Molecular Cell Biology of the Growth and Differentiation of Plant Cells

Molecular Cell Biology of the Growth and Differentiation of Plant Cells

Editor

Ray J. Rose
School of Environmental and Life Sciences
The University of Newcastle
Newcastle, NSW, Australia

CRC Press
Taylor & Francis Group
Boca Raton London New York

CRC Press is an imprint of the
Taylor & Francis Group, an **informa** business
A SCIENCE PUBLISHERS BOOK

CRC Press
Taylor & Francis Group
6000 Broken Sound Parkway NW, Suite 300
Boca Raton, FL 33487-2742

First issued in paperback 2021

© 2016 by Taylor & Francis Group, LLC
CRC Press is an imprint of Taylor & Francis Group, an Informa business

No claim to original U.S. Government works

ISBN-13: 978-0-367-78291-7 (pbk)
ISBN-13: 978-1-4987-2602-3 (hbk)

Visit the Taylor & Francis Web site at
http://www.taylorandfrancis.com

and the CRC Press Web site at
http://www.crcpress.com

Preface

Plants provide humankind with food, fibre and timber products, medicinal and industrial products as well as ecological and climate sustainability. Understanding how a plant grows and develops is central to providing the ability to cultivate plants to provide a sustainable future. 'Molecular Cell Biology of the Growth and Differentiation of Plant Cells' encompasses cell division, cell enlargement and differentiation; which is the cellular basis of plant growth and development. Understanding these developmental processes is fundamental for improving plant growth and the production of special plant products, as well as contributing to biological understanding. The dynamics of cells and cellular organelles are considered in the context of growth and differentiation, made possible particularly by advances in molecular genetics and the visualization of organelles using molecular probes. There is now a much clearer understanding of these basic plant processes of cell division, cell enlargement and differentiation. Each chapter provides a current and conceptual view in the context of the cell cycle (6 chapters), cell enlargement (5 chapters) or cell differentiation (9 chapters).

The cell cycle section examines the regulation of the transitions of the cell cycle phases, proteins of the nucleus which houses most of the genomic information, the division of key energy-related organelles - chloroplasts, mitochondria and peroxisomes and their transmission during cell division. The final chapter in this section deals with the transitioning from cell division to cell enlargement.

The cell enlargement section considers the organisation of the cell wall, the new technical strategies being used, the biosynthesis and assembly of cellulose microfibrils and signaling dependent cytoskeletal dynamics. There are then chapters on the regulation of auxin-induced, turgor driven cell elongation and hormonal interactions in the control of cell enlargement

The cell differentiation section considers the regulation of the cell dynamics of the shoot and root apical meristems, the procambium and cambial lateral meristems as well as nodule ontogeny in the legume-rhizobia symbiosis. There are chapters on asymmetric cell divisions, stem cells, transdifferentiation, genetic reprogramming in cultured cells and the paradox of cell death in differentiation. The final chapter deals with the protein bodies and lipid bodies of storage cells.

Each chapter is written by specialists in the field and the book provides state of the art knowledge (and open questions) set out in a framework that provides a long term reference point. The book is targeted to plant cell biologists, molecular biologists, plant physiologists and biochemists, developmental biologists and those interested in plant growth and development. The chapters are suitable for those already in the field, those plant scientists entering the field and graduate students. The cover images are taken from Chapters 5 and 12.

Ray J. Rose

Contents

The Plant Cell Cycle

1

Plant Cell Cycle Transitions

José Antonio Pedroza-Garcia, Séverine Domenichini and Cécile Raynaud*

Introduction

Plant development is largely post-embryonic, and relies on the proliferative activity of meristematic cells that can form new organs and tissues throughout the life cycle of the plant. Tight control of cell proliferation is therefore instrumental to shape the plant body. In the root meristem, the quiescent centre cells have a low division rate; they play a key role in the self-maintenance of the stem cell pool and function as a reservoir of stem cells that can divide to replace more actively dividing initials (Heyman et al. 2014). The shoot meristem, although less strictly organized than the root meristem, also contains a pool of slowly dividing cells at its centre. On the sides of the meristem, an increase in mitotic index precedes or at least accompanies primordium outgrowth to initiate leaf development (Laufs et al. 1998). Finally, cell proliferation gradually ceases from the tip of the developing leaf to its base as cells progressively differentiate (Andriankaja et al. 2012). This brief summary of the basic mechanisms underlying plant development perfectly illustrates that tight control of the cell cycle plays a central role in this process (Polyn et al. 2015).

Study of the cell cycle began in the second half of the XIXth century with the discovery of cell division and the understanding that cells originate from pre-existing cells. With the identification of chromosomes as the source of genetic information at the beginning of the XXth century, the cell cycle was placed at the centre of the growth, development and heredity for all living organisms (Nurse 2000). Next, in the 1950s the elucidation of the structure of the DNA molecule, and the use of radioactive labelling led to the finding that in eukaryotes, DNA is duplicated during a restricted phase of the cell cycle in interphase that was called S-phase (for synthesis). The cell cycle was thus divided in four phases, S-phase, M-phase or mitosis and two so-called Gap phases, G1 before S-phase and G2 before mitosis. After these crucial

Institute of Plant Sciences Paris-Saclay (IPS2), UMR 9213/UMR1403, CNRS, INRA, Université Paris-Sud, Universitéd'Evry, Université Paris-Diderot,Univeristé Paris-Saclay, Sorbonne Paris-Cité, Bâtiment 630, 91405 Orsay, France.
* Corresponding author : cecile.raynaud@u-psud.fr
  (All three authors have contributed equally in this chapter)

conceptual advances, further dissection of the cell cycle and notably of its regulation had to wait until technical progresses allowed its genetic analysis. This was achieved in the 1970s: combination of genetics, biochemistry and molecular biology allowed the identification of Cyclin Dependent Kinase (CDK)-cyclin complexes as the universal motors of cell cycle regulation in all eukaryotes. CDKs are protein kinases that phosphorylate various substrates to promote transitions from one cell cycle phase to the next. Their activity is modulated by their association with the regulatory subunits called cyclins that are characterized by their cyclic accumulation during the cell cycle. In 2001, L. Hartwell, P. Nurse and T. Hunt were awarded the Nobel prize in Physiology or Medicine for their complementary achievements: their work not only unravelled the role of CDK/cyclin complexes but also introduced the concept of checkpoints to explain the observation that impairing one phase of the cell cycle inhibits subsequent progression.

Basic mechanisms regulating cell cycle progression, DNA replication and mitosis are conserved in all eukaryotes including plants. This high degree of conservation allowed fast progress in the understanding of cell cycle regulation in all organisms. For example, the first plant CDK was isolated by functional complementation of a yeast mutant with an Alfalfa cDNA (Hirt et al. 1991), and considerable progress has been made in the last 35 years in our understanding of plant cell cycle transitions. In spite of this conservation of molecular effectors, the plant cell cycle has a number of specificities. One obvious difference concerns plant mitosis that is characterized by the absence of centrosomes and mechanisms governing cytokinesis. Another hallmark of the plant cell cycle is the relatively frequent occurrence of endoreduplication, a particular type of cell cycle consisting of several rounds of DNA replication without mitosis, and leading to an increase in cell ploidy. Although this process can be found in animals, it is generally restricted to relatively specific cell types such as the salivary glands in Drosophila and hepatocytes in mammals (Fox and Duronio 2013). By contrast in plants, it is widely distributed in various organs such as fruits in tomato, endosperm in cereals or even leaves in plants such as Arabidopsis (Fox and Duronio 2013). In addition, there are also differences in terms of molecular mechanisms regulating cell cycle transitions between plants and other eukaryotes. In the present chapter, we will describe plant cell cycle regulation with a specific emphasis on the molecular mechanisms that control cell cycle transitions, and we will briefly discuss how these basic mechanisms are modulated during plant development or according to external stimuli.

Plant CDKs and Cyclins, Motors of Cell Cycle Progression with an Intriguing Diversity

Core CDK/Cyclin complexes

One feature of plants is the surprisingly high diversity of core cell cycle regulators encompassed by their genomes. Indeed, the Arabidopsis genome encodes 5 CDKs distributed in two sub-classes (a single A-type CDK and four B-type CDKs) and 31 Cyclins belonging to three families (10 CycA, 11 CycB and 10 CycD), whereas *Saccharomyces cerevisiae* has a single CDK and 9 Cyclins, and *Homo sapiens* has

4 CDKs and 9 Cyclins (Van Leene et al. 2010). The number of putative CDK/Cyclin pairs is thus very large in plants, making the elucidation of their role problematic. One important step forward in the understanding of how plant CDK/Cyclin complexes control cell cycle transitions has been the comprehensive analysis of their expression in synchronized cell suspensions (Menges et al. 2005) followed by the systematic analysis of interactions between core cell cycle regulators using Tandem Affinity Purification (Van Leene et al. 2010). These results led to a global picture of CDK/cyclin complexes around the cell cycle (Fig.1). According to these studies, CDKA;1 is expressed throughout the cell cycle and stably associates with D-type cyclins and S-phase expressed A-type cyclins as well as with CYCD3;1 in G2/M, suggesting it could be involved in the control of the G1/S as well as the G2/M transition. Consistently, expression of a dominant negative form of CDKA;1 drastically inhibits cell proliferation (Gaamouche et al. 2010). Likewise, CDKB2s are required for normal cell cycle progression and meristem organisation (Andersen et al. 2008). More recently, analysis of *cdka* and *cdkb* knock-out mutants revealed that CDKA;1 is required for S-phase entry, while it redundantly controls the G2/M transition with B-type CDKs (Nowack et al. 2012).

The large size of Cyclin families complicates the genetic analysis of their respective functions, but as for CDKs, a global view of their respective roles has been obtained by compiling information about their expression during the cell cycle and ability to bind to different CDKs. Very schematically, D-type Cyclins are thought to control cell cycle onset whereas A-type cyclins would be involved at later stages during the S and G2-phases in complex with CDKA1;1 or CDKBs and B-type cyclins bound to CDKBs would control the G2 and M phases [Fig.1, (Van Leene et al. 2010)]. However, Cyclin D3;1 has the particularity of peaking both at the G1/S and at the G2/M transition (Menges et al. 2005), and genetic analysis supports its role

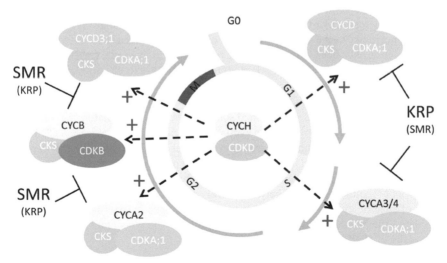

FIGURE 1 Succession of CDK/Cyclin complexes during the cell cycle (adapted from Van Leene et al. 2010). CYCD/CDKA, CYCA/CDKA and CYCB/CDKB sequentially accumulate and are activated to allow progression through the various phases of the cell cycle. CKS sub-units are scaffolding proteins associated with all complexes. Likewise, all CDK/Cyclin complexes are activated by the CYCH/CDKD kinase.

as a positive regulator of both cell cycle transitions (Riou-Khamlichi et al. 1999). Conversely, triple mutants lacking the whole *CYCD3* family show premature exit of cell proliferation towards endoreduplication (Dewitte et al. 2007). Very few genetic studies have been performed on A-type cyclins, and their respective roles are thus largely inferred from expression and interaction data. Nevertheless, the proposed role for Cyclin A3 during S-phase is supported by the observation that down-regulation of CYCA3;2 in Tobacco leads to reduced cell proliferation and endoreduplication (Yu et al. 2003). Two more members of the CycA family have been studied in more detail in Arabidopsis: Cyclin A;1 has thus been shown to be required for the meiotic cell cycle (d'Erfurth et al. 2010), although it could also have functions in vegetative cells (Jha et al. 2014), while Cyclin A2;3 negatively regulates endoreduplication (Imai et al. 2006) by associating with CDKB1;1 and activating cell division (Boudolf et al. 2009). Loss of function studies have allowed this role to be extended to the whole CYCA;2 sub-family: *cycA2;2,3,4* triple mutants show a global reduction of cell proliferation in both shoots and roots (Vanneste et al. 2011). Finally, B-type cyclins are involved in the control of the G2/M transition. This view is supported by their expression pattern that peaks in G2/M, their ability to form complexes with the B-type CDKs, and the observation that ectopic expression of CYCB1;2 is sufficient to induce cell division instead of endoreduplication in developing trichomes (Schnittger et al. 2002). It is worth noting that this model may be over-simplified. For example, CYCD4-1 which has been found by Van Leene et al. (2010) to behave like other D-type cyclins and to bind CDKA;1, has been reported to interact with CDKB2;1 and to be expressed in G2 (Kono et al. 2003). Authors hypothesize that this finding may reflect transient inter-actions due to the ability of CYCD4/CDKA complexes to regulate CDKB-containing complexes, but clearly, more detailed functional analysis of the various Cyclins will be required to reconcile sometimes conflicting experimental data.

As stated above, the large size of Cyclin gene families hampers the genetic dissection of their respective function. In addition, transcriptomic analysis revealed little tissue specificity in the expression pattern of cyclins (Menges et al. 2005). However, a few cyclins have been assigned specific functions. For example, CYCD6;1 has been shown to act downstream of SCARECROW and SHORTROOT to regulate the formative divisions required for root patterning (Sozzani et al. 2006), nevertheless, loss of CYCD6;1 alone is not sufficient to fully compromise these formative divisions, and even triple cyclin mutants still retained some degree of normal patterning, indicating a large level of redundancy between cyclins in this pathway. Likewise, CYCD4-1 and 2 have been involved in stomata formation (Kono et al. 2007), and CYCD4-1 appears to be specifically involved in the regulation of the pericycle cell cycle and during lateral root formation (Nieuwland et al. 2009). Globally, results available so far suggest that a lot of redundancy exists between closely related cyclins. However, the potential role of specific cyclins in response to stress or changes in external conditions have to date little been explored, and could shed light on the physiological role of such a diversity of CDK/cylin complexes.

Atypical CDKs and Cyclins are involved in basal activation of core complexes and in the regulation of gene expression

According to (Menges et al. 2005), the list of Arabidopsis CDKs and Cyclins can be further extended to 29 CDKs and 49 Cyclins by including other sub-groups: CDKC-G and CDK-like (CKL) proteins and CycH, L, P and T. CDKC (in complex with CYCT) and CDKE classes of CDKs are likely involved in the control of gene expression rather than cell cycle progression, and will thus not be further discussed, with the exception of CYCP2;1 (see below) (Barroco et al. 2003, Wang and Chen 2004, Cui et al. 2007, Kitsios et al. 2008). Likewise, CDKG-Cyclin L complexes are involved in chromosome pairing during meiosis, either by directly regulating the meiotic cell cycle or more indirectly by regulating gene expression (Zheng et al. 2014).

By contrast, CDKD-CycH and CDKF are considered as core cell cycle regulators: they are the CDK Activating Kinases (CAK). These proteins can activate CDKs by phosphorylating a conserved threonine in their T-loop. Their accumulation is constant throughout the cell cycle and they are probably not involved in the regulation of one specific cell cycle phase. Consistently, CDKD-deficient mutants show gametophytic lethality, suggesting that CDKD-CycH complexes are required to phosphorylate and activate all core CDKs (Takatsuka et al. 2015). Interestingly, although *cdkf* mutants show reduced cell proliferation, this effect does not seem to be mediated by reduced CDKA or B activity, suggesting that the CDKF could control cell cycle progression via a different pathway that may rely on the regulation of basal transcription (Takatsuka et al. 2009).

Various post-translational mechanisms control CDK/Cyclin complexes activity

In addition to the activating phosphorylation by the CAK, multiple mechanisms acting at the post-translational level modulate CDK/Cyclin activity. The WEE1 protein kinase can inhibit CDKs by phosphorylating them on Tyr15 and Thr14 (Berry and Gould 1996). This phosphorylation plays an important role in the control of the G2/M transition in eukaryotes and functions to avoid premature division of cells that have not sufficiently expanded as well as to delay mitosis after DNA damage. However, in Arabidopsis, the WEE1 kinase seems to be predominantly involved in DNA stress response, and not in growth regulation under normal conditions (De Schutter et al. 2007, Cools et al. 2011). Finally, CDK/cyclin complexes can be inhibited by binding of small proteins called CDK inhibitors (or CKI). In plants they are distributed between two unrelated families: the KRP (for KIP-related Proteins) that share homology with the human cell cycle inhibitor p27 and the SMR (for SMR related) (Van Leene et al. 2010). Like CDKs and cyclins, these inhibitors are extremely diverse: the Arabidopsis genome encompasses 7 KRPs and 14 SMRs. KRPs (also called ICKs for Inhibitors of Cyclin-dependant Kinases) were the first identified plant cell cycle inhibitors (Wang et al. 1998). They associate preferentially with CYCD or CYCA/CDKA;1 complexes (Van Leene et al. 2010). Consistently, over-expression of a number of KRPs induces the same phenotypic defects including reduction of cell division and endoreduplication, reduction of lateral root formation and dramatically enlarged cell size (Wang et al. 2000, Jasinski et al. 2002). The respective roles of the various KRPs remain to be elucidated, and a high level of redundancy between these cell cycle inhibitors is likely to exist. Consistently, quintuple *krp1,2,3,4,7* mutants show only a mild increase in organ size due to the activation of cell proliferation via the E2F pathway (Cheng et al. 2013), however, multi-silenced KRP lines with reduced levels of KRP1-7 show severe developmental defects and ectopic callus formation, providing further evidence for the role of KRPs as negative regulators of cell proliferation (Anzola et al. 2010). Until now, only one member of the KRP family seems to play a distinctive role that cannot be fulfilled by other KRPs: KRP5 is required for the regulation of hypocotyl cell elongation in the dark (Jégu et al. 2013), and cell expansion in the root (Wen et al. 2013). Interestingly, it seems to function at least partly by binding chromatin and regulating the expression of genes involved in cell elongation and endoreduplication, providing evidence for yet unsuspected functions of plant cell cycle inhibitors (Jégu et al. 2013). Whether other KRPs may function as positive regulators of endoreduplication despite their ability to reduce the activity of G1/S CDK-Cyclin complexes remains to be fully established, but this hypothesis is supported by the observation that mild-over-expression of KRP2 results in an increase in endoreduplication (Verkest et al. 2005). SIAMESE (SIM), the founding member of the SMR family, also appears to positively regulate endoreduplication: *sim* mutants display multicellular trichomes, indicating that the SIM protein is required not only to promote endoreduplication but also to inhibit cell proliferation (Churchman et al. 2006). SIM-RELATED proteins (SMRs) have been proposed to play a role in cell cycle arrest during stress response (Peres et al. 2007). Consistently, SMR5 and SMR7 are involved in cell cycle arrest caused by reactive oxygen species, for example during high light stress (Yi et al. 2014), and contribute to the growth reduction caused by chloroplasts dysfunction (Hudik et al. 2014).

Control of the G1/S Transition: The E2F/RBR Pathway

As previously described, CYCD/CDKA complexes are the first CDK/Cyclin complexes activated for cell cycle onset. Consistently, expression of a number of CycDs responds to external cues (see below). In all eukaryotes, CYCD/CDKA complexes promote the G1/S transition by phosphorylating the Retinoblastoma (Rb) protein and alleviating its inhibitory action on E2F transcription factors that can in turn activate genes involved in DNA replication (Berckmans and De Veylder 2009) (Fig. 2). This pathway is conserved in plants, and the Arabidopsis genome encompasses a single Rb homologue (RBR, RetinoBlastoma Related) and six E2Fs (Lammens et al. 2009). Interestingly, most defects of the *cdka;1* mutant are rescued in a *cdka;1 rbr* double mutant, indicating that CDKA;1 regulates cell cycle progression mainly by targeting RBR (Nowack et al. 2012). Plant E2F transcription factors can be divided in two sub-groups: canonical E2Fs (E2Fa, b and c) require a Dimerization Partner (DP) to efficiently bind DNA, whereas atypical E2Fs (E2Fd, e and f) function as monomers. Plant E2Fs also differ by their function in cell cycle regulation, E2Fa and b being activators of the cell cycle whereas E2Fc behaves as a negative regulator (Berckmans and De Veylder 2009). Genome-wide identification of E2F target genes by combining promoter analysis for E2F binding sites and transcriptomic analysis performed on E2F over-expressing lines identified genes involved in DNA replication, DNA repair and chromatin dynamics further supporting the notion that E2Fa and b positively

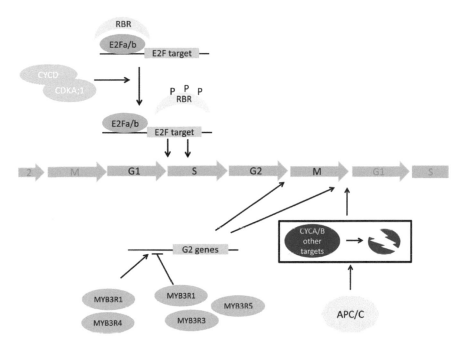

FIGURE 2 Regulation of cell cycle transitions. Activation of CYCD/CDKA complexes leads to phosphorylation of RBR and release of its inhibitory action on E2F factors thereby allowing expression of S-phase genes. G2 and M genes are under the control of MYB3R transcription factors Activation of the APC/C is required to degrade various targets and allow exit from mitosis.

regulate the G1/S transition (Ramirez-Parra et al. 2003, Vandepoele et al. 2005). By contrast, over-expression of E2Fc inhibits cell proliferation (del Pozo et al. 2002) and its down regulation activates cell division (del Pozo et al. 2006), although it is not clear whether E2Fc acts antagonistically to E2Fa and b during S-phase or if it is more specifically involved in regulating the balance between cell proliferation and endoreduplication. This relatively simple model is made complex by the observation that E2Fa also controls a number of genes involved in cell differentiation: the maintenance of proliferative activity in meristems therefore requires partial inactivation of E2Fa by RBR (Magyar et al. 2012, Polyn et al. 2015). Finally, E2Fe and f are involved in the control of cell expansion: E2Fe prevents endocycles onset and thereby delays cell elongation whereas E2Ff is directly involved in cell expansion (Lammens et al. 2009).

Upon RBR release, activating E2Fs stimulate the expression of genes required for DNA replication, including the ones encoding the pre-replication complex (pre-RC). Assembly of the pre-RC on the replication origin and DNA replication licencing are key steps to the regulation of the G1/S transition. ORC (origin replication complex) proteins bind to replication origins and recruit CDC6 and CDT1 that in turn allow binding of MCM proteins that function as helicases to open the replication fork (DePamphilis 2003). All these factors are conserved in Arabidopsis, and interactions between the various constituents of the pre-RC have been observed in the yeast two-hybrid system (Shultz et al. 2007). In addition, there is genetic evidence that the function of CDC6, CDT1, MCM2 and MCM7 in DNA replication is conserved in plants (Springer et al. 2000, Castellano et al. 2001, Castellano et al. 2004, Ni et al. 2009, Domenichini et al. 2012). Licencing of replication origins has to be tightly controlled so that it occurs once and only once per cell cycle in order to avoid incomplete DNA replication or re-replication of fractions of the genome (Xouri et al. 2007). Although these regulatory mechanisms are very well described in animals, it is much less clear how they function in plants. However, CDT1 that is the target of many regulatory pathways in animals also appears to be regulated by proteolysis in plants (Castellano et al. 2004). In addition, plant genomes encode homologues of the CDC7/Dbf4 kinase involved in replication licencing (Shultz et al. 2007), but their function has never been studied. Finally, origin licensing is also regulated between early and late-firing origins, early replicating regions corresponding mainly to euchromatin while heterochromatin is replicated at the end of the cell cycle (Hayashi et al. 2013, Bass et al. 2014). How replication timing is controlled in plants remains to be elucidated, but chromatin modifications such as histone marks are likely to play a role in this process (Raynaud et al. 2014). Consistently, mutants deficient for the deposition of the repressive mark H3K27me1 show re-replication of constitutive heterochromatin regions (Jacob et al. 2010), and this defect is aggravated by the over-expression of the cell cycle inhibitor KRP5 (Jégu et al. 2013), suggesting that heterochromatin not only specifies late replicating regions but could also function as a barrier against endoreduplication. Once pre-RC are activated, CDC6 and CDT1 are released from replication origins and inactivated. MCM proteins open the replication fork bi-directionally and are associated with replicative DNA polymerases via CDC45 and the GINS (go ichini san, also called PSF1, 2, 3 and SLD5), which are instrumental to the stabilization of the replication fork (Friedel et al. 2009). Data regarding the function of these factors in plants is scarce but down-regulation of CDC45 in meiocytes results in DNA fragmentation independently of programmed double-strand breaks that form during meiosis, suggesting that CDC45 is required for DNA replication to proceed normally (Stevens et al. 2004). Although data available so far support the notion that plant DNA replication functions in the same way as what is described in yeast and animals, it is worth noting that CDT1 homologues were found to form complexes with DNA polymerase ε, the replicative polymerase that synthesizes the leading strand (Pursell and Kunkel 2008), suggesting that the molecular events occurring during pre-RC formation or fork progression may differ in plants and other eukaryotes (Domenichini et al. 2012).

Regulation of G2 and Mitosis

Many genes expressed during the G2 and M phases harbour a specific regulatory sequence in their promoter called MSA (mitosis-specific activator) (Ito et al. 1998, Menges et al. 2005) that is recognized by MYB3R transcription factors (Haga et al. 2011). The Arabidopsis genome encodes 5 MYB3R: MYB3R2 which appears to be involved in the control of the circadian clock, but MYB3R1, 3, 4 and 5 have all been reported to control cell cycle progression (Fig. 2). MYB3R1 and 4 activate the expression of G2/M specific genes such as *KNOLLE* to allow proper cytokinesis (Haga et al. 2011); whereas MYB3R3 and 5 are repressors of G2/M genes (Kobayashi et al. 2015). Surprisingly MYB3R1, that was originally thought to function as an activating MYB3R, also functions redundantly with MYB3R 3 and 5: *myb3R1,3,5* triple mutants display hypertrophy of all organs. MYB3R1 thus likely plays opposite roles in different cellular contexts, possibly by binding to different partners. In addition, repressor MYB3Rs play different roles in proliferating and post-mitotic cells: in the former they are required to narrow-down the expression window of their targets to the G2 and M phases while in the latter repressor-MYB3Rs are required to repress cell division genes (Kobayashi et al. 2015). This dual role likely depends on the ability of MYB3R to bind other cell cycle regulators: in mature cells, MYB3R3 can form complexes with the repressor E2Fc and RBR1; consistently, ChIP-seq experiments revealed that MYB3Rs can bind promoters containing MSA sequences as well as E2F targets, indicating that they could play a role in the global repression of cell cycle genes in differentiated cells (Kobayashi et al. 2015).

In addition to the transcriptional regulation of G2/M gene expression, targeted protein degradation plays a pivotal role for progression through mitosis (Fig. 2). The Anaphase Promoting Complex/Cyclosome is a highly conserved E3-ubiquitin ligase specifically targeting cell cycle regulators towards proteolysis (Heyman and De Veylder 2012), that was named for its role in the degradation of the mitosis inhibitor securin (Vodermaier 2004). This complex comprises 11 sub-units [APC1-11, (Van Leene et al. 2010)], some of which are constitutively expressed while others accumulate specifically during G2 and M (Heyman and De Veylder 2012). In addition, several inhibitors and activators of the complex have been identified: CDC20, CCS52 and SAMBA are activators of the complex and OSD1/UVI4-LIKE/GIGAS and UVI4/PIM are inhibitors. A number of APC/C targets have been identified in various plant species, including the expected A and B-type cyclins, but also other cell cycle regulators (for review see Heyman and De Veylder 2012). Loss of function of core sub-units or activators generally results either in defects in gametophyte development in the case of null mutants, or drastic reduction of plant stature in the case of hypomorphic alleles (Heyman and De Veylder 2012). For example, silencing of CDC20-1 and 2 results in severe dwarfism due to reduced cell proliferation (Kevei et al. 2011). Likewise, loss of CCS52, also causes reduced growth, but in this case it is mainly due to defects in cell differentiation and elongation (Lammens et al. 2008, Mathieu-Rivet et al. 2010). By contrast, over-expression of core APC/C sub-units has been reported to increase plant size via an increase in cell proliferation (Rojas et al. 2009, Eloy et al. 2011). However, similar observations have been reported for mutants lacking the APC/C activator SAMBA: *samba* mutants show enhanced cell proliferation and CYCA2; 3 stability (Eloy et al. 2012), illustrating that the outcome of reduced or enhanced APC/C activity cannot be easily predicted and likely depends highly on the affected substrates and cell types.

Modulating Cell Cycle Transitions According to Plant Development and External Cues

Tight regulation of the balance between cell proliferation and differentiation is critical for proper development and to shape the whole plant body according to environmental cues. Entry into the cell cycle requires activation of D-type cyclins, but

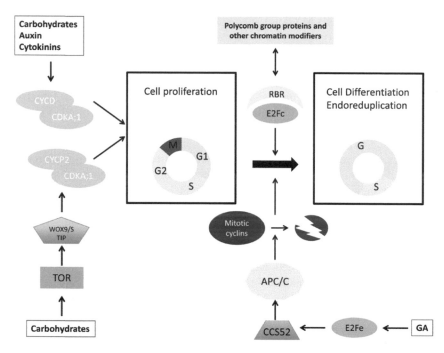

FIGURE 3   Regulation of the balance between proliferation and differentiation. External cues such as carbohydrates or phytohormones activate D-type cyclins and CYCP2 (likely via the TOR kinase and WOX9 transcription factor) to stimulate cell proliferation. By contrast both transcriptional regulation mediated by inhibitor E2Fs and RBR, and ubiquitination by the APC/C act to promote cell cycle exit and cell differentiation.

the signalling events governing their expression are not fully elucidated. Signals known to stimulate the expression of D-type cyclins include sugars (Menges et al. 2006), auxin (Fuerst et al. 1996) and cytokinin (Dewitte et al. 2007) (Fig. 3): all these factors are therefore important to stimulate cell proliferation *in vivo* (Inze and De Veylder 2006). How cells integrate all these signals to modulate cell cycle progression remains largely unknown, but some transcription factors have been identified for their ability to stimulate cell proliferation by activating core cell cycle genes (for review see Berckmans and De Veylder 2009). In the root, mechanisms allowing the reactivation of cell proliferation during germination have been elucidated. Photosynthesis-derived carbohydrates are sensed via the TOR kinase that controls the activity of the *STIMPY/WOX9* transcription factor, which in turn activates the expression of CYCP2;1, which associates with CDKA;1 to activate cell cycle progression (Xiong et al. 2013, Peng et al. 2014) (Fig. 3). In addition, in all proliferating tissues, the OBP1 (OBF binding Protein 1) transcription factor directly regulates core cell cycle genes and its over-expression results in a shortening of the cell cycle (Skirycz et al. 2008), but how the activity of this transcription factor is modulated remains unknown.

Some developmental steps have also been studied in more detail. For example, lateral root formation, that requires reactivation of cell proliferation, has been shown to depend on the concerted action of CYCD2;1, CDKA;1 and KRP2: sucrose would induce CYCD2;1 expression, and it would accumulate in an inactive complex with CDKA;1 and KRP2 in the nucleus. Auxin would subsequently induce KRP2 degradation, leading to the activation of CYCD2;1/CDKA;1 complexes and subsequent cell cycle activation (Sanz et al. 2011). In addition, *e2fa* deficient mutants show fewer lateral root primordia, and the transcription factors LBD (LATERAL ORGAN BOUNDARY DOMAIN) 18 and 33 have been shown to activate the expression of E2Fa upon auxin treatment to stimulate lateral root initiation (Berckmans et al. 2011). How this pathway is connected to the CYCD2;1 pathway has to be clarified, illustrating the extreme complexity of plant cell cycle regulation. Limiting cell proliferation in some specific cells can also be instrumental to plant development: in the root quiescent centre, WOX5 suppresses CYCD3;1 and CYCD3;3 expression to inhibit cell division (Forzani et al. 2014). The possibility to suppress cell division in the QC and to reactivate it when necessary likely plays a central role in the ability of plants to maintain stem cell pools and to protect the integrity of their genome under adverse conditions (Heyman et al. 2014).

Like cell cycle entry, the transition from cell proliferation to differentiation is tightly regulated. Root and leaf development both rely on the gradual cessation of cell proliferation followed by cell elongation, which is often accompanied by endoreduplication in Arabidopsis. As described above, the equilibrium between cell proliferation and cell differentiation depends on the expression of mitotic cyclins and their regulation by the APC/C as well as on the activity of cell cycle inhibitors. This conclusion is largely supported by over-expression experiments that demonstrated the capacity of mitotic cyclins to promote ectopic cell division or the ability of cell cycle inhibitors to stimulate endoreduplication, but other studies have placed these cell cycle regulators in a more physiological context. For example, in roots, cytokinins restrict meristem size by promoting the expression of the APC/C activator CCS52A1 and thus endoreduplication (Takahashi et al. 2013), possibly by targeting CYCA3;2 (Boudolf et al. 2009). Intriguingly, chloroplast differentiation appears to be required for the transition from cell proliferation to differentiation in leaves (Andriankaja et al. 2012), but the underlying mechanisms remain unknown.

Finally, one essential actor of cell cycle modulation during development is the RBR protein: it is involved not only in the G1/S transition as described above, but also probably in the progression through G2/M and it coordinates cell cycle arrest with cell differentiation by associating with a wide range of chromatin modifiers (Kuwabara and Gruissem 2014) (Fig. 3). The developmental roles of RBR are complex and diverse, and therefore cannot be extensively described here.

In addition to the programmed changes in cell proliferation associated with normal plant development, the ability to modulate the cell cycle in response to stress is a key parameter for ability to cope with changing environmental conditions and to adjust their body plan accordingly. This is mediated at least in part by the activation of checkpoints that can block cells in a specific cell cycle phase or reorient cell cycle progression towards endoreduplication or cell death. As a general rule, stress induces cell differentiation, possibly to avoid the transmission of induced mutations to the progeny of the cells (Cools and De Veylder 2009). However, CYCB1;1 has the particularity of being induced by genotoxic stress, and has been proposed to function to block some cells in G2, thereby preserving some proliferative potential until conditions become favourable again (Cools and De Veylder 2009). In the root of Arabidopsis, replenishment of the meristem after initial cell death is achieved by stimulating the division of QC cells that are probably less vulnerable to stress because of their low division rate: when plants are transferred from a medium containing DNA damaging agents back to normal growth medium, the ERF115 transcription factor that is a positive regulator of QC cell division is activated, thereby allowing the replacement of cells that have undergone programmed cell death (Heyman et al. 2013). Yet another mechanism has been described in rice where the RSS1 protein is required to maintain the proliferative capacity of meristematic cells during salt stress (Ogawa et al. 2011), but this factor is not conserved in eudicots. Stress also induces premature cell differentiation in growing organs: in leaves, drought activates gibberellin signalling and thus stabilization of DELLA proteins that in turn activate the atypical E2F factor E2Fe thereby stimulating the expression of CCS52A and triggering early endoreduplication (Claeys et al. 2012) (Fig. 3). High light stress also promotes early cell differentiation by activating the expression of the cell cycle inhibitors SMR5 and SMR7 (Hudik et al. 2014, Yi et al. 2014) and DNA damage causes early differentiation of root meristematic cells (Cools et al. 2011). The analysis of cell cycle progression in response to stress is still in its infancy, but there is also accumulating evidence that biotic stresses also impinge on cell cycle regulation (Reitz et al. 2015), and even that some pathogens modify cell cycle regulation to their advantage (Chandran et al. 2013) opening exciting new research prospects.

Conclusions

Although a very large number of reports have shed light on the mechanisms regulating the plant cell cycle, many questions remain to be addressed. Notably, although several studies have illustrated the role of phytohormones in plant cell cycle regulation, how these signalling pathways are integrated and cooperate to control plant development is far from being fully elucidated. In addition, to get a comprehensive picture of the plant cell cycle, we still need to understand why core cell cycle regulators are so diverse in plants and what the respective roles of the various isoforms can be. One answer to this question probably resides in the plasticity of plant development, and analysis of plant cell cycle regulation in the context of biotic or abiotic stress is likely to reveal the specificities of seemingly redundant factors.

References

Andersen, S.U., S. Buechel, Z. Zhao, K. Ljung, O. Novak, W. Busch, C. Schuster and J.U. Lohmann. 2008. Requirement of B2-type Cyclin-Dependent Kinases for meristem intergrity in *Arabidopsis thaliana*. Plant Cell 20: 88-100.

Andriankaja, M., S. Dhondt, S. De Bodt, H. Vanhaeren, F. Coppens, L. De Milde, P. Muhlenbock, A. Skirycz, N. Gonzalez, G.T. Beemster and D. Inze. 2012. Exit from proliferation during leaf development in *Arabidopsis thaliana*: a not-so-gradual process. Dev. Cell 22: 64-78.

Anzola, J.M., T. Sieberer, M. Ortbauer, H. Butt, B. Korbei, I. Weinhofer, A.E. Mullner and C. Luschnig. 2010. Putative Arabidopsis transcriptional adaptor protein (PROPORZ1) is required to modulate histone acetylation in response to auxin. Proc. Natl. Acad. Sci. U.S.A. 107: 10308-10313.

Barroco, R.M., L. De Veylder, Z. Magyar, G. Engler, D. Inze and V. Mironov. 2003. Novel complexes of cyclin-dependent kinases and a cyclin-like protein from *Arabidopsis thaliana* with a function unrelated to cell division. Cell. Mol. Life Sci. 60: 401-412.

Bass, H.W., E.E. Wear, T.J. Lee, G.G. Hoffman, H.K. Gumber, G.C. Allen, W.F. Thompson and L. Hanley-Bowdoin. 2014. A maize root tip system to study DNA replication programmes in somatic and endocycling nuclei during plant development. J. Exp. Bot. 65: 2747-2756.

Berckmans, B. and L. De Veylder. 2009. Transcriptional control of the cell cycle. Curr. Opin. Plant Biol. 12: 599-605.

Berckmans, B., V. Vassileva, S.P. Schmid, S. Maes, B. Parizot, S. Naramoto, Z. Magyar, C.L. Alvim Kamei, C. Koncz, L. Bogre, G. Persiau, G. De Jaeger, J. Friml, R. Simon, T. Beeckman and L. De Veylder. 2011. Auxin-dependent cell cycle reactivation through transcriptional regulation of Arabidopsis E2Fa by lateral organ boundary proteins. Plant Cell 23: 3671-3683.

Berry, L.D. and K.L. Gould. (1996). Regulation of Cdc2 activity by phosphorylation at T14/Y15. Prog. Cell Cycle Res. 2: 99-105.

Boudolf, V., T. Lammens, J. Boruc, J. Van Leene, H. Van Den Daele, S. Maes, G. Van Isterdael, E. Russinova, E. Kondorosi, E. Witters, G. De Jaeger, D. Inze and L. De Veylder. 2009. CDKB1;1 forms a functional complex with CYCA2;3 to suppress endocycle onset. Plant Physiol. 150: 1482-1493.

Castellano, M.M., M.B. Boniotti, E. Caro, A. Schnittger and C. Gutierrez. 2004. DNA replication licensing affects cell proliferation or endoreplication in a cell type-specific manner. Plant Cell 16: 2380-2393.

Castellano, M.M., J.C. del Pozo, E. Ramirez-Parra, S. Brown and C. Gutierrez. 2001. Expression and stability of Arabidopsis CDC6 are associated with endoreplication. Plant Cell 13: 2671-2686.

Chandran, D., J. Rickert, C. Cherk, B.R. Dotson and M.C. Wildermuth. 2013. Host cell ploidy underlying the fungal feeding site is a determinant of powdery mildew growth and reproduction. Mol. Plant Microbe Interact. 26: 537-545.

Cheng, Y., L. Cao, S. Wang, Y. Li, X. Shi, H. Liu, L. Li, Z. Zhang, L.C. Fowke, H. Wang and Y. Zhou. 2013. Downregulation of multiple CDK inhibitor ICK/KRP genes upregulates the E2F pathway and increases cell proliferation, and organ and seed sizes in Arabidopsis. Plant J. 75: 642-655.

Churchman, M.L., M.L. Brown, N. Kato, V. Kirik, M. Hulskamp, D. Inze, L. De Veylder, J.D. Walker, Z. Zheng, D.G. Oppenheimer, T. Gwin, J. Churchman and J.C. Larkin. 2006. SIAMESE, a plant-specific cell cycle regulator, controls endoreplication onset in *Arabidopsis thaliana*. Plant Cell 18: 3145-3157.

Claeys, H., A. Skirycz, K. Maleux and D. Inze. 2012. DELLA signaling mediates stress-induced cell differentiation in Arabidopsis leaves through modulation of anaphase-promoting complex/cyclosome activity. Plant Physiol. 159: 739-747.

Cools, T. and L. De Veylder. 2009. DNA stress checkpoint control and plant development. Curr. Opin. Plant Biol. 12: 23-28.

Cools, T., A. Iantcheva, A.K. Weimer, S. Boens, N. Takahashi, S. Maes, H. Van den Daele, G. Van Isterdael, A. Schnittger and L. De Veylder. 2011. The *Arabidopsis thaliana* checkpoint kinase WEE1 protects against premature vascular differentiation during replication stress. Plant Cell 23: 1435-1448.

Cui, X., B. Fan, J. Scholz and Z. Chen Z. 2007. Roles of Arabidopsis cyclin-dependent kinase C complexes in cauliflower mosaic virus infection, plant growth, and development. Plant Cell 19: 1388-1402.

d'Erfurth, I., L. Cromer, S. Jolivet, C. Girard, C. Horlow, Y. Sun, J.P. To, L.E. Berchowitz, G.P. Copenhaver and R. Mercier. 2010. The cyclin-A CYCA1;2/TAM is required for the meiosis I to meiosis II transition and cooperates with OSD1 for the prophase to first meiotic division transition. PLoS Genet. 6: e1000989.

De Schutter, K., J. Joubes, T. Cools, A. Verkest, F. Corellou, E. Babiychuk, E. Van Der Schueren, T. Beeckman, S. Kushnir, D. Inze and L. De Veylder. 2007. Arabidopsis WEE1 kinase controls cell cycle arrest in response to activation of the DNA integrity checkpoint. Plant Cell 19: 211-225.

del Pozo, J.C., M.B. Boniotti and C. Gutierrez. 2002. Arabidopsis E2Fc functions in cell division and is degraded by the ubiquitin-SCF(AtSKP2) pathway in response to light. Plant Cell 14: 3057-3071.

del Pozo, J.C., S. Diaz-Trivin, N. Cisneros and C. Gutierrez. 2006. The balance between cell division and endoreplication depends on E2FC-DPB, transcription factors regulated by the ubiquitin-SCFSKP2A pathway in Arabidopsis. Plant Cell 18: 2224-2235.

DePamphilis, M.L. 2003. The 'ORC cycle': a novel pathway for regulating eukaryotic DNA replication. Gene 310: 1-15.

Dewitte, W., S. Scofield, A.A. Alcasabas, S.C. Maughan, M. Menges, N. Braun, C. Collins, J. Nieuwland, E. Prinsen, V. Sundaresan and J.A. Murray.2007. Arabidopsis CYCD3 D-type cyclins link cell proliferation and endocycles and are rate-limiting for cytokinin responses. Proc. Natl. Acad. Sci. U.S.A. 104: 14537-14542.

Domenichini, S., M. Benhamed, G. De Jaeger, E. Van De Slijke, S. Blanchet, M. Bourge, L. De Veylder, C. Bergounioux and C. Raynaud. 2012. Evidence for a role of Arabidopsis CDT1 proteins in gametophyte development and maintenance of genome integrity. Plant Cell 24: 2779-2791.

Eloy, N.B., M. de Freitas Lima, D. Van Damme, H. Vanhaeren, N. Gonzalez, L. De Milde, A.S. Hemerly, G.T. Beemster, D. Inze and P.C. Ferreira. 2011. The APC/C subunit 10 plays an essential role in cell proliferation during leaf development. Plant J. 68: 351-363.

Eloy, N.B., N. Gonzalez, J. Van Leene, K. Maleux, H. Vanhaeren, L. De Milde, S. Dhondt, L. Vercruyssen, E. Witters, R. Mercier, L. Cromer, G. Beemster, H. Remaut, M.C.E. Van Montagu, G. De Jaeger, P.C.G. Ferreira and D. Inze. 2012. SAMBA, a plant-specific anaphase-promoting complex/cyclosome regulator is involved in early development and A-type cyclin stabilization. Proc. Nat. Acad. Sci. U.S.A. 109: 13853-13858.

Forzani, C., E. Aichinger, E. Sornay, V. Willemsen, T. Laux, W. Dewitte and J.A.H. Murray. 2014. WOX5 suppresses CYCLIN D activity to establish quiescence at the center of the root stem cell niche. Curr. Biol. 24: 1939-1944.

Fox, D.T. and R.J. Duronio. 2013. Endoreplication and polyploidy: insights into development and disease. Development 140: 3-12.

Friedel, A.M., B.L. Pike and S.M. Gasser. 2009. ATR/Mec1: coordinating fork stability and repair. Curr. Opin. Cell Biol. 21: 237-244.

Fuerst, R.A., R. Soni, J.A. Murray and K. Lindsey. 1996. Modulation of cyclin transcript levels in cultured cells of *Arabidopsis thaliana*. Plant Physiol. 112: 1023-1033.

Gaamouche, T., C.L. Manes, D. Kwiatkowska, B. Berckmans, R. Koumproglou, S. Maes, T. Beeckman, T. Vernoux, J.H. Doonan, J. Traas, D. Inze and L. De Veylder. 2010. Cyclin-dependent kinase activity maintains the shoot apical meristem cells in an undifferentiated state. Plant J. 64: 26-37.

Haga, N., K. Kobayashi, T. Suzuki, K. Maeo, M. Kubo, M. Ohtani, N. Mitsuda, T. Demura, K. Nakamura, G. Jurgens and M. Ito. 2011. Mutations in MYB3R1 and MYB3R4 cause pleiotropic developmental defects and preferential down-regulation of multiple G2/M-specific genes in Arabidopsis. Plant Physiol. 157: 706-717.

Hayashi, K., J. Hasegawa and S. Matsunaga. 2013. The boundary of the meristematic and elongation zones in roots: endoreduplication precedes rapid cell expansion. Sci. Rep. 3: 2723.

Heyman, J., T. Cools, F. Vandenbussche, K.S. Heyndrickx, J. Van Leene, I. Vercauteren, S. Vanderauwera, K. Vandepoele, G. De Jaeger, D. Van Der Straeten and L. De Veylder. 2013. ERF115 controls root quiescent center cell division and stem cell replenishment. Science 342: 860-863.

Heyman, J. and L. De Veylder. 2012. The anaphase-promoting complex/cyclosome in control of plant development. Mol. Plant 5: 1182-1194.

Heyman, J., R.P. Kumpf and L. De Veylder. 2014. A quiescent path to plant longevity. Trends Cell Biol. 24: 443-448.

Hirt, H., A. Pay, J. Gyorgyey, L. Bako, K. Nemeth, L. Bogre, R.J. Schweyen, E. Heberle-Bors and D. Dudits. 1991. Complementation of a yeast cell cycle mutant by an alfalfa cDNA encoding a protein kinase homologous to p34cdc2. Proc. Natl. Acad. Sci. U.S.A. 88: 1636-1640.

Hudik, E., Y. Yoshioka, S. Domenichini, M. Bourge, L. Soubigout-Taconnat, C. Mazubert, D. Yi, S. Bujaldon, H. Hayashi, L. De Veylder, C. Bergounioux, M. Benhamed and C. Raynaud. 2014. Chloroplast dysfunction causes multiple defects in cell cycle progression in the Arabidopsis *crumpled leaf* mutant. Plant Physiol. 166: 152-167.

Imai, K.K., Y. Ohashi, T. Tsuge, T. Yoshizumi, M. Matsui, A. Oka and T. Aoyama. 2006. The A-type cyclin CYCA2;3 is a key regulator of ploidy levels in Arabidopsis endoreduplication. Plant Cell 18: 382-396.

Inze, D. and L. De Veylder. 2006. Cell cycle regulation in plant development. Annu. Rev. Genet. 40: 77-105.

Ito, M., M. Iwase, H. Kodama, P.Lavisse, A. Komamine, R. Nishihama, Y. Machida and A. Watanabe. 1998. A novel cis-acting element in promoters of plant B-type cyclin genes activates M phase-specific transcription. Plant Cell 10: 331-341.

Jacob, Y., H. Stroud, C. Leblanc, S. Feng, L. Zhuo, E. Caro, C. Hassel, C. Gutierrez, S.D. Michaels and S.E. Jacobsen. 2010. Regulation of heterochromatic DNA replication by histone H3 lysine 27 methyltransferases. Nature 466: 987-991.

Jasinski, S., C. Riou-Khamlichi, O. Roche, C. Perennes, C. Bergounioux and N. Glab. 2002. The CDK inhibitor NtKIS1a is involved in plant development, endoreduplication and restores normal development of cyclin D3; 1-overexpressing plants. J. Cell Sci. 115: 973-982.

Jégu, T., D. Latrasse, M. Delarue, C. Mazubert, M. Bourge, E. Hudik, S. Blanchet, M.N. Soler, C. Charon, L. De Veylder, C. Raynaud, C. Bergounioux and M. Benhamed. 2013. Multiple functions of KRP5 connect endoreduplication and cell elongation. Plant Physiol. 161: 1694-1705.

Jha, A.K., Y. Wang, B.S. Hercyk, H.S. Shin, R. Chen and M. Yang. 2014. The role for CYCLIN A1;2/TARDY ASYNCHRONOUS MEIOSIS in differentiated cells in Arabidopsis. Plant Mol. Biol. 85: 81-94.

Kevei, Z., M. Baloban, O. Da Ines, H. Tiricz, A. Kroll, K. Regulski, P. Mergaert and E. Kondorosi. 2011. Conserved CDC20 cell cycle functions are carried out by two of the five isoforms in *Arabidopsis thaliana*. PLoS One 6: e20618.

Kitsios, G., K.G. Alexiou, M. Bush, P. Shaw and J.H. Doonan. 2008. A cyclin-dependent protein kinase, CDKC2, colocalizes with and modulates the distribution of spliceo-somal components in Arabidopsis. Plant J. 54: 220-235.

Kobayashi, K., T. Suzuki, E. Iwata, N. Nakamichi, P. Chen, M. Ohtani, T. Ishida, H. Hosoya, S. Muller, T. Leviczky, A. Pettko-Szandtner, Z. Darula, A. Iwamoto, M. Nomoto, Y. Tada, T. Higashiyama, T. Demura, J.H. Doonan, M.T. Hauser, K. Sugimoto, M. Umeda, Z. Magyar, L. Bogre and M. Ito. 2015. Transcriptional repression by MYB3R proteins regulates plant organ growth. EMBO J.

Kono, A., C. Umeda-Hara, S. Adachi, N. Nagata, M. Konomi, T. Nakagawa, H. Uchimiya and M. Umeda. 2007. The Arabidopsis D-type cyclin CYCD4 controls cell division in the stomatal lineage of the hypocotyl epidermis. Plant Cell 19: 1265-1277.

Kono, A., C. Umeda-Hara, J. Lee, M. Ito, H. Uchimiya and M. Umeda. 2003. Arabidopsis D-type cyclin CYCD4;1 is a novel cyclin partner of B2-type cyclin-dependent kinase. Plant Physiol. 132: 1315-1321.

Kuwabara, A. and W. Gruissem. 2014. Arabidopsis RETINOBLASTOMA-RELATED and Polycomb group proteins: cooperation during plant cell differentiation and develop-ment. J. Exp. Bot. 65: 2667-2676.

Lammens, T., V. Boudolf, L. Kheibarshekan, L.P. Zalmas, T. Gaamouche, S. Maes, M. Vanstraelen, E. Kondorosi, N.B. La Thangue, W. Govaerts, D. Inze and L. De Veylder. 2008. Atypical E2F activity restrains APC/CCCS52A2 function obligatory for endocycle onset. Proc. Natl. Acad. Sci. U.S.A. 105: 14721-14726.

Lammens, T., J. Li, G. Leone and L. De Veylder. 2009. Atypical E2Fs: new players in the E2F transcription factor family. Trends Cell Biol. 19: 111-118.

Laufs, P., O. Grandjean, C. Jonak, K. Kieu and J. Traas. 1998. Cellular parameters of the shoot apical meristem in Arabidopsis. Plant Cell 10: 1375-1390.

Magyar, Z., B. Horvath, S. Khan, B. Mohammed, R. Henriques, L. De Veylder, L. Bako, B. Scheres and L. Bogre. 2012. Arabidopsis E2FA stimulates proliferation and endocycle separately through RBR-bound and RBR-free complexes. EMBO J. 31: 1480-1493.

Mathieu-Rivet, E., F. Gevaudant, A. Sicard, S. Salar, P.T. Do, A. Mouras, A.R. Fernie, Y. Gibon, C. Rothan, C. Chevalier and M. Hernould. 2010. Functional analysis of the anaphase promoting complex activator CCS52A highlights the crucial role of endo-reduplication for fruit growth in tomato. Plant J. 62: 727-741.

Menges, M., S.M. de Jager, W. Gruissem and J.A. Murray. 2005. Global analysis of the core cell cycle regulators of Arabidopsis identifies novel genes, reveals multiple and highly specific profiles of expression and provides a coherent model for plant cell cycle control. Plant J. 41: 546-566.

Menges, M., A.K. Samland, S. Planchais and J.A. Murray. 2006. The D-type cyclin CYCD3;1 is limiting for the G1-to-S-phase transition in Arabidopsis. Plant Cell 18: 893-906.

Ni, D.A., R. Sozzani, S. Blanchet, S. Domenichini, C. Reuzeau, R. Cella, C. Bergounioux and C. Raynaud. 2009. The Arabidopsis *MCM2* gene is essential to embryo devel-opment and its over-expression alters root meristem function. New Phytol. 184: 311-322.

Nieuwland, J., S. Maughan, W. Dewitte, S. Scofield, L. Sanz and J.A. Murray. 2009. The D-type cyclin CYCD4;1 modulates lateral root density in Arabidopsis by affecting the basal meristem region. Proc. Natl. Acad. Sci. U.S.A. 106: 22528-22533.

Nowack, M.K., H. Harashima, N. Dissmeyer, X. Zhao, D. Bouyer, A.K. Weimer, F. De Winter, F. Yang and A. Schnittger. 2012. Genetic framework of cyclin-dependent kinase function in Arabidopsis. Dev. Cell 22: 1030-1040.

Nurse, P. 2000. A long twentieth century of the cell cycle and beyond. Cell 100: 71-78.

Ogawa, D., K. Abe, A. Miyao, M. Kojima, H. Sakakibara, M. Mizutani, H. Morita, Y. Toda, T. Hobo, Y. Sato, T. Hattori, H. Hirochika and S. Takeda. 2011. RSS1 regulates the cell cycle and maintains meristematic activity under stress conditions in rice. Nat. Commun. 2: 278.

Peng, L., A. Skylar, P.L. Chang, K. Bisova and X. Wu. 2014. CYCP2;1 integrates genetic and nutritional information to promote meristem cell division in Arabidopsis. Dev. Biol. 393: 160-170.

Peres, A., M.L. Churchman, S. Hariharan, K. Himanen, A. Verkest, K. Vandepoele, Z. Magyar, Y. Hatzfeld, E. Van Der Schueren, G.T. Beemster, V. Frankard, J.C. Larkin, D. Inze D and L. De Veylder. 2007. Novel plant-specific cyclin-dependent kinase inhibitors induced by biotic and abiotic stresses. J. Biol. Chem. 282: 25588-25596.

Polyn, S., A. Willems and L. De Veylder. 2015. Cell cycle entry, maintenance, and exit during plant development. Curr. Opin. Plant Biol. 23: 1-7.

Pursell, Z.F. and T.A. Kunkel. 2008. DNA polymerase epsilon: a polymerase of unusual size (and complexity). Prog. Nucleic Acid Res. Mol. Biol. 82: 101-145.

Ramirez-Parra, E., C. Frundt and C. Gutierrez. 2003. A genome-wide identification of E2F-regulated genes in Arabidopsis. Plant J. 33: 801-811.

Raynaud, C., A.C. Mallory, D. Latrasse, T. Jégu, Q. Bruggeman, M. Delarue, C. Bergounioux and M. Benhamed. 2014. Chromatin meets the cell cycle. J. Exp. Bot. 65: 2677-2689.

Reitz, M.U., M.L. Gifford and P. Schafer. 2015. Hormone activities and the cell cycle machinery in immunity-triggered growth inhibition. J. Exp. Bot. 66: 2187-2197.

Riou-Khamlichi, C., R. Huntley, A. Jacqmard and J.A. Murray. 1999. Cytokinin activation of Arabidopsis cell division through a D-type cyclin. Science 283: 1541-1544.

Rojas, C.A., N.B. Eloy, M.D. Lima, R.L. Rodrigues, L.O. Franco, K. Himanen, G.T.S. Beemster, A.S. Hemerly and P.C.G. Ferreira. 2009. Overexpression of the Arabidopsis anaphase promoting complex subunit CDC27a increases growth rate and organ size. Plant Mol. Biol. 71: 307-318.

Sanz, L., W. Dewitte, C. Forzani, F. Patell, J. Nieuwland, B. Wen, P. Quelhas, S. De Jager, C. Titmus, A. Campilho, H. Ren, M. Estelle, H. Wang, J.A.H. Murray. 2011. The Arabidopsis D-Type Cyclin CYCD2;1 and the Inhibitor ICK2/KRP2 Modulate Auxin-Induced Lateral Root Formation. Plant Cell 23: 641-660.

Schnittger, A., U. Schobinger, Y.D. Stierhof and M. Hulskamp. 2002. Ectopic B-type cyclin expression induces mitotic cycles in endoreduplicating Arabidopsis trichomes. Curr. Biol. 12: 415-420.

Shultz, R.W., V.M. Tatineni, L. Hanley-Bowdoin and W.F. Thompson. 2007. Genome-wide analysis of the core DNA replication machinery in the higher plants Arabidopsis and rice. Plant Physiol. 144: 1697-1714.

Skirycz, A., A. Radziejwoski, W. Busch, M.A. Hannah, J. Czeszejko, M. Kwasniewski, M.I. Zanor, J.U. Lohmann, L. De Veylder, I. Witt and B. Mueller-Roeber. 2008. The DOF transcription factor OBP1 is involved in cell cycle regulation in *Arabidopsis thaliana*. Plant J. 56: 779-792.

Sozzani, R., C. Maggio, S. Varotto, S. Canova, C. Bergounioux, D. Albani and R. Cella. 2006. Interplay between Arabidopsis activating factors E2Fb and E2Fa in cell cycle progression and development. Plant Physiol. 140: 1355-1366.

Springer, P.S., D.R. Holding, A. Groover, C. Yordan and R.A. Martienssen. 2000. The essential Mcm7 protein PROLIFERA is localized to the nucleus of dividing cells during the G(1) phase and is required maternally for early Arabidopsis development. Development 127: 1815-1822.

Stevens, R., M. Grelon, D. Vezon, J. Oh, P. Meyer, C. Perennes, S. Domenichini and C. Bergounioux. 2004. A CDC45 homolog in Arabidopsis is essential for meiosis, as shown by RNA interference-induced gene silencing. Plant Cell 16: 99-113.

Takahashi, N., T. Kajihara, C. Okamura, Y. Kim, Y. Katagiri, Y. Okushima, S. Matsunaga, I. Hwang and M. Umeda. 2013. Cytokinins control endocycle onset by promoting the expression of an APC/C Activator in Arabidopsis eoots. Curr. Biol. 23: 1812-1817.

Takatsuka, H., R. Ohno and M. Umeda. 2009. The Arabidopsis cyclin-dependent kinase-activating kinase CDKF;1 is a major regulator of cell proliferation and cell expansion but is dispensable for CDKA activation. Plant J. 59: 475-487.

Takatsuka, H., C. Umeda-Hara and M. Umeda. 2015. Cyclin-dependent kinase-activating kinases CDKD;1 and CDKD;3 are essential for preserving mitotic activity in *Arabidopsis thaliana*. Plant J. 82: 1004-1017.

Van Leene, J., J. Hollunder, D. Eeckhout, G. Persiau, E. Van De Slijke, H. Stals, G. Van Isterdael, A. Verkest, S. Neirynck, Y. Buffel, S. De Bodt, S. Maere, K. Laukens, A. Pharazyn, P.C. Ferreira, N. Eloy, C. Renne, C. Meyer, J.D. Faure, J. Steinbrenner, J. Beynon, J.C. Larkin, Y. Van de Peer, P. Hilson, M. Kuiper, L. De Veylder, H. Van Onckelen, D. Inze, E. Witters and G. De Jaeger. 2010. Targeted interactomics reveals a complex core cell cycle machinery in *Arabidopsis thaliana*. Mol. Syst. Biol. 6: 397.

Vandepoele, K., K. Vlieghe, K. Florquin, L. Hennig, G.T. Beemster, W. Gruissem, Y. Van de Peer, D. Inze and L. De Veylder. 2005. Genome-wide identification of potential plant E2F target genes. Plant Physiol. 139: 316-328.

Vanneste, S., Coppens F., Lee E., Donner T.J., Xie Z., Van Isterdael G., Dhondt S., De Winter F., De Rybel B., Vuylsteke M., De Veylder L., Friml J., Inze D., Grotewold E., Scarpella E., Sack F., Beemster G.T., Beeckman T. 2011. Developmental regulation of CYCA2s contributes to tissue-specific proliferation in Arabidopsis. EMBO J 30: 3430-3441

Verkest, A., C.L. Manes, S. Vercruysse, S. Mae, E. Van Der Schueren, T. Beeckman, P. Genschik, M. Kuiper, D. Inze and L. De Veylder. 2005. The cyclin-dependent kinase inhibitor KRP2 controls the onset of the endoreduplication cycle during Arabidopsis leaf development through inhibition of mitotic CDKA;1 kinase complexes. Plant Cell 17: 1723-1736.

Vodermaier, H.C. 2004. APC/C and SCF: Controlling each other and the cell cycle. Curr. Biol. 14: 787-796.

Wang, H., Q. Qi, P. Schorr, A.J. Cutler, W.L. Crosby and L.C. Fowke. 1998. ICK1, a cyclin-dependent protein kinase inhibitor from *Arabidopsis thaliana* interacts with both Cdc2a and CycD3, and its expression is induced by abscisic acid. Plant J. 15: 501-510.

Wang, H., Y. Zhou, S. Gilmer, S. Whitwill and L.C. Fowke. 2000. Expression of the plant cyclin-dependent kinase inhibitor ICK1 affects cell division, plant growth and morphology. Plant J. 24: 613-623.

Wang, W. and X. Chen. 2004. HUA ENHANCER3 reveals a role for a cyclin-dependent protein kinase in the specification of floral organ identity in Arabidopsis. Development 131: 3147-3156.

Wen, B., J. Nieuwland and J.A. Murray. 2013. The Arabidopsis CDK inhibitor ICK3/KRP5 is rate limiting for primary root growth and promotes growth through cell elongation and endoreduplication. J. Exp. Bot. 64: 1-13.

Xiong, Y., M. McCormack, L. Li, Q. Hall, C. Xiang and J. Sheen. 2013. Glucose-TOR signalling reprograms the transcriptome and activates meristems. Nature 496: 181-186.

Xouri, G., M. Dimaki, P.I. Bastiaens and Z. Lygerou. 2007. Cdt1 interactions in the licensing process: a model for dynamic spatiotemporal control of licensing. Cell Cycle 6: 1549-1552.

Yi, D., C.L.A. Kamei, T. Cools, S. Vanderauwera, N. Takahashi, Y. Okushima, T. Eekhout, K.O. Yoshiyama, J. Larkin, H. Van den Daele, A. Britt, M. Umeda andL. De Veylder. 2014. *The Arabidopsis thaliana* SIAMESE-RELATED cyclin-dependent kinase inhibitors SMR5 and SMR7 control the DNA damage checkpoint in response to reactive oxygen species. Plant Cell 26: 296-309.

Yu, Y., A. Steinmetz, D. Meyer, S. Brown and W.H. Shen. 2003. The tobacco A-type cyclin, Nicta;CYCA3;2, at the nexus of cell division and differentiation. Plant Cell 15: 2763-2777.

Zheng, T., C. Nibau, D.W. Phillips, G. Jenkins, S.J. Armstrong and J.H. Doonan. 2014. CDKG1 protein kinase is essential for synapsis and male meiosis at high ambient temperature in *Arabidopsis thaliana*. Proc.Nat. Acad. Sci.U. S. A. 111: 2182-2187.

2

Discovering the World of Plant Nuclear Proteins

Beáta Petrovská[1,], Marek Šebela[2] and Jaroslav Doležel[1]*

Introduction

Despite the separation of evolutionary lineages many hundred million years ago, cells of all eukaryotic organisms are structurally similar. Their control centre – the nucleus – contains most of the DNA of the cell and regulates the majority of cellular processes. DNA is packed in a small volume of the nucleus after interacting with nuclear proteins. These proteins facilitate DNA folding into a small space; participate in DNA replication, repair and transcription; and help to separate it from the cytoplasm. Additionally, these proteins have a strong impact on the function of the genome. Indeed, the latter cannot be understood without a good knowledge of the composition, structure and behaviour of nuclear proteins, which are the most abundant components of the nucleus (Sutherland et al. 2001). However, little information is available regarding plant nuclear proteins, except for histones and a few other proteins. We are only beginning to understand how the plant genome is organized and how it works. In this chapter, we summarize the current knowledge regarding the plant nucleus and its protein composition, structure and function, with the aim of shedding light on the nature and function of vital components of plant cell nuclei.

Identified Proteins of the Plant Nuclear Envelope

The eukaryotic nucleus is composed of two primary structural parts – the nucleoplasm and nuclear envelope (NE). The nucleoplasm includes chromosomal domains, inter-chromosomal domains, the nucleolus, other nuclear bodies, and nuclear speckles (for review, see Lanctôt et al. 2007). The NE is composed of the inner (INM) and

[1] Institute of Experimental Botany, Centre of the Region Haná for Biotechnological and Agricultural Research, Šlechtitelů 31, 783 71 Olomouc, Czech Republic.

[2] Department of Protein Biochemistry and Proteomics, Centre of the Region Haná for Biotechnological and Agricultural Research, Faculty of Science, Palacký University, Šlechtitelů 11, 783 71 Olomouc, Czech Republic.

[*] Corresponding author : petrovska@ueb.cas.cz

outer (ONM) nuclear membranes, nuclear pore complexes (NPCs), and the nuclear lamina (Hetzer et al. 2005). The NE is a dynamic structure that controls macromolecules trafficking between the nucleoplasm and cytosol and that links chromatin and the cytoskeleton. Silent chromatin associates with the NE (Akhtar and Gasser 2007, Kalverda et al. 2008) and interacts with the nuclear lamina. Active chromatin interacts with nuclear pore proteins at inner parts of the nucleus. In addition, the components of NE participate in mitosis and cell division (Kutay and Hetzer 2008) and function as a microtubule-organizing centre (Stoppin et al. 1994, Murata et al. 2005, Binarová et al. 2006).

A metazoan nuclear lamina is made up of a network of lamin filaments (Krohne and Benavente 1986, Aebi et al. 1986). In contrast, a unique lamina-like structure (Fiserova and Goldberg 2010) has been observed in plant nuclei. Because no lamin homologues have been identified in plant genome sequences (Fiserova and Goldberg 2010), the focus of research has been on the identification and characterization of specific plant lamin-like proteins (for more detail, see Guo and Fang 2014, Ciska and Moreno Díaz de la Espina 2014) (Fig. 1).

The first identified plant protein that specifically binds to the matrix attachment region (MAR) was MAR-binding filament-like protein 1 (MFP1), which was isolated from tomato by Meier et al. (1996) and from onion by Samaniego et al. (2006). In addition to the ability to form filaments typical of structural proteins of the nucleoskeleton, MFP1 is localized in plastids and is associated with thylakoid membranes (Jeong et al. 2003, Samaniego et al. 2005). Gindullis ct al. (1999) identified a protein

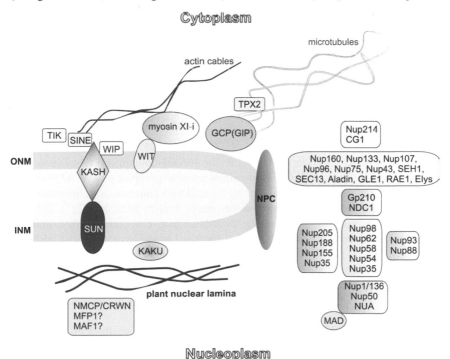

FIGURE 1 Schematic representation of plant nuclear proteins identified to date showing their approximate relative positions.

that specifically interacts with MFP1: the matrix attachment factor 1 (MAF1 or MFP1-associated factor 1). MFP1 and MAF1 localize on the NE (Meier et al. 1996, Gindullis et al. 1999, Samaniego et al. 2006). However, tomato MAF1 was later characterized as a WPP domain protein with a subcellular location outside the nucleus (Patel et al. 2004). Thus, the role of MFP1 and MAF1 as lamin-like proteins has not yet been well defined.

The best candidates for plant lamina proteins are members of the nuclear matrix constituent protein (NMCP) group. Carrot NMCP1 localizes only to the nuclear periphery (Masuda et al. 1993, 1997). Four NMCP1-related proteins were characterized in *Arabidopsis thaliana*; LINC1 (little nuclei 1) localizes to the nuclear periphery, whereas LINC2 (Dittmer et al. 2007) localizes to the nucleoplasm. LINC1 and LINC4 regulate nuclear morphology (Sakamoto and Takagi 2013). All LINC proteins are known as CRWN (crowded nuclei) because of confusion with the linker of nucleoskeleton and cytoskeleton protein (the same acronym, LINC) (Wang et al. 2013). Another NE protein, KAKU4, interacts with CRWN1 and CRWN4, localizes at the INM and functions in nuclear morphology modulation (Goto et al. 2014). Altogether, 97 NMCP proteins have been identified from 37 plant genomes (Kimura et al. 2010, Ciska et al. 2013, Ciska and Moreno Díaz de la Espina 2013, Wang et al. 2013).

The LINC complex (linker of nucleoskeleton and cytoskeleton) connects the nuclear lamina to the cytoskeletal components of cytoplasm (Starr 2009, Tzur et al. 2006, Worman and Gundersen 2006, Tatout et al. 2014). This complex is composed of Sad1/Unc84 (SUN)-domain and Klarsicht/ANC-1/Syne-1 homology (KASH)-domain proteins that associate with the INM and ONM, respectively (Sosa et al. 2012, Zhou et al. 2012).

SUN-domain proteins are the only known INM proteins in plants (Graumann et al. 2010, Oda and Fukuda 2011). In *Arabidopsis*, SUN1 and SUN2 interact with 3 tryptophan-proline-proline (WPP) domain-interacting tail-anchored (WIP) proteins, i.e., WIP1, WIP2 and WIP3 (Meier et al. 2010). Apart from *A. thaliana*, the presence of SUN-domain proteins was also confirmed in other plant species. Five different SUN genes were found in maize: SUN1 and SUN2 are structural homologues of animal SUNs, and SUN3–5 have a predicted specific role at the plant NE (Murphy et al. 2010).

WIP proteins of *Arabidopsis* belong to the KASH-domain proteins (Zhou et al. 2012). Recently, Zhou et al. (2014) identified another group of plant KASH-domain proteins, termed SINE (SUN interacting NE). SINE1 connects with the actin cytoskeleton on the ONM and is required for proper nuclear anchorage in guard cells. SINE2 contributes to innate immunity towards an oomycete pathogen in *A. thaliana*. Another KASH-domain protein, TIK [Toll-Interleukin-Resistance (TIR)-KASH protein], localizes to the NE and controls nuclear morphology in root cells (Graumann et al. 2014). The microtubule-associated protein TPX2, which was identified recently in *A. thaliana* by Petrovská et al. (2013), appears to be a potential member of the LINC complex. TPX2, together with importin, reinforces microtubule formation near chromatin and the NE.

Tamura et al. (2013) showed that myosin XI-i localizes on the ONM and interacts with WPP-domain-interacting tail-anchored (WIT) proteins. WIT proteins are required for myosin XI-i anchorage to the nuclear membrane and for nuclear movement. In addition to previous demonstrations of association between the actin cytoskeleton and the plant NE (Zhou et al. 2014), Dryková et al. (2003) demonstrated that γ-tubulin localizes on the plant NE. Subsequently, demonstrations of the localization of *A. thaliana* gamma-tubulin complex protein 2 (AtGCP2) and AtGCP3, which are γ-tubulin ring complex proteins, on the ONM were published (Seltzer et al. 2007). Batzenschlager et al. (2013) showed a significant role of γ-tubulin complex protein 3 (GCP3)-interacting proteins (GIPs) in both nuclear shaping and NE organization. These authors found that GIP proteins have a role in microtubule nucleation and function as adaptors and/or modulators of NE-associated proteins (Janski et al. 2012, Nakamura et al. 2012, Batzenschlager et al. 2013). Evidence from our laboratory also demonstrates that actin and γ-tubulin associate with the NE (unpublished observation). However, proteins facilitating this association remain unknown.

Plant-specific Elements of Nuclear Pore Complexes

The nuclear pore complexes (NPCs) are the largest cellular multiprotein complexes (40–66 MDa) of approximately 30 different nucleoporin proteins (Nups) (Tamura and Hara-Nishimura 2013). Thirty Nups with a domain organization similar to those of human and yeast origin were identified in *A. thaliana* using proteomic approaches (Tamura et al. 2010). *Arabidopsis* NPCs lack six vertebrate components (Nup358, Nup188, Nup97, Nup 45, Nup37, and a pore membrane protein of 121 kDa; Boruc et al. 2012) and have the Nup136/Nup1 protein. DNA sequence analysis of Nup136/Nup1 did not identify any vertebrate homologue of this protein.

The *Arabidopsis* NUA protein (nuclear pore anchor) represents another NPC protein (Xu et al. 2007) known for its function in SUMOylation (small ubiquitin-like modifier – post-translational modification) and mRNA export. NUA forms a complex with *Arabidopsis* MAD1 and MAD2 (mitotic arrest deficient 1 and 2) proteins (Ding et al. 2012) and plays a role in spindle checkpoint activation during mitosis.

Components of Plant Nuclear Bodies

These membraneless nuclear organelles include the nucleolus, Cajal bodies (CBs), nuclear speckles, cyclophilin-containing speckles, dicing bodies, AKIP1-containing bodies and photobodies.

Nucleoli are the largest nuclear bodies that are primarily composed of proteins (85–90%). Nucleolar RNA represents only 5–10 %, with the least abundant being rDNA (Gerbi 1997, Shaw and Brown 2012). Several nucleolar proteins have already been characterized in plants, e.g., a few homologues of nucleolin (Martin et al. 1992, Minguez and Moreno Díaz de la Espina 1996, Gonzáles-Camacho and Medina 2004, Sobol et al. 2006, Pontvianne et al. 2010, Medina et al. 2010), fibrillarin (Cerdido and Medina 1995, Pih et al. 2000), and many ribosomal proteins (Brown et al. 2005, Brown and Shaw 2008). Pendle et al. (2005) identified 217 nucleolar proteins of *Arabidopsis* using a proteomic approach.

A major component of CBs is the protein coilin (Sleeman et al. 2001, Makarov et al. 2013, Njedojadło et al. 2014). Pontes et al. (2006) showed that ARGONAUTE4 (AGO4) protein colocalizes with CBs in *Arabidopsis*.

The protein composition of plant nuclear speckles, storage sites for splicing factors, has not yet been analysed (Reddy et al. 2012).

Cyclophilin-containing speckles, which are plant-specific nuclear bodies, are composed of the CypRS64 (arginine/serine-rich domain-containing cyclophilin) protein (Lorkovic et al. 2004).

Other plant-specific nuclear bodies – dicing bodies – are composed of DCL1 (dicer-like 1) and HYL1 (hyponastic leaves 1) proteins, which are required for the processing of primary microRNA and/or the storage/assembly of the miRNA processing machinery (Fang and Spector 2007, Song et al. 2007).

AKIP1-containing plant-specific bodies were observed in the guard-cell nuclei in 2002 by Li et al. (2002) and were relocalized to the speckles after treatment with abscisic acid.

Many nuclear proteins may appear to have already been described and even functionally characterized. However, this is not true. As mentioned above, Tamura et al. (2010) used proteomic approaches for the identification of plant NPC proteins. The reason was that most nucleoporin homologues could not be found in plant genomes using homology-based searches (Meier 2006). This study is only one example of how proteomic analyses have slowly entered into the discovery of new plant nuclear proteins.

Unlocking the Nuclear Proteome

Large-scale proteomic approaches not only enable the simultaneous study of numerous proteins but also typically provide their identification. In addition, proteomic data mining leads to descriptions of protein properties such as intracellular distribution, concentration level, turnover dynamics, interaction partners and posttranslational modifications (Trinkle-Mulcahy and Lamond 2007). Despite these opportunities, the nuclear proteome research in plants remains in its infancy. Thus far, no plant nuclear proteome has been characterized completely. The first data regarding the protein composition of plant nuclei were obtained in model organisms with sequenced genomes such as *A. thaliana* (Bae et al. 2003, Jones et al. 2009) and *Oryza sativa* (Tan et al. 2007) and later in other crop species (Table 1).

The nuclear proteome research in plants started when Bae et al. (2003) identified 158 nuclear proteins in the model plant *A. thaliana* with the aim of characterizing the plant nuclear proteome and its response to cold stress. Fifty-four of the identified nuclear proteins were up- or down-regulated by cold treatment. Among these nuclear proteins, six were selected for further functional characterization. Simultaneously, Calikowski et al. (2003) published the results of the first proteomic characterization of the plant nuclear matrix. These authors isolated the nuclear matrix of *Arabidopsis* and performed its characterization by confocal and electron microscopy. Using a proteomic approach, these authors identified 36 proteins, which included known or predicted homologues of nucleolar proteins, e.g., IMP4, Nop56, Nop58, fibrillarins, nucleolin, ribosomal components, histone deacetylase, tubulins, and homologues of eEF-1, HSP/HSC70 and DnaJ, as well as a number of novel proteins with unknown functions. Later, Jones et al. (2009) identified 345 nuclear proteins in *Arabidopsis*. Novel phosphorylation sites and kinase motifs on proteins involved in nuclear transport, such as Ran-associated proteins, as well as on transcription factors, chromatin-remodelling proteins, RNA-silencing components, and the spliceosome, were characterized. In addition, these authors found several proteins involved in Golgi

TABLE 1. A list of plant nuclear proteomes identified thus far.

Organism	No. of nuclear proteins	Growth conditions	Reference
Arabidopsis thaliana	158	cold stress	Bae et al. (2003)
	36	–	Calikowski et al. (2003)
	217 ●	–	Pendle et al. (2005)
	345	–	Jones et al. (2009)
	879	–	Bigeard et al. (2014)
Oryza sativa	190	–	Khan and Komatsu (2004)
	269	–	Tan et al. (2007)
	468	–	Li et al. (2008)
	109	drought stress	Choudhary et al. (2009)
	657	sugar response	Aki and Yanagisawa (2009)
	78	dehydration response	Jaiswal et al. (2013)
	382	cell wall removal response	Mujahid et al. (2013)
Capsicum annuum	6	response to TMV infection	Lee et al. (2006)
Cicer arietinum	150	–	Pandey et al (2006)
	147	dehydration response	Pandey et al. (2008)
	75	dehydration response	Subba et al. (2013)
	107	–	Kumar et al. (2014)
Medicago truncatula	143	seed filling	Repetto et al. (2008)
Zea mays	98	UV-B light treatment	Casati et al. (2008)
	163	heterosis	Guo et al. (2014)
Xerophyta viscosa	18	dehydration response	Abdalla et al. (2010)
	122	dehydration response	Abdalla and Rafudeen (2012)
Glycine max	4975	rust infection	Cooper et al. (2011)
Hordeum vulgare	803 ♦	–	Petrovská et al. (2014)

● proteomic analysis of nucleolus, ♦ proteomic analysis of G1 nuclei, – normal conditions

vesicle trafficking that most likely contribute to cell plate formation during cytokinesis (Jones et al. 2009).

Bigeard et al. (2014) oriented his research on proteomic and phosphoproteomic analyses of chromatin-associated proteins in *Arabidopsis*. These authors identified 879 proteins, of which 198 were phosphoproteins that participate in chromatin remodelling, transcriptional regulation and RNA processing.

The nucleolar proteome of *Arabidopsis* was also characterized. Pendle et al. (2005) identified 217 nucleolar proteins. After a comparison of *Arabidopsis* and human nucleolar proteomes, these authors identified proteins with the same function in humans, proteins that are plant specific, proteins of unknown function and proteins that are nucleolar in plants but non-nucleolar in humans.

The nuclear proteome of rice, which is one of the most important crops, was described for the first time by Khan and Komatsu (2004). These authors identified 190 nuclear proteins, of which a majority are involved in signalling and gene regulation, reflecting the role of the plant nucleus in gene expression and regulation. Later, Tan et al. (2007) identified 269 unique chromatin-associated proteins, such as nucleosome assembly proteins, high-mobility group proteins, histone modification proteins, transcription factors and a large number of unknown proteins. These authors also identified 128 chromatin-associated proteins, including 11 variants of histone H2A.

A nuclei-enriched fraction of rice endosperm was used for the identification of 468 proteins by Li et al. (2008). These proteins included transcription factors, histone modification proteins, kinetochore proteins, centromere/microtubule binding proteins, and transposon proteins.

Aki and Yanagisawa (2009) identified 657 rice nuclear and nucleic acid-associated proteins including novel nuclear factors implicated in evolutionary conserved mechanisms for sugar responses in plants. Simultaneously, Choudhary et al. (2009) detected 109 rice nuclear proteins that displayed changes during drought stress. Their functions include cellular regulation, protein degradation, cellular defence, chromatin remodelling, and transcriptional regulation, supporting the role of the nucleus as the primary cellular regulator.

Jaiswal et al. (2013) characterized the nuclear proteome of rice under water-deficit conditions. These authors identified 78 nuclear dehydration-responsive proteins. Mujahid et al. (2013) analysed changes in the rice nuclear proteome during a response to cell wall removal in suspension-cultured cells. These authors identified 382 nuclear proteins including histone modification proteins, chromatin structure regulatory proteins and transcriptional factors. Gene ontology analysis showed that chromatin and nucleosome assembly proteins, protein-DNA complex assembly, and DNA packaging proteins were associated with the response to cell wall removal.

The nuclear proteome of hot pepper was studied only in 2006 by Lee et al. (2006), who followed changes in nuclear proteins as a response to tobacco mosaic virus (TMV) infection. These authors identified 6 related protein spots representing proteins that may be expressed during the hypersensitive response against the virus infection. The subsequent functional study was performed on the hot pepper 26S proteasome subunit RPN7 (CaRPN7) and demonstrated the possible involvement of CaRPN7 in TMV-induced programmed cell death.

In the first work on chickpea, Pandey et al. (2006) identified 150 nuclear proteins. A comparison of the chickpea nuclear proteome with those of *Arabidopsis* and rice showed only 8 identical proteins in all three organisms. Chickpea and rice shared 11 proteins, while rice and *Arabidopsis* shared only six. These authors showed that 71 % of the chickpea nuclear proteins were novel, highlighting the need for further research regarding the nuclear proteomes of plants.

Dehydration-responsive nuclear proteins of chickpea were studied by Pandey et al. (2008), who identified 147 differentially expressed nuclear proteins. These proteins are involved in gene transcription and replication, molecular chaperones, cell signalling, and chromatin remodelling. Subba et al. (2013) analysed the nuclear proteome of a dehydration-sensitive chickpea cultivar. These authors identified 75 differentially expressed proteins associated with different metabolic and regulatory pathways. Comparing dehydration-sensitive and tolerant cultivars, these authors identified unique proteins and overlapping proteins and indicated their contributions to dehydration tolerance.

An insight into the function of protein phosphorylation in the plant nucleus was obtained recently by Kumar et al. (2014). These authors identified 107 putative nucleus-specific chickpea phosphoproteins with various cellular functions, including protein folding; signalling; gene regulation; DNA replication, repair and modification; and metabolism. Additionally, according to their biological and molecular functions, the two most abundant categories were stress-responsive and nucleotide-binding proteins.

The nuclear proteome of another legume species, barrel clover, was characterized by Repetto et al. (2008). These authors identified 143 proteins in nuclei isolated from seed tissues. A number of proteins that are related to ribosome subunit biogenesis, chromatin structure/organization, transcription, RNA maturation, silencing and transport were identified that could regulate gene expression and prepare seeds for reserve synthesis during the filling stage. In addition, these authors identified novel nuclear proteins involved in the biogenesis of ribosomal subunits (pescadillo-like) and nucleocytoplasmic trafficking (dynamin-like GTPase). These authors hypothesized that the genome architecture could be extensively modified during seed development (e.g., by the timing of the expression of genes encoding chromatin-modifying enzymes, by the presence of RNA interference proteins in seed nuclei).

The first proteomic study on maize nuclei was performed by Casati et al. (2008). These authors compared maize nuclear proteome data before and after exposure to UV-B light and identified 98 proteins showing differences in abundance between the two conditions. Many proteins were classified as DNA binding and chromatin factors, including core histones. More recently, a reference map of the maize nuclear proteome in the basal region of the third seedling leaf was produced by Guo et al. (2014). This work led to the identification of 163 nuclear proteins primarily implicated in RNA and protein-associated functions (including amino acid activation, protein synthesis initiation, protein degradation, protein folding, protein targeting, and post translational modification). Moreover, these authors performed a comparative proteomic analysis between a highly heterotic hybrid, Mo17/B73, and its parental lines. This analysis indicated that hybridization between the two parental lines caused changes in the expression of different nuclear proteins, which could be responsible for leaf size heterosis.

Abdalla et al. (2010) analysed the nuclear proteome of the resurrection plant *Xerophyta viscosa* after its exposure to dehydration stress. In total, 438 protein spots were detected, of which 18 were shown to be up-regulated in response to dehydration. During further research, these authors identified 122 nuclear proteins (Abdalla and Rafudeen 2012) with similarity to proteins identified in the nuclear proteomes of *Arabidopsis* exposed to cold stress (Bae et al. 2003) and in chickpea under dehydration stress (Pandey et al. 2008). Approximately 66 % of *Xerophyta* nuclear proteins did not show changes in their abundance in response to dehydration, and this group included structural proteins and metabolic proteins. Approximately 22 % of the total proteins identified were shown to be more abundant, and up to 10 % proteins were less abundant in response to dehydration stress.

The nuclear proteome of soybean was studied by Cooper et al. (2011), who followed a response to a rust infection. These authors detected approximately 4975 proteins from nuclei preparations of soybean leaves, and proteins with differential accumulation changes between isogenic soybeans susceptible and resistant to the soybean rust fungus were found. However, the identified proteins were not only nuclear and included many cytoplasmic proteins. Nevertheless, this study showed that numerous plant proteins are post-translationally modified in the nucleus after pathogen infection and that some of the proteomic changes most likely reflect defence responses that may confer resistance to soybean rust.

The nuclear proteome of the fourth most important cereal crop, barley, was characterized only recently (Petrovská et al. 2014). These authors developed a novel approach to prevent contamination by cytoplasmic proteins. In addition, these authors were able to discriminate among G1, S and G2 phases of cell cycle nuclei. These authors identified 803 nuclear proteins from G1 phase nuclei of barley.

Conclusions

This chapter summarizes the current knowledge regarding the composition and function of plant nuclear proteins. The primary goal of a majority of nuclear proteomic studies performed thus far was to identify proteins involved in response to stress conditions, and no major effort has been made to characterize the complete plant nuclear proteome. Thus, a large gap in our knowledge of the critical components of the plant hereditary machinery remains, and a greater number of comprehensive and systematic studies of proteins in plant nuclei and their precise functional analyses are needed. When planning future work, keeping in mind that the nuclear proteomics alone is not powerful enough to reveal the functional and structural organization of the plant nucleus is important. Thus, plant nuclear proteomics should be complemented with functional approaches (e.g., biochemical, molecular biological, immunocytochemical, cytological, and reverse genetics approaches) and with the analysis of the three dimensional organization of the nuclear genome to obtain information needed to understand protein functions.

Acknowledgments

This work was supported by the National Program of Sustainability I (LO1204) from the Ministry of Education, Youth and Sports of the Czech Republic. We sincerely apologize to all colleagues whose relevant works could not be cited due to space limitation.

References

Abdalla, K.O. and M.S. Rafudeen. 2012. Analysis of the nuclear proteome of the resurrection plant *Xerophyta viscosa* in response to dehydration stress using iTRAQ with 2DLC and tandem mass spectrometry. J. Proteomics 18: 2361-2374.

Abdalla, K.O., B. Baker and M.S. Rafudeen. 2010. Proteomic analysis of nuclear proteins during dehydration of the resurrection plant *Xerophyta viscosa*. Plant Growth Regul. 62: 279-292.

Aebi, U., J. Cohn, L. Buhle and L. Gerace. 1986. The nuclear lamina is a mesh work of intermediate type filaments. Nature 323: 560.

Akhtar, A. and S.M. Gasser. 2007. The nuclear envelope and transcriptional control. Nat. Rev. Genet. 8: 507-517.

Aki, T. and S. Yanagisawa. 2009. Application of rice nuclear proteome analysis to the identification of evolutionarily conserved and glucose-responsive nuclear proteins. J. Proteome Res. 8: 3912-3924.

Bae, M.S., E.J. Cho, E.Y. Choi and O.K. Park. 2003. Analysis of the *Arabidopsis* nuclear proteome and its response to cold stress. Plant J. 36: 652-663.

Batzenschlager, M., K. Masoud, N. Janski, G. Houlné, E. Herzog, J.-L. Evrard, N. Baumberger, M. Erhardt, Y. Nominé, B. Kieffer, A.-C. Schmit and M-E. Chabouté. 2013. The GIP gamma-tubulin complex-associated proteins are involved in nuclear architecture in *Arabidopsis thaliana*. Front. Plant Sci. 4: 480.

Bigeard, J., N. Rayapuram, D. Pflieger and H. Hirt. 2014. Phosphorylation-dependent regulation of plant chromatin and chromatin-associated proteins. Proteomics 14: 2127-2140.

Binarová, P., V. Cenklová, J. Procházková, A. Doskočilová, J. Volc, M. Vrlík and L. Bögre. 2006. Gamma-tubulin is essential for acentrosomal microtubule nucleation and coordination of late mitotic events in *Arabidopsis*. Plant Cell 18: 1199-1212.

Boruc, J., X. Zhou and I. Meier. 2012. Dynamics of the plant nuclear envelope and nuclear pore. Plant Physiol. 158: 78-86.

Brown, J.W. and P.J. Shaw. 2008. The role of the plant nucleolus in pre-mRNA processing. Curr. Top. Microbiol. Immunol. 326: 291-311.

Brown, J.W.S., P.J. Shaw, P. Shaw and D.F. Marshall. 2005. *Arabidopsis* nucleolar protein database (AtNoPDB). Nucleic Acids Res. 33: 633-636.

Calikowski, T.T., T. Meulia and I. Meier. 2003. A proteomic study of the *Arabidopsis* nuclear matrix. J. Cell. Bioch. 90: 361-378.

Casati, P., M. Campi, F. Chu, N. Suzuki, D. Maltby, S. Guan, A.L. Burlingame and V. Walbot. 2008. Histone acetylation and chromatin remodelling are required for UV-B-dependent transcriptional activation of regulated genes in maize. Plant Cell 20: 827-842.

Cerdido, A. and F.J. Medina. 1995. Subnucleolar location of fibrillarin and variation in its levels during the cell cycle and during differentiation of plant cells. Chromosoma 103: 625-634.

Choudhary, M.K., D. Basu, A. Datta, N. Chakraborty and S. Chakraborty. 2009. Dehydration-responsive nuclear proteome of rice (*Oryza sativa* L.) illustrates protein network, novel regulators of cellular adaptation and evolutionary perspective. Mol. Cell. Proteomics 8: 1579-1598.

Ciska, M. and S. Moreno Díaz de la Espina. 2013. NMCP/LINC proteins: putative lamin analogs in plants? Plant Signal. Behav. 8: e26669.

Ciska, M. and S. Moreno Díaz de la Espina. 2014. The intriguing plant nuclear lamina. Front. Plant Sci. 5: 166.

Ciska, M., K. Masuda and S. Moreno Díaz de la Espina. 2013. Lamin-like analogues in plants: the characterization of NMCP1 in *Allium cepa*. J. Exp. Bot. 64: 1553-1564.

Cooper, B., K.B. Campbell, J. Feng, W.M. Garrett and R. Frederick. 2011. Nuclear proteomic changes linked to soybean rust resistance. Mol. BioSyst. 3: 773-783.

Ding, D., S. Muthuswamy and I. Meier. 2012. Functional interaction between the *Arabidopsis* orthologs of spindle assembly checkpoint proteins MAD1 and MAD2 and the nucleoporin NUA. Plant Mol. Biol. 79: 203-216.

Dittmer, T.A., N.J. Stacey, K. Sugimoto-Shirasu and E.J. Richards. 2007. LITTLE NUCLEI genes affecting nuclear morphology in *Arabidopsis thaliana*. Plant Cell 19: 2793-2803.

Dryková, D., V. Cenklová, V. Sulimenko, J. Volc, P. Dráber and P. Binarová. 2003. Plant gamma-tubulin interacts with alpha/beta-tubulin dimers and forms membrane-associated complexes. Plant Cell 15: 465-480.

Fang, Y. and D.L. Spector. 2007. Identification of nuclear dicing bodies containing proteins for microRNA biogenesis in living *Arabidopsis plants*. Curr. Biol. 17: 818-823.

Fiserova, J. and M.W. Goldberg. 2010. Relationships at the nuclear envelope: lamins and nuclear pore complexes in animals and plants. Biochem. Soc. T. 38: 829-831.

Gerbi, S.A. 1997. The nucleolus: then and now. Chromosoma 105: 385-387.

Gindullis, F., N.J. Peffer and I. Meier. 1999. MAF1, a novel plant protein interacting with matrix attachment region binding protein MFP1, is located at the nuclear envelope. Plant Cell 11: 1755-1768.

Gonzáles-Camacho, F. and F.J. Medina. 2004. Identification of specific plant nucleolar phosphoproteins in a functional proteomic analysis. Proteomics 4: 407-417.

Goto, C., K. Tamura, Y. Fukao, T. Shimada and I. Hara-Nishimura. 2014. The novel nuclear envelope protein KAKU4 modulates nuclear morphology in *Arabidopsis*. Plant Cell 26: 2143-2155.

Graumann, K., E. Vanrobays, S. Tutois, A.V. Probst, E.D. Evans and C. Tatout. 2014. Characterization of two distinct subfamilies of SUN-domain proteins in *Arabidopsis* and their interactions with the novel KASH-domain protein AtTIK. J. Exp. Bot. 65: 6499-6512.

Graumann, K., J. Runions and D.E. Evans. 2010. Characterization of SUN-domain proteins at the higher plant nuclear envelope. Plant J. 61: 134-144.

Guo, T. and Y. Fang. 2014. Functional organization and dynamics of the cell nucleus. Front. Plant Sci. 5: 378.

Guo, B., Y. Chen, C. Li, T. Wang, R. Wang, B. Wang, S. Hu, X. Du, H. Xing, X. Song, Y. Yao, Q. Sun and Z. Ni. 2014. Maize (*Zea mays* L.) seedling leaf nuclear proteome and differentially expressed proteins between a hybrid and its parental lines. Proteomics 14: 1071-1087.

Hetzer, M.W., T.C. Walther and M. IW. 2005. Pushing the envelope: structure, function, and dynamics of the nuclear periphery. Annu. Rev. Cell Dev. Biol. 21: 347-380.

Jaiswal, D.K., D. Ray, M.K. Choudhary, P. Subba, A. Kumar, J. Verma, R. Kumar, A. Datta, S. Chakraborty and N. Chakraborty. 2013. Comparative proteomics of dehydration response in the rice nucleus: new insights into the molecular basis of genotype-specific adaptation. Proteomics 13: 3478-3497.

Janski, N., K. Masoud, M. Batzenschlager, E. Herzog, J.-L. Evrard, G. Houlné, M. Bourge, M.-E. Chabouté and A.-C. Schmit. 2012. The GCP3-interacting proteins GIP1 and GIP2 are required for γ-tubulin complex protein localization, spindle integrity, and chromosomal stability. Plant Cell 24: 1171-1187.

Jeong, S.Y., A. Rose and I. Meier. 2003. MFP1 is a thylakoid-associated, nucleoid-binding protein with a coiled-coil structure. Nucleic Acids Res. 31: 5175-5185.

Jones, A.M., D. MacLean, D.J. Studholme, A. Serna-Sanz, E. Andreasson, J.P. Rathjen and S.C. Peck. 2009. Phosphoproteomic analysis of nuclei-enriched fractions from *Arabidopsis thaliana*. J. Proteomics 72: 439-451.

Kalverda, B., M.D. Röling and M. Fornerod. 2008. Chromatin organization in relation to the nuclear periphery. FEBS Lett. 582: 2017-2022.

Khan, M.K.K. and S. Komatsu. 2004. Rice proteomics: recent developments and analysis of nuclear proteins. Phytochemistry 65: 1671-1681.

Kimura, Y., C. Kuroda and K. Masuda. 2010. Differential nuclear envelope assembly at the end of mitosis in suspension-cultured *Apium graveolens* cells. Chromosoma 119: 195-204.

Krohne, G. and R. Benavente. 1986. The nuclear lamins. A multigene family of proteins in evolution and differentiation. Exp. Cell Res. 162: 1-10.

Kumar, R., A. Kumar, P. Subba, S. Gayali, P. Barua, S. Chakraborty and N. Chakraborty. 2014. Nuclear phosphoproteome of developing chickpea seedlings (*Cicer arietinum* L.) and protein-kinase interaction network. J. Proteomics 105: 58-73.

Kutay, U. and M.W. Hetzer. 2008. Reorganization of the nuclear envelope during open mitosis. Curr. Opin. Cell Biol. 20: 669-677.

Lanctôt, C., T. Cheutin, M. Cremer, G. Cavalli and T. Cremer. 2007. Dynamic genome architecture in the nuclear space: regulation of gene expression in three dimensions. Nature Rev. Genet. 8: 104-115.

Lee, B.J., S.J. Kwon, S.K. Kim, K.J. Kim, C.J. Park, Y.J. Kim, O.K. Park and K.H. Paek. 2006. Functional study of hot pepper 26S proteasome subunit RPN7 induced by tobacco mosaic virus from nuclear proteome analysis. Biochem. Biophys. Res. Commun. 351: 405-411.

Li, J., T. Kinoshita, S. Pandey, C.K.-Y. Ng, S.P. Gygi, K. Shimazaki and S.M. Assmann. 2002. Modulation of an RNA-binding protein by abscisic-acid-activated protein kinase. Nature 418: 793-797.

Li, G, B.R. Nallamilli, F. Tan and Z. Peng. 2008. Removal of high abundance proteins for nuclear subproteome studies in rice (*Oryza sativa*) endosperm. Electrophoresis 29: 604-617.

Lorkovic, Z.J., S. Lopato, M. Pexa, R. Lehner and A. Barta A. 2004. Interactions of *Arabidopsis* RS domain containing cyclophilins with SR proteins and U1 and U11 snRNP-specific proteins suggest their involvement in pre-mRNA splicing. J. Biol. Chem. 279: 33890-33898.

Makarov, V., D. Rakitina, A. Protopopova, I. Yaminsky, A. Arutiunian, A.J. Love, M. Taliansky and N. Kalinina. 2013. Plant Coilin: Structural characteristics and RNA-binding properties. PLoS ONE 8: e53571.

Martin, M., L.F. Garcia-Fernandes, S. Moreno Díaz de la Espina, J. Noaillac-Depeyre, N. Gas and F.J. Medina. 1992. Identification and localization of a nucleolin in onion nucleoli. Exp. Cell Res. 199. 74-84.

Masuda, K., S. Takahashi, K. Nomura, M. Arimoto and M. Inoue M. 1993. Residual structure and constituent proteins of the peripheral framework of the cell nucleus in somatic embryos from *Daucus carota* L. Planta 191: 532-540.

Masuda, K., Z.J. Xu, S. Takahashi, A. Ito, M. Ono, K. Nomura and M. Inoue. 1997. Peripheral framework of carrot cell nucleus contains a novel protein predicted to exhibit a long alpha-helical domain. Exp. Cell Res. 232: 173-181.

Medina, F.J., F. González-Camacho, A.I. Manzano, A. Manrique and R. Herranz. 2010. Nucleolin, a major conserved multifunctional nucleolar phosphoprotein of proliferating cells. J. Appl. Biomed. 8: 141-150.

Meier, I., T. Phelan, W. Gruissem, S. Spiker and D. Schneider. 1996. MFP1, a novel plant filament-like protein with affinity for matrix attachment region DNA. Plant Cell 8: 2105-2115.

Meier, I., X. Zhou, J. Brkljacić, A. Rose, Q. Zhao and X.M. Xu. 2010. Targeting proteins to the plant nuclear envelope. Biochem. Soc. Trans. 38: 733-740.

Meier, I. 2006. Composition of the plant nuclear envelope: theme and variations. J. Exp. Bot. 58: 27-34.

Minguez, A. and S. Moreno Díaz de la Espina. 1996. *In situ* localization of nucleolin in the plant nucleolar matrix. Exp. Cell Res. 222: 171-178.

Mujahid, H., F. Tan, J. Zhang, B.R.R. Nallamilli, K. Pendarvis and Z. Peng. 2013. Nuclear proteome response to cell wall removal in rice (*Oryza sativa*). Proteome Sci. 11: 26.

Murata, T., S. Sonobe, T.I. Baskin, S. Hyodo, S. Hasezawa, T. Nagata, T. Horio and M. Hasebe. 2005. Microtubule-dependent microtubule nucleation based on recruitment of gamma-tubulin in higher plants. Nat. Cell Biol. 7: 961-968.

Murphy, S.P., C.R. Simmons and H.W. Bass. 2010. Structure and expression of the maize (*Zea mays* L.) SUN-domain protein gene family: evidence for the existence of two divergent classes of SUN proteins in plants. BMC Plant Biol. 10: 269.

Nakamura, M., N. Yagi, T. Kato, S. Fujita, N. Kawashima, D. Ehrhardt and T. Hashimoto. 2012. *Arabidopsis* GCP3-interacting protein1/MOZART1 is an integral component of the γ-tubulin-containing microtubule nucleating complex. Plant J. 71: 216-225.

Njedojadło, J., E. Kubicka, B. Kalich and D.J. Smoliński. 2014. Poly(A) RNAs including coding proteins RNAs occur in plant Cajal bodies. PLoS ONE 9: e111780.

Oda, Y. and H. Fukuda. 2011. Dynamics of *Arabidopsis* SUN proteins during mitosis and their involvement in nuclear shaping. Plant J. 66: 629-641.

Pandey, A., S. Chakraborty, A. Datta and N. Chakraborty. 2008. Proteomics approach to identify dehydration responsive nuclear proteins from chickpea (*Cicer arietinum* L.). Mol. Cell. Proteomics 7: 88-107.

Pandey, A., M.K. Choudhary, D. Bhushan, A. Chattopadhyay, S. Chakraborty, A. Datta and N. Chakraborty. 2006. The nuclear proteome of chickpea (*Cicer arietinum* L.) reveals predicted and unexpected proteins. J. Proteome Res. 5: 3301-3311.

Patel, S., A. Rose, T. Meulia, R. Dixit, R.J. Cyr and I. Meier. 2004. *Arabidopsis* WPP-domain proteins are developmentally associated with the nuclear envelope and promote cell division. Plant Cell 16: 3260-3273.

Pendle, A.F., G.P. Clark, R. Boon, D. Lewandowska, Y.M. Lam, J. Andersen, M. Mann, A.I. Lamond, J.W. Brown and P.J. Shaw. 2005. Proteomic analysis of the *Arabidopsis* nucleolus suggests novel nucleolar functions. Mol. Biol. Cell 16: 260-269.

Petrovská, B., H. Jeřábková, I. Chamrád, J. Vrána, R. Lenobel, J. Uřinovská, M. Šebela and J. Doležel. 2014. Proteomic analysis of barley cell nuclei purified by flow sorting. Cytogenet. Genome Res. 143: 78-86.

Petrovská, B., H. Jeřábková, L. Kohoutová, V. Cenklová, Ž. Pochylová, Z. Gelová, G. Kočárová, L. Váchová, M. Kurejová, E. Tomaštíková and P. Binarová. 2013. Overexpressed TPX2 causes ectopic formation of microtubular arrays in the nuclei of acentrosomal plant cells. J. Exp. Bot. 64: 4575-4587.

Pih, K.T., M.J. Yi, Y.S. Liang, B.J. Shin, M.J. Cho, I. Hwang and D. Son. 2000. Molecular cloning and targeting of a fibrillarin homolog from *Arabidopsis*. Plant Physiol. 123: 51-58.

Pontes, O., C.F. Li, P.C. Nunes, J. Haag, T. Ream, A. Vitins, S.E. Jacobsen and C.S. Pikaard. 2006. The *Arabidopsis* chromatin-modifying nuclear siRNA pathway involves a nucleolar RNA processing center. Cell 126: 79-92.

Pontvianne, F., M. Abou-Ellail, J. Douet, P. Comella, I. Matia, C. Chandrasekhara, A. DeBures, T. Blevins, R. Cooke, F.J. Medina, S. Tourmente, C.S. Pikaard and S. Sáez-Vásquez S. 2010. Nucleolin is required for DNA methylation state and the expression of rRNA gene variants in *Arabidopsis thaliana*. PLoS Genet. 6: e1001225.

Reddy, A.S., I.S. Day, J. Gohring and A. Barta. 2012. Localization and dynamics of nuclear speckles in plants. Plant Physiol. 158: 67-77.

Repetto, O., H. Rogniaux, C. Firnhaber, H. Zuber, H. Küster, C. Larré, R. Thompson and K. Gallardo. 2008. Exploring the nuclear proteome of *Medicago truncatula* at the switch towards seed filling. Plant J. 56: 398-410.

Sakamoto, Y. and S. Takagi. 2013. LITTLE NUCLEI 1 and 4 regulate nuclear morphology in *Arabidopsis thaliana*. Plant Cell Physiol. 54: 622-633.

Samaniego, R., S.Y. Jeong, C. de la Torre, I. Meier and S. Moreno Díaz de la Espina. 2006. CK2 phosphorylation weakens 90kDa MFP1 association to the nuclear matrix in *Allium cepa*. J. Exp. Bot. 57: 113-124.

Samaniego, R., S.Y. Jeong, I. Meier and S. Moreno Díaz de la Espina. 2005. Dual location of MAR-binding, filament-like protein 1 in *Arabidopsis*, tobacco, and tomato. Planta 223, 1201-1206.

Seltzer, V., N. Janski, J. Canaday, E. Herzog, M. Erhardt, J. Evrard and A.-C. Schmit. 2007. *Arabidopsis* GCP2 and GCP3 are part of a soluble γ-tubulin complex and have nuclear envelope targeting domains. The Plant Journal 52, 322-331.

Shaw, P. and J.W.S. Brown. 2012. Nucleoli: composition, function and dynamics. Plant Physiol. 158: 44-51.

Sleeman, J.E., P. Ajuh and A.I. Lamond. 2001. snRNP protein expression enhances the formation of Cajal bodies containing p80-coilin and SMN. J. Cell Sci. 114: 4407-4419.

Sobol, M., F. Gonzalez-Camacho, V. Rodríguez-Vilariño, E. Kordyum and F.J. Medina. 2006. Subnucleolar location of fibrillarin and NopA64 in *Lepidium sativum* root meristematic cells is changed in altered gravity. Protoplasma 228: 209-219.

Song, L., M.H. Han, J. Lesicka and N. Fedoroff. 2007. *Arabidopsis* primary microRNA processing proteins HYL1 and DCL1 define a nuclear body distinct from the Cajal body. PNAS 104: 5437-5442.

Sosa, B.A., A. Rothballer, U. Kutay and T.U. Schwartz. 2012. LINC complexes form by binding of three KASH peptides to domain interfaces of trimeric SUN proteins. Cell 149: 1035-1047.

Starr, D.A. 2009. A nuclear-envelope bridge positions nuclei and moves chromosomes. J. Cell Sci. 122: 577-586.

Stoppin, V., M. Vantard, A.-C. Schmit and A.M. Lambert. 1994. Isolated plant nuclei nucleate microtubule assembly: The nuclear surface in higher plants has centrosome-like activity. Plant Cell 6: 1099-1106.

Subba, P., R. Kumar, S. Gayali, S. Shekhar, S. Parveen, A. Pandey, A. Datta, S. Chakraborty and N. Chakraborty. 2013. Characterisation of the nuclear proteome of a dehydration-sensitive cultivar of chickpea and comparative proteomic analysis with a tolerant cultivar. Proteomics 13: 1973-1992.

Sutherland, H.G.E., G.K. Mumford, K. Newton, L.V. Ford, R. Farrall, G. Dellaire, J.F. Cáceres and W.A. Bickmore. 2001. Large-scale identification of mammalian proteins localized to nuclear sub-compartments. Hum. Mol. Gen. 10: 1995-2011.

Tamura, K. and I. Hara-Nishimura. 2013. The molecular architecture of the plant nuclear pore complex. J. Exp. Bot. 64: 823-832.

Tamura, K., K. Iwabuchi, Y. Fukao, M. Kondo, K. Okamoto, H. Ueda, M. Nishimura and I. Hara-Nishimura. 2013. Myosin XI-i links the nuclear membrane to the cytoskeleton to control nuclear movement and shape in *Arabidopsis*. Curr. Biol. 23: 1776-1781.

Tamura, K., Y. Fukao, M. Iwamoto, T. Haraguchi and I. Hara-Nishimura. 2010. Identification and characterization of nuclear pore complex components in *Arabidopsis thaliana*. Plant Cell 22: 4084-4097.

Tan, F., G. Li, B.R. Chitteti and Z. Peng Z. 2007. Proteome and phosphoproteome analysis of chromatin associated proteins in rice (*Oryza sativa*). Proteomics 7: 4511-4527.

Tatout, C., D.E. Evans, E. Vanrobays, A.V. Probst and K. Graumann. 2014. The plant LINC complex at the nuclear envelope. Chromosome Res. 22: 241-252.

Trinkle-Mulcahy, L. and A.I. Lamond. 2007. Towards a high-resolution view of nuclear dynamics. Science 318: 1402-1407.

Tzur, Y.B., K.L. Wilson and Y. Gruenbaum. 2006. SUN-domain proteins: 'Velcro' that links the nucleoskeleton to the cytoskeleton. Nat. Rev. Mol. Cell Biol. 7: 782-788.

Wang, H., T.A. Dittmer and E.J. Richards. 2013. *Arabidopsis* CROWDED NUCLEI (CRWN) proteins are required for nuclear size control and heterochromatin organization. BMC Plant Biol. 13: 200.

Worman, H.J. and G.G. Gundersen. 2006. Here come the SUNs: a nucleocytoskeletal missing link. Trends Cell Biol. 16: 67-69.

Xu, X.M., A. Rose and I. Meier. 2007. NUA activities at the plant nuclear pore. Plant Signal. Behav. 2: 553-555.

Zhou, X., K. Graumann, L. Wirthmueller, J.D.G. Jones and I. Meier. 2014. Identification of unique SUN-interacting nuclear envelope proteins with diverse functions in plants. J. Cell Biol. 205: 677-692.

Zhou, X., K. Graumann, D.E. Evans and I. Meier. 2012. Novel plant SUN-KASH bridges are involved in RanGAP anchoring and nuclear shape determination. J. Cell Biol. 196: 203-211.

3

Plastid Division

Dr Kevin A. Pyke[*]

Introduction

Plastids are a class of organelles resident in plant cells fundamental to plant cell function in a variety of different ways and are one of the major defining features which distinguish plant cells from animal cells. The endosymbiotic evolution of plastids and the uptake of an early photosynthetic single celled organism into early eukaryotic cells represent a fundamental point in the evolution of complex life on the planet (Gray and Archibald 2012). The continued evolution of the endosymbiont plastid in its new environment within the eukaryotic cell required significant modification of plastid function and interaction with its host both at the genetic and biochemical levels (McFadden 2014). The end result is that modern day plant cells are completely dependent on plastid function for an array of metabolic processes and developmental signals (Jarvis and López-Juez 2013) as well as the seminal process of photosynthesis by which the plant can acquire carbon, increase in biomass and grow.

A critical step in the successful integration of the endosymbiont plastid into the cellular function was acquiring the ability to divide in a manner coordinated with the cell cycle and cell division. A mechanism that ensures plastids are present in both daughter cells after cell division is crucial to cell function, since without such a mechanism, plastids would be quickly lost from the cell lineage (Birky 1983). In addition, once plants became multi-cellular containing different types of differentiated cells, the population size of plastids in certain types of cells increased in size in order to fulfil their specific function within those cells. Thus the modern day plant cell has a variety of different plastids performing different metabolic and biochemical functions in different cell types (Pyke 2009).

An understanding of the mechanism by which plastids divide has increased significantly in the last 20 years, primarily through work on chloroplast division and by identifying genes and proteins that function in the chloroplast division process by forward genetic screens, primarily in Arabidopsis (Osteryoung and Pyke 2014). Young chloroplasts divide during the expansion of leaf mesophyll cells during leaf development to generate large populations of green chloroplasts in these mesophyll cells. In most plants, mature mesophyll cells contain between 50-200 chloroplasts

Division of Plant and Crop Sciences, University of Nottingham, Sutton Bonington Campus, Loughborough, Leicestershire, LE12 5RD, UK.
[*] Corresponding author : Kevin.Pyke@Nottingham.ac.uk

per cell, which have differentiated and subsequently divided from 15-20 proplastids, which were present in the post mitotic cell derived from the shoot apical meristem (SAM) prior to mesophyll differentiation. There are exceptions though and in the small mesophyll cells of rice leaves (*Oryza sativa*), there are only 6-10 chloroplasts present (Pyke 2012). Indeed relatively few species of the Angiosperms have been studied in detail for chloroplast number and size (Pyke 1999) and it is highly likely that significant variation exists in this characteristic between different species. Other types of plastids accumulate in different cell types within the plant; for instance ripening fruit cells or petal cells accumulate coloured chromoplasts and cells in storage tissues such as seed endosperm or tubers accumulate amyloplasts which store starch.

At the outset, studies on plastid division have made the slightly presumptuous assumption that different types of plastids used a common mechanism to divide, if they divide at all. This article will consider the current understanding of how plastids divide and will consider whether differences might exist in the way that different plastid types might divide in different cell types.

Chloroplast Division

The classic and best characterised mode of plastid division is that of the chloroplast which typically divide by binary fission, in which a centrally constricting mechanism pulls in the outer and inner chloroplast envelope membranes by force resulting in dumbbell shaped plastids (Leech et al. 1981). This results subsequently in fusion of the inner and outer envelope membranes respectively and the formation of two new daughter chloroplasts. Such constriction sites are normally situated at the mid length of the plastid's longer axis and thus the two daughter plastids are approximately evenly sized. Stages of this constriction process can be easily viewed in expanding leaf cells and gave rise to the first evidence that plastids actually divided (Leech et al.1981). By screening for and identifying Arabidopsis chloroplast division mutants and considering analogies of chloroplast division to bacterial cell division from the chloroplast's prokaryotic evolutionary past, an array of genes and their proteins have now been identified that function in a chloroplast division mechanism (Osteryoung and Pyke 2014). The mechanism consists of concentric rings of proteins both on the inside of the envelope membrane in the stroma and on the cytosolic external surface, along with proteins, which interconnect the concentric rings through the envelope membranes and its lumen. Central in this process is a ring of FtsZ protein, composed of FtsZ1 and FtsZ2 monomers, which forms on the stromal surface of the inner envelope membrane. In a co-ordinated manner, on the outside of the outer envelope membrane, a ring of DR5B/ARC5 protein, related to dynamin, also forms and together these two rings generate a co-ordinated constrictive force which is the major source of energy driving the whole binary fission process. The two rings are co-ordinated in their construction and their synchronised constriction by several other proteins including ARC6, PDV1, PDV2 and PARC1 which either play a role in recruiting proteins to form ring structures or facilitate linking them together through the envelope lumen (Osteryoung and Pyke 2014). During the division process, the constriction mechanism pulls in the membranes until the latter stages of the constriction where both envelope membranes fuse with their counter parts to produce two separated chloroplasts. Detail of the mechanism by which the envelope membranes fuse and how the internal thylakoid membranes are divided and segregated into the two new daughter chloroplasts are unknown. Based on genetic and biochemical experiments, a working model of how this chloroplast division mechanism works has been proposed (Osteryoung and Pyke 2014). The energy for the constriction is derived from the hydrolysis of GTP since both FtsZ and dynamin are GTPases. Isolation of intact constriction complexes and measurement of the force generated, showed clearly that the major driver in this system is the dynamin ring which pushes the membrane in by constriction as the FtsZ ring pulls on the inside (Yoshida et al. 2006).

The placement of the ring at the midpoint of the chloroplast's axis is important since highly asymmetric divisions would result in mini chloroplasts which may well be non-functional. A system using Min proteins, derived from that which directs the placement of the division ring in bacterial cell division, also functions in chloroplast division where by the proteins minD, minE and ARC3 direct the FtsZ ring to be constructed centrally at the midpoint of the long axis of the chloroplast and actively inhibit it forming near the end poles of the chloroplast thereby ensuring that the constriction proceeds at a central point (Basak and Møller 2013, Osteryoung and Pyke 2014).

In addition, other factors have also been shown to influence chloroplast division in an as yet unknown capacity. Somewhere in the control of the central placement of division, membrane tension and mechano-sensing appears to play a role since mutations in the MSL2 and MSL3 mechano-sensing proteins in Arabidopsis dramatically alter chloroplast division by changing the placement of the FtsZ ring (Wilson et al. 2011). It is possible that changes in the tension of the chloroplast envelope through the division process may be monitored by such a mechano-sensing system, giving rise to an element of control of the division process (Pyke 2006). Furthermore, a role for lipids in control of chloroplast division is also implicated, evidence coming from mutants in which KASI, the elongation factor in fatty acid synthesis is mutant and in which FtsZ ring placement and FtsZ polymerization are both perturbed (Wu and Xue 2010, Fan and Xu 2011). A consideration of lipids and membrane function in chloroplast division has been, until now, lacking and this may be a fruitful area for coming research in the plastid division field. Indeed a recent study (Okazaki et al. 2015) clearly shows that PDV1 and PDV2 proteins interact with the plastid envelope lipid phosphatidylinositol 4-phosphate to regulate the extent of chloroplast division in the cell.

One pertinent question here is how quickly the constriction process leading to final separation of the two daughter chloroplasts happens. Few people have observed chloroplast division within a cell in real time, although final plastid separation has been observed in hanging drops from isolated chloroplasts (Ridley and Leech 1970). Most recent research has focussed more on the molecular and genetic aspects of the process rather than the cell biology and older studies followed the morphology of the chloroplast division process in great detail (Leech et al.1981). In young wheat leaf cells, young chloroplasts in constriction are observable for 20 minutes (Boffey et al.1979) suggesting that the entire division process may last around one hour. However it remains unclear how the process is controlled in terms of initiation, progression through ring recruitment, motile constriction and membrane fusion and one suspects that in different situations the division process may take much longer.

The experiments leading up to our current understanding of the way chloroplasts divide has been well documented and the reader is directed to several recent excellent reviews (Miyagishima 2011, Miyagishima et al. 2011, Pyke 2013, Osteryoung and Pyke 2014). We will consider questions here now which build on this current knowledge. Although early workers on plastid division considered that a standard division mechanism would work in all types of plastids, evidence is accumulating that this may not be so and that variations on this FtsZ/Dr5B constriction mechanism may occur in different plastid types.

Proplastid Division in the Shoot Apical Meristem

Let us consider proplastids, which are the progenitors of all plastids within the plant. Proplastids are resident in meristem cells in both the shoot apical meristem (SAM) and the root apical meristem (RAM) and are required to divide and segregate at cell division to ensure continuity of plastids in cell lineages within the meristem cells. Similarly proplastid division is necessary in associated tissues in which extensive cell division occurs, such as leaf primordia and developing embryos. Classic electron micrographs of proplastids in shoot and root meristem cells reveal them as small plastids with minimal thylakoid membrane and pigment and pleomorphic in shape (Chaley and Possingham 1981, Robertson et al. 1995), although more recent elegant studies using 3D electron tomography reveal variation in proplastid structure within different layers and zones of the SAM with more extensive thylakoid biogenesis in proplastids in distinct regions (Charuvi et al. 2012). Most estimates of proplastid numbers in SAM cells suggest between 10-20 proplastids are present in each cell but must be required to replicate in the face of extensive cell divisions within the meristem, giving rise to leaf primordia and subsequent leaf expansion. Indeed modern imaging of the SAM proplastids reveals that proplastid number per cell increases from 9-12 in the G1 and S stages of the cell cycle to 23-25 proplastids per cell in prometaphase at the start of mitosis; a doubling in numbers which presumably pre-empts the subsequent cell division (Segui-Simarro and Staehlin 2009). Little is known about proplastid division itself but several lines of evidence suggest that perturbation of the chloroplast division machinery by mutation does not necessarily affect the division of proplastids in the SAM, since cells lacking plastids are not observed and seedling function and plant growth in such chloroplast division mutants is relatively normal (Robertson et al. 1995). The DR5B/ARC5 gene was shown not to be expressed in the SAM (Okazaki et al. 2009) and when mutated has no effect on proplastid number in the SAM (Robertson et al. 1996), suggesting that it does not function in proplastid division in the SAM. In contrast, *arc6* mutants do have fewer greatly enlarged proplastids in their SAM cells (Robertson et al. 1995). Surprisingly, Arabidopsis plants in which all FtsZ proteins have been knocked out by mutation are still viable and grow normally (Schmitz et al. 2009). Consequently an alternative or moderated mechanisms for proplastid division may exist. What might this be? Analysis of the transcriptome of the SAM and associated tissues maybe a useful tool to look for expression of known plastid division related genes. Such an analysis of the transcriptome of the Arabidopsis shoot apex containing the SAM and leaf primordia clearly shows expression of the key chloroplast division genes *FtsZ, ARC3, DR5B/ARC5, ARC6, MinD, MinE* (López-Juez et al. 2008). Furthermore, a transcriptome analysis of very young Arabidopsis embryos shows transcription of chloroplast division genes (Slane et al. 2014). Most plastids in young developing embryos are likely to be proplastids although some evidently contain chlorophyll and maybe chloroplast-like even at this early stage (Tejos et al. 2010). Together the evidence suggests that machinery that performs chloroplast division in leaf mesophyll cells may operate in proplastid division but possibly in a slightly different manner. No proplastid specific division mutants have been identified to date, although they may well confer embryo lethality. It is possible that a secondary "backup" division mechanism is exists in proplastids, as suggested by the *arc6* mutant phenotype in SAM cells.

Whatever the mechanism to proliferate proplastids, a given proplastid will need to replicate through several cycles of cell division and subsequent segregation into newly divided cells. Indeed the control of proplastid segregation, which is critical to ensure plastid continuity in cell populations, is also poorly understood. Experiments addressing organelle segregation at cytokinesis (Sheahan et al. 2004) suggest that an even distribution of proplastids around the cell or association of proplastids with the nuclear membrane would be enough to position proplastids in such a way that at cell plate formation there are always sufficient proplastids positioned into each daughter cell.

Root Plastid Division

Proplastid division also occurs at several other points during plant development including embryo development in the developing seed and in the meristem in the root apical meristem (RAM). Few studies have examined plastids in the RAM or indeed in the root as a whole in terms of plastid division, even though the RAM is much easier to image than the SAM. Unlike the SAM, plastids in cells leaving the RAM do not differentiate significantly as do chloroplasts in the stems and leaves of above ground organs. Root plastids are generally small, highly variable in their morphology and appear not to accumulate in large populations in root cells (Osteryoung and Pyke 2014). Knowledge of the cellular dynamics of root plastids is surprisingly scant probably because root plastids have no pigment and are difficult to see other than by electron microscopy. Using plastid targeted green fluorescent protein allows them to imaged relatively easily (Pyke 2013) and analysis of populations of root plastids in cells along the length of the growing root suggests that numbers of plastids per cell shows little change, implying that root plastids do not divide (Pyke unpublished). In general root cells have around 15-25 plastids per cell and even through the region of major cell expansion in the root, this number does not increase resulting in a decline in the proportion of plastid coverage of the expanded cells. Root plastids are crucial to root cell function and metabolism in many different ways although establishment of large populations covering much of the cell's surface is obviously not required in roots. Interestingly no mutants have been identified thus far which enable plastid replication in roots cells resulting in root cells full of plastids, suggesting that plastid division in root cells is heavily repressed. This feature may be part of the general repression of chloroplast development in roots which is controlled by auxin flow from the shoot and the subsequent down-regulation of HY5 and GLK transcription factors in root cells (Kobayashi et al. 2012) Obviously proplastids must divide in the RAM to generate the 15-25 plastids observed in root cells further up the root. It is of note that transcriptomics of the soybean root do not show expression of the chloroplast division genes discussed earlier (Haerizadeh et al. 2011), suggesting that root proplastids may well replicate by a mechanism different to that of chloroplasts.

Division of Plastids Accumulating Starch

Amyloplasts are an important type of plastid that are long-term cellular stores of starch and are found primarily in plant storage tissues such as seed endosperm and storage tissues such as potatoes. It is salient to remember that all of the starch synthesised in all of the plants in the world is made in amyloplasts. The best analysis of amyloplast development and division has been carried out in the endosperm of cereal seeds in which amyloplasts differentiate from proplastids early in the development of the storage structure and synthesise grains of starch. Different species have differing complexities of starch grain structure being simple or complex and varying in number per amyloplast. In wheat seed endosperm, amyloplasts divide by binary fission a process involving a central constriction (Langeveld et al. 2000, Bechtel and Wilson 2003). In rice seed endosperm during the early stages of amyloplast formation, several division sites within an amyloplasts can be initiated simultaneously, resulting in a "beads on a string" appearance of dividing amyloplasts before they finally separate, a process in which the ARC5/DR5BP protein is involved (Yun and Kawagoe 2009, Langeveld et al. 2000). This is a similar division phenotype to that of *minD* chloroplast mutants and suggests that the *min* system of constriction positioning is modified during amyloplast division as well as FtsZ and PDV proteins playing a role in this process (Yun and Kawagoe 2010). Certainly FtsZ can be manipulated to change starch grain size in potato tubers (de Pater et al. 2006). Once amyloplasts approach maturity, various septa form within the amyloplasts demarcating regions of the amyloplast stroma, which eventually contain compound starch granules (Yun and Kawagoe 2010). Detailed analysis of septum positioning and growth suggests that they may initiate the synthesis of compound starch granules. It is clear however that these septa do not facilitate actual physical separation of amyloplasts. It seems likely that amyloplast division has had to be adapted from that of chloroplast division simply because the amyloplast is full of starch grains (Crumpton-Taylor et al. 2012) which would hinder a simple binary fission and the mechanism of septal growth within the amyloplast may reflect this.

Chloroplasts also exhibit amyloplast features in that in many species transitory starch granules are synthesised during the light period and are degraded through the dark period (Crumpton-Taylor et al. 2012). It appears that a precise number of granule initiation sites exist in these chloroplasts. During the rounds of chloroplast division occurring during leaf development, new initiation sites for starch grain synthesis must be generated in daughter chloroplasts to maintain starch grain number and such synthesis could be induced by the subsequent expansion of the chloroplast after division, since starch grain number per unit chloroplast volume is tightly maintained (Crumpton-Taylor et al. 2012).

Plastid Division Producing Chromoplasts

Chromoplasts are differentiated plastids which accumulate coloured pigments, normally a spectrum of carotenoid molecules, and are found in coloured plant structures such as petals and fruits. Studies of chromoplast differentiation have been largely focussed on fruit ripening and in particular tomato and peppers, in which coloured chromoplasts differentiate from populations of green chloroplasts, which is the process underlying the green to red transition of ripening fruit (Egea et al. 2010, Pesarasi et al. 2014). There is little evidence that chromoplasts themselves are capable of extensive division, although they can differentiate from a variety of other plastid types (Egea et al. 2010). The large populations of chromoplasts in tomato fruit cells differentiate from large populations of green chloroplasts which appear to replicate during development of the young green fruit by binary fission in a similar way to leaf mesophyll chloroplasts (Waters et al. 2004, Forth and Pyke 2006). The bulk of this chloroplast division seems to happen as chloroplast division in green fruit up to the point at breaker, when ripening changes become visible in terms of chlorophyll loss and carotenoid biosynthesis (Cookson et al. 2003, Pan et al. 2013). Several studies analysing the transcriptome and proteome of the fruit chloroplast development followed by chromoplast differentiation show that plastid division genes are expressed in immature green fruit when most chloroplast division is occurring (Tang et al. 2013) whereas once chromoplasts have differentiated, proteins associated with plastid division are mostly lost (Barsan et al. 2010, Wang et al. 2013a), although the presence of FtsZ protein in mature chromoplasts may reflect a secondary structural role for this protein in chromoplast function (Barsan et al. 2010). Many mutants and transgenic lines exist which perturb the normal tomato fruit ripening process and some have been characterised for their effect on chloroplast accumulation and subsequent chromoplast differentiation from the point of view of increasing total plastid content of tomato fruit cells to enhance nutritional quality. The *hp-1* mutant, which is mutant in the *UV-DAMGED DNA BINDING PROTEIN-1* (DDB1) shows increased chloroplast division and subsequent chromoplast compartment size in mature tomato fruit (Cookson et al. 2003) and transcriptome analysis shows a marked increase in the transcription of the chloroplast division genes *FtsZ-1* and *ARC3* (Tang et al. 2013). Both the *hp-2* (Kolotilin et al. 2007) and the *hp-3* (Galpaz et al. 2008) tomato mutants also show an increase in chloroplast number and at maturity, carotenoid content. Such lines are of major interest for those trying to increase carotenoid content of mature ripe tomato fruit or tomato fruit photosynthesis (Cocaliadis et al. 2014). Unfortunately both encode photomorphic control proteins which have a more global effect on cellular development than plastid division alone. It remains unclear exactly how these three *hp* mutants manifest a greater plastid compartment size in either green fruit cells or mature ripe fruit cells. Another gene, *ARRP2-LIKE*, also plays a role in accumulation of plastids in tomato pericarp cells since changes in the level of expression of *ARRP2-LIKE* are able to manipulate the plastid compartment size in both green and red tomato fruit (Pan et al. 2013).

Division of Other Differentiated Plastid Types

In addition to chloroplast, amyloplasts, chromoplasts and root plastids, plants contain other general types of plastids normally referred to as non-green plastids since they largely lack chlorophyll. Indeed throughout the Angiosperms there will be many types of plastids which will have features of chloroplasts, amyloplasts, chromoplasts and others principally because plastids cannot really be pigeonholed as to specific types, principally on the spectrum of pigments or storage molecules which they accumulate. Whether other non-green plastids divide is unclear. Few cells other than fully green photosynthetic tissues such as leaves, cotyledons, sepals, outer parts of stems or some floral tissues seem to accumulate large populations of plastids which dominate within the cell as occurs during leaf mesophyll development. However some of the components of the chloroplast division machinery have been shown to be functionally active in other plastid types in different cells, such as FtsZ1 function in pollen grain plastids (Tang et al. 2009) and PARC6 and ARC12 in etioplasts. (Wang et al. 2013b). Analysis of the min system in non-green etioplasts for constriction site placement suggests that division of such plastids does not require the full complement of min components to facilitate constriction positioning as occurs in chloroplasts (Wang et al. 2013a).

Plastid Dividing Ring

Throughout the history of plastid division research, a key feature has been the presence of two plastid division rings (PD), which are in addition to the FtsZ and ARC5/DR5B rings already discussed. These rings have been observed extensively by electron microscopy in chloroplasts from a range of species, especially algae and they have been implicated as playing a major role in the chloroplast division process (Kuroiwa et al. 1998, Miyagishima et al. 2001). The innermost PD is constructed on the stromal surface of the inner envelope membrane in coordination with the FtsZ and ARC5/DR5B rings and the outer PD ring forms in a similar way on the cytoplasmic surface of the outer envelope membrane giving rise to four concentric rings at the midpoint of the chloroplast (Osteryoung and Pyke 2014). The outer PD ring in the red alga *Cyanidioschyzon merolae* is composed of filaments of polyglucans along with the enzyme PDR1 which is glycogenin-like in nature and is likely to be involved in polyglucan synthesis (Yoshida et al. 2010). The gene encoding PDR1 has homologs widely conserved across the plant kingdom including higher plants (Yoshida et al. 2010) so it is generally considered that the PD rings function in higher plant chloroplast division, although most of the evidence for such a role has come from electron micrographs and clear molecular genetic evidence is lacking (Kuroiwa et al. 2002; Yoshida et al. 2012). If PD rings do indeed function in higher plant chloroplast division then one would expect a mutant in the *PDR1* gene to impair chloroplast division but such mutants have not been reported. Thus exactly how PD rings might function in higher plant chloroplasts remains an open question.

Alternative Mechanisms for Plastid Replication

There are some lines of evidence that suggest plastids might replicate by means other than central binary fission. A budding type of replication by which bits of plastid are pinched off of the parent plastid has been suggested (Pyke 2010, 2012) and several clear images of higher plant chloroplasts in this form have been shown (Kulandaivelu and Gnanam 1985, Forth and Pyke 2006, Pyke 2010). Large chloroplasts of the fruit of the *suffulta* mutant of tomato clearly undergo a budding type of process, which can also be observed in wild type tomato chloroplasts (Forth and Pyke 2006, Pyke 2010). Such a mechanism is not dissimilar to the beads on a string phenotype shown by amyloplasts in division (Yun and Kawogoe 2009) nor dissimilar to the phenotype of Arabidopsis *minD* mutants in which plastids divide asymmetrically (Colletti et al. 2000). Thus a budding mechanism may simply be an adaptation of the conventional FtsZ ARC5/DR5B system except physically displaced.

Control of Plastid Division and Populations of Plastids in Cells

Most of the studies to date on plastid division have focussed on the mechanism by which binary fission occurs. An equally important point is the question of how the cell controls the number and size of the plastid population within it, which is part of the bigger question of how a cell knows how big it is and how many different organelles it contains and how big they are (Chan and Marshall 2012). It is clear that any control system that occurs in different cells within a developing plant results in markedly different sizes of plastid populations in the mature cells. For instance, a comparison of leaf mesophyll cells, leaf epidermal cells and root cells reveals a big difference in the plastid population size and the total plastid compartment size in relation to the size of the cell is massively different. One clear characteristic of chloroplast compartment size however in leaf mesophyll cells is that there is a major compensatory mechanism that trades off chloroplast number and chloroplast size such that the total amount of chloroplast material in relation to cell size is relatively constant (Osteryoung and Pyke 2014). This is clearly shown by *arc* mutants of Arabidopsis, since perturbing plastid division and reducing the number of chloroplasts per cell is compensated for by chloroplast enlargement thus the total compartment size remains similar (Pyke and Leech 1992, 1994). In several cell types, which undergo rapid cell expansion, it is commonly observed that the plastids initiate division but apparently fail to complete it and accumulate as dumbbell shaped plastids. This has been observed in expanding petals cells in Arabidopsis (Pyke and Page 1998), expanding epidermal cells in cotyledons (Pyke 1997) and leaf trichome cells (Pyke unpublished), which suggests that some signal either emanating from the cell expansion process or as a result of a reduction in plastid density within the cell can initiate the division process, at least in the early stages but which apparently fails to complete. The cell specific nature of how plastid populations are controlled is also well illustrated by neighbouring cell layers in the SAM. The L1 layer gives rise to the epidermal cells in which plastids are low in number and are small, poorly differentiated chloroplasts with low levels of chlorophyll whereas in cells derived from the neighbouring L2 layer which differentiate into proper mesophyll cells with large populations of fully developed functional chloroplasts. Certainly a repression of plastid division must play a role in epidermal cells in preventing full chloroplast development and division although the nature of such cell specific control has yet to be discovered. Another question relating to how cells produce large populations of individual chloroplasts is why do they and is it necessary? Many photosynthetic organisms such as algae only contain one or two chloroplasts per cell yet the evolution of higher plants appears to have facilitated a complex control system enabling extensive chloroplasts division in its primary photosynthetic cells, the mesophyll. It is generally considered that the production of large populations of small chloroplasts in these cells optimises various photosynthetic parameters to enhance photosynthetic potential of the leaf. This has been shown by analysing *arc* mutants of Arabidopsis which generally have fewer larger chloroplasts in these cells and show reduced mesophyll conductance to

CO_2 (Weise et al. 2015). In addition the mechanism by which chloroplasts relocate within mesophyll cells in relation to light intensity (Morita and Nakamura 2012) is also compromised in *arc* mutants (Jeong et al. 2002). This suggests that the replication and expansion of chloroplasts in leaf mesophyll cells has produced the optimal arrangement to maximise photosynthesis of the leaf as a whole.

Hints at the underlying control of chloroplast division have started to be revealed by the characterisation of mutant genes which alter chloroplast division but turn out to be homologues of other known control pathways in plants. Specifically, there appears to be cross over between far red light signalling in plant development and chloroplast division since the chloroplast division mutant CDP45 is also FHY3 which functions in far red light signalling pathways (Gao et al. 2013, Chang et al. 2015). Exactly how these two different pathways interact is unclear but it strongly suggests that any control of chloroplast division in particular will have its roots deeply embedded within more general developmental control system within plant development.

References

Barsan, C., P. Sanchez-Bel, C. Rombaldi, I. Egea, M. Rossignol, M. Kuntz, M. Zouine, A. Latche, M. Bouzayen and J.C. Pech. 2010. Characteristics of the tomato chromoplast revealed by proteomic analysis. J. Exp. Bot. 61: 2413-2431.

Basak, I. and S.G. Møller. 2013. Emerging facets of plastid division regulation. Planta 237: 389-98.

Bechtel, D.B. and J.D. Wilson. 2003. Amyloplast formation and starch grain development in hard winter wheat. Cereal Chem. 80: 175-183.

Boffey, S.A., J.R. Ellis, G. Selldén and R.M. Leech. 1979. Chloroplast division and DNA synthesis in light-grown wheat leaves. Plant Physiol. 64: 502-505.

Birky, C.W. Jr. 1983. The partitioning of cytoplasmic organelles at cell division. Int. Rev. Cytol. Suppl. 15: 49-89.

Chaley, N. and J.V. Possingham. 1981. Structure of constricted proplastids in meristematic plant tissues. Biologie Cellulaire 41: 203-210.

Chan, Y-H. M. and W.F. Marshall. 2012. How cells know the size of their organelles. Science 337: 1186-1189.

Chang, N., Gao, Y.F., Zhao, L., Liu, X.M. and Gao, H.B. 2015. Arabidopsis FHY3/CPD45 regulates far-red light signaling and chloroplast division in parallel. Sci. Reports 5 Article Number: 9612.

Charuvi, D., V. Kiss, R. Nevo, E. Shimoni, Z. Adam, Z. and Z. Reich. 2012. Gain and loss of photosynthetic membranes during plastid differentiation in the shoot apex of Arabidopsis. Plant Cell 24: 1143-1157.

Cocaliadis, M.F., R. Fernandez-Munoz, C. Pons, D. Orzaez and A. Granell. 2014. Increasing tomato fruit quality by enhancing fruit chloroplast function. A double-edged sword? J. Exp. Bot. 65: 4589-4598.

Colletti, K.S., E.A. Tattersall, K.A. Pyke, J.E. Froelich, K.D. Stokes and K.W. Osteryoung. 2000. A homologue of the bacterial cell division site-determining factor MinD mediates placement of the chloroplast division apparatus. Curr. Biol 10: 507-516.

Cookson, P.J., J.W. Kiano, C.A. Shipton, P.D. Fraser, S. Romer, W. Schuch, P.M. Bramley and K.A. Pyke. 2003. Increases in cell elongation, plastid compartment size and phytoene synthase activity underlie the phenotype of the *high pigment-1* mutant of tomato. Planta 217: 896-903.

Crumpton-Taylor, M., S. Grandison, K.M.Y. Png, A.J. Bushby and A.M. Smith. 2012. Control of starch granule numbers in Arabidopsis chloroplasts. Plant Physiol. 158: 905-916.

de Pater S., M. Caspers, M. Kottenhagen, H. Meima, R. ter Stege and N. de Vetten. 2006. Manipulation of starch granule size distribution in potato tubers by modulation of plastid division. Plant Biotech. J. 4: 123-34.

Egea, I., C. Barsan, W. Bian, E. Purgatto, A. Latché, C. Chervin, M. Bouzayen and J.C. Pech. 2010. Chromoplast differentiation: current status and perspectives. Plant Cell Physiol. 51: 1601-11.

Fan, J. and C. Xu. 2011. Genetic analysis of Arabidopsis mutants impaired in plastid lipid import reveals a role of membrane lipids in chloroplast division. Plant Signal. Behav. 6: 458-460.

Forth, D. and K.A. Pyke. 2006. The *suffulta* mutation in tomato reveals a novel method of plastid replication during fruit ripening. J. Exp. Bot. 57: 1971-1979.

Galpaz, N., Q. Wang, N. Menda, D. Zamir and J. Hirschberg. 2008. Abscisic acid deficiency in the tomato mutant *high-pigment 3* leading to increased plastid number and higher fruit lycopene content. Plant J. 53: 717-30.

Gao, Y.F., H., Liu, C.J., An, Y.H., Shi, X., Liu, W.Q., Yuan, B., Zhang, J., Yang, C.X. Yu, and H.B. Gao, 2013. Arabidopsis FRS4/CPD25 and FHY3/CPD45 work cooperatively to promote the expression of the chloroplast division gene ARC5 and chloroplast division. Plant J. 75: 795-807.

Gray, M.W. and J.M. Archibald. 2012. Origins of mitochondria and plastids. pp. 1-30. *In* Govindjee and T.D. Sharkey (eds). Advances in Photosynthesis and Respiration Vol. 35 Springer, New York.

Haerizadeh, F., M.B. Singh and P.L. Bhalla. 2011. Transcriptome profiling of soybean root tips. Funct. Pl. Biol. 38: 451-461.

Jarvis, P., and E. López-Juez. 2013. Biogenesis and homeostasis of chloroplasts and other plastids. Nat. Rev. Mol. Cell Biol. 14: 787-802.

Jeong W.J., Y-I. Park, K.H. Suh, J.A. Raven, O.J. Yoo and Y.R. Liu. 2002. A large population of small chloroplasts in tobacco leaf cells allows more effective chloroplast movement than a few enlarged chloroplasts. Plant Physiol. 129: 112-121.

Kobayashi, K., S. Baba, T. Obayashi, M. Sato, K. Toyooka, M. Keranen, E.M. Aro, H. Fukaki, H. Ohta, K. Sugimoto and T. Masuda. 2012. Regulation of root greening by light and auxin/cytokinin signaling in Arabidopsis. Plant Cell 24: 1081-1095.

Kolotilin, I., H. Koltai, Y. Tadmor, C. Bar-Or, M. Reuveni, A. Meir, S. Nahon, H. Shlomo, L. Chen and I. Levin. 2007. Transcriptional profiling of *high pigment-2(dg)* tomato mutant links early fruit plastid biogenesis with its overproduction of phytonutrients. Plant Physiol. 145: 389-401.

Kulandaivelu, G. and A. Gnanam. 1985. Scanning electron microscopic evidence for a budding mode of chloroplast multiplication in higher plants. Physiol. Plant. 63: 299-302.

Kuroiwa, T., H. Kuroiwa, A. Sakai, H. Takahashi, K. Toda and R. Itoh. 1998. The division apparatus of plastids and mitochondria. Int. Rev. Cytol. 181: 1-41.

Kuroiwa, H., T. Mori, M. Takahara, S-Y. Miyagishima and T. Kuroiwa T. 2002. Chloroplast division machinery as revealed by immunofluorescence and electron microscopy. Planta 215: 185-190.

Langeveld, S.M.J., R. Van Wijk, N. Stuurman, J.W. Kijne and S. de Pater. 2000. B-type granule containing protrusions and interconnections between amyloplasts in developing wheat endosperm revealed by transmission electron microscopy and GFP expression. J. Exp. Bot. 51: 1357-1361.

Leech, R.M., W.W. Thomson and K.A. Platt-Aloia. 1981. Observations of the mechanism of chloroplast division in higher plants. New Phytol. 87: 1-9.

López-Juez, E., E. Dillon, Z. Magyar, S. Khan, S, Hazeldine, S.M. de Jager, J.A. Murray, G.T. Beemster, L. Bögre and H. Shanahan. 2008. Distinct light-initiated gene expression and cell cycle programs in the shoot apex and cotyledons of Arabidopsis. Plant Cell 20: 947-968.

Morita M.T. and M. Nakamura. 2012. Dynamic behaviour of plastids related to environmental response. Curr. Opin. Plant Biol. 15: 722-728.

McFadden, G.I. 2014. Origin and evolution of plastids and photosynthesis in eukaryotes. Cold Spring Harb. Perspect. Biol. 6: a016105.

Miyagishima, S-Y., S. Takahara and T. Kuroiwa. 2001. Novel filaments 5 nm in diameter constitute the cytosolic ring of the plastid division apparatus. Plant Cell 13: 707-721.

Miyagishima, S-Y. 2011. Mechanism of plastid division: from a bacterium to an organelle. Plant Physiol. 155: 1533-1544.

Miyagishima, S-Y., H. Nakanishi and Y. Kabeya. 2011. Structure, regulation and evolution of the plastid division machinery. Int. Rev. Cell Mol. Biol. 291: 115-153.

Okazaki, K., Y. Kabeya, K. Suzuki, T. Moria, T. Ichikawa, M. Matsui, H. Nakanishi and S-Y. Miyagishima. 2009. The PLASTID DIVISION1 and 2 components of the chloroplast division machinery determine the rate of chloroplast division in land plant cell differentiation. Plant Cell 21: 1769-1780.

Osteryoung, K.W. and K.A. Pyke. 2014. Division and dynamic morphology of plastids. Ann. Rev. Plant Biol. 65: 443-472.

Okazaki, K., S-Y. Miyagishima and H. Wada. 2015. Phosphatidylinositol 4-phosphate negatively regulates chloroplast division in Arabidopsis. Plant Cell 27: 663-674.

Pan Y., G. Bradley, K. Pyke, G. Ball, C. Lu, R. Fray, A. Marshall, S. Jayasuta, C. Baxter, R. van Wijk, L. Boyden, R. Cade, N.H. Chapman, P.D. Fraser, C. Hodgman and G. B. Seymour. 2013. Network inference analysis identifies an *APRR2-Like* Gene linked to pigment accumulation in tomato and pepper fruits. Plant Physiol 161: 1476-1485.

Pesaresi P., C. Mizzotti, M. Colombo and S. Masiero. 2014. Genetic regulation and structural changes during tomato fruit development and ripening. Front. Plant Sci. 23: 124.

Pyke, K.A. 1997. The genetic control of plastid division in higher plants. Am. J. Bot. 84: 1017-1027.

Pyke, K.A. 1999. Plastid division and development. Plant Cell 11: 549-556.

Pyke, K.A. 2006. Plastid division: the squeezing gets tense. Curr. Biol. 16: R60-R62.

Pyke, K.A. 2009. Plastid Biology. First Edition. Cambridge University Press, Cambridge, UK.

Pyke, K.A. 2010. Plastid Division. AoB Plants. AoB Plants doi: 10.1093/aobpla/plq016.

Pyke, K. 2012. Mesophyll. In: eLS. John Wiley & Sons Ltd, Chichester. http://www.els.net [doi: 10.1002/9780470015902.a0002081.pub2].

Pyke, K.A. 2013. Divide and shape: an endosymbiont in action. Planta 237: 381-387.

Pyke, K.A. and A.M. Page. 1998. Plastid ontogeny during petal development in Arabidopsis. Plant Physiol. 116: 797-803.

Pyke K.A. and R.M. Leech. 1992. Nuclear mutations radically alter chloroplast division and expansion in *Arabidopsis thaliana*. Plant Physiol. 99: 1005-1008.

Pyke K.A. and R.M. Leech. 1994. A genetic analysis of chloroplast division in *Arabidopsis thaliana*. Plant Physiol. 104: 201-207.

Ridley, S.M. and R.M. Leech. 1970. Division of chloroplasts in an artificial environment. Nature 227: 463-465.

Robertson, E.J., K.A. Pyke and R.M. Leech. 1995. *arc*6, a radical chloroplast division mutant of Arabidopsis also alters proplastid proliferation and morphology in shoot and root apices. J. Cell Sci. 108: 2937-2944.

Robertson, E.J., S.M. Rutherford and R.M. Leech. 1996. Characterisation of chloroplast division using the Arabidopsis mutant *arc5*. Plant Physiol. 112: 149-159.

Schmitz, A.J., J.M. Glynn, B.J.S.C. Olson, K.D. Stokes and K.W. Osteryoung. 2009. Arabidopsis *FtsZ2-1* and *FtsZ2-2* are functionally redundant, but FtsZ-based plastid division is not essential for chloroplast partitioning or plant growth and development. Mol. Plant 2: 1211-1222.

Segui-Simarro, J.M. and L.A. Staehelin. 2009. Mitochondrial reticulation in shoot apical meristem cells of Arabidopsis provides a mechanism for homogenization of mtDNA prior to gamete formation. Plant Signal. Behav. 4: 168-171.

Sheahan, M.B., R.J. Rose and D.W. McCurdy. 2004. Organelle inheritance in plant cell division: the actin cytoskeleton is required for unbiased inheritance of chloroplasts, mitochondria and endoplasmic reticulum in dividing protoplasts. Plant J. 37: 379-390.

Slane D., J. Kong, K.W. Berendzen, J, Kilian, A. Henschen, M. Kolb, M. Schmid, K. Harter, U. Mayer, I. De Smet, M. Bayer and G. Jürgens. 2014. Cell type-specific transcriptome analysis in the early *Arabidopsis thaliana* embryo. Development 141: 4831-4840.

Tang, L.Y., N. Nagata, R. Matsushima, Y.L. Chen, Y. Yoshioka and W. Sakamoto. 2009. Visualization of plastids in pollen grains: Involvement of FtsZ1 in pollen plastid division. Plant Cell Physiol. 50: 904-908.

Tang X., Z. Tang, S. Huang, J. Liu, J. Lium, W. Shi, X. Tian, Y. Li, D. Zhang, J. Yang, Y. Gao, D. Zeng, P. Hou, X. Niu, Y. Cao, G. Li, X. Li, F. Xiao and Y. Liu. 2013. Whole transcriptome sequencing reveals genes involved in plastid/chloroplast division and development are regulated by the *HP1/DDB1* at an early stage of tomato fruit development. Planta 238: 923-936.

Tejos, R.I., A.V. Mercado and L.A. Meisel. 2010. Analysis of chlorophyll fluorescence reveals stage specific patterns of chloroplast-containing cells during Arabidopsis embryogenesis. Biol. Res 43: 99-111.

Wang, Y-Q., Y. Yang, Z. Fei, H. Yuan, T. Fish, T.W. Thannhauser, M. Mazourek, L.V. Kochian, X. Wang and L. Lil. 2013a. Proteomic analysis of chromoplasts from six crop species reveals insights into chromoplast function and development. J. Exp. Bot. 64: 949-961.

Wang P., J. Zhang, J.B. Su, P. Wang, J. Liu, B. Liu, D.R. Feng, J.F. Wang and H.B. Wang HB. 2013b. The chloroplast Min system functions differentially in two specific nongreen plastids in *Arabidopsis thaliana*. PLoS ONE 8: e71190.

Waters, M.T., R.G. Fray and K.A. Pyke. 2004. Stromule formation is dependent upon plastid size, plastid differentiation status and the density of plastids within the cell. Plant J. 39: 655-667.

Weise, S.E., D.J. Carr, A.M. Bourke, D.T. Hanson, D. Swarthout and T.D. Sharkey. 2015. The arc mutants of Arabidopsis with fewer large chloroplasts have a lower mesophyll conductance. Photosynth. Res. 124: 117-126.

Wilson, M.E., G.S. Jensen and E.S. Haswell. 2011. Two mechanosensitive channel homologs influence division ring placement in Arabidopsis chloroplasts. Plant Cell 23: 2939-2949.

Wu, G.Z. and H.W. Xue. 2010. Arabidopsis β-ketoacyl-[acyl carrier protein] synthase is crucial for fatty acid synthesis and plays a role in chloroplast division and embryo development. Plant Cell 22: 3726-44.

Yoshida Y, Kuroiwa H, Misumi O, Nishida K, Yagisawa F, Fujiwara T, Nanamiya H, Kawamura F, Kuroiwa T. 2006. Isolated chloroplast division machinery can actively constrict after stretching. Science 313: 1435-1438.

Yoshida Y., H. Kuroiwa, O. Misumi, M. Yoshida, M. Ohnuma, T. Fujiwara, F. Yagisawa, S. Hirooka, Y. Imoto, K. Matsushita, S. Kawano and T. Kuroiwa. 2010. Chloroplasts divide by contraction of a bundle of nanofilaments consisting of polyglucan. Science 329: 949-953.

Yoshida T., S-Y. Miyagishima, H. Kuroiwa, T. Kuroiwa. 2012. The plastid-dividing machinery: formation, constriction and fission. Curr. Opin. Plant Biol. 15: 714 -721.

Yun M-S. and Y. Kawagoe. 2009. Amyloplast division progresses simultaneously at multiple sites in the endosperm of rice. Plant Cell Physiol. 50: 1617-1626.

Yun M-S. and Y. Kawagoe. 2010. Septum formation in amyloplasts produces compound granules in the rice endosperm and is regulated by plastid division proteins. Plant Cell Physiol. 51: 1469-1479.

4

Mitochondrial and Peroxisomal Division

Shin-ichi Arimura[1,2,] and Nobuhiro Tsutsumi[1,3]*

Introduction

Mitochondria and peroxisomes are essential organelles in eukaryotic cells. They divide to maintain themselves during cell division. The division is also important for controlling the numbers and shapes of the organelles in order to meet the needs of the cell. Although mitochondria and peroxisomes have different numbers, shapes and physiological functions, they share some components involved in division. Here we review recent studies on the components, mechanisms and regulation of division of these organelles.

Mitochondrial Division

Plant mitochondria

Plant mitochondria are usually smaller (about 0.5 μm in width and from 0.5 to 2 μm in length) and more numerous than mitochondria in animals and fungi. Each plant cell has hundreds to more than ten thousand mitochondria with small particulate shapes (Fig. 1A). They move around the cell at speeds of several μm/sec, which is about ten times faster than the movement of mitochondria in mammals and yeast (Fig. 1B). Plant mitochondrial genomes, which range in size from 200 Kbp to more than 10 Mbp, are much larger than those in mammals, which have a constant size of around 16 Kbp (reviewed in Kubo and Newton 2008, Sloan 2013). The large plant mitochondrial genomes are usually depicted as one or more master-circle(s), that contain various repeat sequences of various lengths. Microscopic observations of plant mitochondrial DNA reveal that it is a mixture of linear, branched (probably recombining or replicating form) and circular structures of various sizes (Oldenburg and Bendich 1996). The DNA is packaged with proteins into nucleoids in the mitochondrial matrix.

---

[1] Graduate School of Agricultural and Life Sciences, The University of Tokyo, 1-1-1 Yayoi, Bunkyo-ku, Tokyo 113-8657, Japan,

[2] PRESTO, Japan Science and Technology Agency, 4-1-8, Honcho, Kawaguchi, Saitama 332-0012, Japan,

[3] CREST, Japan Science and Technology Agency, 4-1-8, Honcho, Kawaguchi, Saitama 332-0012, Japan

[*] Corresponding author : arimura@mail.ecc.u-tokyo.ac.jp,

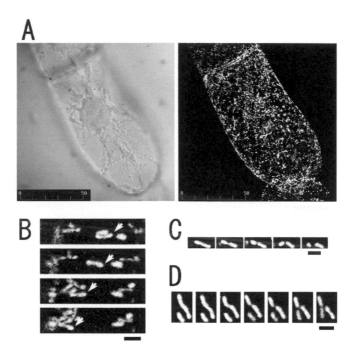

FIGURE 1   Mitochondria in living tobacco suspension cultured (BY-2) cells.

A: Transmission (left) and confocal laser scanning microscopic (right) images of the same tobacco BY-2 cell. Mitochondria are stained by Mito Tracker Orange. Scale unit is 50 μm.

B: Moving mitochondria imaged at 3-sec intervals. Scale bar, 2 μm.

C: Mitochondrial fission imaged at 10-sec intervals. Scale bar, 2 μm.

D: Mitochondrial fission and probable fusion imaged at 10-sec intervals. The long mitochondrion at right divides into two mitochondria, and the upper one becomes attached to the mitochondrion at lower left. Scale bar, 2 μm. Figure is reproduced from Arimura and Tsutsumi (2006).

The roles of plant mitochondria include not only oxidative respiration but also the metabolisms of amino acids, lipids and nucleotides. Plant mitochondria also have plant-specific roles in photorespiration and balancing the redox state in response to over-reduction by photosynthesis (Noguchi and Yoshida 2008, Foyer and Noctor 2013). Mitochondria change their shape by constant fission and fusion (Fig. 1C, D), not only to maintain or to increase their numbers but also to make their shapes fit the physiological needs of the cell. The following sections describe the similarities and differences in mitochondrial division in plants and other organisms.

Mitochondrial Division in Budding Yeast

Mitochondrial movement and division have been studied since the early 20th century (Lewis and Lewis, 1914). Starting around 1990, budding yeast *Saccharomyces cerevisiae* has been used as a model organism for mitochondrial dynamics because much is known about its genetics and because it is well suited for microscopic observation. Analyses of yeast mutants with defects in mitochondrial distribution and morphology (*mdm* mutants) have identified several genes involved in mitochondrial dynamics

TABLE. Homologous relationship of mitochondrial fission and fusion factors in diversified eukaryotes

Process	Organism				Features, functions and localization (estimated)	Involvement in peroxisomal division
	Yeast *S. cerevisiae*	Mammal *H. sapiens*	Green plant *A. thaliana*	Red algae *C. merolae*		
Fission	Dnm1	Drp1	DRP3A, DRP3B	Dnm1	Cytosol, mt & px division ring, GTPase	+
	Fis1	Fis1	AtFIS1A, AtFIS1B	CMQ197C	OM, for localization of DRPs? (through Mdv1 in yeast)	+
	Mdv1, Caf4	–	–	Mdal	OM outersurface, interact with DRP and Fis1 (in yeast)	+
	–	Mff	–	–	OM, for localization of DRP to fission sites	+
	–	Mid49&51	–	–	OM, for localization of DRP to fission sites	–
	–	–	ELM1, At5g06180	–	OM outersurface, interact with DRP3,	–
	–	–	PMD1, PMD2	–	OM, independent of DRP3, Rosidae specific,	+ (PMD1)
	–	–	–	FtsZ1	Matrix, GTPase, polymerizes into ftsZ ring	–
	–	–	–	ZED	Matrix, bacterial cell division ZapA homologue, forms ring	–
Fusion	Fzo1	Mfn1, Mfn2	–	–	OM, for OM fusion, GTPase	–
	Mgm1	Opa1	–	–	IM and intermembrane space, for IM fusion, GTPase	–
	Ugo1	–	–	–	OM, for OM fusion Interact with Fzo1 and Mgm1. OM	–

including division (McConnell et al. 1990, reviewed in Yaffe 1995, Okamoto and Shaw 2005, Shaw and Nunnari 2002). Dnm1/Mdm37, a GTPase dynamin-related protein, is a central factor in mitochondrial division in yeast. At the same time that Dnm1 was identified, two independent groups identified a mitochondrial fusion, factor, Fzo1 which is another dynamin-related protein that counterbalances mitochondrial division (Bleazard et al. 1999, Sesaki and Jensen 1999). Two other mitochondrial division factors, Mdv1 (and its paralogue Caf4) and Fis1, were identified by searching for mutants that suppress the defects of mitochondrial morphology in the mitochondrial fusion mutant *fzo1* (reviewed in Shaw and Nunnari 2002). Dnm1 polymerizes into a ring-like helical structure surrounding the dividing site of mitochondria on the outside of the outer membrane. It constricts and divides mitochondria by its GTPase activity (Ingerman et al. 2005). The function of the other two factors, Fis1 and Mdv1, is to localize the cytosolic Dnm1 to the mitochondrial division sites. Fis1 localizes to the mitochondrial outer membrane by a transmembrane domain in its C-terminus so that almost all of its N-terminus extends into the cytosol. Mdv1 and its paralogue Caf4 interact with Fis1 by their N-terminal extensions and interact with Dnm1 by their C-terminal beta propeller regions (Tieu and Nunnari 2000, Griffin et al. 2005). Therefore Mdv1 and Caf4 are thought to act as molecular adaptors between these two proteins. Dnm1 and Fis1 are well conserved among eukaryotes, but Mdv1 and Caf4 are not (table). Therefore, Dnm1 and Fis1 are considered to be universal factors involved in mitochondrial division. Much is known about the roles of Dnm1 and its orthologous dynamin-related proteins in mitochondrial division, but there is some controversy about the role of Fis1 in mitochondrial division in other eukaryotes (Otera et al. 2010, Loson et al. 2013, Palmer et al. 2013). In addition, yeast Fis1 was shown to be an interchangeable factor for mitochondrial fission in transgenic yeast expressing the chimeric Mdv1 that directly localized to the outer surface of mitochondria without Fis1 (Koirala et al. 2013). In transgenic yeast lacking Fis1 but expressing the chimeric Mdv1, Mdv1 directly localized to the outer surface of mitochondria without Fis1, allowing Dnm1 to localize to mitochondria and initiate mitochondrial division (Koirala et al. 2013). Further studies are needed to determine whether Fis1 has a common role in eukaryotes that is not related to mitochondrial division.

Mitochondrial Division in Mammals

Dynamin-related proteins (DRPs) that function in mitochondrial division (MtDRPs) are more conserved than other dynamin-related proteins including authentic dynamins which are involved in animal endocytosis (Leger et al. 2015). Like yeast Dnm1, the cytosolic mammalian and nematode MtDRPs localize to the mitochondrial division sites where they form a ring that constricts the mitochondria (Smirnova et al. 2001, Labrousse et al. 1999). The expression levels of mammalian *Fis1* affect the morphology of mitochondria, but the molecular role of Fis1 in mitochondrial division is controversial. If it has a role similar to that of yeast Fis1, it seems to be relatively weak (Otera et al. 2010, Loson et al. 2013, Palmer et al. 2013). Because no animal orthologues of Mdv1 or Caf4 were found in the databases, the molecular mechanisms for recruiting MtDRP seem to be different in yeasts and animals. Studies of *Drosophila* and mouse cells identified two proteins (Mff and MiD49/ MiD51) embedded in the mitochondrial outer-membrane of both species that act as mitochondrial adaptors for MtDRPs in place of Fis1 (Gandre-Babbe and van der

Bliek 2008, Palmer 2011). Fis1, Mff and MiD49/51 all seem to localize MtDrp1 but in distinct ways, and with different efficacies (Otera et al. 2010, Loson et al. 2013, Palmer et al. 2013, Shen et al. 2014).

Post-translational modifications of MtDRP, including phosphorylation, ubiquitination, SUMOylation and S-nitrosylation, have several roles, such as in localization, protein stability, oligomerization and GTPase activity (Willems et al. 2015). In yeast and mammals, the endoplasmic reticulum (ER) also contacts or surrounds mitochondrial division sites before the recruitment of MtDRPs (Friedman et al. 2011).

There is increasing evidence that defects in mitochondrial division cause disease or affect development in mammals (Ishihara et al. 2009, Waterham et al. 2007, Wakabayashi et al. 2009, reviewed in Mishra and Chan 2014). Defects in mitochondrial division disturb the mitochondrial distribution in neuronal cells (Li et al. 2004) and degrade the integrity of the mitochondrial genome passed on to the daughter cells (Osman et al. 2015, Ishihara et al. 2015). During apoptosis in some types of mammalian cells, mitochondria release cytochrome c just after they divide (Frank et al. 2001). The multiple defects caused by disruption of mitochondrial division reflect the multifaceted importance of mitochondrial division in mammalian cells and tissues.

Mitochondrial Division in Red Algae

Much has been learned about the mechanisms of organelle division in the unicellular red alga, *Cyanidioschyzon merolae*, by Dr. Kuroiwa and colleagues (reviewed in Kuroiwa et al. 2008, Kuroiwa 2010). *C. merolae* is a good experimental model because its genome (16.5Mb) has been completely sequenced (Matsuzaki et al. 2004) and because each cell has a single plastid, a single mitochondrion and a single microbody (peroxisome), while the cells of other eukaryotes typically have several organelles of each type. Therefore, organellar divisions occur only once per cell division and are tightly linked to the host cell division cycle. The cell cycle in *C. merolae* cells can be highly synchronized by diurnal light/dark conditions. Plastid and mitochondrial division rings have been purified from synchronized *C. merolae* cells and the proteins involved have been identified (Yoshida et al. 2006, Nishida et al. 2007). Isolated plastid rings with a dynamin-related protein (CmDnm2) were shown to have some elasticity by using optic tweezers (Yoshida et al. 2006). *C. merolae* has two mitochondrial division proteins, a FtsZ-homologue (CmFtsZ1) and a ZapA-like protein (ZED), that are very similar to bacterial cell division proteins (Takahara et al. 2000, Yoshida et al. 2009). Both of these genes have been lost in the genomes of yeasts, animals and green plants. In the first step of mitochondrial division, which occurs at the end of the G1/S phase, ZED and FtsZ1 make ring-like structures in succession in the matrix beneath the inner membrane at the division site. In the second step, another protein (Mda1) and an MD (Mitochondrial Dividing) ring localize to the outer surface and start to constrict the mitochondrion to give it a dumbbell-like shape. Mda1 is weakly similar to yeast Mdv1 at the amino acid sequence level. Both proteins have coiled coil structures in the middle and C-terminal WD40 domains. In the third and last step, CmDnm1, which acts as a MtDRP in *C. merolae*, forms a ring around the constricted site, and the CmDnm1 ring with another outer membrane MD ring cut the mitochondrion into two pieces (Nishida et al. 2003). The two MD rings are visible in electron micrographs (reviewed in Kuroiwa et al. 2008, Kuroiwa 2010).

Mitochondrial Division in Arabidopsis

BLAST searches of the nuclear genome of Arabidopsis for homologs of yeast Dnm1 and other MtDRPs yielded two MtDRPs (DRP3A and DRP3B) (renamed from the original names ADL2a and ADL2b, Hong et al. 2003) (Arimura and Tsutsumi 2002, Arimura et al. 2004a, Logan et al. 2004). Although DRP3A and DRP3B function redundantly in mitochondrial division, DRP3A seems to function more than DRP3B because *drp3a* mutant has more defects than *drp3b* mutant (Fujimoto et al. 2009). Cells of *drp3a* mutants and *drp3b* mutants have longer and fewer mitochondria than do wild-type cells. The mitochondria in cells of double knockout *drp3a drp3b* Arabidopsis plants are more severely elongated and more interconnected than are mitochondria in the single mutants (Fujimoto et al. 2009). Fluorescently-labeled DRP3A and DRP3B localize to the mitochondrial division sites before the divisions occur (Fig. 2A).

Arabidopsis has two FIS homologues, called FIS1A (also called BYGYIN) and FIS1B. Mitochondria of Arabidopsis mutants that express little or no *FIS1A* or *FIS1B* transcripts are slightly elongated (Scott et al. 2006, Zhang and Hu 2008). FIS1A and FIS1B localize to the mitochondrial outer membranes and peroxisomal membranes (Zhang and Hu 2008, Lingard et al. 2008) and chloroplast membranes (Ruberti et al. 2014).

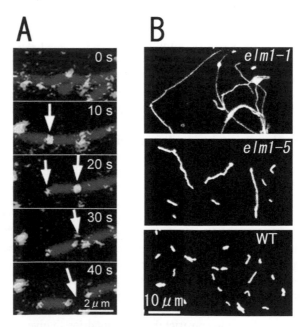

FIGURE 2 Mitochondrial division and mitochondrial morphology.

A. Time course observation of mitochondrial division in tobacco BY-2 cells transiently expressing GFP-DRP3A with mitochondrial marker MitoTracker, examined by confocal laser scanning microscopy. GFP-DRP3A localized to the sites where mitochondrial divisions occur (Arrows).

B. Mitochondria in the wild type (WT) and mitochondrial division mutants, *elm1-1* (strong allele) and *elm1-5* (weak allele), of A. thaliana. Mitochondria are visualized by the stable transformation with expressing GFP localized to the matrix. B is reproduced from Arimura et al. 2008.

Screenings of Arabidopsis mutants for mutants with aberrant mitochondrial morphology revealed genes controlling mitochondrial morphology (Logan et al. 2003, Feng et al. 2004, Arimura et al. 2008) In addition to *DRP3A*, a plant-specific gene *ELM1* (e̲longate m̲itochondria) was identified as a factor in mitochondrial division (Fig. 2B, Arimura et al. 2008). Although ELM1 does not have a transmembrane domain and does not interact with FIS1, it localizes to the outer surface of the mitochondrial outer membrane by an unknown mechanism. Localization of DRP3A to the mitochondrial division sites was found to require ELM1. The Arabidopsis genome has a paralogue of *ELM1* (At5g06180), but it has little or no role in mitochondrial division (Arimura et al. unpublished data).

Other factors that affect mitochondrial morphology in Arabidopsis, and they may also be involved in mitochondrial division, include PMD1 and PMD2 (P̲eroxisomal and M̲itochondrial D̲ivision factors 1 and 2) (Aung and Hu 2011) and UBP27 (U̲biquitin b̲inding p̲rotein 27) (Pan et al. 2014). The authors of these studies suggest that these factors may also be involved in mitochondrial division. PMD1 and PMD2 each have a coiled coil domain extending into the cytosol and a transmembrane domain in its C-terminus to anchor it in the mitochondrial outer membrane. Homologues of PMD1 and PMD2 have so far been found only in Rosids. The loss of *UBP27* in Arabidopsis did not change the morphology of mitochondria, but its overexpression changed mitochondrial morphology from rod-shaped to spherical and reduced the association of DRP3A with mitochondria. Although it is unclear whether UBP27 interacts with DRP3A/DRP3B, it may affect their posttranslational modifications or degradations (Pan et al. 2014). DRP3A and DRP3B have been shown to be phosphorylated (Heazlewood et al. 2008; Sugiyama et al. 2008, Wang et al. 2012).

Mitochondrial Fusion

A gene necessary for mitochondrial fusion (*Fzo1*) was first identified in sperm cells from a sterile male *Drosophila* mutant (*fuzzy onion*) in which mitochondria cannot properly fuse (Hales and Fuller 1997). Fzo1 has a GTPase domain in its N-terminal part that extends into the cytosol and a transmembrane domain at its C-terminus that localizes to the mitochondrial outer membrane. Fzo1 orthologues are also present in yeast (Hermann et al. 1998) and human (Santel and Fuller 2001), and are distantly related to proteins in the dynamin super family. Another dynamin-related GTPase, OPA1 (Mgm1 in yeast), which is the inter-membrane space protein attached to the inner membrane, is required for the fusion of the inner membranes of two mitochondria (Wong et al. 2000). Human homologues of *Fzo1* and *OPA1* have been identified as the causative genes in Charcot-Marie-Tooth disease and Optic Atrophy, respectively (reviewed in Mishra and Chan 2014).

Although no homologues of these genes or homologs of other genes involved in mitochondrial fusion have been found in plant genomes, plant mitochondria frequently and repeatedly undergo fusion and fission (Fig. 3, Arimura et al. 2004b, Sheahan et al. 2005). Observations of onion epidermal cells reveal the fusion and fission of mitochondria but not the fusion and fission of plastids and peroxisomes (Arimura et al. 2004b). The frequent fission and fusion of plant mitochondria may help to explain why more than half of the mitochondria in plant somatic cells have no DNA at all (Sheahan et al. 2005, Takanashi et al. 2006, Preuten et al. 2010, Wang et

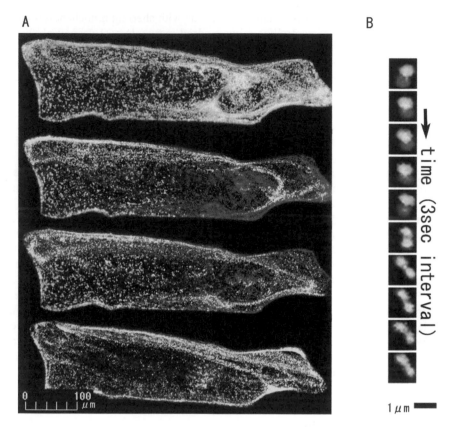

FIGURE 3 Times courses of mitochondrial fusions in an onion bulb epidermal cell transformed with mitochondria-localized Kaede fusion proteins.

A. Demonstration of the movement and fusion of mitochondria. The cell has more than 10,000 mitochondria (green dots; top panel). The right half of the cell was exposed to 400-nm light to photoconvert most of the mitochondria to red (second panel). Thirty minutes later, many of the green and red mitochondria have migrated into the other half of the cell (third panel). After 2 hours, the green and red Kaede were completely mixed by mitochondrial fusions and redistributed to all mitochondria in the cell (bottom panel).

B. Magnified image of mitochondrial fusion between two (red and green) mitochondria. Images were acquired at 3-sec intervals. Mitochondrial fusion occurs between the fifth and sixth rows, resulting in sharing the green and red signals between the two mitochondria. Images are reproduced from Arimura et al. 2004b.

al. 2010), as it enables all mitochondria to share internal proteins or small molecules in each cell. It may also explain how DNA recombination frequently occurs between two mitochondrial genomes (but not plastid genomes) from two different cells after protoplast fusion (Belliard et al. 1979).

Mitochondrial fusion and fission help to determine mitochondrial shape. Fusion may also regulate genetic inheritance between mitochondria. Although plant mitochondria are usually numerous and small, in shoot apical meristem cells, they are interconnected around the nucleus (Segui-Simarro et al. 2008). Mitochondrial fusion and division work together to determine mitochondrial shape. Fusion may also assure genetic inheritance between mitochondria and during cellular development.

Peroxisomal Division

Division of plant peroxisomes

Peroxisomes are single membrane-bound, usually round-shaped, DNA-less organelles that can grow and divide. They have 0.5-1 μm diameters similar to those of mitochondria, but are less numerous than mitochondria. Peroxisomes have roles in photorespiration, the metabolism of lipids and the synthesis of plant hormones, such as auxin and jasmonic acid (Hu et al. 2012). There has been speculation that peroxisomes had an endosymbiotic origin but analyses of peroxisomal proteins do not seem to support this idea (Gabaldón et al. 2006, Gabaldón 2010). Many genes involved in generating peroxisomes (peroxins, *PEXs*) have been identified in yeast, mammals and plants (Nito et al. 2007). One of them, the highly conserved *PEX11*, has been implicated in peroxisomal division (Marshall et al. 1995, Lingard and Trelease 2006, Schrader et al. 1998, Williams et al. 2015). Overexpression of *PEX11* elongates peroxisomes. Arabidopsis has five copies of *PEX11*, which are expressed under different conditions and which have slightly different functions (Lingard and Trelease 2006, Orth et al. 2007, Lingard et al. 2008). Expression of one of the copies (*AtPEX11b*) is induced by light through the photoreceptor phytochrome A and involves HY5, a transcription factor (Desai and Hu 2008).

Other genes involved in peroxisomal division include MtDRPs and closely related DRPs such as *DRP3A* (Mano et al. 2004, Aung and Hu 2009, Zhang and Hu 2009) and DRP5B, which is also involved in chloroplast division (Zhang and Hu 2010), Fis1a and Fis1b (Zhang and Hu 2008, 2009, Lingard et al. 2008) and PMD1 (Aung and Hu 2011). The fact that these factors participate in the divisions of multiple organelle divisions is reasonable in view of the limited numbers of organelle division genes in the nucleus. Further studies are needed to determine how the frequency of division in these organelles is regulated. The frequency of mitochondrial fission must be higher than that of peroxisome divisions because mitochondrial fission occurs frequently to counter balance fusion, which occurs frequently in mitochondria but rarely (or not at all) in peroxisomes. In addition, peroxisomes are formed *de novo* from the ER (Hoepfner et al. 2005, Mullen and Trelease 2006), while mitochondria are formed only by division of existing mitochondria. Therefore, the mechanisms regulating the number and shapes of these organelles are expected to be quite different, even though they use some common factors.

Conclusions

Although mitochondrial and peroxisomal divisions are fundamental cellular processes, only a limited number of factors related to these processes have so far been identified. It is likely that new factors, both those that are widely conserved in eukaryotes and those that are lineage specific, will be uncovered in the future. Identification of such division-related factors will greatly improve our understanding of the development and growth of plants.

Acknowledgments

We apologize that we were not able to cite all of the relevant publications due to space limitations. This work was supported by grants from the Japan Society for the Promotion of Science and the Ministry of Education, Culture, Sports, Science, and Technology of JAPAN (grant24248001 to N.T. and grants24380814 to S.A.) and JST (CREST to N.T. and PERSTO to S.A.).

References

Arimura, S., G.P. Aida, M. Fujimoto, M. Nakazono and N. Tsutsumi. 2004a. Arabidopsis dynamin-like protein 2a (ADL2a), like ADL2b, is involved in plant mitochondrial division. Plant Cell Physiol. 45: 236-242.

Arimura, S., M. Fujimoto, Y. Doniwa, N. Kadoya, M. Nakazono, W. Sakamoto and N. Tsutsumi. 2008. Arabidopsis ELONGATED MITOCHONDRIA1 is required for localization of DYNAMIN-RELATED PROTEIN3A to mitochondrial fission sites. Plant Cell 20: 1555-1566.

Arimura, S. and N. Tsutsumi. 2002. A dynamin-like protein (ADL2b), rather than FtsZ, is involved in Arabidopsis mitochondrial division. Proc. Natl. Acad. Sci. USA. 99: 5727-5731.

Arimura, S., J. Yamamoto, G.P. Aida, M. Nakazono and N. Tsutsumi. 2004b. Frequent fusion and fission of plant mitochondria with unequal nucleoid distribution. Proc. Natl. Acad. Sci. USA. 101: 7805-7808.

Arimura, D. and N. Tsutsumi. 2006. Dynamic mitochondria, their fission and fusion in higher plants. pp. 225-240. *In*: T. Nagata , K. Matsuoka and D. Inzé (eds.). Tobacco BY-2 Cells: From Cellular Dynamics to Omics. Springer, New York.

Aung, K. and J. Hu. 2009. The Arabidopsis peroxisome division mutant pdd2 is defective in the DYNAMIN-RELATED PROTEIN3A (DRP3A) gene. Plant Signal. & Behaviour 4: 542-544.

Aung, K. and J. Hu. 2011. The Arabidopsis tail-anchored protein PEROXISOMAL AND MITOCHONDRIAL DIVISION FACTOR1 is involved in the morphogenesis and proliferation of peroxisomes and mitochondria. Plant Cell 23: 4446-4461.

Belliard G., F. Vedal and G. Pelleiter. 1979. Mitochondrial recombination in cytoplasmic hybrids of *Nicotiana tabacum* by protoplast fusion. Nature 281: 401-403.

Bleazard, W., J.M. McCaffery, E.J. King, S. Bale, A. Mozdy, Q. Tieu, J. Nunnari and J.M. Shaw.1999. The dynamin-related GTPase Dnm1 regulates mitochondrial fission in yeast. Nat. Cell. Biol. 1: 298-304.

Desai, M. and J. Hu. 2008. Light induces peroxisome proliferation in Arabidopsis seedlings through the photoreceptor phytochrome A, the transcription factor HY5 HOMOLOG, and the peroxisomal protein PEROXIN11b. Plant Physiol. 146: 1117-1127.

Feng, X.G., S. Arimura, H.Y. Hirano, W. Sakamoto and N. Tsutsumi. 2004. Isolation of mutants with aberrant mitochondrial morphology from *Arabidopsis thaliana*. Genes Genet. Syst. 79: 301-305.

Foyer, C.H. and G. Noctor. 2013. Redox signaling in plants. Antioxid. Redox Signal. 18: 2087-2090.

Frank, S., B. Gaume, E.S. Bergmann-Leitner, W.W. Leitner, E.G. Robert, F. Catez, C.L. Smith and R.J. Youle. 2001. The role of dynamin-related protein 1, a mediator of mitochondrial fission, in apoptsis. Dev. Cell 1: 515-525.

Friedman, J.R., L.L. Lackner, M. West, J.R. DiBenedetto, J. Nunnari and G.K. Voeltz. 2011. ER tubules mark sites of mitochondrial division. Science 334: 358-362.

Fujimoto, M., S. Arimura, S. Mano, M. Kondo, C. Saito, T. Ueda, M. Nakazono, A. Nakano, M. Nishimura and N. Tsutsumi. 2009. Arabidopsis dynamin-related proteins DRP3A and DRP3B are functionally redundant in mitochondrial fission, but have distinct roles in peroxisomal fission. Plant J. 58: 388-400.

Gabaldón, T. and S. Capella-Gutierrez. 2010. Lack of phylogenetic support for a supposed actinobacterial origin of peroxisomes. Gene 465: 61-65.

Gabaldón, T., B. Snel, F. van Zimmeren, W. Hemrika, H. Tabak and M.A. Huynen. 2006. Origin and evolution of the peroxisomal proteome. Biol. Direct 1: 8.

Gandre-Babbe, S. and A.M. van der Bliek. 2008. The novel tail-anchored membrane protein Mff controls mitochondrial and peroxisomal fission in mammalian cells. Mol. Biol. Cell 19: 2402-2412.

Griffin E.E., J. Graumann and D.C. Chan. 2005. The WD40 protein Caf4p is a component of the mitochondrial fission machinery and recruits Dnm1p to mitochondria. J Cell Biol. 170:237-48.

Hales, K.G. and M.T. Fuller. 1997. Developmentally regulated mitochondrial fusion mediated by a conserved, novel, predicted GTPase. Cell 90: 121-129.

Heazlewood, J.L., P. Durek, J. Hummel, J. Selbig, W. Weckwerth, D. Walther and W.X. Schulze. 2008. PhosPhAt: a database of phosphorylation sites in Arabidopsis thaliana and a plant-specific phosphorylation site predictor. Nucleic Acids Res. 36: D1015-1021.

Hermann, G.J., J.W. Thatcher, J.P. Mills, K.G. Hales, M.T. Fuller, J. Nunnari and J.M. Shaw. 1998. Mitochondrial fusion in yeast requires the transmembrane GTPase Fzo1p. J. Cell Biol. 143: 359-373.

Hoepfner, D., D. Schildknegt, I. Braakman, P. Philippsen and H.F. Tabak. 2005. Contribution of the endoplasmic reticulum to peroxisome formation. Cell 122: 85-95.

Hong, Z., S.Y. Bednarek, E. Blumwald, I. Hwang, G. Jurgens, D. Menzel, K.W. Osteryoung, N.V. Raikhel, K. Shinozaki, N. Tsutsumi and D. P. Verma. 2003. A unified nomenclature for Arabidopsis dynamin-related large GTPases based on homology and possible functions. Plant Mol. Biol. 53: 261-265.

Hu, J., A. Baker, B. Bartel, N. Linka, R.T. Mullen, S. Reumann and B.K. Zolman. 2012. Plant peroxisomes: biogenesis and function. Plant Cell 24: 2279-2303.

Ingerman, E., E.M. Perkins, M. Marino, J.A. Mears, J.M. McCaffery, J.E. Hinshaw and J. Nunnari. 2005. Dnm1 forms spirals that are structurally tailored to fit mitochondria. J. Cell Biol. 170: 1021-1027.

Ishihara, N., M. Nomura, A. Jofuku, H. Kato, S.O. Suzuki, K. Masuda, H. Otera, Y. Nakanishi, I. Nonaka, Y. Goto., N. Taguchi, H. Morinaga, M. Maeda, R. Takayanagi, S. Yokota and K. Mihara. 2009. Mitochondrial fission factor Drp1 is essential for embryonic development and synapse formation in mice. Nat. Cell Biol 11: 958-966.

Ishihara, T., R. Ban-Ishihara, M. Maeda, Y. Matsunaga, A. Ichimura, S. Kyogoku, H. Aoki, S. Katada, K. Nakada, M. Nomura, N. Mizushima, K. Mihara, N. Ishihara. 2015. Dynamics of mitochondrial DNA nucleoids regulated by mitochondrial fission is essential for maintenance of homogeneously active mitochondria during neonatal heart development. Mol. Cell Biol. 35: 211-223.

Koirala, S., Q. Guo, R. Kalia, H.T. Bui, D.M. Eckert, A. Frost and J.M. Shaw. 2013. Interchangeable adaptors regulate mitochondrial dynamin assembly for membrane scission. Proc Natl Acad Sci USA 110: E1342-1351.

Kubo, T. and K.J. Newton. 2008. Angiosperm mitochondrial genomes and mutations. Mitochondrion 8, 5-14.

Kuroiwa, T. 2010. Mechanisms of organelle division and inheritance and their implications regarding the origin of eukaryotic cells. Proceedings of the Japan Academy Series B, Physical and biological sciences 86: 455-471.

Kuroiwa, T., O. Misumi, K. Nishida, F. Yagisawa, Y. Yoshida, T. Fujiwara and H. Kuroiwa. 2008. Vesicle, mitochondrial, and plastid division machineries with emphasis on dynamin and electron-dense rings. Int. Rev. Cell Mol. Biol. 271: 97-152.

Labrousse, A.M., M.D. Zappaterra, D. A. Rube and A.M. van der Bliek. 1999. *C. elegans* dynamin-related protein DRP-1 controls severing of the mitochondrial outer membrane. Mol. Cell 4: 815-826.

Leger, M.M., M. Petru, V. Zarsky, L. Eme, C. Vlcek, T. Harding, B.F. Lang, M. Elias, P. Dolezal and A.J. Roger. 2015. An ancestral bacterial division system is widespread in eukaryotic mitochondria. Proc. Natl. Acad. Sci. USA. 112: 10239-10246.

Lewis, M.R. and W.H. Lewis.1914. Mitochondria in tissue culture. Science 39: 330-333.

Li, Z., K. Okamoto, Y. Hayashi and M. Sheng. 2004. The importance of dendritic mitochondria in the morphogenesis and plasticity of spines and synapses. Cell 119: 873-887.

Lingard, M.J., S.K. Gidda, S. Bingham, S.J. Rothstein, R.T. Mullen and R.N. Trelease. 2008. Arabidopsis PEROXIN11c-e, FISSION1b, and DYNAMIN-RELATED PROTEIN3A cooperate in cell cycle-associated replication of peroxisomes. Plant Cell 20: 1567-1585.

Lingard, M.J. and R.N. Trelease. 2006. Five Arabidopsis peroxin 11 homologs individually promote peroxisome elongation, duplication or aggregation. J. Cell Sci. 119: 1961-1972.

Logan, D.C., I. Scott and A.K. Tobin. 2003. The genetic control of plant mitochondrial morphology and dynamics. Plant J. 36: 500-509.

Logan, D.C., I. Scott and A.K. Tobin. 2004. ADL2a, like ADL2b, is involved in the control of higher plant mitochondrial morphology. J. Exp. Bot. 55: 783-785.

Loson, O.C., Z. Song, H. Chen, and D.C. Chan. 2013. Fis1, Mff, MiD49, and MiD51 mediate Drp1 recruitment in mitochondrial fission. Mol. Biol. Cell 24: 659-667.

Mano, S., C. Nakamori, M. Kondo, M. Hayashi and M. Nishimura. 2004. An Arabidopsis dynamin-related protein, DRP3A, controls both peroxisomal and mitochondrial division. Plant J. 38: 487-498.

Marshall, P.A., Y.I. Krimkevich, R.H. Lark, J.M. Dyer, M. Veenhuis and J.M. Goodman. 1995. Pmp27 promotes peroxisomal proliferation. J. Cell Biol. 129: 345-355.

Matsuzaki, M., O. Misumi, I.T. Shin, S. Maruyama, M. Takahara, S.Y. Miyagishima, T. Mori, K. Nishida, F. Yagisawa, K. Nishida, Y. Yoshida, Y. Nishimura, S. Nakao, T. Kobayashi, Y. Momoyama, T. Higashiyama, A. Minoda, M. Sano, H. Nomoto, K. Oishi, H. Hayashi, F. Ohta, S. Nishizaka, S. Haga, S. Miura, T. Morishita, Y. Kabeya, K. Terasawa, Y. Suzuki, Y. Ishii, S. Asakawa, H. Takano, N. Ohta, H. Kuroiwa, K. Tanaka, N. Shimizu, S. Sugano, N. Sato, H. Nozaki, N. Ogasawara, Y. Kohara and T. Kuroiwa. 2004. Genome sequence of the ultrasmall unicellular red alga *Cyanidioschyzon merolae* 10D. Nature 428: 653-657.

McConnell, S.J., L.C. Stewart, A. Talin and M.P. Yaffe. 1990. Temperature-sensitive yeast mutants defective in mitochondrial inheritance. J. Cell Biol. 111: 967-976.

Mishra, P. and D.C. Chan. 2014. Mitochondrial dynamics and inheritance during cell division, development and disease. Nature Reviews: Molec. Cell Biol. 15: 634-646.

Mozdy, A.D., M.A. Karren, L. Martin and J.M. Shaw. 2001. The role of the Fzo1 GTPase in mitochondrial fusion. Mol. Biol. Cell 12: 3A-3A.

Mullen, R.T. and R.N. Trelease. 2006. The ER-peroxisome connection in plants: development of the "ER semi-autonomous peroxisome maturation and replication" model for plant peroxisome biogenesis. Biochim. Biophys. Acta. 1763: 1655-1668.

Nishida, K., M. Takahara, S. Miyagishima, H. Kuroiwa, M. Matsuzaki and T. Kuroiwa. 2003. Dynamic recruitment of dynamin for final mitochondrial severance in a primitive red alga. Proc Natl Acad Sci USA 100: 2146-2151.

Nishida, K., F. Yagisawa, H. Kuroiwa, Y. Yoshida and T. Kuroiwa. 2007. WD40 protein Mda1 is purified with Dnm1 and forms a dividing ring for mitochondria before Dnm1 in Cyanidioschyzon merolae. Proc Natl Acad Sci USA 104: 4736-4741.

Nito, K., A. Kamigaki, M. Kondo, M. Hayashi and M. Nishimura. 2007. Functional classification of Arabidopsis peroxisome biogenesis factors proposed from analyses of knockdown mutants. Plant Cell Physiol. 48: 763-774.

Noguchi, K. and K. Yoshida. 2008. Interaction between photosynthesis and respiration in illuminated leaves. Mitochondrion 8: 87-99.

Okamoto, K. and J.M. Shaw. 2005. Mitochondrial morphology and dynamics in yeast and multicellular eukaryotes. Annu. Rev. Genet. 39: 503-536.

Oldenburg, D.J. and A.J. Bendich. 1996. Size and Structure of Replicating Mitochondrial DNA in Cultured Tobacco Cells. Plant Cell 8: 447-461.

Orth, T., S. Reumann, X. Zhang, J. Fan, D. Wenzel, S. Quan and J. Hu. 2007. The PEROXIN11 protein family controls peroxisome proliferation in Arabidopsis. Plant Cell 19: 333-350.

Osman, C., T.R. Noriega, V. Okreglak, J.C. Fung and P. Walter. 2015. Integrity of the yeast mitochondrial genome, but not its distribution and inheritance, relies on mitochondrial fission and fusion. Proc. Natl. Acad. Sci. USA. 112: E947-956.

Otera, H., C. Wang, M.M. Cleland, K. Setoguchi, S. Yokota, R.J. Youle and K. Mihara. 2010. Mff is an essential factor for mitochondrial recruitment of Drp1 during mitochondrial fission in mammalian cells. J. Cell Biol. 191: 1141-1158.

Palmer, C.S., L.D. Osellame, D. Laine, O.S. Koutsopoulos, A.E. Frazier and M.T. Ryan. 2011. MiD49 and MiD51, new components of the mitochondrial fission machinery. EMBO Rep. 12: 565-573.

Palmer, C.S., K.D. Elgass, R.G. Parton, L.D. Osellame, D. Stojanovski and M.T. Ryan. 2013. Adaptor proteins MiD49 and MiD51 can act independently of Mff and Fis1 in Drp1 recruitment and are specific for mitochondrial fission. J. Biol. Chem. 288: 27584-27593.

Pan, R., N. Kaur and J. Hu. 2014. The Arabidopsis mitochondrial membrane-bound ubiquitin protease UBP27 contributes to mitochondrial morphogenesis. Plant J. 78: 1047-1059.

Preuten, T., E. Cincu, J. Fuchs, R. Zoschke, K. Liere and T. Borner. 2010. Fewer genes than organelles: extremely low and variable gene copy numbers in mitochondria of somatic plant cells. Plant J. 64: 948-959.

Ruberti, C., A. Costa, E. Pedrazzini, F. Lo Schiavo and M. Zottini. 2014. FISSION1A, an Arabidopsis tail-anchored protein, is localized to three subcellular compartments. Molec. Plant 7: 1393-1396.

Santel, A. and M.T. Fuller. 2001. Control of mitochondrial morphology by a human mitofusin. J. Cell Sci. 114: 867-874.

Schrader, M., B.E. Reuber, J.C. Morrell, G. Jimenez-Sanchez, C. Obie, T.A. Stroh, D. Valle, T.A. Schroer and S.J. Gould. 1998. Expression of PEX11beta mediates peroxisome proliferation in the absence of extracellular stimuli. J. Biol. Chem. 273: 29607-29614.

Scott, I., A.K. Tobin and D.C. Logan. 2006. BIGYIN, an orthologue of human and yeast FIS1 genes functions in the control of mitochondrial size and number in Arabidopsis thaliana. J. Exp. Bot. 57: 1275-1280.

Segui-Simarro, J.M., M.J. Coronado and L.A. Staehelin. 2008. The mitochondrial cycle of Arabidopsis shoot apical meristem and leaf primordium meristematic cells is defined by a perinuclear tentaculate/cage-like mitochondrion. Plant Physiol. 148: 1380-1393.

Sesaki, H. and R.E. Jensen. 1999. Division versus fusion: Dnm1p and Fzo1p antagonistically regulate mitochondrial shape. J. Cell Biol. 147: 699-706.

Shaw, J.M. and J. Nunnari. 2002. Mitochondrial dynamics and division in budding yeast. Trends Cell Biol. 12: 178-184.

Sheahan, M.B., D.W. McCurdy and R.J. Rose. 2005. Mitochondria as a connected population: ensuring continuity of the mitochondrial genome during plant cell dedifferentiation through massive mitochondrial fusion. Plant J. 44: 744-755.

Shen, Q., K. Yamano, B.P. Head, S. Kawajiri, J.T. Cheung, C. Wang, J.H. Cho, N. Hattori, R.J. Youle and A.M. van der Bliek. 2014. Mutations in Fis1 disrupt orderly disposal of defective mitochondria. Mol. Biol. Cell 25: 145-159.

Sloan, D.B. 2013. One ring to rule them all? Genome sequencing provides new insights into the 'master circle' model of plant mitochondrial DNA structure. New Phytol. 200: 978-985.

Smirnova, E., L. Griparic, D.L. Shurland and A.M. van der Bliek. 2001. Dynamin-related protein Drp1 is required for mitochondrial division in mammalian cells. Mol. Biol. Cell 12: 2245-2256.

Sugiyama, N., H. Nakagami, K. Mochida, A. Daudi, M. Tomita, K. Shirasu and Y. Ishihama. 2008. Large-scale phosphorylation mapping reveals the extent of tyrosine phosphorylation in Arabidopsis. Mol. Syst. Biol. 4: 193.

Takahara, M., H. Takahashi, S. Matsunaga, S. Miyagishima, H Takano, A. Sakai, S. Kawano and T. Kuroiwa. 2000. A putative mitochondrial ftsZ gene is present in the unicellular primitive red alga Cyanidioschyzon merolae. Mol. Gen. Genet. 264: 452-460.

Takanashi, H., S. Arimura, W. Sakamoto and N. Tsutsumi. 2006. Different amounts of DNA in each mitochondrion in rice root. Genes Genet. Syst. 81: 215-218.

Tieu, Q. and J. Nunnari. 2000. Mdv1p is a WD repeat protein that interacts with the dynamin-related GTPase, Dnm1p, to trigger mitochondrial division. J. Cell Biol. 151: 353-365.

Wakabayashi, J., Z. Zhang, N. Wakabayashi, Y. Tamura, M. Fukaya, T.W. Kensler, M. Iijima and H. Sesaki. 2009. The dynamin-related GTPase Drp1 is required for embryonic and brain development in mice. J. Cell Biol. 186: 805-816.

Wang, D.Y., Q. Zhang, Y. Liu, Z.F. Lin, S.X. Zhang, M.X. Sun and Sodmergen 2010. The levels of male gametic mitochondrial DNA are highly regulated in angiosperms with regard to mitochondrial inheritance. Plant Cell 22: 2402-2416.

Wang, F., P. Liu, Q. Zhang, J. Zhu, T. Chen, S. Arimura, N. Tsutsumi and J. Lin. 2012. Phosphorylation and ubiquitination of dynamin-related proteins (AtDRP3A/3B) synergically regulate mitochondrial proliferation during mitosis. Plant J. 72: 43-56.

Waterham, H.R., J. Koster, C.W. van Roermund, P.A. Mooyer, R.J. Wanders and J.V. Leonard. 2007. A lethal defect of mitochondrial and peroxisomal fission. New England J. Med. 356: 1736-1741.

Willems, P.H., R.Rossignol, C.E. Dieteren, M.P. Murphy and W.J. Koopman. 2015. Redox Homeostasis and Mitochondrial Dynamics. Cell Metabolism 22: 207-218.

Williams, C., L. Opalinski, C. Landgraf, J. Costello, M. Schrader, A.M. Krikken, K. Knoops, A.M. Kram, R. Volkmer and I.J. van der Klei. 2015. The membrane remodeling protein Pex11p activates the GTPase Dnm1p during peroxisomal fission. Proc. Natl. Acad. Sci. USA .112: 6377-6382.

Wong, E.D., J.A. Wagner, S.W. Gorsich, J.M. McCaffery, J.M. Shaw and J. Nunnari. 2000. The dynamin-related GTPase, Mgm1p, is an intermembrane space protein required for maintenance of fusion competent mitochondria. J. Cell Biol. 151: 341-352.

Yaffe, M.P. 1995. Isolation and analysis of mitochondrial inheritance mutants from *Saccharomyces cerevisiae*. pp. 447- 453. *In*: G.M. Attardi and A. Chomyn (eds.). Mitochondrial Biogenesis and Genetics, Pt A , Methods in Enzymology Vol. 260. Academic Press Inc, San diego,

Yoshida, Y., H. Kuroiwa, S. Hirooka, T. Fujiwara, M. Ohnuma, M. Yoshida, O. Misumi, S. Kawano and T. Kuroiwa. 2009. The bacterial ZapA-like protein ZED is required for mitochondrial division. Curr. Biol. 19: 1491-1497.

Yoshida, Y., H. Kuroiwa, O. Misumi, K. Nishida, F. Yagisawa, T. Fujiwara, H. Nanamiya, F. Kawamura and T. Kuroiwa. 2006. Isolated chloroplast division machinery can actively constrict after stretching. Science 313: 1435-1438.

Zhang, X. and J. Hu. 2009. Two small protein families, DYNAMIN-RELATED PROTEIN3 and FISSION1, are required for peroxisome fission in Arabidopsis. Plant J. 57: 146-159.

Zhang, X. and J. Hu. 2010. The Arabidopsis chloroplast division protein DYNAMIN-RELATED PROTEIN5B also mediates peroxisome division. Plant Cell 22: 431-442.

Zhang, X.C. and J.P. Hu. 2008. FISSION1A and FISSION1B proteins mediate the fission of peroxisomes and mitochondria in Arabidopsis. Molec. Plant 1: 1036-1047.

5

Mechanisms of Organelle Inheritance in Dividing Plant Cells

Michael B. Sheahan, David W. McCurdy and Ray J. Rose*

Introduction

Membrane-bound organelles are a characteristic feature of eukaryotic cells, providing intracellular compartments that concentrate enzymes, separate distinct biochemical pathways and generate chemical and electrical gradients – crucial to the production of energy and cell signaling processes (Cavalier-Smith 2002). Organelles, in particular plastids, mitochondria, peroxisomes, Golgi and the endoplasmic reticulum (ER), must be present throughout the cell cycle to ensure a continuity of organelles and organelle-derived functions for future generations. During the cell cycle each organelle (or the organelle population) must double in size, divide, and be delivered to its proper location in the daughter cells (Warren and Wickner 1996).

For the nucleus, highly organized and elaborate inheritance processes exist, ensuring stringently accurate partitioning of the nuclear genome to each daughter cell. It had been common belief that organelle inheritance was merely the result of stochastic dispersion processes (Huh and Paulsson 2011), and consequently little investigation was conducted into the cellular and molecular mechanisms that may be acting to ensure the continuity of organellar information systems. Evidence from a number of experimental organisms including yeast (Fagarasanu et al. 2010) and plants (Sheahan et.al. 2007), however, indicate that ordered mechanisms exist to ensure organelle continuity during cell division. A growing recognition of organelle association with the cytoskeleton and coordinated movements of organelles during cell division implicates cellular machinery in the mediation of unbiased organelle inheritance (Sheahan et al. 2004).

In this Chapter, we examine current understanding of organelle partitioning to maintain organelle continuity to ensure cellular function in future generations. After a brief introduction to organelle evolution, we focus our discussion on inheritance mechanisms of plastids, mitochondria, peroxisomes, the Golgi and ER. Investigations on the developing and reproducing flowering plant as well as on cell suspension cultures and regenerating protoplasts are considered.

School of Environmental and Life Sciences, The University of Newcastle, NSW 2308, Australia
* Corresponding author: Ray.Rose@newcastle.edu.au

The Origin of Eukaryotic Organelles

It is widely accepted that mitochondria and plastids descended from eubacterial symbionts (Gray 2004). Strong molecular evidence supports the claim that the proto-mitochondrion evolved following the phagocytosis of a eubacterium from within the a-proteobacteria (Cavalier-Smith 2009, Gray 2012), more than 1.5 billion years ago (Dyall et al. 2004, Martin and Mentel 2010). Mitochondria likely predated the advent of plastids (Raven and Allen 2003, Dyall et al. 2004) with the cyanobacterial endosymbiosis occurring more than 1.2 billion years ago (Dyall et al. 2004). The identity of the host organism that captured these free-living prokaryotes that evolved into mitochondria and chloroplasts is less certain but is likely to be a member of the Archae (Spang et al. 2015). While during the early stages of endosymbiosis, proto-mitochondria and proto-plastids are likely to have functioned and replicated autonomously, the eventual insertion of a host ATP/ADP exchange protein into the proto-organelle membranes would have enslaved each to the host eukaryote, thus rendering the proto-organelles obligate endosymbionts (Cavalier-Smith 2002). The transfer of vast amounts of genetic material from these proto-organelles to the host cell genome and the insertion of protein-translocation machinery further cemented the fate of the proto-organelles.

The origins of the endomembrane systems are less certain. However, recent results with the archaeal phylum Lokiarchaeota with its actin homologues, GTPases and potentially endocytic and/or phagocytic capacities (Spang et al. 2015), suggests the feasibility of the invasion of endosymbionts and the development of endomembrane systems by the Archae progenitors of eukaryotes. Cavalier–Smith (2002, 2009) has argued that phagotrophy would enable invagination and vesiculation of the plasma membrane with subsequent vesicle fusion eventually generating an internal membrane system. Specialization of this internal membrane system could ultimately lead to the development of the nuclear membrane, endoplasmic reticulum, Golgi apparatus, peroxisome and vacuolar membranes found in present-day eukaryotic cells.

When a eukaryote cell divides, its organellar capacity in the daughter cells needs to be maintained. The most basic requirement for unbiased organelle inheritance is the duplication of organelle numbers (and volume) and distribution of the resulting populations, such that populations partition to daughter cells with minimal bias. Thus, organelle membranes and genomes must be synthesized and for some organelles (plastids, mitochondria, peroxisomes), organelle division must also occur. Consequently, a distribution strategy needs to be implemented in order to achieve unbiased partitioning of the organelle population to each daughter cell.

Organelle Inheritance in the Germline

Much research into organelle inheritance has focused on the DNA-containing organelles, namely plastids (chloroplasts) and mitochondria, and on the events that occur in reproductive meristems during gamete formation. Nuclear genes usually display Mendelian segregation while plastids and mitochondria show non-Mendelian inheritance. Within the plant kingdom, patterns of organelle inheritance can be uniparental-maternal, biparental or uniparental-paternal. However, the predominant pattern of inheritance in angiosperms, like almost all eukaryotes, is uniparental-maternal (Russell 1993, Greiner et al. 2015).

The mechanisms underlying uniparental-maternal inheritance vary and in the male gamete include organelle nucleoid degradation, degradation of generative cell cytoplasm during fertilization or exclusion of organelles from the generative cell (Birky 2001). In cases of gamete exclusion in the generative cell, asymmetric division of the haploid microspore occurs with the larger vegetative cell containing all the plastids (Greiner et al. 2015). Interestingly, during microsporogenesis in *Lilium longiflorum*, a plant that displays strict uniparental-maternal inheritance, plastids develop a striking polar distribution during microspore mitotic division. MTs radiating from the haploid nuclei appear to enforce this pattern of distribution, and disruption of MTs with colchicine causes randomization of plastid distribution (Tanaka 1991). In this system, no evidence for the involvement of actin filaments was observed. The exclusion of plastids from around the nucleus plus the asymmetric nuclear location of the haploid microspore may facilitate the exclusion of plastids from the generative cell. Thus, although converse to the situation in megagametogenesis, where organelle continuity must be maintained, the MT cytoskeleton clearly plays a role in driving and maintaining organelle distributions during male gamete formation. In the *Lilium longiflorum* study, mitochondria, unlike the plastids, tended to concentrate around the nucleus rather than displaying a polar distribution (Tanaka 1991).

The egg cell in most angiosperms is highly polarized with the nucleus residing at the chalazal end of the cell and a large vacuole or several smaller vacuoles at the micropylar end (Huang and Russell 1992). During megasporogenesis, particularly during prophase of meiosis I, the DNA-containing organelles within the megasporocyte frequently undergo replication, generating large populations of small organelles. Because the majority of angiosperms have uniparental inheritance, the events that occur during formation of the egg cell determines the organelle complement in the filial generation. The patterns of organelle replication and distribution during egg cell formation vary considerably between plant species, however, some generalizations can be made. Most egg cells are poor in Golgi bodies with variable plastid and mitochondrial numbers and amounts of endoplasmic reticulum (Russell 1993). The number of DNA-containing organelles varies according to species and size of the egg cell. For example, in *Impatiens glandulifera*, the egg cell contains between 150 and 480 plastids and 1000-2500 mitochondria, whereas the larger egg cells of *Plumbago zeylanica* contain, on average, 730 plastids and the astonishingly large number of 39900 mitochondria per cell (Huang and Russell 1992). Thus, large increases in organelle number appear to contribute to the maintenance of plastid and mitochondrial continuity during this phase of female gamete development. Plastids at this stage tend to be variable in size and shape, while mitochondria are normally spheroidal (Dickinson and Li 1988, Huang and Russell 1992, Russell 1993). The DNA-containing organelles also tend to dedifferentiate, resulting in plastids and mitochondria that possess only rudimentary internal membrane systems at the completion of megagametogenesis (Dickinson and Li 1988, Huang and Russell 1992). Depending on species, the distribution of organelles within the egg cell cytoplasm varies. Although the DNA-containing organelles can distribute throughout the gamete cytoplasm, it is more common for these organelles to associate with the egg cell nucleus (Dickinson and Li 1988, Huang and Russell 1992). MTs and AFs are reported to have a random orientation (Russell 1993).

The ER is typically scarce in the egg cells of most angiosperms, but again, varies according to species. For example, the ER is abundant in egg cells from *Glycine max*, *Beta vulgaris* and *Epidendrum* (Huang and Russell 1992), which in the latter species is typically appressed to the plasma membrane. Egg cells are also typically deficient in Golgi, and peroxisomes in egg cells have only been observed in *Plumbago* (Huang and Russell 1992, Russell 1993). Although the egg cells of most species appear meta-bolically quiescent and have limited endomembrane architecture, most appear to be poised for rapid protein synthesis, possessing large nucleoli and many ribosomes (Huang and Russell 1992). Considering that endomembrane components are numer-ous in the zygote, the ER must undergo extensive proliferation following fertilization while components of the endomembrane system downstream of the ER are presum-ably synthesized largely *de novo* (Kuroiwa et al. 2002).

Maintaining the continuity of organelle genes between successive generations of plants thus appears to employ mechanisms that act to increase the number of trans-missible units. The association of organelles with the maternal nucleus may enhance unbiased partitioning of organelles during the first zygotic divisions by close associa-tion with the spindle apparatus. Indeed, in *Pelargonium*, this perinuclear distribution of plastids and mitochondria is maintained during the first zygotic division (Kuroiwa et al. 2002). In addition, the clustering of plastids and mitochondria around the egg cell nucleus may play an important role in keeping the large numbers of DNA-containing organelles within the cytoplasm of the nascent egg cell during the phase of cellularization that occurs in late megagametogenesis. What drives and maintains such associations between the DNA-containing organelles and the maternal nucleus is uncertain however.

The first cell divisions following gamete fertilization occur in the zygote and during embryogenesis. Although only a few studies examining organelles in devel-oping zygotes exist, studies by Singh and Mogensen (1975) and Kuroiwa et al. (2002) in *Quercus* and *Pelargonium*, respectively, indicate that strategies similar to those employed during female gametogenesis maintain organelle continuity in the developing zygote. Before the first zygotic division in *Pelargonium*, for example, mitochondrial number increases about 10-fold. Unbiased partitioning of this large mitochondrial population to the apical and basal cells appears to be achieved by the close association of mitochondria with the zygotic nucleus. Again, however, what drives and maintains such an association is uncertain.

Organelle Inheritance in Continually Dividing Cells

Organelle inheritance in vegetative meristems

To ensure appropriate plastid segregation during cell division, plastids must divide before cytokinesis. In meristematic cells, proplastid division keeps pace with cell division. In cells of young primary leaves in *Phaseolus vulgaris*, the continuity of plastid numbers appears to be maintained by mechanisms that entrain plastid divi-sion with the cell cycle, such that, when the cell divides, plastids have also divided (Whatley 1980). The densely cytoplasmic cells of the shoot apical meristem (SAM) contain 10-20 plastids per cell (Possingham 1980), often forming a distinct ring around the nucleus (Robertson et al. 1995). During mitosis in the meristematic cells of *Isoetes lacustris* shoots, which contain a single plastid, the plastid divides before the onset of metaphase and the daughter plastids, remaining closely associated with the nucleus, partition to opposite poles of the forming daughter cells before mitosis completes (Whatley 1974). Thus, although present in low copy number, inheritance appears to be ensured by an ordered partitioning mechanism involving close association with the nucleus. The mechanisms that drive such partitioning processes are unknown, however, in the root meristematic cells of *Isoetes*, the plane of plastid division orientates precisely with the preprophase band of MTs (Brown and Lemmon 1984). Thus, there is a potential role for MTs in the positioning of plastids in the cells of this species. In cells of the root apical meristem (RAM) in tobacco, plastids display a perinuclear distribution, organizing in a large circle around what appears to be the nucleus (Köhler and Hanson 2000).

Mitochondria have been studied in the SAM and leaf primordium meristematic cells in Arabidopsis and show quite striking changes across the cell cycle (Seguí-Simarro et al. 2008). Using serial thin section reconstruction and high-pressure-freezing, Seguí-Simarro and co-workers showed that meristematic cells develop a large mitochondrion that wraps around the nucleus to form a reticulate (tentaculate/cage-like) mitochondrion as the cell cycle progresses. Small mitochondria occur at the periphery and undergo dynamic fusion/fission with the large tentaculate-like mitochondrion. This structure and its dynamics ensures mixing of nucleoids and subsequent roughly equal partitioning to daughter cells. In the same study, Seguí-Simarro et al. (2008) reported that other cell types had no similar nuclear-associated mitochondrial mass but instead contained mitochondria with oval or sausage-like morphology that were distributed throughout the cell. These authors argued that cells of the SAM are special in that they are precursors to all the cells of the aerial parts of the plant, and this unique mitochondrial morphology enables passing of homogeneous, recombined mtDNA to the next generation. In animal cells, a recent study emphasizes the subtleties of mitochondrial behaviour during cell division of stem cells. The continuing stem cells inherit the newer mitochondria while the cells heading for differentiation inherit the older mitochondria (Katajisto et al. 2015). Age was determined by using different flurophores to mark mitochondria at different times. One suggestion was that old mitochondria were nuclear located while new ones were peripheral (Katajisto et al. 2015, Suomalainen 2015).

Other studies have focused on the endomembrane system, with particular attention paid to Golgi. The requirement for cell wall and cell plate synthesis in continuously dividing plant cells requires an active secretory pathway, which in turn influences the inheritance strategy, particularly the distribution, of Golgi bodies. In dividing cells of the onion RAM, the volume of Golgi bodies approximately doubles between prophase and telophase of the cell cycle (Garcia-Herdugo et al. 1988). Similar findings were made more recently in cells of the Arabidopsis RAM, with the number of Golgi stacks found to double in the G2 phase of the cell cycle (Seguí-Simarro and Staehelin 2006). During interphase, Golgi bodies randomly distribute throughout the cytoplasm, but during early cytokinesis, most (around 80%) of the Golgi bodies accumulate in a ring-like configuration around the forming cell plate (Seguí-Simarro and Staehelin 2006). Such a distribution invariably relates to the function of Golgi bodies in delivery of cell plate-forming components. However, such a doubling of Golgi numbers and distribution between the forming daughter nuclei also perhaps facilitates unbiased inheritance, with each daughter cell inheriting between 45 and 55% of the mother cell's Golgi population (Seguí-Simarro and Staehelin 2006).

Collings et al. (2003) investigated the distribution of peroxisomes in dividing cells of the onion RAM and found that peroxisomes distribute randomly throughout the cytoplasm during interphase, but by late anaphase, accumulate at the division plane. Initially, peroxisomes occur within the phragmoplast in two zones either side of the developing cell plate. However, as the phragmoplast expands outwards to form an annulus, peroxisomes redistribute into a ring circumscribing the inner edge of the expanding phragmoplast. The redistribution of peroxisomes, which becomes most apparent during telophase, may relate to formation of the cell plate, with peroxisomes perhaps recycling excess lipid or regulating hydrogen peroxide levels (Collings et al. 2003). Disrupting AFs or the AF-based motor protein, myosin, with chemical inhibitors prevents peroxisome aggregation, whereas disruption of MTs has little effect, indicating that peroxisome aggregation is actomyosin and not a MT-dependent process. Examination of the images in the Collings et al. (2003) paper suggests that peroxisome number increases before cell division, similarly to other organelles.

Examination of the distribution of the ER in dividing cells from the RAM and dividing leaf cells (stomata, mesophyll) of barley indicates that ER aggregates at the polar regions of the nucleus following prometaphase breakdown of the nuclear envelope (Hepler, 1980). During metaphase, tubes of ER surround the mitotic apparatus, protrude into its interior and encompass the chromosomes. That the ER closely associates with kinetochore MTs in these cells suggests that MTs may also have a role in this ER redistribution (Hepler, 1980). Recent studies in HeLa cells show that nuclear pore complexes disassemble and the nuclear membrane disperses into the ER then reemerges as the nuclear envelope reassembles. MT breakage and nuclear envelope resealing is accomplished by the enzyme spastin and the ESCRT-III endosomal sorting complex (Olmos et al. 2015, Vietri et al. 2015).

Organelle inheritance in cultured, constitutively dividing cells

In sycamore suspension cultured cells, the plastid population closely associates with the nucleus – these numerous plastids segregate at cell division with roughly half congregating at each end of the nucleus (Roberts and Northcote 1970). Plastids in tobacco suspension-cultured cells show a similar distribution, with most of the highly pleomorphically-shaped plastids tending to associate closely with the nucleus (Köhler and Hanson 2000). These plastids have an extensive network of stromules that can radiate large distances into the cell. Whether the association of these plastids with the nucleus and their large network of stromules play important functional roles in these cells is uncertain. However, such a distribution would enhance unbiased plastid inheritance.

In tobacco BY-2 cell cultures, as in the RAM, the number of Golgi bodies increase during the cell cycle (Nebenführ et al. 2000). In interphase, Golgi bodies disperse in an apparently random but AF dependent manner throughout the cytoplasm. In preparation for cell division, Golgi redistribute to the perinuclear cytoplasm, whereas by metaphase a large proportion of Golgi bodies accumulate around the spindle MTs and spindle poles (Nebenführ et al. 2000). This distribution appears to be maintained until late cytokinesis. Clearly, the important role for Golgi bodies in the synthesis of the growing phragmoplast influences the distribution of these organelles. However, such a distribution also enables Golgi bodies to partition to opposing faces of the forming cell plate and segregate in roughly equal numbers to the daughter cells.

Interestingly, application of inhibitors of either MT or AF polymerization has little effect on the maintenance of this distribution, but this is not to say movement to the perinuclear region is not MT or AF dependent. Importantly here, the distribution of Golgi during mitosis and cytokinesis was separate from plastids and mitochondria. The partitioning of the Golgi apparatus in mammalian cells is rather different, possibly because of the absence of a phragmoplast. Golgi inheritance is ensured by fragmentation of the Golgi apparatus into numerous dispersed small vesicles in late G2 and early in mitosis (Colenzi et al. 2003, Carcedo at al. 2004, Diao and Lowe 2004). A protein, CtBP3/BARS, which remodels lipid bilayers, is an essential regulator of Golgi fragmentation after the Golgi stacks are disassembled into a tubular network (Carcedo et al. 2004). The BARS-induced fission of the Golgi membranes, and associated signaling, is necessary for entry into prophase (Carcedo et al. 2004, Cervigni et al. 2015). Following cytokinesis, these Golgi vesicles fuse to reform a functional Golgi apparatus in each daughter cell.

During division of tobacco NT-1 cells, the perinuclear ER undergoes a dramatic redistribution (Gupton et al. 2006). In these cells, the ER concentrates at the nuclear envelope before prophase and as division progresses, an increasing number of ER tubules enter the nucleus. Following breakdown of the nuclear envelope in prophase, the density of ER at the spindle poles increases, with ER tubules accumulating around the chromosomes and apparently aligning parallel to the orientation of MTs in the spindle apparatus. With the commencement of anaphase, the ER rapidly fills the space between the separating chromosomes with the phragmoplast becoming enriched in ER during telophase (Gupton et al. 2006). Throughout the process of cell division, the cortical ER in contrast undergoes little change (Gupton et al. 2006). As cells progress from interphase to cell division, the cytoskeletal factors that organize the ER switch from being AF-based to MT-based. When MTs begin directing ER arrangement at nuclear envelope dissipation, inhibition of MT or AF polymerization indicated that the distribution of ER was a MT-, and not an AF-dependent phenomenon (Gupton et al. 2006). It is likely that this redistribution of ER has a functional role, with ER derived vesicles being important for cell plate formation. However, the colocalization of the ER with chromosomes and the spindle apparatus may concomitantly ensure that approximately half the ER volume partitions to each daughter cell. The relationship between the nuclear membrane and the ER is consistent with the recent studies on mammalian cells (Olmos et al. 2015, Vietri et al. 2015).

The division of peroxisomes has been studied in Arabidopsis suspension cells (Lingard et al. 2008). Data are consistent with a doubling of peroxisome numbers in G2 from division of preexisting peroxisomes and then their partitioning into daughter cells to maintain the peroxisome population. In this study, however, the partitioning mechanism was not investigated.

Organelle Inheritance in Cultured Protoplasts: A Cell Division Re-Initiation System

Plants, possibly due to their sessile nature, have an unparalleled ability to recruit somatic cells to re-enter the cell cycle. The ability of most somatic plant cells to regenerate from stems cells formed via dedifferentiation, or from pre-existing stem cells, constitutes an important regenerative and reproductive strategy, enhancing the ability of plants to survive under adverse conditions from which they cannot escape. A requirement for unbiased organelle inheritance is particularly evident in cells capable of totipotency. Experimental systems that examine the changes that occur during the transition from somatic cell to stem cell provide an excellent means to assess the processes that facilitate unbiased organelle inheritance. Cultured meso-phyll protoplasts can undergo the transition from fully differentiated mesophyll cell to a totipotent stem cell that will undergo division and ultimately reform an entire plant (Sheahan et al. 2004). The changes that occur to chloroplasts, mitochondria, ER and peroxisomes in these cell types have provided novel insights into the underlying cellular mechanisms that drive unbiased organelle inheritance.

In freshly prepared protoplasts (0 h), chloroplasts reside in the cortical cytoplasm, as is the case for their mesophyll cell progenitors, but during culture and before cell division, chloroplasts reposition to the perinuclear region (Fig. 1 and 2). The propor-tion of cells with perinuclear-positioned chloroplasts increases with further culture, such that in divided protoplasts, almost all chloroplasts cluster tightly in the peri-nuclear region of either daughter cell (Sheahan et al. 2004). Interestingly, the number of chloroplasts approximately halves during both the first and second cell divisions; thereafter, however, chloroplast number remains reasonably constant at around 11 chloroplasts per cell. Thus, chloroplast replication in this developmental context is uncoupled from cell division, until the cell reaches a steady-state chloroplast number (Thomas and Rose, 1983). Experiments with inhibitors of either MT or AF polym-erization indicates that this repositioning is an AF- and not MT-dependent phenom-enon (Sheahan et al. 2004). How chloroplast perinuclear clustering is maintained,

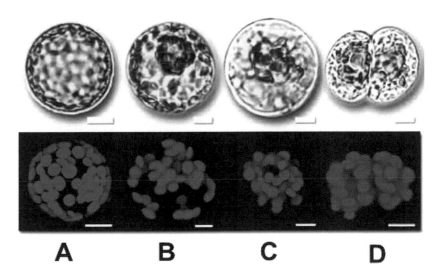

A   B   C   D

FIGURE 1   Chloroplast redistribution during culture of *Nicotiana tabacum* mesophyll protoplasts. Chloroplasts become localised to the perinuclear cytoplasm prior to cell division. Top images are in brightfield, bottom images show chloroplasts visualised by chlorophyll autofluorescence. (A) Cortical chloroplast distribution; 0 h. (B) Transitional chloroplast distribution; 48 h. (C) Perinuclear chloroplast distribution; 72 h. (D) Perinuclear chloroplast distribution in a divided cell; 96 h. Chloroplasts move from cortical to perinuclear cytoplasm during culture. Bars = 10 μm.

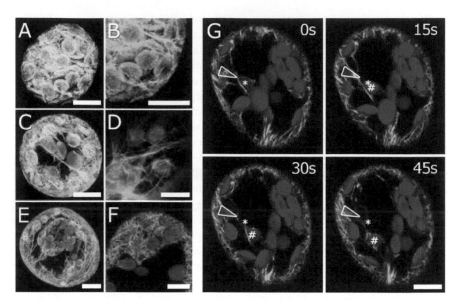

FIGURE 2 Chloroplast motility and chloroplast-AF interaction in cultured protoplasts. (A-F) Confocal projections of *Nicotiana tabacum* mesophyll protoplasts during culture showing the interaction of chloroplasts with GFP-fABD2-labelled AFs. Confocal projections of top cell half (A, B), middle third (C, E), or single optical sections (D, F). (A, B) Freshly isolated protoplasts displayed prominent chloroplast AF-baskets. Image in (B) is a higher magnification image of that in (A) showing AF-baskets. (C, D) Chloroplast-AF interaction in protoplasts cultured for 48 h. Chloroplast AF-baskets were lost during culture as chloroplasts became enmeshed within and moved by the actin network. (E, F) Protoplasts cultured for 96 h. Chloroplasts in the perinuclear region were enmeshed by the perinuclear AF array. Chloroplasts elsewhere in the cell remained enmeshed by a dense network of AFs. (G) Movement of chloroplasts with the AF network. Time series composed of single 2.5-μm optical sections. Arrowhead indicates initial position of AF bundle, while '*' indicates the initial position of chloroplast centroid (0s) and '#' the final position (45s). Note how the chloroplast moves with the AF bundle rather than along it. Bars = 10 μm.

however, is uncertain; maybe the chloroplasts are "trapped" in some way in close association with the nuclear membrane, possibly by interactions with the nuclear membrane itself and/or by a perinuclear cytoskeletal network.

Mitochondria in freshly prepared protoplasts are present as small, coccoid (approximately 0.7-μm diameter) organelles in close association with chloroplasts and the nucleus. During early culture (4–24 h), however, mitochondria fuse to produce tubular structures with a highly plastic morphology. During this phase of culture, the number of mitochondria decreases slightly; however, the total amount of mitochondrial volume per cell remains relatively constant, consistent with net mitochondrial fusion (Sheahan et al. 2004, 2005). Following this phase, fission predominates and tubular mitochondria give rise to numerous smaller mitochondria that disperse uniformly throughout the cytoplasm (Figs. 3 and 4). By 72 h culture, mitochondria are present as punctate spheres (approximately 0.4 μm diameter), with a corresponding doubling in the number of mitochondria per cell. In subsequent divisions, mitochondria remain as small, punctate organelles, uniformly distributed throughout the cytoplasm (Sheahan et al. 2004). Similarly to chloroplasts, mitochondrial repositioning (dispersion in this case) is an entirely AF-dependent phenomenon (Figs. 2 and 4) with little or no role for MTs (Sheahan et al. 2004).

Peroxisome partitioning is similar to mitochondria with large numbers distributing throughout the cytoplasm by an AF-dependent mechanism, with little involvement of MTs (Tiew et al. 2015). Again, this phenomenon ensures that daughter cells inherit similar peroxisome number. In freshly isolated protoplasts the average protoplast area was 2.2 μm^2 with 62 peroxisome per cell. These numbers increased more than tenfold per cell prior to cell division; individual area of peroxisomes decreased but total area (from the sum of peroxisome surface area for each 1μm section of a cell) increased. These changes in protoplast numbers and surface area, as well as contributing to peroxisome continuity, help maintain ROS homeostasis and suitable redox for cell division induction (Tiew et al. 2015).

Much of the ER volume in freshly isolated protoplasts localizes to perinuclear and transvacuolar regions of the cell. During protoplast culture, however, both the volume and density of total cell ER increases, while the volume of perinuclear ER

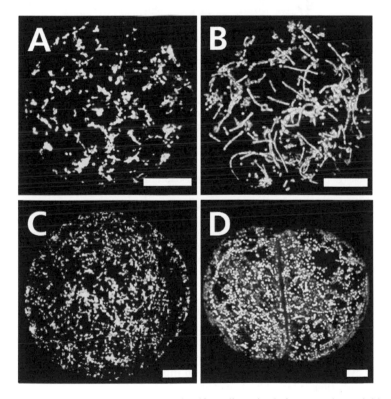

FIGURE 3 Mitochondrial fusion, fission and uniform dispersion before protoplasts reinitiate cell division. *Nicotiana tabacum* mesophyll protoplasts expressing GFP targeted to mitochondria by the cytochrome oxidase IV signal peptide from yeast. (A) Mitochondria are initially small, oval-shaped organelles. (B) Early in protoplast culture mitochondria undergo a phase of massive elongation caused by mitochondrial fusion. (C) Following the fusion phase, mitochondria undergo fission, generating large numbers of small mitochondria. (D) Large numbers of uniformly dispersed mitochondria enhance unbiased inheritance of the organelle at cell division. Bars = 10 μm. Based on Sheahan et al. (2004), with permission of John Wiley and Sons Publishers.

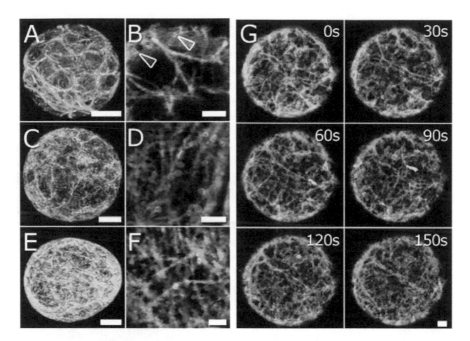

FIGURE 4 Confocal projections of tobacco mesophyll protoplasts during culture showing the interaction of mitochondria with AFs. Confocal projections of the top half (A, C, E) and single optical sections (B, D, F) of protoplasts at higher magnification. (A, B) Freshly isolated protoplasts. (A) Mitochondria associated with chloroplast actin-baskets. (B) Motile mitochondria appear in direct contact with AF bundles. (C, D) Mitochondrial-AF interaction in protoplasts cultured for 48 h. Mitochondria increased in number but remained closely associated with AFs. (E, F) Protoplasts cultured for 96h. (E) Large numbers of punctate mitochondria dispersed throughout the cell remaining in close association with AFs. (F) Motile mitochondria observed to be travelling along sub-cortical AFs. (G) Time-lapse imaging of erratic and vectorial mitochondrial movements. Mitochondria undergoing erratic motion (*magenta*) appear to follow the path of movement taken by underlying AFs. Mitochondria undergoing rapid vectorial movements (*cyan*) appear to move along static AFs or are moved by rapidly sliding AFs. Bars = 10 μm (A, C, E), 2.5 μm (B, D, F, G). From Sheahan et al. 2007, with permission of John Wiley and Sons Publishers.

decreases. Thus, redistribution occurs before cell division and results in the almost complete transfer of ER volume from the perinuclear region into the cell cortex, eventually stabilizing with an 85% reduction in perinuclear-associated ER (Sheahan et al. 2004). Redistribution of the ER to the cell cortex strictly requires the actin cytoskeleton, with no role for MTs in this process. Notably, disruption of these partitioning events by treating protoplasts with inhibitors of AF polymerization greatly increases inheritance bias (Sheahan et al. 2004).

Coordination of Organelle Replication with Cell Division

Although organelle division can be uncoupled from cell division, for example, as occurs with plastid division during mesophyll development, there is a clear need to maintain organelle numbers in dividing cells to ensure organelle continuity (Boffey et al. 1979, Rose 1988). In dividing cells, therefore, organelle replication must coordinate with progression through the cell cycle, ensuring functional requirements and accurate transmission to daughter cells. Checkpoints that convey information about the status of total organelle volume and number must exist before cell division commences. Chloroplast dysfunction is known to influence cell cycle progression (Hudik et al. 2014).

Evidence from synchronized BY-2 cell cultures indicates that expression of the plastid division protein, FtsZ, peaks during mitosis, suggesting that cell and plastid division are coordinated in these cells (El-Shami et al. 2002). Microarray analysis of gene expression in synchronized cultures of dividing Arabidopsis cells provides further support for the coordination of cell and plastid division. In these cells, plastid-targeted homologues of the yeast protein, ABF2 (involved in mitochondrial DNA segregation and copy number control), show strong regulation of expression during mitosis (Menges et al. 2002). More evidence of the molecular mechanisms that entrain organelle replication with the cell cycle comes from a study of *AtCDT1* genes in Arabidopsis (Raynaud et al. 2005). AtCDT1 forms a component of the nuclear DNA pre-replication complex, which functions to ensure DNA synthesis occurs only once before entry into mitosis (DePamphilis, 2005). In Arabidopsis, AtCDT1 localizes to the nucleus, but also interacts with ARC6, a component of the plastid division machinery. Silencing of AtCDT1 leads to endoreduplication and reduced cell and plastid division. Thus, AtCDT1 represents an important coordinator of cell and plastid division (Raynaud et al. 2005).

DNA replication in the DNA-containing organelles must occur before cell division and coordinate with the cell cycle. Synthesis of plastid DNA precedes both plastid division and synthesis of nuclear DNA (Yasuda et al. 1988, Suzuki et al. 1992). Transfer of stationary BY-2 cells to fresh medium, a treatment which stimulates cell division, results in a similar dramatic increase in the amount of plastid DNA per cell within 24 h. Interestingly, levels of plastid DNA per cell are relatively low in cells of the SAM in Arabidopsis, but as plastid numbers increase as cells move through the leaf primordia, plastid DNA synthesis increases dramatically and precedes the differentiation of proplastids into chloroplasts (Fujie et al. 1994). Examination of plastid DNA in young and fully expanded leaves of a number of plant species indicates that during leaf development, when chloroplast division ceases, a significant decline in the levels of plastid DNA occurs (Shaver et al. 2006). This means that plastid DNA synthesis is not necessarily co-ordinated with chloroplast division once cell division ceases, due to extensive prior plastid DNA replication.

The replication of mitochondrial DNA in cultured BY-2 cells parallels that of plastids, with a greater than four-fold increase in mitochondrial DNA within 24 h of transferring stationary phase BY-2 cells to fresh medium. Again, as with plastids, mitochondrial DNA synthesis precedes nuclear DNA synthesis and cell division (Suzuki et al. 1992). In contrast to plastid DNA synthesis in the meristematic cells, mitochondrial DNA synthesis is restricted to cells immediately surrounding the quiescent center in the RAM (Suzuki et al. 1992, Fujie et al. 1994). In cells leaving the SAM, the amount of mitochondrial DNA per mitochondria drops considerably, as occurs in cells distal to the RAM (Suzuki et al. 1992, Fujie et al.1994, Takanashi et al. 2006). Such findings suggest that mitochondrial DNA is synthesized in large quantities in the meristematic cells to support subsequent mitochondrial division. Mitochondrial DNA can also be distributed throughout the population because of mitochondrial fusion. Unlike plastids, mitochondria undergo fusion (Arimura et al. 2004, Sheahan et al. 2005) possibly to compensate for variable DNA levels per mitochondrion (Takanashi et al. 2006).

Although the situation is dynamic and dependent on plastid type and developmental context, plastid nucleoids in dicotyledons are typically scattered throughout the plastid interior and associate with thylakoids, whereas in some monocotyledons nucleoids disperse around the plastid periphery and probably associate with the inner plastid envelope (Rose and Possingham 1976, Rose 1979, Selldén and Leech 1981). Several studies indicate that plastid DNA is physically attached to plastid membranes by DNA-binding proteins such as PEND and MFP1 (Sato et al. 1993, Jeong et al. 2003, Krupinska et al. 2013). By virtue of attachment of the plastid DNA to either membrane system, plastid nucleoids can thus be distributed and segregated by membrane growth and separation of membranes into daughter plastids following plastid division (Rose 1988). Less is known about the segregation mechanism of mitochondrial DNA; however, DNA-binding proteins associated with the inner mitochondrial membrane, such as Yhm2, exist in yeast and thus a similar mechanism of mitochondrial DNA segregation as occurs with plastids may occur in plant mitochondria (Cho et al. 1998).

As discussed earlier non-DNA containing organelles such as peroxisomes (Lingard et al. 2008) and Golgi (Seguí-Simarro and Staehelin 2006) approximately double in number during the cell cycle prior to partitioning to daughter cells.

Cytoskeletal Basis for Organelle Repositioning in Plant Cells

The key criterion for unbiased organelle inheritance following on from organelle replication is distribution of organelles in such a manner that facilitates their unbiased partitioning to daughter cells. Although such distributions may appear stochastic or enforced by steric constraints, for example in the cytoplasmically dense cells of the germline or meristem, as seen from the above considerations there is an increasing recognition that the AF and MT cytoskeleton plays an important role in mediating these distributions. Distribution of organelles to specific subcellular locations thus often requires complex, regulated interactions between organelles and the cytoskeleton.

Organelle motility in interphase plant cells

Although plants are sessile, their cellular cytoplasm displays a great deal of dynamism. Organelle dynamics are driven by the cytoskeleton, either indirectly, as occurs in cytoplasmic streaming, or directly, for example during light-responsive chloroplast repositioning. Unlike animal cells, which use MTs, plants use the actin cytoskeleton for long distance transport (Williamson 1993, Wada and Suetsugu, 2004, Avisar et al. 2008, Cai et al. 2014). AFs or bundles provide the tracks for myosin, utilizing ATP hydrolysis, to power organelle movement (Avisar et al. 2008, Cai et al. 2014). Mitochondria (Van Gestel et al. 2002), peroxisomes (Collings et al. 2002, Jedd and Chua 2002, Mathur et al. 2002), Golgi (Boevink et al. 1998, Nebenführ et al. 1999) and the ER (Quader et al. 1987, Williamson 1993, Sparkes et al. 2009) undergo AF-dependent movement, powered by myosin X1 (Hashimoto et al. 2005, Avisar et al. 2008, Cai et al. 2014). These conclusions are based on studies using inhibitors,

RNAi (Avisar et al. 2008) and T-DNA knockout lines (Peremyslov et al. 2008, Cai et al. 2014). While fast mitochondrial movement depends on an AF-myosin system, immobile mitochondrial positioning in the cortical cytoplasm seems to be dependent on both AFs and MTs (Van Gestel et al. 2002).

Plastid motility appears to be driven by AFs while MTs act to restrain movement (Kandasamy and Meagher 1999, Wada et al. 2003, Kwok and Hansen 2003, Wada and Suetsugu 2004). The chloroplast positioning mutant, *chup1*, was isolated from Arabidopsis using a screen for a defective chloroplast light-avoidance response (Oikawa et al. 2003). The CHUP protein binds to F-actin and the outer chloroplast membrane and may function as a chloroplast actin anchoring protein (Wada and Suetsugu 2004). Two kinesin-like proteins, KAC1 and KAC2 are essential for chloroplasts to move and anchor to the plasma membrane (Suetsugu et al. 2010). These proteins are AF rather than MT based. It does not appear that myosins are involved in chloroplast movements (Avisar et al. 2008, Peremyslov et al. 2008).

Organelle repositioning during cell division

There have been fewer investigations about the cellular and molecular mechanisms of organelle motility and repositioning in dividing plant cells. The major forms of organelle distribution observed in dividing cells, or cells preparing to enter division, are cytoplasmic dispersion (Fig.4, Sheahan et al. 2004, Tiew et. al. 2015), perinuclear clustering (Fig 2, Sheahan et al. 2004) or circumscribing the phragmoplast (Nebenführ et al. 2000). In some cases, the cellular mechanisms that achieve these organelle distributions are similar to interphase cells. For example, peroxisome repositioning in cells of the RAM in onion is AF- and myosin-dependent (Collings et al. 2003). Similarly, repositioning of chloroplasts, mitochondria, ER (Sheahan et al. 2004) and peroxisomes (Tiew et al. 2015) in protoplasts preparing to enter cell division are AF-dependent. In protoplasts initiating cell division, mitochondria track along individual AF bundles (Fig. 4) while chloroplasts appear to be surrounded by AF baskets as they translocate with a network of actin arrays (Fig. 2). However, in some cell types, organelle distributions appear to be cytoskeleton-independent, for example the maintenance of Golgi positioning in cultured BY-2 cells (Nebenführ et al. 2000). There are also observations indicating that MTs may play a bigger role in organelle repositioning in dividing compared to interphase cells. Examples in this case are MT-dependent ER repositioning in dividing tobacco NT-1 cells (Gupton et al. 2006), and MT-dependent plastid redistribution during *Lillium* microsporogenesis (Tanaka 1991). Weaker evidence also exists to suggest that MTs may also be involved in driving plastid and mitochondrial association with the nucleus in egg cells (Huang and Russell 1992), in dividing barley cells (Hepler 1980) and possibly in positioning of the plastid in *Isoetes* meristematic cells (Whatley 1974).

In budding yeast each organelle in the cell division process is anchored to actin-linked myosin motors by a specific protein factor (Fagarasanu et al. 2010). In fission yeast, mitochondria partition to opposing poles of the mother cell but use forces generated from the fast-growing ends of dynamic MTs that emanate from the mitotic spindle rather than kinesin motor proteins (Yaffe et al. 2003). Thus, attachment to MTs and dynamic MT polymerization processes within the spindle drive mitochondria to opposing poles of the cell. These yeast data are consistent with diverse cytoskeletal mechanisms used in eukaryotes for organelle inheritance. Even though

AFs or MTs may be major drivers in different eukaryote cell types it is increasingly apparent that AFs and MTs can interact via discrete proteins (Petrášek and Schwarzerová 2009, Schneider and Persson 2015) such as kinesin 14 motor proteins (Schneider and Persson 2015).

Conclusions and Future Prospects

Considerable variation exists in the patterns of organelle replication and distribution before plant cells divide, depending largely on cell type and developmental context. However, two basic distribution strategies appear to have evolved to ensure unbiased organelle partitioning; firstly, a stochastic, but uniform dispersion of organelles throughout the cytoplasm, and secondly, a close association of organelles with the nucleus. Such strategies enhance the probability that each daughter cell will inherit approximately half the organelle population. In the case of clustering in the mid-plane of a dividing cell, unbiased inheritance occurs as a consequence of the placement of an internal cell plate.

As discussed in this Chapter, in almost all dividing plant cells, organelle number (and typically volume as well) increases by two-fold or more (Possingham 1980, Sheahan et al. 2004, Seguí-Simarro and Staehelin 2006, Lingard et al. 2008) before cytokinesis separates the cytoplasm of the mother cell into two daughter cells. This increase in organelle number is particularly evident when quiescent cells prepare to enter an extended phase of proliferation, for example, as occurs with the marked increase in mitochondrial number during megasporogenesis (Huang and Russell 1992). In such cases, an increased importance is placed on organelle distributions that permit unbiased partitioning of organelles at cytokinesis. Thus, organelles with a low copy number are often found in close association with the nucleus, for example plastids in shoot and root meristems. Such a distribution allows organelles to take advantage of the highly accurate partitioning properties of the mitotic spindle apparatus. Organelles with a high copy number may also associate with the nucleus (sometimes only because of steric constraints), but more often, disperse uniformly throughout the cytoplasm, for example, as occurs with mitochondria in cultured protoplasts. The distribution of organelles can change considerably during mitosis and cytokinesis, which is particularly evident for ER, Golgi bodies and perhaps peroxisomes, which move to locations between segregating chromosomes, primarily because of functional requirements for these organelles during the process of cell plate formation.

The partitioning mechanisms are predominantly AF dependent but with MTs being important in some cases and in circumscribing the phragmoplast. AFs can serve as a track upon which myosin drives the organelle movement (Williamson 1993, Wada and Suetsugu 2004, Avisar et al. 2008, Cai et al. 2014). Although the basic requirements for either AFs or MTs in organelle repositioning have been elucidated in some specific developmental contexts, very little is known about the detailed molecular regulation of such processes. For example, which signals cause organelles to cluster around the nucleus, disperse throughout the cytoplasm or circumscribe the phragmoplast? What is the nature of the components that tether organelles such as plastids and mitochondria to the cytoskeleton and maintain an association with the nucleus? Information obtained from the careful analysis of cytoskeletal architecture with appropriate tags and the mutational dissection of cytoskeleton-interacting proteins in mutant plants are being used to address these questions. The study of the detailed mechanisms of organelle inheritance in plants is still at an early stage and further study is clearly required in order to reveal the gamut of organelle inheritance processes and the relationship to the cytoskeleton and cell cycle that have evolved in different developmental contexts and cell types.

Acknowledgments

This work was supported by the Australian Research Council project number CEO348212. David McCurdy and Ray Rose particularly acknowledge the contributions of our late colleague Dr Michael Sheahan, who pioneered the study of organelle inheritance and its mechanisms in our laboratories and passed away after initiating this review.

References

Arimura, S., J. Yamamoto, G.P. Aida, M. Nakazono and N. Tsutsumi. 2004. Frequent fusion and fission of plant mitochondria with unequal nucleoid distribution. Proc. Natl. Acad. Sci. USA. 101: 7805-7808.

Avisar, D., A.I. Prokhnevsky, K.S. Makarova, E.V. Koonon and V.V. Dolja. 2008. Myosin XI-K is required for rapid trafficking of Golgi stacks, peroxisomes, and mitochondria in leaf cells of *Nicotiana benthamiana*. Plant Physiol. 146: 1098-1108.

Birky, C.W. Jr. 2001. The inheritance of genes in mitochondria and chloroplasts: laws, mechanisms, and models. Annu. Rev. Genet. 35: 125-148.

Boevink, P., K. Oparka, S. Santa Cruz, B. Martin, A. Betteridge and C. Hawes. 1998. Stacks on tracks: the plant Golgi apparatus traffics on an actin/ER network. Plant J. 15: 441-447.

Boffey, S. A., J.R. Ellis, G. Sellden and R.M. Leech. 1979. Chloroplast division and DNA synthesis in light-grown wheat leaves. Plant Physiol. 64: 502-505.

Brown, R.C. and B.E. Lemmon. 1984. Plastid apportionment and preprophase microtubule bands in monoplastidic root meristem cells of *Isoetes* and *Selaginella*. Protoplasma 123: 95-103.

Cai, C., J.L. Henty-Ridilla, D.B. Szymanski and C.J. Staiger. 2014. Arabidopsis myosin XI: a motor rules the tracks. Plant Physiol. 166: 1359-1370.

Carcedo, C.H., M. Bonazzi, S. Spanò, G. Turacchio, A. Colanzi, A. Luini and D. Corda. 2004. Mitotic Golgi partitioning is driven by the membrane-fissioning protein CtBP3/BARS. Science 305: 93-96.

Cavalier-Smith, T. 2002. The phagotrophic origin of eukaryotes and phylogenetic classification of Protozoa. Int. J. Syst. Evol. Microbiol. 52: 297-354.

Cavalier-Smith, T. 2009. Predation and eukaryote cell origins: A coevolutionary perspective. Int. J. Biochem. and Cell. Biol. 41: 307-322.

Cervigni, R.I., R. Bonavita, M.L. Barretta, D. Spano, I. Ayala, N. Nakamura, D. Corda and A. Colanzi. 2015. JNK2 controls fragmentation of the Golgi complex and the G2/M transition through phosphorylation of GRASP65. J. Cell Sci. 128: 2249-2260.

Cho, J.H., S.J. Ha, L.R. Kao, T.L. Megraw and C-B. Chae. 1998. A novel DNA-binding protein bound to the mitochondrial inner membrane restores the null mutation of mitochondrial histone abf2p in *Saccharomyces cerevisiae*. Mol. Cell Biol. 18: 5712-5723.

Colanzi, A., C. Süetterlin and V. Malhotra. 2003. Cell-cycle-specific Golgi fragmentation: how and why? Curr. Opin. Cell Biol. 15: 462-467.

Collings, D.A., J.D.I. Harper, J. Marc, R.L. Overall and R.T. Mullen. 2002. Life in the fast lane: actin-based motility of plant peroxisomes. Can. J. Bot. 80: 430-441.

Collings, D.A., J.D.I. Harper and K.C. Vaughn. 2003. The association of peroxisomes with the developing cell plate in dividing onion root cells depends on actin microfilaments and myosin. Planta 218: 204-216.

DePamphilis, M.L. 2005. Cell cycle dependent regulation of the origin recognition complex. Cell Cycle 4: 70-79.

Diao, A. and M. Lowe 2004. The Golgi goes fission. Science 305: 48-49.

Dickinson, H.G. and F.L. Li. 1988. Organelle behaviour during higher plant gametogenesis. pp.131-148. *In*: S.A. Boffey and D. Lloyd (eds.). Division and Segregation of Organelles. Cambridge University Press, Cambridge.

Dyall, S.D., M.T.Brown and P.J. Johnson. 2004. Ancient invasions: from endosymbionts to organelles. Science 304: 253-257.

El-Shami, M., S. El-Kafafi, D. Falconet and S. Lerbs-Mache. 2002. Cell cycle-dependent modulation of FtsZ expression in synchronized tobacco BY2 cells. Mol. Genet. Genom. 267: 254-261.

Fagarasanu, A., F.D. Mast, B. Knoblach and R.A. Rachubinski. 2010. Molecular mechanisms of organelle inheritance: lessons from peroxisomes in yeast. Nature Rev. Mol. Cell Biol. 11: 644-654.

Fujie, M., H. Kuroiwa, S. Kawano, S. Mutoh and T. Kuroiwa. 1994. Behavior of organelles and their nucleoids in the shoot apical meristem during leaf development in *Arabidopsis thaliana* L. Planta 194: 395-405.

Garcia-Herdugo, G., J.A. Gonzáles-Reyes, F. Gracia-Navarro and P. Navas. 1988. Growth kinetics of the Golgi apparatus during the cell cycle in onion root meristems. Planta 175: 305-312.

Gray, M.W. 2004. The evolutionary origins of plant organelles. pp. 15-36. *In*: H. Daniell and C.D. Chase, (eds.). Molecular Biology and Biotechnology of Plant Organelles: Chloroplasts and Mitochondria. Springer, Dordrecht.

Gray, M.W. 2012. Mitochondrial evolution. Cold Spring Harb. Perspect. Biol. 4: a011403.

Greiner, S., J. Sobanski and R. Bock. 2015. Why are most organelle genomes transmitted maternally? Bioessays 37: 80-94.

Gupton, S.L., D.A. Collings and N.S. Allen. 2006. Endoplasmic reticulum targeted GFP reveals ER organization in tobacco NT-1 cells during cell division. Plant Physiol. Biochem. 44: 95-105.

Hashimoto, K., H. Igarashi, S. Mano, M. Nishimura, T. Shimmen and E. Yokota. 2005. Peroxisomal localization of a myosin XI isoform in *Arabidopsis thaliana*. Plant Cell Physiol. 46: 782-789.

Hepler, P. K. 1980. Membranes in the mitotic apparatus of barley cells. J. Cell Biol. 86: 490-499.

Huang, B-Q. and S.D. Russell. 1992. Female germ unit: organization, reconstruction and isolation. Int. Rev. Cytol. 140: 233-293.

Hudik, E., Y. Yoshioka, S. Domenichini, M. Bourge, L. Soubigout-Taconnat, C. Mazubert, D. Yi, S. Bujaldon, H. Hayashi, L. De Veylder, C. Bergounioux, M. Behamed and C. Raynaud. 2014. Chloroplast dysfunction causes multiple defects in cell cycle progression in the Arabidopsis *crumpled leaf* mutant. Plant Physiol. 166: 152-167.

Huh, D. and J. Paulsson. 2011. Random partitioning of molecules at cell division. Proc. Natl. Acad. Sci. USA. 108: 15004-15009.

Jedd, G. and N.H. Chua. 2002. Visualization of peroxisomes in living plant cells reveals acto-myosin-dependent cytoplasmic streaming and peroxisome budding. Plant Cell Physiol. 43: 384-392.

Jeong, S.Y., A. Rose, I. Meier. 2003. MFP1 is a thylakoid-associated, nucleoid-binding protein with a coiled-coil structure. Nucl. Acids Res. 31: 5175-5185.

Kandasamy, M.K. and R.B. Meagher. 1999. Actin-organelle interaction: association with chloroplast in Arabidopsis leaf mesophyll cells. Cell Motil. Cytoskeleton 44: 110-118.

Katajisto, P., J. Döhla, C.L. Chaffer, N. Pentinmikko, N. Marjanovic, S. Iqbal, R. Zoncu, W.Chen, R.A. Weinberg and D.M. Sabitini. 2015. Asymmetric apportioning of aged mitochondria between daughter cells is required for stemness. Science 348: 340-344.

Köhler, R. and M. Hanson. 2000. Plastid tubules of higher plants are tissue-specific and developmentally regulated. J. Cell Sci. 113, 81-89.

Krupinska, K., J. Melonek and K. Krause. 2013. New insights into plastid nucleoid structure and functionality. Planta 237: 653-664.

Kuroiwa, H., Y. Nishimura, T. Higashiyama and T. Kuroiwa. 2002. *Pelargonium* embryogenesis: cytological investigations of organelles in early embryogenesis from the egg to the two-celled embryo. Sex. Plant Reprod. 15: 1-12.

Kwok, E.Y. and M.R. Hanson. 2003. Microfilaments and microtubules control the morphology and movement of non-green plastids and stromules in *Nicotiana tabacum*. Plant J. 35: 16-26.

Lingard, M.J., S.K. Gidda, S. Bingham, S.J. Rothstein, R.T. Mullen and R.N. Trelease. 2008. Arabidopsis PEROXIN11c-e, FISSION1b, and DYNAMIN-RELATED PROTEIN3A cooperate in cell cycle associated replication of peroxisomes. Plant Cell 20: 1567-1585.

Martin, W.F. and M. Mentel. 2010. The origin of mitochondria. Nature Education 3: 58.

Mathur, J., N. Mathur and M. Hülskamp. 2002. Simultaneous visualization of peroxisomes and cytoskeletal elements reveals actin and not microtubule-based peroxisome motility in plants. Plant Physiol. 128: 1031-1045.

Menges, M., L. Hennig, W. Gruissem and J.A.H. Murray. 2002. Cell cycle-regulated gene expression in *Arabidopsis*. J. Biol. Chem. 277: 41987-42002.

Nebenführ, A., L.A. Gallagher, T.G. Dunahay, J.A. Frohlick, A.M. Mazurkiewicz, J.B. Meehl and L.A. Staehelin. 1999. Stop-and-go movements of plant Golgi stacks are mediated by the acto-myosin system. Plant Physiol. 121: 1127-1142.

Nebenführ, A., J.A. Frohlick and L.A. Staehelin. 2000. Redistribution of Golgi stacks and other organelles during mitosis and cytokinesis in plant cells. Plant Physiol. 124: 135-151.

Oikawa,K., M. Kasahara, T. Kiyosue, T. Kagawa, N. Suetsugu, F. Takahashi, T. Kanagae, Y. Niwa, A. Kadota and M. Wada. 2003. CHLOROPLAST UNUSUAL POSITIONING1 is essential for proper chloroplast positioning. Plant Cell 15: 2805-2815.

Olmos, Y., L. Hodgson, J. Mantell, P. Verkade and J.G. Carlton. 2015. ESCRT-III controls nuclear envelope reformation. Nature 522: 236-239.

Peremyslov, V.V., A.I. Prokhnevsky, D. Avisar and V.V. Dolja. 2008. Two class XI myosins function in organelle trafficking and root hair development in Arabidopsis. Plant Physiol. 146: 1109-1116.

Petrášek, J. and K. Schwarzerová. 2009. Actin and microtubule cytoskeleton interactions. Curr. Opin. Plant Biol. 12: 728-734.

Possingham, J.V. 1980. Plastid replication and development in the life cycle of higher plants. Annu. Rev. Plant Physiol. 31: 113-129.

Quader H, A. Hofmann and E. Schnepf. 1987. Shape and movement of the endoplasmic reticulum in onion bulb epidermis cells: possible involvement of actin. Eur. J. Cell Biol. 44: 17-26.

Raven, J.A. and J.F. Allen. 2003. Genomics and chloroplast evolution: what did cyanobacteria do for plants? Genome Biol. 4: 2009.

Raynaud C, C. Perennes, C. Reuzeau, O. Catrice, S. Brown and C. Bergounioux. 2005. Cell and plastid division are coordinated through the prereplication factor AtCDT1. Proc. Natl. Acad. Sci. USA. 102: 8216-8221.

Roberts, K. and D.H. Northcote. 1970. The structure of *Sycamore* callus cells during division in a partially synchronized suspension culture. J. Cell Sci. 6: 299-321.

Robertson, E.J., K.A. Pyke and R.M. Leech. 1995. *arc6*, an extreme chloroplast division mutant of Arabidopsis also alters proplastid proliferation and morphology in shoot and root apices. J. Cell Sci. 108: 2937-2944.

Rose, R.J. 1979. The association of chloroplast DNA with photosynthetic membrane vesicles from spinach chloroplasts. J. Cell Sci. 36: 169-193.

Rose, R.J. 1988. The role of membranes in the segregation of plastid DNA. pp 171-195. *In*: S.A. Boffey and D. Lloyd (eds.) Division and Segregation of Organelles. Cambridge University Press, Cambridge.

Rose, R.J. and J.V. Possingham. 1976. The localization of (^{3}H) thymidine incorporation in the DNA of replicating spinach chloroplasts by electron-microscope autoradiography. J. Cell Sci. 20: 341-355.

Russell, S.D. 1993. The egg cell: development and role in fertilization and early embryogenesis. Plant Cell 5: 1349-1359.

Sato, N., C. Albrieux, J. Joyard, R. Douce and T. Kuroiwa. 1993. Detection and characterization of a plastid envelope DNA-binding protein which may anchor plastid nucleoids. EMBO J. 12: 555-561.

Schneider R and S. Persson. 2015. Connecting two arrays: the emerging role of actin-microtubule cross-linking motor proteins. Front. Plant Sci. 6: 415.

Seguí-Simarro, J.M. and L.A. Staehelin. 2006. Cell cycle-dependent changes in Golgi stacks, vacuoles, clathrin-coated vesicles and multivesicular bodies in meristematic cells of *Arabidopsis thaliana*: a quantitative and spatial analysis. Planta 223: 223-236.

Seguí-Simarro, J.M., M.J. Coronado and L.A. Staehelin. 2008. The mitochondrial cycle of Arabidopsis shoot apical meristem and leaf primordium meristematic cells is defined by a perinuclear tentaculate/cage-like mitochondrion. Plant Physiol. 148: 1380-1393.

Selldén, G. and R.M. Leech. 1981. Localization of DNA in mature and young wheat chloroplasts using the fluorescent probe 4'-6-diamidino-2-phenylindole. Plant Physiol. 68: 731-734.

Shaver, J.M., D.J. Oldenburg and A.J. Bendich. 2006. Changes in chloroplast DNA during development in tobacco, *Medicago truncatula*, pea, and maize. Planta 224: 72-82.

Sheahan, M.B., D.W. McCurdy and R.J. Rose. 2005. Mitochondria as a connected population: ensuring continuity of the mitochondrial genome during plant cell dedifferentiation through massive mitochondrial fusion. Plant J. 44: 744-755.

Sheahan, M.B., R.J. Rose and D.W. McCurdy. 2004. Organelle inheritance in plant cell division: the actin cytoskeleton is required for unbiased inheritance of chloroplasts, mitochondria and endoplasmic reticulum in dividing protoplasts. Plant J. 37: 379-390.

Sheahan, M.B., R.J. Rose and D.W. McCurdy. 2007. Mechanisms of organelle inheritance in dividing plant cells. J. Int. Plant Biol. 49: 1208-1218.

Singh, A.P. and H.L. Mogensen. 1975. Fine structure of the zygote and early embryo in *Quercus gambelii*. Am. J. Botany 62: 105-115.

Spang, A., J.H. Saw, S.L. Jorgensen, K. Zaremba-Niedzwiedzka, J. Martijn, A.E. Lind, R. van Eijk, C. Schleper, L. Guy and T.J.G. Ettema. 2015. Complex archae that bridge the gap between prokaryotes and eukaryotes. Nature 521: 173-179.

Sparkes, I., J. Runions, C. Hawes and L. Griffing. 2009. Movement and remodelling of the endoplasmic reticulum in non-dividing cells of tobacco leaves. Plant Cell 21: 3937-3948.

Suetsugu N., N. Yamada, T. Kagawa, H. Yonekura, T.Q.P. Uyeda and A. Kadota. 2010. Two kinesin-like proteins mediate actin-based chloroplast movement in *Arabidopsis thaliana.* Proc. Natl. Acad. Sci. USA 101: 7805-7808.

Suomalainen, A. 2015. Asymmetric rejuvenation. Nature 521: 296-297.

Suzuki, T., S. Kawano, A. Sakai, M. Fujie, H. Kuroiwa, H. Nakamura and T. Kuroiwa T. 1992. Preferential mitochondrial and plastid DNA synthesis before multiple cell divisions in *Nicotiana tabacum.* J. Cell Sci. 103: 831-837.

Takanashi, H., S. Arimura, W. Sakamoto and N. Tsutsumi. 2006. Different amounts of DNA in each mitochondrion in rice root. Genes Genet. Syst. 81: 215-218.

Tanaka, H. 1991. Microtubule-determined plastid distribution during microsporogenesis in *Lilium longiflorum.* J. Cell Sci. 99: 21-31.

Thomas, M.R. and R.J. Rose. 1983. Plastid number and plastid structural changes associated with tobacco mesophyll protoplast culture and plant regeneration. Planta 158: 329-338.

Tiew, T.W-Y., M.B. Sheahan and R.J. Rose. 2015. Peroxisomes contribute to reactive oxygen species homeostasis and cell division induction in *Arabidopsis* protoplasts. Front. Plant Sci. 6: 658.

Van Gestel, K., R.H. Köhler and J-P. Verbelen. 2002. Plant mitochondria move on F-actin, but their positioning in the cortical cytoplasm depends on both F-actin and microtubule. J. Exp. Bot. 53: 659-667.

Vietri, M., K. O. Schink, C. Campsteijn, C.S. Wegner, S.W. Schultz, L. Christ, S.B. Thoresen, A. Brech, C. Raiborg and H. Stenmark. 2015. Spastin and ESCRT-III coordinate mitiotic spindle disassembly and nuclear envelope scaling. Nature 522: 231-235.

Wada, M., T. Kagawa and Y. Sato. 2003. Chloroplast movement. Annu. Rev. Plant Biol. 54: 455-468.

Wada, M. and N. Suetsugu. 2004. Plant organelle positioning. Curr Opin Plant Biol 7: 626-631.

Warren, G. and W. Wickner. 1996. Organelle inheritance. Cell 84: 395-400.

Whatley, J. M. 1974. The behaviour of chloroplasts during cell division of *Isoetes lacustris* L. New Phytol. 73: 139-142.

Whatley, J.M. 1980. Plastid growth and division in *Phaseolus vulgaris*. New Phytol. 86: 1-16.

Williamson, R. E. 1993. Organelle movements. Annu. Rev. Plant Physiol. and Plant. Mol. Biol. 44: 181-202.

Yaffe, M.P., N. Stuurman and R.D. Vale. 2003. Mitochondrial positioning in fission yeast is driven by association with dynamic microtubule and mitotic spindle poles. Proc. Natl. Acad. Sci. USA. 100: 11424-11428.

Yasuda, T., T. Kuroiwa and T. Nagata. 1988. Preferential synthesis of plastid DNA and increased replication of plastids in cultured tobacco cells following medium renewal. Planta 174: 235-241.

6

Cell Division and Cell Growth

Takuya Sakamoto, Yuki Sakamoto and Sachihiro Matsunaga*

Introduction

Plant morphogenesis and differentiation are strongly dependent on the coordination between cell division and cell growth. In many plant species, cell growth is occasionally accompanied by endoreduplication (also referred to as endoreplication and the endocycle), which arises from multiple DNA replications without cell division, although endoreduplication in most cereal plants is limited to endosperm development (Sabelli and Larkins 2009). In endoreduplication, DNA replication is not followed by subsequent mitosis and cytokinesis. Endoreduplication precedes rapid cell elongation in the boundary region between the meristematic and elongation regions of roots (Hayashi et al. 2013), trichome formation on the leaf epidermis (Melaragno et al. 1993), extensive elongation of hypocotyls in the dark (Jakoby and Schnittger 2004), and the differentiation of giant cells in the sepal epidermis (Roeder et al. 2010).

Recently, regulatory proteins that control the switch to endoreduplication from mitosis have been identified (De Veylder et al. 2011, Matsunaga et al. 2013). For example, entry into the S-phase of the plant cell cycle is regulated by the E2F-RETINOBLASTOMA-RELATED1 (RBR1) pathway like in animals (Berckmans and De Veylder 2009). More than 300 E2F-target genes have been identified, including regulators for DNA replication and chromatin structure (Naouar et al. 2009). The switching mechanism between cell division and cell growth is influenced by environmental stress. In fact, endoreduplication induced by numerous stresses results in cell expansion, meristem size reduction and activation of metabolic pathways (Scholes and Paige 2014). The transition from mitosis to endoreduplication increases the transcriptional activity of stress-responsive genes and production of stress-mitigating compounds.

Leaf size is also determined by the balance between cell division and cell growth by expansion. After leaf cells proliferate through mitosis, they transition to expansion with an increase in vacuole size and endoreduplication in epidermal cells (Johnson and Lenhard 2011). The transition occurring through proliferation arrest, called the "arrest front", starts at the leaf tip and moves to the basal sides corresponding with

Department of Applied Biological Science, Faculty of Science and Technology, Tokyo University of Science, 2641 Yamazaki, Noda, Chiba 278-8510, Japan.
* Corresponding author : sachi@rs.tus.ac.jp

an increase in leaf size (Powell and Lenhard 2012, Hepworth and Lenhard 2014). A great transcriptome shift in the arrest front induces the exit from cell proliferation and transition to cell growth (Andriankaja et al. 2012).

In this chapter, we discuss the mechanism and describe examples of the coordination between cell division and cell growth in plant roots and leaves.

Transition from Cell Division to Cell Growth in Roots

Plant roots consist of three general regions containing cells in different developmental stages, the meristematic, elongation, and differentiation zones, which are sequentially ordered from the tip to the base of the root. The meristematic zone, where cell division occurs randomly, is composed of mitotic cells. Conversely, in the elongation zone, cells can rapidly expand with endoreduplication. Cell growth in the meristematic region is achieved by turgor-driven cell wall extension with an increase in cytoplasmic growth, whereas in the differentiation region, cell growth is based on turgor-driven cell wall extension dependent on vacuolar expansion. Thus, the control system for coordination between cell division and cell growth is fundamental for creating the highly ordered structure of plant roots (Fig. 1).

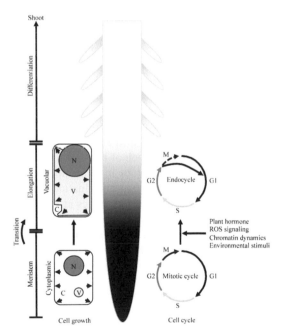

FIGURE 1 Cell division and cell growth transition in roots. Cell cycle and cell growth transitions occur at regions located between the meristem and elongation zones. The cell cycle transition is regulated by hormone-based signaling and reactive oxygen species (ROS) signaling, and affected by environmental stimuli such as DNA damage induction. The cell cycle transition from mitosis to the endocycle occurs before the cell growth transition from cytoplasmic growth to vacuolar growth. Arrows indicate turgor-driven outward pressure. N, nucleus; V, vacuole; C, cytoplasm.

Endoreduplication Precedes Cell Expansion

For the past two decades, it has remained unclear whether endoreduplication occurs prior to cell enlargement. However, recent studies on *Arabidopsis thaliana* roots using advanced imaging techniques have answered this question. Visualization of DNA-replicating cells with 5-ethynyl-2'-deoxyuridine (EdU) revealed that boundary cells in the transition zone between the meristematic and elongation zones contain duplicated genomes and replicate their DNA at S phase (Hayashi et al. 2013). Similarly, an S/G2 phase reporter, pHTR2: CDT1a (C3)-RFP, was steadily expressed in 4-5 consecutive cells preceding the first dramatically elongated cells (Yin et al. 2014). These findings established that endoreduplication precedes rapid cell expansion at the boundary where the transition from cell division to cell growth occurs. This is consistent with previous suggestions of the physiological significance of endoreduplication in rapid cell expansion (Scholes and Paige 2014). It is possible that a lack of necessity for cytoskeleton reconstitution may allow the faster growth of endore duplicating cells (Kondorosi and Kondorosi 2004). In line with this, adequate amounts of metabolites supplemented through enhanced transcriptional activities accompanied by increased gene dosage could contribute to dynamic turgor-driven cell expansion. In this scenario, it is speculated that ribosome synthesis could be the limiting factor related to the amount of DNA available (Sablowski and Dornelas 2014).

Chromatin Dynamics for the Transition

In terms of cell cycle regulation, the molecular mechanism for the transition from mitosis to endoreduplication has been well characterized (Heyman and De Veylder, 2012). The inhibition of G2/M progression is established by a specific distribution of plant hormones, mainly auxin and cytokinin (Fig. 1) (Takatsuka and Umeda 2014, Sablowski and Dornelas 2014, Scholes and Paige 2014). DNA replication and transcriptional regulation are highly associated with chromatin structure. Chromatin is a high-order and highly compact macromolecular structure formed of repeating units called nucleosomes, which consist of a segment of DNA wrapped around a nucleosome core of eight histones (two each of H2A, H2B, H3, and H4). Chromatin modifiers such as histone chaperones, histone modification enzymes, and chromatin remodeling factors mediate the construction and fluctuations (open and closed) of chromatin structure during the cell cycle (Desvoyes et al. 2010). These chromatin modifiers are essential for maintaining a foothold for appropriate gene regulation in the cell cycle transition. Although little information is available, a few chromatin modifiers and histone modification states have been shown to be involved in the transition switch from cell proliferation to cell expansion based on endoreduplication control in roots (Fig. 1).

Chromatin Assembly Factor-1 (CAF-1) is a heterotrimeric complex conserved among all eukaryotes that is composed of three subunits, FASCIATA1 (FAS1), FAS2, and MULTICOPY SUPPRESSOR OF IRA1 in *A. thaliana* (Polo and Almouzni 2006). CAF-1 facilitates *de novo* deposition of H3.1/H4 histone dimers onto DNA in response to replication-dependent nucleosome assembly (Kaya et al. 2001, Hennig et al. 2003). The higher ploidy in various organs including roots caused by defects in CAF-1 points to the involvement of this complex in transition control (Ramirez-Parra and Gutierrez 2007). CAF-1 mutants suffer from a constitutively high amount of endogenous double-strand breaks (DSBs) and show increased expression of G2-specific DNA damage checkpoint genes (Endo et al. 2006, Kirik et al. 2006, Ramirez-Parra and Gutierrez 2007, Hisanaga et al. 2013). Consistently, the high ploidy phenotype of *fas1* was shown to be dependent on ATM activity, a master kinase for DNA damage signaling (Hisanaga et al. 2013). These observations suggest that CAF-1 activity in chromatin assembly is necessary to avoid premature switching of the transition, possibly by maintaining genomic stability, resulting in the avoidance of excessive DNA damage, a trigger of endoreduplication.

Chromatin events set up by histone modifications that have extensive and diverse effects on transcriptional patterns occur at defined times during the cell cycle and control the transition through specific cell cycle stages (Desvoyes et al. 2014). One of the E2F-target genes, ASH1-RELATED3 (ASHR3), is a histone H3 Lys-36 (H3K36) monomethyltransferase. ASHR3 plays an important role in the maintenance of mitotic cell division in the root meristem of *A. thaliana* (Kumpf et al. 2014). The meristem size of the *ashr3-1* mutant is reduced by disruption of coordination between DNA replication and cell division. Chromatin immunoprecipitation analysis indicated that ASHR3-dependent histone methylation contributes to the synchronization of DNA replication and cell division in the root meristem. HISTONE MONOUBIQUITINATION 1 (HUB1), a homolog of yeast BRE1, is a RING E3 ligase that mono-ubiquitinates H2B at residue K143 in plants (Desvoyes et al. 2014) and is involved in the transcriptional control of key regulators of G2/M transition. The absence of HUB1 triggers a prolonged G2 phase and consequent early start of endoreduplication in *A. thaliana* (Fleury et al. 2007). These data indicate that H2B mono-ubiquitination (H2Bub) by HUB1 can cause negative transcriptional regulation for transition from mitosis to endoreduplication, although the link between H2Bub and transcriptional regulation still needs to be further investigated (Fen and Shen 2014).

Histone acetylation opens (loosens) the chromatin structure by modulating the interactions between the N-terminal histone tails and the DNA, resulting in increased accessibility of target genes to transcription. Histone acetylation plays a significant role in *A. thaliana* root development (Rosa et al. 2014). The level of histone acetylation is strongly associated with global changes in histone-DNA interactions, whose extent is stronger in dividing cells and weaker in differentiated cells. Interestingly, treatment with trichostatin A, an inhibitor of histone deacetylases, leads to histone hyperacetylation and reduces the cell division rate but increases cell number in the meristem, suggesting that increased levels of histone acetylation delay cell differentiation in the root meristem. Similarly, the differentiation of human embryonic stem cells is associated with reduced histone acetylation (Legartová et al. 2014). Therefore, dynamic changes of transcriptional patterns around the boundary regulated by histone acetylation could be an upstream trigger of endoreduplication, although the differences in histone acetylation levels during the cell cycle transition need to be determined.

Transition in Response to Environmental Stimuli

Endoreduplication has been shown to be adaptive and responsive to environmental stimuli (Fig. 1). In roots, abiotic stresses and biotic stresses related to pathogenic and symbiotic biotrophs are often accompanied by enhanced endoreduplication (Scholes and Paige 2014). Growing evidence suggests that genotoxicity including DNA DSBs is an active stimulus of endoreduplication in roots. It has been established that endoreduplication is systematically induced in roots subjected to DNA damage (Ramirez-Parra and Gutierrez 2007, Adachi et al. 2011) and a putative transcription factor, SOG1, which functions analogously to mammalian p53, contributes in large part to this genetic program (Yoshiyama et al. 2013). Interestingly, salinity, excess boron, and cadmium stress have been shown to induce endoreduplication as well as increased accumulation of reactive oxygen species (ROS), which consequently increases levels of DSBs (Sakamoto et al. 2011, Sharma et al. 2012, Scholes and Paige 2014). Therefore, under these environmental conditions, it is plausible that DNA damage may be a common factor that determines the transition from mitosis to endo-reduplication. DNA damage could prompt rapid remodeling of chromatin or DNA topology relevant to gene expression that results in immediate and extensive changes in the expression profile. Interestingly, rapid resolution of DNA topology through active DSB formation in promoter regions was shown to be crucial for immediate gene expression in neurons (Madabhushi et al. 2015). A recent study found that ROS signaling under the control of the transcription factor UPBEAT1 (UPB1) functions in the transition to differentiation (Tsukagoshi et al. 2010), suggesting the possible mechanism that active accumulation of endogenous DSBs through the control of ROS homeostasis drives systemically programmed endoreduplication at the bound-ary between the meristematic and elongation zones. Practically, a similar system has already been shown in mammalian cells. Caspase 3/caspase-activated DNase depen-dent DNA damage promotes cell differentiation by directly modifying genomic organization, which affects the expression of critical regulatory genes (Larsen et al. 2010). Although caspase is not encoded in plant genomes, it is interesting to explore and focus on factors that are dependent on ROS signaling. UPB1 acts through a pathway independent of the auxin/cytokinin hormonal signaling pathway, although it remains possible that these two signaling systems converge at some level. One pos-sible explanation for the functional difference between these two signaling pathways is that the ROS pathway plays a key role in the local response, whereas hormones mediate long-distance communication (Tsukagoshi et al. 2010).

Transition of Cell Expansion Pathway

Cell expansion is achieved through the increased growth of intracellular compo-nents, mainly the cytoplasm/vacuole, and subsequent turgor-driven oriented cell wall extension. In meristems, the increased cell volume is mostly attributable to cytoplas-mic growth, whereas vacuolar enlargement plays a major role in the expansion of differentiated cells (Fig. 1). The central regulator of cytoplasmic growth is TARGET OF RAPAMYCIN (TOR), a large Ser/Thr protein kinase. The activation of genes related to ribosome biogenesis could be a key function of TOR for the production of metabolites involved in the increase of cytoplasmic volume (Xiong et al. 2013). In contrast, vacuolar enlargement is promoted by efficient influx of water driven by the accumulation of osmotic solutes. Therefore, it is thought that vacuolar invertase (VIN), which catalyzes various carbon hydrate metabolic pathways in concert with sucrose synthase, may have a key role in vacuolar expansion (Wang et al. 2010). Cotton GhVIN1 complemented the short-root phenotype of *vin* knockout *A. thaliana* plants, suggesting that *A .thaliana* root growth is limited by VIN activity. However, in this case, it was shown that the mechanism of VIN-mediated *A. thaliana* root elongation was independent of osmotic regulation (Wang et al. 2010). Alternatively, it has been suggested that VIN-mediated *A. thaliana* root elongation possibly occurs through glucose-mediated effects on auxin signaling (Mishra et al. 2009). As mentioned above, there is also a transition in the way that cells expand, from cytoplasmic to vacuolar growth, in roots. One possibility, considering that endoreduplication precedes rapid cell expansion, is that genomic reorganization accompanied by a conversion from mitosis to endoreduplication could extensively affect gene sets relevant to the growth of intracellular components.

Transition from Cell Division to Cell Growth in Shoots

The leaf developmental process can be divided into three steps. In the first step, primary cells are provided from the shoot apical meristem to leaf primordia. In the second step, cell proliferation occurs throughout the leaf primordia (cell division and elongation are coupled). In the third step, cell expansion begins (only cell elongation occurs). As there is a positive correlation between cell number and organ size, the number of cells created by cell proliferation is a critical factor in determining the final leaf size and shape. For example, Japanese *Bonsai* is an art, in which plants, such as pine, are grown in a container. Usually, Bonsai plants are much smaller than wild plants because their leaves have fewer but not smaller cells than those of wild plants (Korner et al. 1989). The rate and duration of cell proliferation are important factors that decide the cell number.

After cell proliferation, cells start expanding. This transition is critical for deciding the cell number. If the transition is too late, extra cell proliferation proceeds; in contrast, if the transition is too early, cell proliferation is insufficient. The proliferating cells do not arrest their cell cycles simultaneously throughout the leaf blade, but sequentially from the tip to base. This differential timing of mitosis arrest produces the boundary called the "cell cycle arrest front", "cell division arrest front" or "proliferation-arrest front" between proliferating and expanding cell populations (Kazama et al. 2010, Powell and Lenhard 2012, Hepworth and Lenhard 2014). The arrest front, which can be visualized using a β-glucuronidase (GUS) driven by cyclin promoters such as *pCyclinB1;1*::GUS, remains at a constant position for a particular period and then moves rapidly toward the base of the leaf blade. Since the arrest front indicates the switch from proliferation to expansion, several factors regulating the transition have been identified by observation of the arrest front. Observations and molecular biological analyses using *A. thaliana* have revealed that the transition is regulated by phytohormone signaling, post-transcriptional, and protein degradation pathways.

Auxin Pathway

The phytohormone auxin plays a role in regulating the transition from the cell proliferation stage to the cell expansion stage. Auxin induces the expression of *AUXIN REGULATED GENE INVOLVED IN ORGAN SIZE (ARGOS)* encoding an unknown function protein localized on the ER (Hu et al. 2003, Feng et al. 2011). The up- and down-regulation of ARGOS causes an increase and decrease of cell number and final leaf size, respectively. Over expression of ARGOS prolongs cell proliferation,

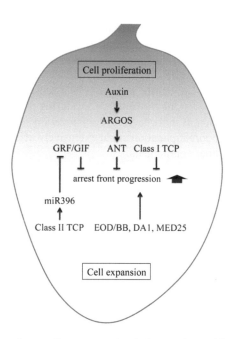

FIGURE 2 Regulation of arrest front progression in leaves. A transition between cell division and cell growth occurs at the arrest front in leaves. ARGOS and ANT are expressed in proliferating cells and inhibit arrest front progression. Class I TEOSINTE-BRANCHED1/CYCLOI-DEA/ PCFs (TCPs) are expressed in proliferating cells and inhibit arrest front progression while Class II TCPs are expressed in expanding cells and promote the progression. ENHANCER OF DA1-1/BIG BROTHER, DA1, and MEDIATOR COMPLEX SUBUNIT 25 are expressed in expanding cells and promote arrest front progression.

through extended AINTEGUMENTA (ANT) expression (Hu et al. 2003). ANT belonging to the APETALA 2 (AP2) family is a DNA binding protein and critical regulator of organ size. Over expression of ANT induces large leaves, whereas *ant* mutants exhibit small leaves. The ectopic expression of ANT induces extra cell division and the *ant* mutation reduces the cell division rate. Over expression of ANT induces prolonged expression of a G1 type cyclin, CYCD3, which is involved in the G1/S phase transition (Mizukami and Fischer 2000), suggesting that ANT delays the progression of the arrest front by prolonging cell proliferation. The loss of ANT function represses the leaf overgrowth phenotype in ARGOS over expression plants (Hu et al. 2003). These results suggest that ARGOS positively regulates cell proliferation activity and inhibits the transition from cell proliferation to cell expansion under the auxin signaling pathway via ANT (Fig. 2).

TCP and MicroRNA Pathway

The TEOSINTE-BRANCHED1/CYCLOI-DEA/PCF (TCP) family comprises plant-specific transcription factors that can be classified into two classes. Class I TCPs, including 13 members in *A. thaliana* such as TCP14 and TCP15, promote cell proliferation. TCP14 Over expression induces CYCB1;2 expression, and *tcp14/tcp15* double mutation down-regulates CYCA1;1, CYCB1;1, CYCB1;2, CDC20, and CDKB2;1

expression, suggesting that class I TCPs enhance cell proliferation (Kieffer et al. 2011, Steiner et al. 2012). The function of class II TCPs during leaf development has been much more extensively studied than that of class I. Class II TCPs include 11 members and can be subdivided into the CIN and CYC/TB1 subfamilies (Uberti Manassero et al. 2013). The CIN-TCP subfamily containing eight members promotes cell differentiation and progression of the arrest front. Five of the eight CIN-TCP members were post-transcriptionally down-regulated by the microRNA miR319 (Palatnik et al. 2003). CIN-TCPs up-regulate the expression level of miR396, which in turn down-regulates the GROWTH-REGULATING FACTOR (GRF) family of transcription factors. There are nine GRF members in *A. thaliana* and all are regulated by miR396 (Palatnik et al. 2003, Wang et al. 2011). Over expression of GRF5 induces large leaves and *grf5* mutants exhibit small leaves, because the cell numbers in plants overexpressing GRF5 and *grf5* mutants are increased and decreased, respectively (Horiguchi et al. 2005). ANGUSTIFOLIA3 (AN3), also known as GRF-INTERACTING FACTOR1 (GIF1), interacts with GRF5. Over expression of AN3 also induces large leaves with many cells and *an3* mutation induces small leaves (Lee et al. 2009). These results suggest that Class II CIN-TCPs promote the arrest front and the transition from cell proliferation to cell expansion via miR396 and GRF/GIFs (Fig. 2).

Proteasome Pathway

Proteasomes are also involved in leaf development. The proteasome is a large protein complex that degrades polyubiquitinated proteins in cells. A predicted ubiquitin binding protein, DA1, and an ubiquitin E3 ligase, BIG BROTHER (BB), also called ENHANCER OF DA1-1 (EOD1), regulate the duration of cell proliferation and control organ size. The *da1-1* mutants, expressing dominant negative proteins affecting both DA1 and DA1-related proteins (DAR), showed increased leaf size and cell number (Li et al. 2008). In observations of the mitotic index and a cell cycle marker, *da1-1* showed more prolonged cell proliferation than the wild-type. EOD1/BB was identified as an enhancer of the *da1-1* phenotype (Li et al. 2008). The *eod1* mutants exhibit large organs with increased cell number and *eod1/da1-1* double mutants demonstrate synergistically larger organs, which suggests that DA1 and EOD/BB act in parallel pathways to negatively control cell proliferation in leaves. DA1 functions together with MEDIATOR COMPLEX SUBUNIT 25 (MED25), which is also an enhancer of *da1-1*. MED25 controls organ size by extending the duration of cell proliferation and expansion (Xu and Li 2011). Over expression of MED25 decreases the cell size and cell number; in contrast, loss of MED25 function increases these. The *da1/med25* double mutant exhibits synergistic enhancement of cell number in the petal. These analyses suggest that DA1, MED25, and EOD1/BB enhance the transition from cell proliferation to expansion via proteasome degradation of unknown proteins that repress the transition (Fig. 2).

Cell Proliferation and Cell Expansion Cooperate in a Compensation-Dependent Manner

In leaves, there are relationships unexplained by the timing of cell cycle arrest between cell size and cell number. When a sufficient number of cells cannot be provided by cell proliferation, cell expansion is enhanced to maintain leaf size. This phenomenon is known as 'compensation' and is conserved in seed plants such as *A. thaliana* and *Oryza sativa*. Compensation seems to occur in leaves and cotyledons. A compensation-like phenotype was first observed in gamma ray-irradiated wheat more than 50 years ago (Haber 1962). Recently, it was reported that several mutants and transgenic plants of *A. thaliana* exhibited the compensation phenotype (Hisanaga et al. 2015).

Compensation is observed in *an3* mutants. The cell number in *an3* is about 0.3 times that in the WT whereas the cell size is about 1.5 times larger (Horiguchi et al. 2005). Careful observation of the cell expansion process in *an3* revealed that cell size was normal in the proliferation stage, but enlarged in the expansion stage. As the duration of cell expansion was not prolonged, *an3* mutation enhanced the expansion rate during the expansion step. Since chimeric expression of AN3 could not recover the cell size of AN3-expressing cells in *an3*, AN3-dependent compensation is a non-cell-autonomous process (Kawade et al. 2010). Compensation is also observed in *ant*, *erecta*, *struwwelpeter*, and *fugu1-fugu5* mutants, and KRP2 Over expression plants (KRP2-ox) (Mizukami and Fischer 2000, De Veylder et al. 2001, Autran et al. 2002, Ferjani et al. 2007, Tisné et al. 2011). The cell size of *fugu5* and KRP2-ox is about 1.3 times and 2.4 times that of the WT, respectively (Ferjani et al. 2007). The causes of enlarged cell size are prolonged cell expansion in *fugu5* and extra cell growth during the cell proliferation step in KRP2-ox plants. It was also confirmed that KRP2-dependent compensation is a cell-autonomous process using chimeric expression plants (Kawade et al. 2010). A positive correlation between DNA content and cell size was reported (Sablowski and Dornelas 2014). The ploidy levels of *an3* and *fugu5* were mildly increased, while those of KRP-ox were similar to those of the WT, which suggests that the enlargement of cell size through compensation is not positively correlated with ploidy levels. These results indicate that several independent systems regulate the cooperation between cell proliferation and cell expansion during leaf development; however, the molecular mechanisms of these systems remain unclear.

Acknowledgments

This research was supported by CREST grants from the Japan Science and Technology Agency to S.M., and MEXT/JSPS KAKENHI to S.M.

References

Adachi, S., K. Minamisawa, Y. Okushima, S. Inagaki, K. Yoshiyama, Y. Kondou, E. Kaminuma, M. Kawashima, T. Toyoda, M. Matsui, D. Kurihara, S. Matsunaga and M. Umeda. 2011. Programmed induction of endoreduplication by DNA double-strand breaks in Arabidopsis. Proc. Natl. Acad. Sci. USA 108: 10004-10009.

Andriankaja, M., S. Dhondt, S. De Bodt, H. Vanhaeren, F. Coppens, L. De Milde, P. Mühlenbock, A. Skirycz, N. Gonzalez, G.T. Beemster and D. Inzé. 2012. Exit from proliferation during leaf development in Arabidopsis thaliana: a not-so-gradual process. Dev. Cell. 22: 64-78.

Autran, D., C. Jonak, K. Belcram, G.T. Beemster, J. Kronenberger, O. Grandjean, D. Inzé and J. Traas. 2002. Cell numbers and leaf development in Arabidopsis: a functional analysis of the STRUWWELPETER gene. EMBO J. 21: 6036-6049.

Berckmans, B and L. De Veylder. 2009. Transcriptional control of the cell cycle. Curr. Opin. Plant Biol. 12: 599-605.

De Veylder, L., J.C. Larkin and A. Schnittger. 2011. Molecular control and function of endoreplication in development and physiology. Trends Plant Sci. 16: 624-34.

De Veylder, L., T. Beeckman, G.T. Beemster, L. Krols, F. Terras, I. Landrieu, E. van der Schueren, S. Maes, M. Naudts and D. Inzé. 2001. Functional analysis of cyclin-dependent kinase inhibitors of Arabidopsis. Plant Cell 13: 1653-1668.

Desvoyes, B., M. Fernández-Marcos, J. Sequeira-Mendes, S. Otero, Z. Vergara and C. Gutierrez. 2014. Looking at plant cell cycle from the chromatin window. Frontiers in Plant Sci. 5: 369.

Desvoyes, B., M.P. Sanchez, E. Ramirez-Parra and C. Gutierrez. 2010. Impact of nucleosome dynamics and histone modification on cell proliferation during Arabidopsis development. Heredity. 105: 80-91.

Endo, M., Y. Ishikawa, K. Osakabe, S. Nakayama, H. Kaya, T. Araki, K. Shibahara, K. Abe, H. Ichikawa, L. Valentine, B. Hohn and S. Toki. 2006. Increased frequency of homologous recombination and T-DNA integration in Arabidopsis CAF-1 mutants. EMBO J. 25: 5579-5590.

Fen, J., and W.H. Shen. 2014. Dynamic regulation and function of histone monoubiquitination in plants. Frontiers in Plant Sci. 5: 83.

Feng, G., Z. Qin, J. Yan, X. Zhang and Y. Hu. 2011. Arabidopsis ORGAN SIZE RELATED1 regulates organ growth and final organ size in orchestration with ARGOS and ARL. New Phytol. 191: 635-646.

Ferjani, A., G. Horiguchi, S. Yano and H. Tsukaya. 2007. Analysis of leaf development in fugu mutants of Arabidopsis reveals three compensation modes that modulate cell expansion in determinate organs. Plant Physiol. 144: 988-999.

Fleury, D., K. Himanen, G. Cnops, H. Nelissen, T. M. Boccardi, S. Maere, G.T.S. Beemster, P. Neyt, S. Anami, P. Robles, J.L. Micol, D. Inzé and M. Van Lijsebettens. 2007. The *Arabidopsis thaliana* homolog of yeast BRE1 has a function in ell cycle regulation during early leaf and root growth. Plant Cell 19: 417-432.

Haber, A.H. 1962. Nonessentiality of concurrent cell divisions for degree of polarization of leaf growth. I. Studies with radiation-induced mitotic inhibition. Am. J. Bot. 49: 583-589.

Hayashi, K., J. Hasegawa and S. Matsunaga. 2013. The boundary of the meristematic and elongation zones in roots: endoreduplication precedes rapid cell expansion. Scientific Rep. 3: 2723.

Hennig, L., P. Taranto, M. Walser, N. Schonrock and W. Gruissem. 2003. Arabidopsis MSI1 is required for epigenetic maintenance of reproductive development. Development 130: 2555-2565.

Hepworth, J. and M. Lenhard. 2014. Regulation of plant lateral-organ growth by modulating cell number and size. Curr. Opin. Plant Biol. 17: 36-42.

Heyman, J. and L. De Veylder. 2012. The anaphase-promoting complex/cyclosome in control of plant development. Mol. Plant 5: 1182-1194.

Hisanaga, T., A. Ferjani, G. Horiguchi, N. Ishikawa, U. Fujikura, M. Kubo, T. Demura, H. Fukuda, T. Ishida, K. Sugimoto and H. Tsukaya. 2013. The ATM-dependent DNA damage response acts as an upstream trigger for compensation in the *fas1* mutation during Arabidopsis leaf development. Plant Physiol. 162: 831-841.

Hisanaga, T., K. Kawade and H. Tsukaya. 2015. Compensation: a key to clarifying the organ-level regulation of lateral organ size in plants. J. Exp. Bot. 66: 1055-1063.

Horiguchi, G., G.T. Kim and H. Tsukaya. 2005. The transcription factor AtGRF5 and the transcription coactivator AN3 regulate cell proliferation in leaf primordia of *Arabidopsis thaliana*. Plant J. 43: 68-78.

Hu, Y., Q. Xie and N.H. Chua. 2003. The Arabidopsis auxin-inducible gene ARGOS controls lateral organ size. Plant Cell 15: 1951-1961.

Jakoby, M. and A. Schnittger. 2004. Cell cycle and differentiation. Curr. Opin. Plant Biol. 7: 661-669.

Johnson, K. and M. Lenhard. 2011. Genetic control of plant organ growth. New Phytol. 191: 319-333.

Kawade, K., G. Horiguchi and H. Tsukaya. 2010. Non-cell-autonomously coordinated organ size regulation in leaf development. Development 137: 4221-4227.

Kaya, H., K.I. Shibahara, K.I. Taoka, M. Iwabuchi, B. Stillman and T. Araki. 2001. FASCIATA genes for chromatin assembly factor-1 in Arabidopsis maintain the cellular organization of apical meristems. Cell 104: 131-142.

Kazama, T., Y. Ichihashi, S. Murata and H. Tsukaya. 2010. The mechanism of cell cycle arrest front progression explained by a KLUH/CYP78A5-dependent mobile growth factor in developing leaves of *Arabidopsis thaliana*. Plant Cell Physiol. 51: 1046-54.

Kieffer, M., V. Master, R. Waites and B. Davies. 2011. TCP14 and TCP15 affect internode length and leaf shape in Arabidopsis. Plant J. 68: 147-158.

Kirik, A., A. Pecinka, E. Wendeler and B. Reiss. 2006. The chromatin assembly factor subunit FASCIATA1 is involved in homologous recombination in plants. Plant Cell 18: 2431-2442.

Kondorosi, E. and A. Kondorosi. 2004. Endoreduplication and activation of the anaphase-promoting complex during symbiotic cell development. FEBS Letters 562: 153-157.

Korner, C., S.P. Menendez-Riedl and P.C.L. John. 1989. Why are bonsai plants small? A consideration of cell size. Aust. J. Plant Physiol. 16: 443-448.

Kumpf, R., T. Thorstensen, M.A. Rahman, J. Heyman, H.Z. Nenseth, T. Lammens, U. Herrmann, R. Swarup, S.V. Veiseth, G. Emberland, M.J. Bennett, L. De Veylder and R.B. Aalen. 2014. The ASH1-RELATED3 SET-domain protein controls cell division competence of the meristem and the quiescent center of the *Arabidopsis* primary root. Plant Physiol. 166: 632-43.

Larsen, B.D., S. Rampallia, L.E. Burnsa, S. Brunettea, F.J. Dilwortha and L.A. Megeneya. 2010. Caspase 3/caspase-activated DNase promote cell differentiation by inducing DNA strand breaks. Proc. Natl. Acad. Sci. USA 108: 4230-4235.

Lee, B.H., J.H. Ko, S. Lee, Y. Lee, J.H. Pak and J.H. Kim. 2009. The Arabidopsis GRF-INTERACTING FACTOR gene family performs an overlapping function in determining organ size as well as multiple developmental properties. Plant Physiol. 151: 655-668.

Legartová, S., S. Kozubek, M. Franek, Z. Zdráhal, G. Lochmanová, N. Martinet and E. Bártová. 2014. Cell differentiation along multiple pathways accompanied by changes in histone acetylation status. Biochem. Cell Biol. 92: 85-93.

Li, Y., L. Zheng, F. Corke, C. Smith and M.W. Bevan. 2008. Control of final seed and organ size by the DA1 gene family in *Arabidopsis thaliana*. Gene Dev. 22: 1331-1336.

Madabhushi, R., F. Gao, A.R. Pfenning, L. Pan, S. Yamakawa, J. Seo, R. Rueda, T.X. Phan, H. Yamakawa, P. Pao, R.T. Stott, E. Gjoneska, A. Nott, S. Cho, M. Kellis and L. Tsai. 2015. Activity-Induced DNA breaks govern the expression of neuronal early-response genes. Cell 161: 1592-1605.

Matsunaga, S., Y. Katagiri, Y. Nagashima, T. Sugiyama, J. Hasegawa, K. Hayashi and T. Sakamoto. 2013. New insights into the dynamics of plant cell nuclei and chromosomes. Int. Rev. Cell Mol. Biol. 305: 253-301.

Melaragno, J.E., B. Mehrotra and A.W. Coleman. 1993. Relationship between endopolyploidy and cell size in epidermal tissue of Arabidopsis. Plant Cell 5: 1661-1668.

Mishra, B.S., M. Singh, P. Aggrawal and A. Laxmi. 2009. Glucose and auxin signaling interaction in controlling *Arabidopis thaliana* seedlings root growth and development. PLoS ONE 4: e4502.

Mizukami, Y. and R.L. Fischer. 2000. Plant organ size control: AINTEGUMENTA regulates growth and cell numbers during organogenesis. Proc. Natl. Acad. Sci. USA 97: 942-947.

Naouar, N., K. Vandepoele, T. Lammens, T. Casneuf, G. Zeller, P. Van Hummelen, D. Weigel ,G. Rätsch, D. Inzé, M. Kuiper, L. De Veylder and M. Vuylsteke. 2009. Quantitative RNA expression analysis with Affymetrix Tiling 1.0R arrays identifies new E2F target genes. Plant J. 57:184-194.

Palatnik, J.F., E. Allen, X. Wu, C. Schommer, R. Schwab, J.C. Carrington and D. Weigel. 2003. Control of leaf morphogenesis by microRNAs. Nature 425: 257-263.

Polo, S.E. and G. Almouzni. 2006. Chromatin assembly: a basic recipe with various flavours. Curr. Opin. Genet. Dev. 16: 104-111.

Powell, A.E. and M. Lenhard. 2012. Control of organ size in plants. Curr. Biol. 22: R360-367.

Ramirez-Parra, E. and C. Gutierrez. 2007. E2F regulates *FASCIATA1*, a chromatin assembly gene whose loss switches on the endocycle and activates gene expression by changing the epigenetic status. Plant Physiol. 144:105-120.

Roeder, A.H.K., V. Chickarmane, A. Cunha, B. Obara, B.S. Manjunath and E.M. Meyerowitz. 2010. Variability in the control of cell division underlies sepal epidermal patterning in *Arabidopsis thaliana*. Plos Biol. 8: e1000367.

Rosa, S., V. Ntoukakis, N. Ohmido, A. Pendle, R. Abranches and P. Shaw. 2014. Cell differentiation and development in Arabidopsis are associated with changes in histone dynamics at the single-cell level. Plant Cell 26: 4821-4833.

Sabelli, P.A. and B.A. Larkins. 2009. The development of endosperm in grasses. Plant Physiol. 149: 14-26.

Sablowski, R. and M.C. Dornelas. 2014. Interplay between cell growth and cell cycle in plants. J. Exp. Bot. 65: 2703-2714.

Sakamoto, T., Y.T. Inui, S. Uraguchi, T. Yoshizumi, S. Matsunaga, M. Mastui, M. Umeda, K. Fukui and T. Fujiwara. 2011. Condensin II alleviates DNA damage and is essential for tolerance of boron overload stress in Arabidopsis. Plant Cell 23: 3533-3546.

Scholes, D.R. and K.N. Paige. 2014. Plasticity in ploidy: a generalized response to stress. Trend Plant Sci. 20:165-175.

Sharma, P., A.B. Jha, R.S. Dubey and M. Pessarakli. 2012. Reactive oxygen species, oxidative damage, and antioxidative defense mechanism in plants under stressful conditions. J. Bot. 2012: 217037.

Steiner, E., I. Efroni, M. Gopalraj, K. Saathoff, T.S. Tseng, M. Kieffer, Y. Eshed, N. Olszewski and D. Weiss. 2012. The Arabidopsis O-linked N-acetylglucosamine transferase SPINDLY interacts with class I TCPs to facilitate cytokinin responses in leaves and flowers. Plant Cell 24: 96-108.

Takatsuka, H. and M. Umeda. 2014. Hormonal control of cell division and elongation along differentiation trajectories in roots. J. Exp. Bot. 65: 2633-2643.

Tisné, S., F. Barbier and C. Granier. 2011. The ERECTA gene controls spatial and temporal patterns of epidermal cell number and size in successive developing leaves of *Arabidopsis thaliana*. Ann. Bot. 108: 159-168.

Tsukagoshi, H., W. Busch and P.N. Benfey. 2010. Transcriptional regulation of ROS controls transition from proliferation to differentiation in the root. Cell 19: 644-646.

Uberti Manassero, N.G., I.L. Viola, E. Welchen and D.H. Gonzalez. 2013. TCP transcription factors: architectures of plant form. Biomol. Concepts. 4: 111-127.

Wang, L., X. Gu, D. Xu, W. Wang, H. Wang, M. Zeng, Z. Chang, H. Huang and X. Cui. 2011. miR396-targeted AtGRF transcription factors are required for coordination of cell division and differentiation during leaf development in Arabidopsis. J. Exp. Bot. 62: 761-773.

Wang, L., X.R. Li, H. Lian, D.A. Ni, Y.K. He, X.Y. Chen and Y.L. Ruan. 2010. Evidence that high activity of vacuolar invertase is required for cotton fiber and Arabidopsis root elongation through osmotic dependent and independent pathways, respectively. Plant Physiol. 154: 744-756.

Xiong, Y., M. McCormack, L. Li, Q. Hall, C. Xiang and J. Sheen. 2013. Glucose–TOR signalling reprograms the transcriptome and activates meristems. Nature 496: 181-186.

Xu, R. and L. Li. 2011. Control of final organ size by mediator complex subunit 25 in *Arabidopsis thaliana*. Development 138: 4545-4554.

Yin, K., M. Ueda, H. Takagi, T. Kajihara, S. Sugamata Aki, T Nobusawa, C. Umeda-Hara and M. Umeda. 2014. A dual-color marker system for *in vivo* visualization of cell cycle progression in Arabidopsis. Plant J. 80: 541-552.

Yoshiyama, K.O., J. Kobayashi, N. Ogita, M. Ueda, S. Kimura, H. Maki and M. Umeda. 2013. ATM-mediated phosphorylation of SOG1 is essential for the DNA damage response in Arabidopsis. EMBO rep. 14: 817-822.

Plant Cell Enlargement

7

Organization of the Plant Cell Wall

Purbasha Sarkar and Manfred Auer*

Introduction

The plant body is distinct from that of animals in having highly specialized organs optimized for water and nutrient uptake (roots), water and nutrient transport (stems) and photosynthesis (leaves). Since plants are autotrophs they do not rely on predation and thus have not evolved locomotive mechanisms. They employ sophisticated strategies for survival, including protection from herbivore animals and microbes as well as unfavorable environmental conditions, including heat, cold, wind, drought and wildfire. The uptake of large amounts of water poses an osmotic challenge to the cells, with the resulting turgor pressure threatening their integrity. A complex extracellular matrix, known as the cell wall, surrounds all plant cells and provides a counterforce to the turgor pressure. It is this combination of turgor pressure and firm cell wall that provides rigidity to plants and allows them to grow tall. The ability to grow tall allows the leaves to be out of reach for many animals and also to optimally compete for sunlight, (e.g. in a dense forest). However tall trees often require a massive trunk that supports the tree crown with its leaves. Such a body plan however results in an immense weight, imposing significant axial compression forces on the trunk, as well as lateral shear forces caused by wind, which in turn leads to bending of the trunk and branches. In other words plant cell walls have evolved to allow swift axial elongation, gradual radial growth, high axial compressibility, high rigidity, and flexibility - characteristics that make wood a widely used and ideal building material.

The 3D organization of cell walls, once experimentally determined in complete detail, will likely be very complex yet elegant, reflecting the different static and dynamic requirements imposed onto its 3D architecture. It should come as no surprise that biologists have been fascinated by the problem of plant cell wall architecture for many decades. We are currently on the verge of understanding the plant cell wall 3D architecture at macromolecular resolution given recent advances in sample preparation, advanced cryogenic 3D imaging technologies and sophisticated statistical image analysis, model building and computational simulations. This chapter lays out a framework in which the emerging 3D architectural data will be interpreted

Life Science Division, Lawrence Berkeley National Laboratory, Berkeley, CA 94720, USA.
* Corresponding author : mauer@lbl.gov

and understood. We will summarize some of the ideas that decades of research have yielded on cell wall formation, the biochemistry and biosynthesis of the main cell wall components (which we will call the wall building blocks), their respective mode of deposition, their interactions and assemblies into larger entities (which we will call the wall modular units), their organization in the 3D space with respect to the cellular elongation axis (which we will call the cell wall complex), as well as discuss whether there exists a quaternary structure, a supramolecular organization in different regions of the cell wall. We will discuss some of the imaging modalities that have been chosen for cell wall architectural studies and discuss their potential and shortcomings, as well as possible sample preparation artifacts associated with them. We will sketch a path forward towards an experimentally determined cell wall, based on cryo electron tomography of ultrarapid-frozen, frozen-hydrated vitrified tissue sections, followed by statistical volumetric analysis, model building and computational mechanical simulations. We will end with a discussion of the different scales of length and resolution that will be needed to truly predict mechanical behavior of plants e.g. for genetic mutants. Plant cell wall reengineering is of interest not only in agriculture but also in the context of lignocellulosic biofuel and high-value chemical production.

Plant Cell Wall Formation

Plant cell walls are formed at an early stage of cell life in most cell types. During the late stage of cell division (cytokinesis), a complex cytoskeletal structure called the phragmoplast is formed between the two newly formed cells and serves as a scaffold for deposition of plasma membrane and cell wall material. The newly deposited material forms a disc-shaped partitioning structure between the two newly forming cells called the cell plate, which at the completion of cell division fuses with the existing plasma membrane and cell wall of the mother cell, and forms continuous membranes surrounding each of the two new cells with a maturing cell wall between them (Segui-Simmaro et al. 2004).

Plant cell walls mature differently depending on the type of the cell it surrounds. A mature plant cell wall may consist of multiple distinct layers, commonly distinguished as middle lamella, primary cell wall, and secondary cell wall. The middle lamella is a thin cementing layer formed between two adjacent cells, mainly composed of pectic polysaccharides (pectins) and some hemicelluloses in the growing cells with additional lignin deposited in the mature cells. A primary cell wall is laid down between the middle lamella and the plasma membrane, and may be defined as the extendable wall layer deposited in most growing cells. Secondary cell wall may be defined as the strong, non-extendable wall layers deposited between primary cell wall and plasma membrane in cell types such as, xylem tracheary element, fibers, and sclereids, after these cells stop growing. A gradation of wall properties between these two extremes can also be found in different cell types (Albersheim et al. 2010, Cosgrove and Jarvis 2012, Lee et al. 2011, Leroux 2012). The primary cell wall is mainly composed of three groups of polysaccharides - cellulose, hemicelluloses and pectins, with some glycoproteins and enzymes also being present, while the secondary cell walls are predominantly composed of cellulose, hemicelluloses and large quantities of lignin (Albersheim et al. 2010).

Cell Wall Building Blocks

Plant cell walls are predominantly polymers of the neutral carbohydrates glucose, xylose, galactose, arabinose, mannose, fucose, rhamnose, as well as charged molecules such as galacturonic acid, glucuronic acid and mannuronic acids. While cellulose forms unbranched linear strands, most hemicellulose and pectin polymers are branched, with the notable exception of mixed-linkage-glucan and glucomannan as

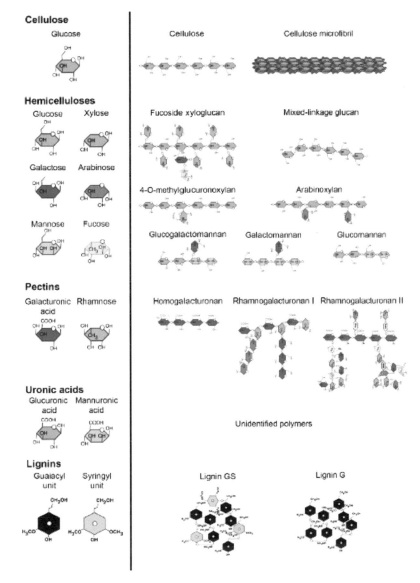

FIGURE 1 Molecular structure of the predominant building blocks of plant cell walls. Left panel: monomers. Right panel: subunit of respective polymers. (From Sarkar et al. 2009.)

well as homogalacturonan, respectively (Sarkar et al. 2009, Fig. 1). Branching allows for a higher degree of complexity with respect to inter-strand association and thus network connectivity.

Cellulose microfibrils

Multiple unbranched strands of cellulose are synthesized by cellulose synthase complexes embedded in the plasma membrane, and then held together by hydrogen bonds to form thicker fibrous structures called cellulose microfibrils - the main load-bearing framework of plant cell wall ultrastructure. Individual cellulose microfibrils are several microns long and ~5 nm in diameter, and are considered to contain a crystalline cellulose core that is surrounded by paracrystalline cellulose and hemicelluloses (Somerville et al. 2004, Ding and Himmel 2006, Sarkar et al. 2014). The number of cellulose chains within a microfibril is debatable and has been reported to be 36 in older literature (Somerville et al. 2004, Ding and Himmel 2006) and 18–24 in recent years (Newman et al. 2013, Thomas et al. 2013). The microfibrils are very closely packed in complex networks within the cell walls with only a few nanometer gaps between them that are filled with the non-cellulosic components of the wall. A multiple of 6 strands is thought to make up a microfibril strand owing to the presumed six-fold-symmetry of the cellulose synthase. The size of this multisubunit protein is somewhat controversial with some reports estimating cellulose synthase to extend up to 50 nm beyond the membrane, allowing it to physically interact with the microtubule network. However all identified subunits to date are transmembrane proteins, that based on their molecular weight and intracellular primary sequence distribution can be compared to the well-characterized P-type ATPases, which extend no further than ~7 nm beyond the plasma membrane. Attempts in our laboratory to detect a ~50 nm sized large protein complex have failed, and while we cannot rule out the existence of a 50 nm sized structure, it would seem that either cellulose synthase must be significantly smaller than 50 nm or to date only a small subset of its subunits have identified.

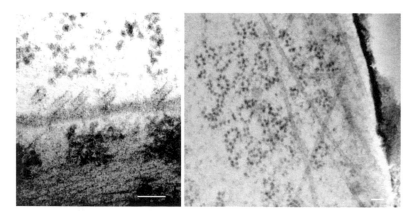

FIGURE 2   Microtubule found underneath the plasma membranes are thought to guide the cellulose synthase and thus orient microfibril fibers in the cell wall. Left: oblique orientation revealing densely spaced microtubule in close proximity with plasma membrane, Right: microtubule network inside the cytoplasm in an en-face view orientation. Scale bars = 50 nm.

In any case, cellulose synthases are thought to be guided in their directionality by the organization of microtubules that resides in close proximity to the plasma membrane (Fig. 2). The microtubules either serve as attachment sites for cellulose synthase or as "bumpers" constricting the direction of cellulose deposition.

Non-cellulosic components

Hemicelluloses are mostly branched polysaccharides composed of neutral pentoses and hexoses that strengthen the cell wall by binding to cellulose via hydrogen bonds, and also by binding with lignins in some cases (Scheller and Ulvskov 2010). Xyloglucan, a predominant hemicellulose is widely believed to bind and surround the paracrystalline cellulose in the outer layer of microfibrils and function as a tether between two neighboring microfibrils (McCann and Roberts 1991, Rose and Bennett 1999, Cosgrove 2005, Ding and Himmel 2006); however, the tether model has been challenged recently (Park and Cosgrove 2012).

Pectins, like hemicelluloses, are branched polysaccharides containing high proportions of D-galacturonic acids, however their precise organization and respective functions are not well known (McCann and Roberts 1991, Cosgrove 2000, Somerville et al. 2004). Pectins are particularly prominent in the middle lamella, the site of interaction between the cell walls of two adjacent cells, and they are thought of as glue holding adjacent cells together.

Both hemicelluloses and pectins are synthesized in the Golgi apparatus (Fig. 3) as smaller subunits through the activity of glycosyltransferases and transported through vesicles that then fuse with the plasma membrane to release their content into the extracellular cell wall matrix space, where subunits are further cross-connected through cell wall-residing enzymes. What makes this process even more

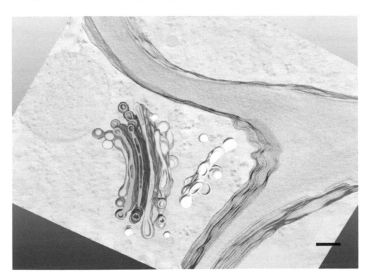

FIGURE 3 Golgi apparatus with its 3D membrane stacks (artificially colored in blue, pink, red, salmon, orange and yellow) in proximity to plant cell walls (lavender) as revealed by segmentation of electron tomograms of high-pressure frozen, freeze-substituted *Arabidopsis* samples. Scale bar = 100 nm.

complex to study is that these extracellular matrix polymers are a product of multi-step enzymatic catalysis and transport, with specific products that cannot be easily traced throughout the entire pathway, unlike the deposition of components of animal extracellular matrices such as collagen or fibronectin.

Lignins are thought to be the complex polymeric result of gualacyl and/or syringyl aromatic radical chemistry. These macromolecules are thought to fill in the space between cellulose microfibrils and hemicellulose strands in mature secondary cell walls, making them rigid and impervious. Their biosynthesis, transport to cell wall and polymerization mechanisms still remains unclear (Vanholme et al. 2010).

Proteins of around 1,000–2,500 types are involved in the synthesis, transport, secretion of cell wall components, and in the remodeling of plant cell walls (Somerville et al. 2004, Yong et al. 2005, Geisler-Lee et al. 2006). A variety of glycoproteins and enzymes are embedded in the cell walls, though their exact location and often their precise function is unclear (Somerville et al. 2004, Cosgrove 2005). Many of these proteins display functional redundancies that make elucidation of their roles by classical genetic knockout approaches challenging.

Polysaccharide Interactions & Cell Wall Modular Units

Having discussed the primary sequence of a polysaccharide as its structural building blocks, this section will explore how the various cell wall components can interact with one another and form "modular units", meaning how individual polysaccharide

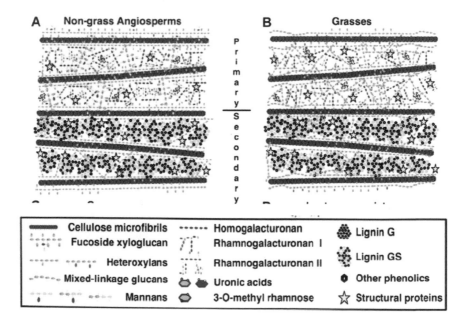

FIGURE 4 Cell wall Modular Units. Cellulose microfibrils are deposited in a direction pattern, whereas hemicellulose (and possibly pectin) cross-connects the microfibrils into a 2D (and ultimately 3D network). Secondary cell walls feature additional lignin polymers in the space left by the cellulose microfibrils and its cross-connecting hemicelluloses. (From Sarkar et al. 2009).

strands can interact with one another to form a network. There are differences in the composition of cell wall components between non-grass angiosperm and grasses (Fig. 4) leading to changes in the way cell wall components can interact with one another, hence the modular units may be different between different plant phyla. But the more basic question on how hemicelluloses interact with one another and with cellulose microfibrils is widely debated; some cell wall models depicting individual hemicellulose to loosely cross-connect multiple adjacent microfibrils with a rather low cross-section of interaction, whereas others imply much stronger interaction with microfibrils through close molecular interactions for the majority of the hemicellulose strands. Indeed experimental determination of the degree of hemicellulose-cellulose interaction will be important to understand the role of hemicellulose to form bridges between microfibrils. Likewise, it is currently unknown whether hemicellulose form bundles or function as individual strands.

Architecture of the Cell Wall Complex

Cell Wall Models

Since the 1930s various groups have described the organization of cell wall components, especially cellulose microfibrils based on optical and electron microscopy. Resolving the complex spatial organization of these nanometer-scale structures within cell walls has been extremely challenging due to poor sample preservation and the resolution limits of available microscopy techniques. Hence the precise 3D organization has remained a debated issue for decades. Various organizations of cellulose microfibrils within plant cell walls have been proposed over the years based on studies done on diverse plant samples, which can be categorized into three distinct organization patterns. However, it needs to be noted that most published studies were carried out on samples that had undergone substantial enzymatic and/or otherwise chemical treatment, hence there remains some doubt about the cell wall's true native 3D organization.

Cell Wall Model 1: Transverse-to-Longitudinal Organization

The earliest observations on cellulose microfibril organization were made using polarized light microscopy on cell walls of long tubular cells of *Tradescantia* stamen hairs (Van Iterson 1937) that always have transversely arranged cellulose microfibrils with respect to the growth axis of the cells, closest to the plasma membrane. Similar microfibril organization patterns were also detected by electron microscopy in these stamen hairs (Roelofsen and Houwink 1951) as well as in walls of filamentous alga *Nitella* (Green 1960). Such observations gave rise to the earliest and most predominant model of cell wall development, commonly known as the multinet growth hypothesis model. This model suggests that newly synthesized cellulose microfibrils are deposited continuously to the surface of the cell wall adjacent to the plasma membrane (denoted as inner surface of cell wall) in transverse orientation (radially in a pattern perpendicular to the cellular elongation axis and parallel to the plasma membrane). Microfibrils in this orientation resist the dominant transverse stress encountered by longitudinally growing cells. As newer layers of transversely oriented microfibrils are deposited on the inner surface, older layers get pushed away from the plasma membrane and hence, away from the dominant transverse stress as well. The older microfibrils were hypothesized to gradually reorient to parallel longitudinal orientation with the changing force gradient, and upon reorganization, become responsible for resisting the longitudinal pull of cell wall extension. Variations within this organization model can be seen in the literature, where the earlier studies show reorientation of microfibril layers within single layers (Van Iterson 1937, Roelofsen and Houwink 1951, Green 1960), and later studies show reorientation of sheets (lamellae) or bundles of microfibrils (Preston 1982, Anderson et al. 2010).

Cell Wall Model 2: Criss-cross Organization

The second major organization pattern reported widely shows microfibrils alternating between two sharply distinct orientations without any gradual change in orientation. Variations of this 'criss-cross' model depict either alternating individual layers (Roelofsen 1958, McCann and Roberts 1991, Somerville et al. 2004) or alternating lamellae (Chafe and Wardrop 1972, Itoh and Shimaji 1976) of either orthogonally arranged-transverse and longitudinal (Roelofsen 1958) or criss-crossing helices of microfibrils with two distinct pitches that are oriented not exactly transverse or longitudinal (i.e. parallel) to the growth axis of the cells (Chafe and Wardrop 1972, Itoh and Shimaji 1976).

However, the observation of non-transverse microfibrils closer to the plasma membrane and transverse or near-transverse microfibrils away from the plasma membrane has challenged the predominant multinet growth hypothesis model of wall development, and resulted in an alternative cell wall development model known as the ordered fibril hypothesis (Lloyd 2011). This hypothesis suggests that new microfibril layers are not always deposited in transverse orientation at the plasma membrane, but each new layer is deposited in a rotated orientation relative to the previous layer.

Cell Wall Model 3: Arc Organization

With respect to the cell wall organization pattern, the ordered fibril hypothesis is expanded on by another less frequently discussed model. Early literature on both higher plants and algae shows alternating orthogonal layers of microfibrils separated by thin layers that have microfibrils with an intermediate orientation (Setterfield and Bayley 1958, Frei and Preston 1961, Roland et al. 1977). While the predominant orthogonal layers could be transverse and longitudinal to the growth axis of the cell, or two other distinct angles, this pattern of organization suggested a gradual change in microfibril orientation similar to the transverse-to-longitudinal model, but also accounted for observation of non-transverse microfibrils near the plasma membrane.

Crucial Parameters of Cell Wall Models

Cellulose microfibril organization

These cell wall models differ starkly with respect to the **microfibril orientation**, i.e. transverse versus longitudinal orientation, gradual changes versus abrupt and repeated changes with respect to microfibril orientation in adjacent microfibril layers. This has been the focus for over 4 decades now, whereas other parameters are less frequently mentioned.

Microfibril diameter is relatively well established to be ~3–5 nm (Somerville et al. 2004) based on scanning electron microscopy measurements, which was recently further confirmed by transmission electron microscopy tomographic 3D imaging or stained and unstained samples (Sarkar et al. 2014). However, microfibrils with a diameter of ~50 nm have been reported, based on atomic force microscopy (AFM) images, but none of the published SEM nor our own extensive TEM 2D and 3D tomographic analysis supports the presence of such large fibers, which after all would be readily visible. Either 50 nm sized microfibrils stem from the specifics of the technique used, and thus could be a potential imaging artifact (AFM), or these large microfibrils must be very rare to escape detection by electron microscopy, or they are specific to the system that was investigated.

Microfibril spacing is yet another important parameter. Our recent study comparing (1) benchtop processed, (2) high-pressure frozen, freeze-substituted and resin embedded as well as (3) ultrarapid frozen, cryo-sectioned plant stem samples revealed that microfibril spacing is highly susceptible to the sample preparation protocols, with extensive chemical treatment protocols yielding about twice the spacing, when compared to cryogenically processed samples that yielded an average microfibril spacing of ~12 nm (mid-axis to mid-axis), with spaces between adjacent microfibrils typically being ~ 6.5 nm (Sarkar et al. 2014).

Hemicellulose cross-link organization

Most models do not explicitly state how they envision the hemicellulose to connect adjacent microfibrils, however the drawings usually depict them as individual strands that just connect two adjacent microfibrils with little interaction between hemicellulose strands and the microfibrils (McCann and Roberts 1991, Cosgrove 2005). So any comprehensive model will need to provide an estimate of the **average number of cross-links** per a given unit length of microfibrils (e.g. 100 nm), thus their respective **average spacing**, as well as the **orientation** within each microfibril layer and between different adjacent microfibril layers (and/or even possibly showing larger range interactions). Are the hemicellulose cross-links randomly positioned or do they follow more regular patterns of organization, both radially and axially? Last but not least, do hemicellulose filaments connect microfibrils as individual strands or do they associate into bundles, which would provide somewhat stiffer cross-bridges but given the finite amount of hemicellulose available would result in fewer bridges. Also if there were loose (or rigid) bridges, what footprint of interaction would the hemicellulose and cellulose have?

All the above mentioned parameters are important for realistic simulation of mechanical properties of cell walls and thus need to be determined experimentally, most likely through a combination of sophisticated high resolution 3D imaging of faithfully preserved samples, subsequent model building and quantitative statistical analysis. A recent study using cryo-tomography has set the stage for an experimentally determined plant cell wall model that can be expected within the near future.

Supra-molecular Cell Wall Organization

So far we have treated cell walls as homogeneous structures, assuming that their composition (building blocks), association into larger entities such as microfibrils (modular units), and their spatial 3D organization of microfibrils and hemicellulose polysaccharides are the same, at least within the borders of the primary and secondary cell walls, respectively. While to date this question has not been systematically asked, it should be noted that secondary cell walls in transmission electron microscopy have been subdivided into S1, S2 and S3 segments due to differences in texture observed in ultrathin section. Hence a comprehensive model of cell walls should also be able to account for such differences in texture and provide a rationale for how these different segments of the secondary cell wall differ from one another.

Imaging Technologies For Plant Cell Wall Determination

Over the last 40 years, microscopy at different levels of resolution has been used for model formation, with different electron microscopy techniques (scanning and transmission electron microscopy), atomic force microscopy and optical microcopy (fluorescence light and Raman microscopy) playing a dominant role. It is worth examining the different imaging approaches with their possible sample preparation and imaging artifacts.

Imaging of maize parenchyma cell wall surface by Atomic Force Microscopy (AFM) in combination with earlier immuno-transmission electron microscopy (TEM) data suggested an elementary microfibril composed of 36 glucan chains with a crystalline core (Somerville et al. 2004, Ding and Himmel 2006), in line with diameter estimates based on electron microscopy and molecular modeling of hydrogen-bonding of adjacent cellulose strands. Multiple of 6 have been widely assumed to be the correct cellulose strand number in a microfibril based on an observation of hexagonally symmetrical rosettes, which were interpreted to be cellulose synthase residing in the plasma membrane. The number of cellulose strands has more recently been reduced to 18–24 cellulose strands. This example illustrates the power of utilizing multiple imaging approaches but also illustrates how rather far-reaching conclusions can be based on very few observations.

Fluorescence light microscopy has been used extensively to study the dynamics of plant cell walls, although mostly indirectly, i.e. by studying the microtubule network underneath the plasma membrane, which is thought to guide the cellulose synthase-mediated cellulose microfibril deposition. In the conventional, diffraction-limited optical imaging scenario, the resolution is at best ~200 nm, a distance that is ~20 times larger than the average cellulose microfibril spacing in cell walls. Yet, even larger distances have been reported to be of significance, as parallel microtubule tracks of ~1μm spacing was altered with microtubule-perturbing drugs, implying an altered architecture of the cell walls (Sommerville 2006). More recently super-resolution light microscopy in their various forms such as PALM, STORM and STED dramatically improve the resolution, but require either genetically encoded tags or affinity tags that serve as reporter molecules (Rust et al. 2006, Ji et al. 2008). Genetically encoded tags, while routine for proteins, are very challenging for carbohydrate, since carbohydrate chains are the result of enzymatic activity and not of ribosomal translation of fluorescence-tag-encoding gene sequence. Affinity probes on the other hand need to have ready access to their target and bind with high specificity. Antibodies are often not specific enough and plagued with cross-reactivity and thus false-positive detection, yet at the same time they are typically too bulky

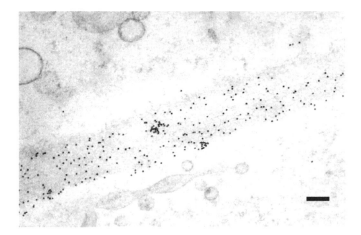

FIGURE 5   Colloidal-gold Cellulose-Binding Module protein detecting cellulose microfibrils in the *Arabidopsis thaliana* primary cell wall. Scale bar = 100 nm.

to penetrate cell walls leading to false-negative results. This problem is independent of the mode of imaging utilized. Both genetically encoded tag and affinity probes can also be visualized with an electron microscope at nanometer resolution. Certain protein probes bind with high affinity, such as cellulose binding modules, but even with ready access on a section, only a fraction of the binding sites are occupied and thus detected (Fig. 5), presumably due to sterical hindrance, epiotope/recognition site preservation and/or accessibility.

Furthermore when applying super-resolution light microscopy to biological samples it is key to suppress nonspecific background fluorescence in order to allow for maximal sensitivity. This requirement can be difficult to fulfill for intact cell walls, which can show a significant amount of background fluorescence particularly due to the lignin polymers. Also while cellulose is a well-defined homopolymer and cellulose-binding proteins are well characterized, detection of other cell wall components is

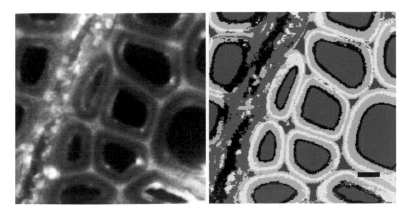

FIGURE 6   Labeling of carbohydrates and lignins within *Arabidopsis* cell wall by Raman microspectroscopy mapping. Left: Optical microscopy image of cell wall; Right: Raman map. Scale bar = 1 µm.

much more challenging and mostly attempted using immuno-fluorescence histochemistry and immunogold cytochemistry labeling. Ideally one would like to image "label-free", and gain specificity by exploiting the differences in the chemical vibrational pattern of the various components, e.g. in the form of infrared (IR) microscopy and/ or Raman microscopy. Raman imaging yields higher resolution than IR and is comparable to optical imaging. In particular, cellulose has a well-characterized spectrum that can be utilized, whereas hemicelluloses are more heterogeneous and much less defined. While specific hemicellulose and pectin molecules cannot easily be assigned to respective regions, at least one can determine areas that display similar vibrational spectra, and thus the regions will have a similar macromolecular composition (Fig. 6).

Focusing on two cellulose-specific Raman spectrum peaks, we have recently developed a Polarized Raman Microspectroscopy imaging approach that allows one to assess the directionality and order of cellulose microfibril orientation in plant cell wall sections by systematically changing the polarization plane of the incoming light and observe the Raman signal as a function of the polarization plane (Sun et al. 2016, Fig. 7). This new approach is fairly sensitive and can even detect the subtle changes in the cellulose directional organization in the brittle culm mutant compared to wild-type *Arabidopsis thaliana* samples (Sun et al. 2016). Interestingly unlike cellulose microfibrils, lignin did not show preferred orientation. This new Raman imaging approach can be used in the imaging mode, e.g. by rastering across the cell wall, however it should be noted that this imaging approach still requires significant expert knowledge, but it constitutes a sensitive, non-invasive and relatively fast method to address microfibril orientation as a function of genetic engineering. In principle, Raman imaging could be extended to very high resolution (~10 nm) by using near-field optical microscopy approaches, e.g. realized through a cantilever typically used for atomic force microscopy (AFM). However the success of this near-field imaging approach still needs to be demonstrated for biological samples.

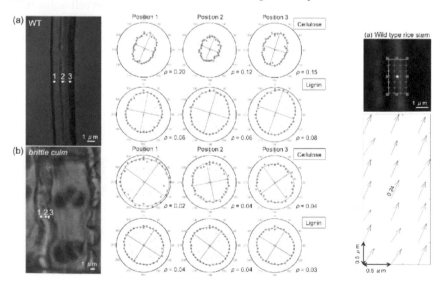

FIGURE 7   Polarized Raman Microspectroscopy of poplar cell wall showing orientation of cellulose microfibrils (from Sun et al. 2016)

Atomic Force Microscopy (AFM) has been successfully used in cell wall research. AFM has the advantage that samples can stay fully hydrated, but it is confined to a surface and can only provide sample surface topology. It should be noted that the AFM signal is a convolution of the tip geometry with the surface topology. Thus is it crucial to ensure that tip geometry does not exceed the space between microfibrils, as otherwise one could easily overestimate an object's dimensions. Success depends on completely immobilizing the tissue samples in order to avoid pushing the biological sample, which would result in a "lateral smearing" of the signal, which would in turn also overestimate the width of certain features. However the fact that AFM can image in solution, albeit only the very surface layer, makes this approach a very interesting one in a broader imaging portfolio.

Electron microscopy in many ways is thought of as the gold standard in high resolution imaging for its resolving power of one to a few nanometers, however, this technology is more than any other approach vulnerable to sample preparation artifacts. Since electrons rather strongly interact with matter only very thin objects can be penetrated by the electron beam, realistically samples require to be less than 0.5 to 1 micron, with thinner samples being more favorable due to better penetration and fewer superposition of macromolecular objects along the path of the electron, rendering the images more interpretable. While transmission electron microscopy (TEM) fully penetrates samples, scanning electron microscopy (SEM) typically detects surface features through the detection of back-scattered electron and secondary electrons being emitted from the sample surface (and a small region underneath). SEM images are inherently surface views of an object and they can provide spectacularly beautiful depictions of a 3D object including surface highlights and shadows, not unlike a 2D photograph of a 3D object. However, they are somewhat hard to quantify, despite surface features that are usually well depicted. But it is important to keep in mind that one does not see far into the specimen, so any observations of the surface cannot necessarily be extrapolated beyond the very surface. Moreover SEM images typically require the deposition of a conducting layer of heavy metal stain and/or carbon and while this layer can be rather small (1–2 nm) its additional coating needs to be taken into consideration when measuring dimensions of biological objects, and the entire process of sample preparation, including critical point drying can lead to artifacts and may alter the 3D organization. Furthermore given the vacuum of electron microscopes, samples need to be dehydrated, either through critical point drying (SEM) or organic solvent dehydration before being sputter-coated or resin-embedded, respectively. This step in SEM sample preparation can be detrimental for biological samples, and thus is ideally carried out at very low temperature (so-called freeze substitution) to minimize sample preparation artifacts (McDonald and Auer 2006). Freeze-etching, as is sometimes carried out for SEM imaging, has similar benefits to freeze-substitution for TEM imaging, but it is an art only mastered by very few labs in the world, and rarely administered these days, possibly due to the superior power of 3D electron microscopy tomographic imaging of cryogenically immobilized samples, widely seen as the gold standard in 3D electron microscopic imaging of macromolecular complexes within their cell and tissue context. Regarding TEM sample preparation, traditional protocols chemically fix, stain and dehydrate samples at room temperature, which can result in alterations of the cellular scenery through aggregation and extraction. High-pressure freezing mostly avoids such problems, as samples are immobilized within milliseconds and subsequently dehydrated at low temperature. Heavy metal staining with uranyl ions and osmium tetroxide is likely to be somewhat less effective for carbohydrates, except for pectins, which are negatively charged and thus bind uranyl ions more effectively. Additional contrast can be achieved by ruthenium red, however a delicate balance needs to be struck to obtain sufficient contrast for imaging while not obscuring fine ultrastructural details. Secondary cell walls can be particularly difficult to study as the hydrophobic lignin prevents heavy metal stain penetration, resulting in a featureless region comprising the secondary cell wall, whereas the primary cell wall shows adequate contrast and texture (Fig. 8).

Cryo-electron microscopy (cryo-TEM) has the advantage that no stain is necessary and samples can be examined in their native frozen-hydrated state, but even in cryo-EM samples need to be thin, ideally not more than 100 nm, as thicker samples mean lower obtainable 3D resolution. TEM by its nature is a projection technique where 3D objects are projected onto the 2D plane of a detector.

All objects in the path of the electron will superimpose onto one another, and can only be separated in Z by recording the objects in a so-called tilt-series that allows a 3D reconstruction. Due to the limitations of tilt angles, the limited total electron dose that needs to be spread over the entire data set and computational 3D reconstruction artifacts resulting from coarsely sampled 3D space, cryo-tomograms are often noisy and anisotropic. Therefore they are challenging to visualize and quantitatively interpret, however, the tedious work pays off as model building and quantitative analysis can reveal a variety of parameters crucial for cell wall model formation. Tomographic imaging can be applied to any biological specimen, and it is useful not only for cryogenically prepared samples (although that may be necessary if one wants to determine the in-vivo native structure of plant cell walls), but

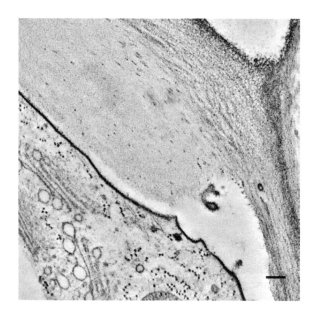

FIGURE 8 High pressure frozen, freeze substitued, resin embedded *Arabidopsis* cell wall showing microfibrils in primary cell walls as well as the homogeneous texture of the lignin-rich secondary cell wall. Scale bar = 100 nm. (From Sarkar et al. 2014.)

also for conventionally processed samples. A recent study from our lab (Sarkar et al. 2014) used electron tomography to measure microfibril diameter and spacing from conventionally and cryogenically processed samples, and found the diameter to be comparable in all three sample preparation approaches, but the spacing differed significantly for the conventionally processed samples, whereas both the high-pressure frozen, freeze-substituted samples as well as the cryo-sectioned, frozen-hydrated samples displayed a nearly identical yet much smaller spacing (Fig. 9).

Naturally there is great interest in methods to determine 3D cell wall architecture e.g. in the context of cell wall maturation upon ripening for agriculture and/

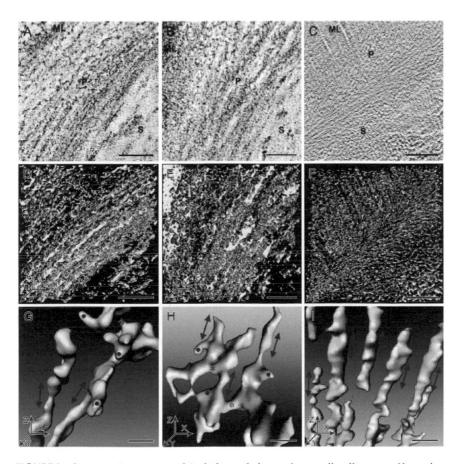

FIGURE 9 Segmented tomograms of *Arabidopsis thaliana* primary cell wall processed by various sample preparation methods showing qualitative comparison of wall preservation quality. A–C. 2D slices taken from the 3D electron tomograms showing overview of cell wall ultrastructure in transverse sections. Scale bars = 100 nm. D–F. Overview of segmented cell wall volumes. Scale bars = 100 nm. G–I. Close-up of segmented primary cell wall volumes showing arrangement of cellulose microfibrils (bidirectional arrow), cross-links (*), and additional density artifacts (●) from a side view (XZ plane) of the tomograms. Z-axis is approximately the axis of cell elongation. Scale bars = 10 nm. A, D, G. Chemically prepared samples (conventional method). B, E, H. High pressure frozen, freeze substituted samples. C, F, I. Vitreous sections of self-pressurized, rapid frozen samples ML = Middle lamella, P = Primary cell wall. S = Secondary cell wall. (From Sarkar et al. 2014.)

FIGURE 10   Segmented tomograms of chemically extracted *Arabidopsis* primary cell wall show-ing the evident loss of polysaccharides. Left: Untreated; Center: Ammonium oxalate treated to remove pectins; Right: Ammonium oxalate and sodium hydroxide treated to remove pectin and hemicelluloses. Scale bar = 100 nm. (From Loss et al. 2012.)

or re-engineering the paper or lignocellulosic biofuel industry. As a proof-of-concept we employed electron tomography to compare plant biomass samples that have been treated to remove specific cell wall components, i.e. hemicellulose and pectin (Fig. 10). Not too surprisingly, and in agreement with biochemical data, we found an increased loss of material resulting in larger porosity of the cell wall material (Loss et al. 2012). Tomographic imaging of plant cell walls will also be great value to study the effect of mutations in genes known to be involved in cell wall synthesis or remodeling.

Future Directions

Determining the plant cell wall 3D architecture will allow detailed mechanical strength simulation and thus an understanding of how the experimentally deter-mined cell wall design is superior over other possibilities. Ideally such mechani-cal predictions can be compared with actual mechanical stress/strain experiments of plant tissues (Varanasi et al. 2011). However, one of the unexplored areas of research is the mesoscale that bridges the macroscale with the micro-/nano-scale. Knowing the 3D architecture of a plant cell wall, as exciting as it will be, can only be the first step in a multi-scale imaging and modeling integration effort. In order to truly pre-dict mechanical properties, we will need to use large volume imaging tools at ide-ally subcellular resolution, and utilize the high-resolution model and adapt it to the mesoscale 3D organization of diverse cell walls of individual cell types in the various tissues in the stem.

None of the methods described in this chapter are well suited for this task, as we need to be able to penetrate samples of several millimeters in size. X-ray microscopy (X-ray tomography) on the other hand is an ideal experimental approach, as it can penetrate entire small animals such as zebrafish larvae or mouse embryos. In this sce-nario, samples are worked up into resin blocks, as is typical for electron microscopy, including heavy metal staining. The entire resin-block containing the plant organ of interest is then subjected to X-ray tomographic imaging, first at the 5–10 micron reso-lution range for the entire sample, and then again at a ~0.6–0.8 micron resolution range for ~100–200 microns fields of view. Since X-ray imaging of heavy-metal contrasted, resin-embedded samples is non-destructive, stitching together the information from smaller volumes can collect larger fields of views. Depending on the instrument used, and the field of view examined, a resolution as high as 50 nm is achievable, although that will require physical trimming of the samples to a field of view of ~65 µm. An ideal workflow would entail studying the samples at different scales and resolution, where possible to obtain fluorescence and/or other optical imaging approaches (such as IR or Raman microspectrometry), and superimpose such patterns onto the X-ray microscopy 3D data sets. Regions of interest would then be chosen for fine-scale X-ray tomography and Focused Ion Beam Scanning Electron Microscopy (FIBSEM) analysis. FIBSEM, another extremely powerful novel advanced imaging approach, which is based on iterative sequential imaging of the block surface of resin-embedded samples and milling of a 5–10 nm surface-slice off the block, allows ~10 nm resolution imaging of volumes of tens to potentially hundreds of microns. Most importantly, all these imaging modalities can be superimposed onto a 3D multiscale coordinate frame and the information can be integrated at the model space. We believe that such a multiscale, multimodal integrated bioimaging approach will constitute the future for a true multiscale understanding of the mechanical properties at the macroscopic, mesoscopic, microscopic and nanoscopic scale.

References

Albersheim, P., A. Darvill, K. Roberts, R. Sederoff and A. Staehelin. 2010. Plant Cell Walls. Garland Science, New York, USA.

Anderson, C.T., A. Carroll, L. Akhmetova and C. Somerville. 2010. Real-time imaging of cellulose reorientation during cell wall expansion in Arabidopsis roots. Plant Physiol. 152: 787-796.

Cosgrove, D.J. 2000. Expansive growth of plant cell walls. Plant Physiol. Biochem. 38: 109-124.

Cosgrove, D.J. 2005. Growth of the plant cell wall. Nature Rev. Molec. Cell Biol. 6: 850-861.

Cosgrove, D.J. and M.C. Jarvis. 2012. Comparative structure and biomechanics of plant primary and secondary cell walls. Front. Plant Sci. 3: 204.

Chafe, S.C. and A.B. Wardrop. 1972. Fine structural observations on the epidermis. I. The epidermal cell wall. Planta 107: 269-278.

Ding, S-Y. and M.E. Himmel. 2006. The maize primary cell wall microfibril: a new model derived from direct visualization. J. Agri. Food Chem. 54: 597-606.

Frei, E. and R.D. Preston. 1961. Cell wall organization and wall growth in the filamentous green algae Cladophora and Chaetomorpha. II. Spiral structure and spiral growth. Proc. Roy. Soc. Lond. B 155: 55-77.

Ji, N., H. Shroff, H. Zhong and E. Betzig. 2008. Advances in the speed and resolution of light microscopy. Curr. Opin. in Neurobiol. 18: 605-616.

Geisler-Lee, J., M. Geisler, P.M. Coutinho, B. Segerman, N. Nishikubo, J. Takahashi, H. Aspeborg, S. Djerbi, E. Master, S. Andersson-Gunnerås, B. Sundberg, S. Karpinski, T.T. Teeri, L.A. Kleczkowski, B. Henrissat, and E.J. Mellerowicz. 2006. Poplar carbohydrate-active enzymes. Gene identification and expression analyses. Plant Physiol. 140: 946-962.

Green, P.B. 1960. Multinet growth of the cell wall of *Nitella*. J. Biophys. Biochem. Cytol. 7: 289-296.

Itoh, T. and K. Shimaji. 1976. Orientation of microfibrils and microtubules in cortical parenchyma cells of poplar during elongation growth. Bot. Mag. Tokyo 89: 291-308.

Lee, K.J., S.E. Marcus and J.P. Knox. 2011. Cell wall biology: perspectives from cell wall imaging. Mol. Plant 4: 212-219.

Leroux, O. 2012. Collenchyma: a versatile mechanical tissue with dynamic cell walls. Ann Bot 110: 1083-1098.

Llyod, C. 2011. Dynamic microtubules and the texture of plant cell walls. Int. Rev. Cell Mol. Biol. 287: 287-329.

Loss, L.A., G. Bebis, H. Chang, M. Auer, P. Sarkar and P. Parvin. 2012. Automatic segmentation and quantification of filamentous structures in electron tomography. Proceedings of the ACM Conference on Bioinformatics, Computational Biology and Biomedicine: 170-177.

McCann, M.C. and K. Roberts. 1991. Architecture of the Primary Cell Wall. Academica Press, London, UK.

McDonald, K. and M. Auer. 2006. High-pressure freezing, cellular tomography, and structural cell biology. Biotechniques 41: 137-143.

Newman, R.H., S.J. Hill and P.J. Harris. 2013. Wide-angle x-ray scattering and solid state nuclear magnetic resonance data combined to test models for cellulose microfibrils in mung bean cell walls. Plant Physiol. 163: 1558-1567.

Preston, R.D. 1982. The case for multinet growth in growing walls of plant cells. Planta 155: 356-363.

Park, Y.B. and D.J. Cosgrove. 2012. A revised architecture of primary cell walls based on biomechanical changes induced by substrate-specific endoglucanases. Plant Physiol. 158: 1933-1943.

Roelofsen, P.A. 1958. Cell-wall structure as related to surface growth. Some supplementary remarks on multinet growth. Acta. Bot. Neerl. 7: 77-89.

Roelofsen, P.A. and A.L. Houwink. 1951. Cell wall structure of staminal hairs of *Tradescantia virginica* and its relation with growth. Protoplasma 40: 1-22.

Roland, J.-C., B. Vian and D. Reis. 1977. Further observations on cell wall morphogenesis and polysaccharide arrangement during plant growth. Protoplasma 91: 125-141.

Rose, J.K.C. and A.B. Bennett. 1999. Cooperative disassembly of the cellulose xyloglucan network of plant cell walls: parallels between cell expansion and fruit ripening. Trends Plant Sci. 4: 176-183.

Rust, M.J., M. Bates and X. Zhuang. 2006. Sub-diffraction-limit imaging by stochastic optical reconstruction microscopy (STORM). Nature Methods 3: 793-796.

Sarkar, P., E. Bosneaga and M. Auer. 2009. Plant cell walls throughout evolution: towards a molecular understanding of their design principles. J. Expt'l. Bot. 60: 3615-3635.

Sarkar, P., E. Bosneaga, E.G. Yap Jr., J. Das, W-T. Tsai, A. Cabal, E. Neuhaus, D. Maji, S. Kumar, M. Joo, S. Yakovlev, R. Csencsits, Z. Yu, C. Bajaj, K. H. Downing and M. Auer. 2014. Electron Tomography of Cryo-Immobilized Plant Tissue: a novel approach to studying 3d macromolecular architecture of mature plant cell walls in situ. *PLoS ONE,* 9: e106928.

Scheller, H.V. and P. Ulvskov. 2010. Hemicelluloses. Ann. Rev. of Plant Biol. 61: 263-289.

Segui-Simarro, J.M., J.R. Austin 2nd, E.A White and L.A. Staehelin. 2004. Electron tomographic analysis of somatic cell plate formation in meristematic cells of Arabidopsis preserved by high-pressure freezing. Plant Cell 16: 836-856.

Setterfield, G. and S.T. Bayley. 1958. Arrangement of cellulose microfibrils in walls of elongating parenchyma cells. J. Biophys. Biochem. Cytol. 4: 377-382.

Somerville, C. 2006. Cellulose synthesis in higher plants. Ann. Rev of Cell and Dev. Biol. 22: 53-78

Somerville, C., S. Bauer, G. Brininstool, M. Facette, T. Hamann, J. Milne, E. Osborne, A. Paredez, S. Persson, T. Raab, S. Vorwerk and H. Youngs. 2004. Toward a systems approach to understanding plant cell walls. Science 306: 2206-2211.

Sun, L., S. Singh, M. Joo, M. Vega-Sanchez, P. Ronald, B.A. Simmons, P. Adams and M. Auer. 2016. Non-invasive imaging of cellulose microfibril orientation within plant cell walls by polarized Raman microspectroscopy. Biotech. and Bioeng. 113: 82-90.

Thomas, L.H., V.T. Forsyth, A. Sturcová, C.J. Kennedy, R.P. May, C.M. Altaner, D.C. Apperley, T.J. Wess and M. Jarvis. C. 2013. Structure of cellulose microfibrils in primary cell walls from collenchyma. Plant Physiol. 161: 465-476.

Vanholme, R., B. Demedts, K. Morreel, J. Ralph and W. Boerjan. 2010. Lignin biosynthesis and structure. Plant Physiol. 153: 895-905.

Van Iterson, G. 1937. A few observations on the hairs of the stamens of *Tradescantia virginica*. Protoplasma 27: 190-211.

Varanasi, P., L. Sun, B. Knierim, E. Bosneaga, P. Sarkar, S. Singh and M. Auer. 2011. Quantifying bio-engineering: the importance of biophysics in biofuel research. pp. 493-520. *In*: M.A.dos Santos Bernardes (ed.). Biofuel's Engineering Process Technology, InTech: Open Access Publisher.

Yong, W., B. Link, R.O'Malley, J. Tewari, C.T. Hunter, C.A. Lu, X. Li, A.B. Bleecker, K.E. Koch, M.C. McCann, D.R. McCarty, S.E. Patterson, W.D. Reiter, C. Staiger, S.R. Thomas, W. Vermerris and N.C. Carpita. 2005. Plant cell wall genomics. Planta 221: 747-751.

8

Biosynthesis and Assembly of Cellulose

Candace H. Haigler[1], Jonathan K. Davis[1],*
Erin Slabaugh[1] and James D. Kubicki[2]

Introduction

Cellulose microfibrils are an essential structural component of plant cell walls. This chapter is focused on cellulose synthesis in expanding primary walls and sclerenchyma cells in angiosperms, which have historically been the major focus of research on cellulose synthesis. However, cellulose is synthesized by many organisms including: some species of eubacteria, cyanobacteria that were involved in the endosymbiotic origin of plant cells, diverse single-celled eukaryotes (such as the social amoeba *Dictyostelium discoideum* and the oomycetes), unicellular and multicellular algae, all land plants including non-vascular bryophytes, and even in an animal—the tunicates. Selected features of cellulose synthesis in these organisms will be mentioned below when they have important implications for understanding cellulose synthesis in angiosperms. In general, we will emphasize cellulose synthesis within canonical primary and secondary cell walls. These terms can only be loosely applied to the diversity that exists in nature. 'Primary wall' typically refers to thin extensible walls with a lesser amount of cellulose around expanding cells. 'Secondary wall' typically refers to thick cellulose-rich walls that are deposited after expansion has stopped. Primary walls are typically less than 0.5 μm thick and contain 20 – 30% cellulose. Secondary walls may be about 2 to 6 μm thick and they often contain 40 – 50% cellulose (Albersheim et al. 2011).

Vascular plant morphogenesis and function absolutely depend on the synthesis of strong cellulose microfibrils. As examples, cellulose microfibrils help to govern the behavior of the apical meristem, participating in a feedback network involving multiple cellular components (Sampathkumar et al. 2014, Boudon et al. 2015). In the primary and secondary cell walls of the plant body, the cellulose microfibrils are

[1] Department of Crop Science and Department of Plant Biology, Campus Box 7620, North Carolina State University, Raleigh, NC 27695, USA.
[2] Department of Geosciences, The Pennsylvania State University, 503 Deike, University Park, PA, 16802, USA.
* Corresponding author : candace_haigler@ncsu.edu

essential, high-strength components (Nishiyama 2009). Organized cellulose micro-fibrils constrain cellular expansion to occur in the direction perpendicular to their net orientation, which causes cells to elongate when the cellulose microfibrils are oriented transversely to the cellular axis (Szymanski and Cosgrove 2009). When cellulose synthesis in primary walls is compromised by mutations in plant cellulose synthases (called CESAs) or exposure to cellulose synthesis inhibitors, organ elongation is typically inhibited as shown by numerous experiments (summarized in McFarlane et al. 2014). Defects in primary wall cellulose synthesis impact the morphogenesis of the whole plant (Landrein et al. 2013). Growth of plants past the first-leaf stage requires the differentiation of xylem tracheary elements with cellulose-rich secondary walls (Zhao et al. 2007). When the normal amount of secondary wall cellulose is reduced, xylem vessels collapse under the stress of water transpiration and stems may be short or weak (Turner et al. 2001). Correspondingly, a dwarfed and weeping growth habit occurred in transgenic aspen trees with impaired secondary wall cellulose synthesis in the woody tissue (Joshi et al. 2011).

Cellulose microfibril formation occurs as the culmination of a fascinating multi-step process. First, the glucose moiety of an activated sugar substrate (uridine diphosphate-alpha-D-glucose, UDP-glucose) in the cytosol is polymerized to form a beta-1,4-linked glucan homopolymer by a plasma membrane localized CESA. Intramolecular interactions cause each glucose to be inverted 180° relative to the previously added monomer, which causes the disaccharide cellobiose to be the <u>structural</u> repeating unit of cellulose. Then each polymer is translocated to the extracytoplasmic side of the plasma membrane, apparently through a pore formed by the transmembrane helices (TMH) of CESA. Finally, the adjacent polymers interact and co-crystallize in a parallel extended conformation to form strong microfibrils. Plant cells assemble numerous CESAs in close proximity within a cellulose synthesis complex (CSC) in order to promote the coalescence of multiple glucan chains into structural microfibrils. The forces associated with cellulose microfibril formation cause the CSC to move in the plane of the plasma membrane as the long cellulose microfibrils form continuously behind it. This remarkable conversion of soluble sugar substrate to a structural microfibril is accomplished in a space of about 100 nm, inclusive of the membrane-bound CSC and the zone near the surface of the plasma membrane where microfibril formation occurs.

To complement the summary of cellulose synthesis in this chapter, more detailed information can be found in recent reviews on cellulose synthesis including Kumar and Turner, 2014, Li et al. 2014, McFarlane et al. 2014, Slabaugh et al. 2014, McNamara et al. 2015, Wallace and Somerville 2015). In addition, these reviews are a source of the primary references for the older well-established facts about cellulose biosynthesis mentioned below. Primary citations are most often provided for points made by single research papers, newer discoveries about cellulose biosynthesis, and other types of information.

Characteristics and Variation in Native Celluloses

Both the primary and secondary walls of the seed plants contain an allomorph of cellulose called cellulose I beta, which has glucan chains oriented in parallel (Nishiyama 2009). Cellulose microfibrils within plant cell walls are too long to be measured directly, but their length can be estimated. The number of glucose residues incorporated into a single cellulose polymer varies from 2,000–4,000 in primary walls to 14,000–15,000 in secondary walls (Brett 2000). Given the 5.2 Å length of a single glucose unit within the beta-1, 4-linked glucan chain (Fig. 1A), the maximum length of a cellulose polymer would be 2 μm or 8 μm in primary or secondary walls,

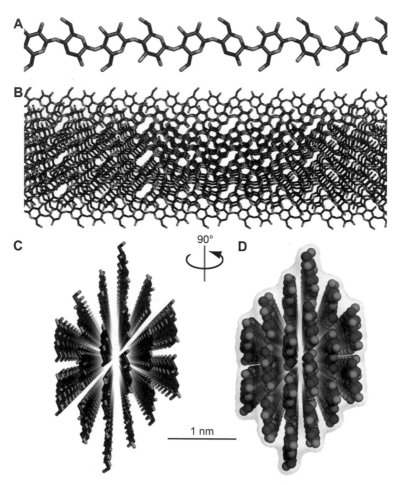

FIGURE 1 Molecular structure of cellulose. (A) Part of a single ß-1,4-linked glucan chain with the linear ribbon-like conformation that facilitates microfibril formation. Carbon and oxygen bonds are portrayed as sticks, while hydrogen bonds are not shown. (B) A segment of an 18-chain microfibril viewed from the side, with structural coordinates corresponding to published data (Fernandes et al. 2011). This microfibril was rotated 90° to generate the end-on views. (C) End-on stick representation of the 18-chain microfibril with only carbon and oxygen bonds represented. (D) End-on molecular surface representation of the 18-chain microfibril to convey its solid nature. The van der Waals radii of carbon and oxygen atoms are 1.7 and 1.62 Å, respectively, and these are portrayed as spheres with half these values. The molecular surface of the fibril was calculated with a solvent radius of 1.4 angstrom and full van der Waals radii for carbon and oxygen, with hydrogen atoms omitted. The 1 nm scale bar refers only to the microfibril cross-sections in (C and D). Molecular representations and solvent shells were determined with The PyMOL Molecular Graphics System (Version 1.7.4 Schrödinger, LLC).

respectively. If the polymer length approximately defines the microfibril length as expected, then the microfibril length is equivalent to 10 – 40% of the diameter of a small plant cell.

Typically the term microfibril is used to describe the smallest structural unit of cellulose in plant cell walls (Zhang et al. 2014; Fig. 1B, C, D). A single CSC in land plants is thought to synthesize this smallest microfibril (Fig. 2A), and the number of glucan chains within it has been debated extensively. Currently, the smallest microfibril is predicted to contain 18 chains and to be 2 to 3 nm wide (Newman et al. 2013). Even so, larger cellulose macrofibrils may form as a result of pairs, triplets, and other small groups of CSCs operating near each other in the plasma membrane (Herth 1985, Haigler et al. 2014). The occurrence and extent of cellulose macrofibril formation varies in different developmental contexts, which is one key regulator of the diverse biophysical properties of cells and tissues (Donaldson 2007, Nishiyama 2009, Newman et al. 2013, Thomas et al. 2014). The mode of assembly of the macrofibrils can also vary, for example including co-crystallization of smaller cellulose microfibrils still in a nascent state of ordering between polymers or bundling along the surfaces of distinct microfibrils that have already crystallized independently (Zhang et al. 2014; Thomas et al. 2014). The hierarchical assembly of a large extracellular macrofibril of cellulose by a bacterium clearly illustrates these principles (Haigler and Benziman 1982).

The control of cellulose assembly in native plant cell walls is an active area of research with implications for the useful properties of biomaterials. In addition to control by the number of active CESAs in the CSC, cellulose macrofibril diameter can be affected by the amount and composition of the cell wall matrix polymers. Potentially, cellulose-interacting matrix polymers such as xyloglucan interact with the surfaces of nascent cellulose microfibrils and alter their higher order aggregation and co-crystallization (Park and Cosgrove 2015). This effect is maximized during primary wall cellulose synthesis when matrix polymers are more abundant. Conversely, the increased proportion of cellulose as compared to the matrix components in secondary walls would promote more extensive microfibril coalescence, co-crystallization, and the formation of cellulose macrofibrils (Thomas et al. 2014). An extreme case is the secondary wall of cotton fibers, which contain about 95% cellulose with a high crystallite size of 5 nm (Lee et al. 2015) as contrasted with the less than 3 nm crystallites in woody cell walls (Thomas et al. 2014) containing 40 – 50% cellulose.

The CESA Gene Family in Plants

CESAs are classified in Glycosyltransferase Family2 (GT-2) (Cantarel et al. 2009) (Fig. 2 B, C). Like all GT-2 family proteins, CESAs are inverting glycosyltransferases that form a glycosidic bond in the opposite anomeric configuration relative to the donor substrate. In the case of cellulose, the alpha linkage of glucose to UDP within UDP-glucose (Fig. 2B) is inverted during glycosyltransfer to the beta linkage within the beta-1, 4-linked cellulose homopolymer (Fig. 1A). CESAs are also processive glycosyltransferases, which means they add successive glucoses without releasing the elongating glucan chain (Fig. 2C). In plants, CESA proteins have about 1000 amino acids and are around 125 kDa in size. A cytoplasmic region of

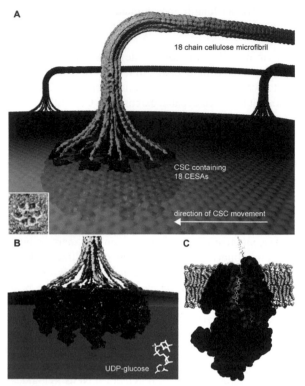

FIGURE 2   Representations of the structure of the rosette CSC and CESA. (A) The inset in the lower left shows a rosette CSC (without the associated cellulose) as viewed by transmission electron microscopy in a replica of the plasma membrane of a plant cell. This structure is the signature of the TMH of multiple CESAs crossing the plasma membrane. The scale bar, which applies only to this inset, equals 20 nm.

The other images are computationally assembled cartoons, representing approximate structures, that were created with the Blender 3D modelling and rendering package (www.blender.org). The view in (A) is from outside the cell, with one CSC highlighted and two others in the background. The cytosolic domains of CESA are below the plasma membrane, whereas the TMH pass through the membrane and form the channel through which the glucan chains are extruded. Near the surface of the plasma membrane the glucan chains coalesce to form the microfibril as the CSC moves within the plasma membrane. The nascent microfibril probably turns earlier and lies closer to the membrane than is depicted here for clarity. The cartoon structures depict 18 CESAs within one rosette CSC and an 18 chain cellulose microfibril, consistent with the latest evidence. (B) A cartoon of the rosette CSC viewed from the side, along with the cytosolic UDP-glucose substrate for cellulose synthesis. The location of the plasma membrane is portrayed at the top of the grey cytosolic region. (C) A cartoon of a single CESA protein embedded in the plasma membrane bilayer and an associated glucan chain. The glucan chain, which originates at the catalytic site and passes through the predicted TMH channel of CESA into the apoplastic space, is viewed through a semi-transparent 'window', whereas in reality it would be buried within CESA assuming similarity with BcsA. The molecular structure of the membrane was published previously (Heller et al. 1993).

Sources of information used to produce these cartoons were: (a) the eight TMH region and translocating glucan chain of bacterial BcsA (Morgan et al. 2013; (b) the large cytosolic loop of the predicted plant CESA structure, which did not include the N- or C-terminal ends of the protein (Sethaphong et al. 2013); and (c) cellulose structure parameters similar to prior publications (Fernandes et al. 2011) without accounting for the hydrogens. The assembled structure representing the CESA monomer was replicated and rotated at 0°, 60°, or 120° prior to creating symmetric homotrimers with close but non-overlapping packing of the cytosolic domain. The synthetic trimers were replicated and assembled by hand to form a six-lobed structure similar to the rosette CSC. The emerging glucan chains were placed by matching the coordinates of the first glucose residue of each glucan chain to the location where the chain would be just emerging from the TMH region of BcsA into the periplasmic space of the bacterial cell (Morgan et al. 2013), which corresponds to the extracytoplasmic region of a plant cell.

approximately 500 amino acids contains the active site where glucan chain elongation occurs. Computational methods most commonly predict eight transmembrane helices (TMH) within CESAs (Fig. 2 B,C), but the number remains ambiguous due to no solved structure for a plant CESA being available. Possibly, one or more of the predicted TMHs lies on the interior surface of the plasma membrane (an interfacial helix) or enters and exits the membrane on the same side (a reentrant helix) (Slabaugh et al. 2014). Biochemical analysis of the topology of CESA in the plasma membrane is needed, along with continuing efforts to crystallize a CESA.

The vascular plants have a multi-gene CESA family with six major groups (or clades) as determined by phylogenetic sequence analysis. Given the homopolymeric nature of cellulose, researchers were surprised to find that CESAs in the different groups were usually required for cellulose synthesis. CESAs in three of the groups are required for synthesis of expanding primary walls, and three other isoforms are required during the synthesis of thickening secondary walls in sclerenchyma cells. In Arabidopsis hypocotyls and roots, AtCESA1, 3, and 6 (or its other clade members) are necessary for normal growth and organogenesis. The AtCESA6 clade includes AtCESA2, 5, and 9, and these may substitute for AtCESA6 when its function is compromised (for an example, see Bischoff et al. 2011). In Arabidopsis xylem and interfascicular fibers, AtCESA4, 7, and 8 are required for normal secondary cell wall thickening (Turner et al. 2001). Cellulose synthesis within secondary walls in wood, cotton fibers, and stalks of grains and grasses depends on the orthologs of AtCESA4, 7, and 8 (Zhong and Ye 2015). There are exceptions to these typical patterns. For example, Arabidopsis leaf trichomes use a 'primary wall' set of CESAs to synthesize their thick 'secondary' walls, which contain a mixture of typical primary and secondary wall polymers (Betancur et al. 2010). Similarly, an unconventional set of CESA isoforms synthesizes the cellulose within the long rays of Arabidopsis seed mucilage, which expands from the seed surface along with uptake of water during germination (Griffiths et al. 2015).

There are sequence variations within five regions of CESA that confer the class specificity defining the six major groups (Carroll and Specht 2011). However, the functional differentiation, if any, of the isoforms remains largely uncharacterized. Although all the isoforms are likely to polymerize cellulose, they may do so at different rates due to inherent biochemical differences or different interactions with other proteins. The CSCs moved slower in dark-grown hypocotyls when even another CESA in the same clade, AtCESA5, was experimentally integrated into the CSC of an AtCESA6 knock-out line (Bischoff et al. 2011). Conversely, the CSC traveled more rapidly when a 'secondary wall' CESA was used to complement a mutation in a 'primary wall' CESA (Carroll et al. 2012).

CESA Structure and Function Relationships

Plant cellulose synthesis usually occurs at the plasma membrane just prior to the incorporation of microfibrils into the cell wall. In an exceptional case, crystalline cellulose is extruded within the endomembrane system during the early stages of cell plate formation during plant cell cytokinesis (Miart et al. 2014). Synthesis of cellulose at the plasma membrane poses questions including: (a) how is the catalytic site defined and what is the mechanism of catalysis; (b) what drives the rotation of each successive glucose so that cellobiose becomes the structural repeating unit; and

(c) how is the glucan chain translocated across the plasma membrane? Revolutionary insight into these questions was derived from the first crystal structure of any cellulose synthase, BcsA from the bacterium *Rhodobacter sphaeroides* (Morgan et al. 2013). In this case, the BcsA cellulose synthase catalytic subunit co-crystallized in a heterodimer with BcsB. The crystallized BcsA also contained a UDP molecule, presumably reflecting the prior transfer of a glucose unit from the UDP-glucose substrate to the elongating chain. Remarkably, an 18-glucose fragment of cellulose was included in the crystal, providing direct insight into the translocation process. The solved structure of BcsA-BcsB can provide strong clues about cellulose polymerization in CESA, as demonstrated by the structural similarity of a computationally folded model of the CESA catalytic domain (Sethaphong et al. 2013). The similarities of the catalytic region of BcsA observed by X-ray diffraction and a plant CESA structure predicted via *ab initio* structure simulation provided confidence about broadly conserved aspects of cellulose polymerization over long evolutionary time.

The common catalytic domain in cellulose synthases is called a GT-A fold, which has a six-stranded beta-sheet surrounded by five core alpha helices. This region contains distinct sites for binding the elongating cellulose polymer (the acceptor substrate) and UDP-glucose (the donor substrate) in close proximity. The similarity of the structure of the predicted CESA catalytic domain with BcsA allows insights into the probable function of amino acid motifs that are conserved between the two proteins (Sethaphong et al. 2013). As occurs in BcsA, the DDG and DCD motifs in CESA are likely involved in coordinating the UDP moiety of UDP-glucose and the required Mg++ cofactor, respectively. The aspartate residue in a conserved TED motif likely acts as the catalytic base, and the glutamate cooperates with a conserved HxKAG motif to accommodate glucose within the substrate-binding pocket. A Q(Q/R)xRW motif near the N-terminus of the cytosolic domain lies in an interfacial alpha helix, and its tryptophan is likely to interact with the terminal glucose of the acceptor glucan in CESA as it does in BcsA.

BcsA and the predicted CESA cytosolic domain each contain only one active site, which is deeply buried in BcsA, indicating that polymerization occurs one glucose residue at a time (Morgan et al. 2013). This makes it necessary to explain how cellobiose becomes the structural repeating unit of cellulose. For glycosyltransfer to occur, the donor substrate must be positioned precisely in the active site to present the anomeric carbon of the donor sugar for nucleophilic attack by the 4-hydroxyl of the terminal glucose of the acceptor substrate. The active site architectures in several different crystals of the BcsA-BcsB dimer suggest that UDP-glucose binds identically during each catalytic cycle, regardless of the orientation of the terminal glucose on the acceptor. Therefore, it is likely that each glucose residue rotates around the glycosidic bond following glycosyltransfer so that the newly added residue adopts a sterically favorable inverted orientation relative to the previously added one (reviewed in McNamara et al. 2015). In the translocating glucan chain within the TMH pore in BcsA, the adjacent glucoses already have inverted orientations. Therefore, glucose rotation occurs shortly after beta-1, 4-glycosidic bond formation, which is consistent with the weak coordination of the terminal acceptor glucose in the BcsA crystal structure that could allow the terminal glucose to rotate freely (Morgan et al. 2013). The inversion of adjacent glucose residues within the elongating glucan chain is favorable because the beta-1, 4 linkage is equatorial relative to the glucopyranose ring. This allows steric interactions (between the hydroxyl groups) and hydrogen bonding (between the 3-hydroxyl and the hemiacetal oxygen and between the 2 and 6 hydroxyls of adjacent glucose residues) to favor and stabilize the molecular structure of the cellobiose structural repeating unit within the polymer. Consequently, the cellulose polymer assumes a linear conformation with hydrophobic and hydrophilic faces, and these properties are essential for cellulose microfibril formation.

The structural information supported modeling of the active site and the polymerization and translocation processes in BcsA using a hybrid quantum mechanical/molecular mechanical (QM/MM) method (Yang et al. 2015). The calculations showed that: (a) the theoretical rate of addition of one glucose residue to the cellulose oligomer was in reasonable agreement with experimental data on the rate of cellulose polymerization *in vitro* by the BcsA-BcsB dimer (Omadjela et al. 2013); and (b) the inversion of conformation of each glucose residue around the glycosidic bond occurs spontaneously during polymerization from one catalytic site. Due to the similar predicted structure of the catalytic site in CESA (Sethaphong et al. 2013), cellulose polymerization in plants is likely to occur in the same way.

Nonetheless, there are differences in other aspects of the bacterial and plant cellulose synthases that are currently being explored (Slabaugh et al. 2014). For example, CESA lacks a regulatory domain involved in activation of bacterial cellulose synthesis by 3', 5'-cyclic diguanylic acid (c-di-GMP; Morgan et al. 2014), an allosteric regulator that is not found in plants. Conversely, several aspects of CESA structure are unique to plants. As compared to BcsA, CESAs have: (a) a longer N-terminal region containing a Zn-finger motif; and (b) two insertions in the catalytic domain (called the plant conserved region, or P-CR, and the class-specific region, or CSR). The P-CR and CSR are about 100 to 125 amino acids long, and they fold independently on the periphery of the cytosolic/catalytic domain (Sethaphong et al. 2013). As reviewed previously (Kumar and Turner 2014), the Zn-finger domain may participate in oxidative crosslinking leading to CESA dimerization, while not being essential for that process, and/or mediate ubiquitination to regulate CESA activity or proteolytic destruction. Presumably the P-CR and CSR mediate plant-specific aspects of CSC assembly and/or cellulose synthesis, but their roles are unknown. Interestingly, no missense mutations have been identified in the CSR through forward-genetic screens, and only one has been identified in the P-CR (Zhong et al. 2003). However, missense mutations made at putative phosphorylation sites in the CSR partially compromise AtCESA1 function (Chen et al. 2010).

The UDP-glucose donor substrate exists in the cytoplasm where the catalytic domain of CESA is located (Fig. 2B). This presents the necessity of translocating the forming glucan chain across the membrane, which is accomplished through a pore formed by the TMH of BcsA (Morgan et al. 2013). The cellulose oligosaccharide within the BcsA-BcsB crystal originated near the catalytic pocket and 10 glucose units lay within in an approximately 8Å diameter pore formed by TMH3–8 of BcsA. In BcsA, hydrophobic and hydrophilic amino acids alternate within the glucan translocating channel and interact either with the glycopyranose ring or the hydroxyl groups of adjacent glucose residues in cellulose, respectively (Morgan et al. 2013). Each glucose would interact with these different types of amino acids successively as the cellulose chain moves through the channel. Presumably the same mechanism operates in CESA (Fig. 2C) despite differences in the TMH regions of the two proteins (Slabaugh et al. 2014). Extrapolating evidence from *in vitro* cellulose synthesis by the purified BcsA-BcsB dimer, the energy provided by glycotransfer and subsequent relaxation of the newly formed beta-glycosidic bond into the energetically favorable equatorial conformation are likely sufficient to drive the continuous translocation of the chain through the channel (McNamara et al. 2015, Yang et al. 2015).

Forward genetic screens based on mutant phenotypes or altered responses to cellulose synthesis inhibitors have been pivotal in identifying plant CESAs and understanding their roles in development. The CESA mutations identified by both strategies have been summarized (McFarlane et al. 2014). The structural information derived from BcsA and the modeled CESA cytosolic domain structure allow hypotheses to be made about how cellulose synthesis may be adversely affected by missense mutations in various CESA isoforms (Morgan et al. 2013, Sethaphong et al. 2013, Slabaugh et al. 2014), as illustrated by the following examples. In the TMH region, the mutation of alanine in TMH4 of AtCESA1 (residue 903) to valine was associated with resistance to a cellulose inhibitor, faster CSC movement, and lower cellulose crystallinity (Harris et al. 2012). Based on sequence alignments, this residue may be orthologous to a tyrosine in TMH6 of BcsA that forms a hydrogen bond with the translocating glucan, supporting the idea that glucan translocation can affect cellulose properties (Morgan et al. 2013). In the cytosolic domain, the change of alanine 522 to valine in the *eli1-2* mutant of AtCESA3, which has stunted growth (Caño-Delgado et al. 2000, Caño-Delgado et al. 2003), is located in what appears to be a flexible loop bordering the substrate binding pocket so that substrate acquisition may be altered. Three other mutations that cause cellulose-deficient phenotypes occur at the site of a proline in a beta strand within the GT-A catalytic fold. These amino acid substitutions reflect a change of proline to: (a) threonine in the *fra5* mutant of AtCESA7 (residue 557; Zhong et al. 2003); (b) lysine in the *repp3* mutant of AtCESA3 (residue 578; Feraru et al. 2011); or (c) serine in the *thanatos* mutant of AtCESA3 (Daras et al. 2009). None of these substituted amino acids has the rigidity of proline and the *than* mutation is semi-dominant with a constitutive root swelling phenotype, supporting the possibility that the catalytic site is destablized by modification of the native proline.

Assembly of Multiple CESAs to Form the CSC

The strength of cellulose in plant cell walls derives from the coalescence of multiple glucan chains to form a strong microfibril. A linear aggregate of glucan chains, without chain folding, occurs due to the assembly of multiple CESAs into a CSC so that many glucan chains are extruded in close proximity (Fig. 2 A, B). In some algae and all of the characterized land plants, including bryophtyes and ferns, the assembled CSC has six lobes as visualized by freeze fracture transmission electron microscopy (FF-TEM). The lobes form a circle, leading to the description as a 'rosette' (Fig. 1A, inset). The presence of CESA in this protein complex was supported by immunolabeling of FF-TEM replicas (Kimura et al. 1999). The average diameter of the rosette CSC is about 25 nm, as visualized where the CESA TMH cross the plasma membrane. Using refined FF-TEM methods, the six-lobed rosette CSCs can also be detected on the surface of the plasma membrane, where only the 'tops' of the TMH emerge (Haigler et al. 2014). A larger structure possibly representing the cytosolic side of the CSC was visualized by use of another TEM technique (Bowling and Brown 2008), but there was no molecular identification provided. Why rosette CSC morphology only appears in land plants and their closest algal relatives, as contrasted with linear or rectangular CSCs in other organisms that synthesize cellulose microfibrils (Tsekos 1999), is still not understood. It is often presumed that the plant-specific regions of CESA are involved in rosette CSC formation, but this remains to be proven.

Several types of evidence support the existence of three CESA isoforms with the rosette CSC. Biochemical analyses have shown 1:1:1 stoichiometry for the CESA isoforms engaged in primary and secondary wall synthesis in Arabidopsis. This ratio occurred for AtCESA1, 3, and 6 within CSCs that were co-immunoprecipitated from extracts of dark-grown hypocotyls using isoform specific CESA antibodies (Gonneau et al. 2014). The 1:1:1 ratio also occurred for AtCESA4, 7, and 8 as assayed by quantitative immunoblotting of protein extracts from stems engaged in secondary wall cellulose synthesis (Hill et al. 2014). Several types of data show that CESA isoforms interact with each other within the primary wall CSC (Desprez et al. 2007, Wang et al. 2008), the secondary wall CSC (Atanassov et al. 2009, Timmers et al. 2009), or in general without regard to this distinction at least under experimental conditions (Carroll et al. 2012). Common models assume that each lobe of the rosette CSC is heteromeric, although available data do not exclude the possibility of the lobes containing only one or two CESA isoforms. Phylogenetic analyses have shown that a single ancestral CESA evolved to generate 'primary' and 'secondary' wall CESAs, then heteromeric CSCs evolved separately in each case (Roberts et al. 2012). The theory of constructive neutral evolution explains how the accumulation of mutations over long evolutionary time can change an initially homomeric protein complex into an obligatorily heteromeric one (Finnigan et al. 2012). Therefore, the current CESA isoforms could possibly only differ by their obligatory position in a heteromeric CSC. Specific isoforms could also modulate interactions with other proteins that regulate CSC activity or embody particular regulatory protein modification sites as discussed further below.

Cellulose Microfibril Formation

Fluorescently labeled CESAs move along linear tracks at about 300 nm/min during primary wall synthesis (Paradez et al. 2006, Desprez et al. 2007). The polymers continually emanating from the moving CSCs must assemble and crystallize efficiently in order for linear cellulose microfibrils with smooth surfaces to be generated (Zhang et al. 2014; Figure 2A). How can individual glucan chains coalesce with extended conformation and in register so that long paracrystalline cellulose microfibrils form? The term paracrystalline acknowledges that cellulose in land plants is a biopolymer with limited width and regions of crystalline imperfection along the length of the microfibrils, so its crystalline order is less precise than in mineral crystals. The biophysical nature of the disordered regions in cellulose, either on the surface or possibly at intervals along the length of the microfibrils, is an active area in cellulose research. Potentially the disordered regions of cellulose microfibrils are a focal point for cellulose microfibril bundling in interaction with xyloglucan, which in turn has a key role in the control of cell wall biomechanical properties (Park and Cosgrove 2015). Therefore, understanding the mechanisms behind the formation of cellulose microfibril properties is key to understanding the cell wall network.

To understand the forces related to cellulose microfibril formation, a pioneering study employed coarse-grained Monte Carlo simulations of the synthesis of a presumed 36-chain cellulose microfibril (Diotallevi and Mulder 2007). This work was done based on a speculative model for 36 CESAs existing in the CSC, before more recent evidence for 18 CESAs. The simulations showed the emerging glucan chains turning at approximately 90° just above the CSC, resulting in microfibrils forming parallel to the plane of the plasma membrane (Diotallevi and Mulder, 2007). This computational prediction helps to explain how microfibrils are able to form within a small space outside the plasma membrane from glucan chains that are extruded approximately upwards through a (putative) pore in the CESA TMH region. The coarse-grained modeling further suggested that the forces inherent in cellulose polymerization and crystallization were sufficient to propel the rosette-type CSC forward within the plane of the plasma membrane (Diotallevi and Mulder, 2007). Structures consistent with the coarse-grained model have been visualized on the surface of the plasma membrane where microfibrils are forming, and atomistic modeling of six cellulose chains (with 60 glucose residues each) also supported spontaneous formation of a small cellulose protofibril from a basal pool of relatively disorganized glucan chains followed by bending toward the modeled membrane (Haigler et al. 2014). A pool of glucan chains at the base would allow a linear paracrystalline cellulose microfibril to form even if the rates of polymerization from individual CESAs were slightly different. This putative pool of basal glucan chains and the adjacent zone where the glucan chains first become linear may also explain the temporal separation between the processes of cellulose polymerization and microfibril crystallization, as demonstrated by addition of dyes that interfere with cellulose crystallization in bacterial and plant cells (Haigler 1991, Roberts et al. 2004).

In the computational simulations, beta-1, 4-glucan chains coalesced and microfibrils formed spontaneously without any helper molecule (Diotallevi and Mulder 2007, Haigler et al. 2014). However, it is possible that a plant-specific plasma membrane-anchored protein binds to and guides the beta-1, 4-glucan chains in their initial coalescence for microfibril formation (Sorek et al. 2014). Similarly, a glucan-binding (chitinase-like) protein positively affects cellulose crystallization, although how it exerts this effect is unknown (Sanchez-Rodriguez 2012).

Regulation of CSC function

A discussion of the many types of higher-level regulation of cellulose synthesis, for example through hormones, light, and temperature is beyond the scope of this chapter. Regulatory factors operating at the level of CESA and the CSC are discussed below. At the first level, cellulose synthesis is strongly influenced by transcriptional control. As examples, the stage of predominant secondary cell wall thickening in the Arabidopsis inflorescence stem is associated with up-regulated expression of the *AtCESA4, 7,* and *8* genes (Hall and Ellis 2013) and secondary wall synthesis in cotton fibers begins with strongly up-regulated expression of the Gossypium orthologs of these Arabidopsis genes (Tuttle et al. 2015). The operation of the CSC to produce cellulose microfibrils occurs in the complex cellular context of the plasma membrane cell-wall continuum, inclusive of the cortical cytoskeleton and endomembrane system (Liu et al. 2015). This complexity may help to explain the difficulty with synthesizing cellulose *in vitro* from plant preparations, although cellulose I microfibrils can be formed in a mixture with beta-1, 3-glucan (callose) synthesis using solubilized plant cell extracts (Lai Kee Him et al. 2002).

The rosette CSC complex is assembled in the endomembrane system and delivered to the plasma membrane within small Golgi-derived secretory vesicles (Bashline et al. 2014). An unknown activation process must occur in the plasma membrane since cellulose is not typically synthesized within Golgi vesicles. A primary wall CSC persists in the plasma membrane for about 7 minutes, and this lifetime is in part regulated via actin (Sampathkumar et al. 2014). The operational time of a CSC in the plasma membrane can logically control the degree of cellulose polymerization and microfibril length. Given CESA movement at about 300 nm/min, a 7 min persistence time would generate a 2.1 µm long cellulose microfibril consistent with the estimate based on the degree of polymerization of primary wall cellulose. Secondary wall CSCs may remain in the plasma membrane longer to synthesize cellulose with higher degree of polymerization or they could synthesize cellulose faster during a similar active lifetime as compared to primary wall CSCs.

CESA activity can be modulated by phosphorylation, as occurs for many proteins. Wallace and Somerville (2015) provide a useful summary of phosphorylation events in AtCESA1, 2, 3, 4, 5 and 7 that have been supported by phosphoproteomics. The phosphorylation sites are in the N-terminal domain before the first TMH or in the CSR. Both of these regions are uniquely found in CESA as compared to BcsA, so the presence of phosphorylation sites that can regulate factors such as CESA velocity and stability is reasonable (Chen et al. 2010, Bischoff et al. 2011, Taylor 2007). Reflecting the integration of cellulose synthesis into the plant system, the phosphorylation of AtCESA1, AtCESA3, a CSC-associated protein (called CSI1; see below) and numerous enzymes in carbohydrate metabolism, were positively correlated with daytime photosynthesis and metabolic flux to cellulose (Boex-Fontvieille et al. 2014).

Stressful conditions often cause the regulated clearance of CESAs from the plasma membrane and accumulation in endomembrane compartments (Gutierrez et al. 2009, Crowell et al. 2010, Bashline et al. 2013, Fujimoto et al. 2015). This may be a means of rapidly stopping cell growth under adverse conditions while preserving the biochemical investment in the synthesis and assembly of the CSC. The CSCs may be stored in intracellular compartments until they can be re-exported to the plasma membrane to once again support cellulose synthesis and plant growth once conditions improve. Long term plant success would be aided by the ability to rapidly stop the synthesis of cellulose, which is a large and irreversible carbon sink required only for plant growth and not short-term survival (Haigler et al. 2001).

In terms of the cellular control of CSC activity, the relationship between the CSC and the microtubules has been a focus of research for several decades. There is a complex relationship between biophysical forces and the organization of cellulose microfibrils and cortical microtubules that is only beginning to be understood at the mechanistic level (Baskin 2001, Paredez et al. 2008, Landrein and Hamant 2013, Lei et al. 2013, Worden et al. 2015). Fluorescently labeled CSCs often travel along the path of microtubules, and two proteins named cellulose synthase interacting 1 and 3 (CSI1 and CSI3), bind both CESA and microtubules. These are large proteins with about 2100 amino acids containing multiple Armadillo repeats, a protein motif present in several microtubule-binding proteins (Tewari et al. 2010). Therefore, the CSI proteins may function to tether CSCs to microtubules (Li et al. 2012, Bringmann et al. 2012). However, CESAs can maintain their trajectory independent of the orientation of microtubules (Sugimoto et al. 2000), and the debate continues about how microtubules within a complex cellular network mechanistically impact cellulose synthesis in various cell types. For example, the properties of the incipient cell wall, inclusive of pectin distribution and extent of methylation, also impact cellulose microfibril orientation (Yoneda et al. 2010).

Finally, a cellulase called Korrigan, with an unknown but required role in cellulose synthesis, is physically associated with the CSC in the plasma membrane and in endomembrane compartments (Lei et al. 2014, Vain et al. 2014, Worden et al. 2015). This signifies the many things that remain to be learned about this process of pivotal importance in nature and industry. Currently there is only partial insight available about how the CSC, as a remarkable protein nanomachine, synthesizes cellulose microfibrils and modulates their biophysical properties. Some major unanswered questions are listed below.

Major Unanswered Questions

What is the atomic structure of CESA and its topology in the membrane? Do plant CESAs within the six major phylogenetic clades have structural differences that impact potential isoform-specific roles in cellulose synthesis? What are the structural and energetic determinants of the processivity of glucan formation and translocation within CESA? How is the predicted TMH pore gated so that plasma membrane electrostatic potential is maintained when cellulose synthesis is not active? Or, are inactive CSCs immediately removed from the plasma membrane? How do the plant specific domains of CESA modulate CSC assembly and/or cellulose synthesis? What molecular mechanisms regulate and constrain the assembly of the rosette-type CSC? What is the complete group of proteins that associate with the CSC as part of its core structure or as regulatory partners? How is cellulose synthesis in the vascular plants activated in the plasma membrane? Does the structure of CESA and/or the CSC aid microfibril formation, for example through guiding the angle of chain extrusion? How does the cellulose bend so much in a short distance within the apoplastic space after it exits the CSC? Do smaller groups of cellulose chain pre-assemble, or do all chains from one CSC assemble at the same time? What is the time lag between initial coalescence of the glucan chains and microfibril crystallization? Do other proteins assist the process of cellulose crystallization, and, if it occurs, what is the mechanism? What can be learned about glucan chain assembly from the future atomistic modeling of the interactions of at least 18 chains in the presence of water? Are the polymer assembly thermodynamics sufficient to overcome the entropy of the initial state and drive the movement of the CSC within the plane of the plasma membrane? In addition to the path of travel of the CSC in the plasma membrane, how do other factors influence the orientation of newly synthesized cellulose microfibrils? How does cellulose interact with diverse polymers in the apoplast to generate cell walls with the greatly variable structures, biophysical properties, and functions observed in plants?

Acknowledgements

The authors acknowledge the support of their related research and the writing of this article as part of The Center for LignoCellulose Structure and Formation, an

Energy Frontier Research Center funded by the U.S. Department of Energy, Office of Science, Basic Energy Sciences under Award # DE-SC0001090.

References

Albersheim, P., A. Darvill, K. Roberts, R. Sederoff and A. Staehelin. 2011. Plant Cell Walls. Garland Science, New York.

Atanassov, I., J.K. Pittmann and S.R. Turner. 2009. Elucidating the mechanisms of assembly and subunit interaction of the cellulose synthase complex of Arabidopsis secondary walls. J. Biol. Chem. 284: 3833-3841.

Bashline, L., S. Li, C.T. Anderson, L. Lei and Y. Gu. 2013. The endocytosis of cellulose synthase in Arabidopsis is dependent on mu2, a clathrin mediated endocytosis adaptin. Plant Physiol. 163: 150-160.

Bashline, L., S. Li and Y. Gu. 2014. The trafficking of cellulose synthase complex in higher plants. Ann Bot, doi: 10.1093/aob/mcu040.

Baskin, T.I. 2001. On the alignment of cellulose microfibrils by cortical microtubules: a review and a model. Protoplasma 215: 150-171.

Betancur, L., B. Singh, R.A. Rapp, J.F. Wendel, M.D. Marks, A.R. Roberts and C.H. Haigler. 2010. Phylogenetically distinct cellulose synthase genes support secondary wall thickening in Arabidopsis shoot trichomes and cotton fiber. J. Integr. Plant Biol. 52: 205-220.

Bischoff, V., R. Desprez, G. Mouille, S. Vernhettes, M. Gonneau and H. Hofte. 2011. Phytochrome regulation of cellulose synthesis in Arabidopsis. Curr. Biol. 21: 1822-1827.

Boex-Fontvieille, E., M. Davanture, M. Jossier, M. Zivy, M. Hodges and G. Tcherkez. 2014. Photosynthetic activity influences cellulose biosynthesis and phosphorylation of proteins involved therein in Arabidopsis leaves. J. Expt. Bot. 65: 4997-5010.

Boudon F, J. Chopard, O. Ali, B. Gilles, O. Hamant, A. Boudaoud, J. Traas and C. Godin. 2015. A computational framework for 3D mechanical modeling of plant morphogenesis with cellular resolution. PLOS Comp. Biol. 11: e1003950.

Bowling, A.J. and R.M. Brown Jr. 2008. The cytoplasmic domain of the cellulose-synthesizing complex in vascular plants. Protoplasma 233: 115-127.

Brett, C.T. 2000. Cellulose microfibrils in plants: biosynthesis, deposition, and integration into the cell wall. Int. Rev. Cytol. 199: 161-99.

Bringmann, M., E. Li, A. Sampathkumar, T. Kocabek, M.-T. Hauser and S. Persson. 2012. POMPOM2/ CELLULOSE SYNTHASE INTERACTING1 is essential for the functional association of cellulose synthase and microtubules in Arabidopsis. Plant Cell 24: 163-77.

Caño-Delgado, A.I., K. Metzlaff and M.W. Bevan. 2000. The eli1 mutation reveals a link between cell expansion and secondary cell wall formation in *Arabidopsis thaliana*. Development 127: 3395-3405.

Caño-Delgado, A., S. Penfield, C. Smith, M. Catley and M. Bevan. 2003. Reduced cellulose synthesis invokes lignification and defense responses in *Arabidopsis thaliana*. Plant J. 34: 351-362.

Cantarel, B.L., P.M. Coutinh., D. Rancurel, T. Bernard, V. Lombard and B. Henrissat. 2009. The Carbohydrate-Active EnZymes database (CAZy): an expert resource for glycogenomics. Nucleic Acids Res. 37: D233-38.

Carroll, A. and C.D. Specht. 2011. Understanding plant cellulose synthases through a comprehensive investigation of the cellulose synthase family sequences. Front. Plant Sci. 2: 5.

Carroll, A., N. Mansoori, S. Li, L. Lei, S. Vernhettes, R.G.F. Visser, C. Somerville, Y. Gu and L.M. Trindade. 2012. Complexes with mixed primary and secondary cellulose synthases are functional in Arabidopsis plants. Plant Physiol. 160: 726-737.

Chen, S., D.W. Ehrhardt and C.R. Somerville. 2010. Mutations of cellulose synthase (CesA1) phosphorylation sites modulate anisotropic cell expansion and bidirectional mobility of cellulose synthase. Proc. Natl. Acad. Sci. USA 107: 17188-17193.

Crowell, E.F., M. Gonneau, Y.D. Stierhof, H. Hofte and S. Vernhettes. 2010. Regulated trafficking of cellulose synthases. Curr. Opin. Plant Biol. 13: 700-705.

Daras, G., S. Rigas, B. Penning, D. Milioni, M.C. McCann, N.C. Carpita, C. Fasseas and P. Hatzopoulos. 2009. The thanatos mutation in *Arabidopsis thaliana* cellulose synthase 3 (AtCesA3) has a dominant-negative effect on cellulose synthesis and plant growth. New Phytol. 184: 114-126.

Desprez, T., M. Juraniec, E.F. Crowell, H. Jouy, Z. Pochylova, F. Parcy, H. Höfte, M. Gonneau and S. Vernhettes. 2007. Organization of cellulose synthase complexes involved in primary cell wall synthesis in *Arabidopsis thaliana*. Proc. Natl. Acad. Sci. USA 104: 15572-15577.

Diotallevi F. and B. Mulder. 2007. Cellulose synthase complex: A polymerization driven supramolecular motor. Biophys. J. 92: 2666-2673.

Donaldson L. 2007. Cellulose microfibril aggregates and their size variation with cell wall type. Wood Sci. Technol. 41: 443-460.

Feraru, Elena, M.I. Feraru, J. Kleine-Vehn, A. Martiniere, G. Mouille, S. Vanneste, S. Vernhettes, J. Runions and J. Friml. 2011. PIN polarity maintenance by the cell wall in Arabidopsis. Curr. Biol. 21: 338-343.

Fernandes, A.N., L.H. Thomas, C.M. Altaner, P. Callow, V.T. Forsyth, D.C. Apperley, C.J. Kennedy and M.C. Jarvis. 2011. Nanostructure of cellulose microfibrils in spruce wood. Proc. Natl. Acad. Sci. USA 108: E1195-E1203.

Finnigan, G.C., V. Hanson-Smith, T.H. Stevens and J.W. Thornton. 2012. Evolution of increased complexity in a molecular machine. Nature 481: 360-364.

Fujimoto, M., Y. Suda, S. Vernhettes, A. Nakano and T. Ueda. 2015. Phosphatidylinositol 3-kinase and 4-kinase have distinct roles in intracellular trafficking of cellulose synthase complexes in Arabidopsis thaliana. Plant Cell Physiol. 56: 287-298.

Gonneau, M., T. Desprez, A. Guillot, S. Vernhettes and H. Höfte. 2014. Catalytic subunit stoichiometry within the cellulose synthase complex. Plant Physiol. 166: 1709-1712.

Griffiths J.S.K. Sola, R. Kushwaha, P. Lam, M. Tateno, R. Young, C. Voiniciuc, G. Dean, S.D. Mansfield, S. DeBolt and G.W. Haughn. 2015. Unidirectional movement of cellulose synthase complexes in Arabidopsis seed coat epidermal cells deposit cellulose involved in mucilage extrusion, adherence, and ray formation. Plant Physiology Preview, doi: 10.1104/pp.15.00478.

Gutierrez, R., J.J. Lindeboom, A.R. Paredez, A.M. Emons and D.W. Ehrhardt. 2009. Arabidopsis cortical microtubules position cellulose synthase delivery to the plasma membrane and interact with cellulose synthase trafficking compartments. Nat. Cell Biol. 11: 797-806.

Haigler, C.H. and M. Benziman. 1982. Biogenesis of cellulose I microfibrils occurs by cell-directed self-assembly in Acetobacter xylinum. pp. 273-296. *In:* R. M Brown, Jr. (ed.). Cellulose and Other Natural Polymer Systems. Plenum, New York.

Haigler, C.H. 1991. The relationship between polymerization and crystallization in cellulose biogenesis. pp. 99-124. *In:* C.H. Haigler and P. Weimer (eds.). Biosynthesis and Biodegradation of Cellulose. Marcel Dekker, New York.

Haigler, C.H., M. Ivanova-Datcheva, P.S. Hogan, V.V. Salnikov, S. Hwang, L.K. Martin and D.P. Delmer. 2001. Carbon partitioning to cellulose synthesis. Plant Mol. Biol. 47: 29-51.-

Haigler, C.H., M.J. Grimson, J. Gervais, N. Le Moigne, H. Hofte, B. Monasse and P. Navard. 2014. Molecular modeling and imaging of initial stages of cellulose fibril assembly: Evidence for a disordered intermediate stage. PLOS ONE 9: e93981.

Hall, H. and B. Ellis. 2013. Transcriptional programming during cell wall maturation in the expanding Arabidopsis stem. BMC Plant Biol. 13: 14.

Harris, D. M., K. Corbin, T. Wang, R. Gutierrez, A.L. Bertolo, C. Petti, D.-M. Smilgiese, J.M. Estevez, D. Bonetta, B.R. Urbanowicz, D.W. Ehrhardt, C.R. Somerville, J.K.C. Rose, M. Hong and S. DeBolt. 2012. Cellulose microfibril crystallinity is reduced by mutating C-terminal transmembrane region residues CESA1A903V and CESA3T942I of cellulose synthase. Proc. Natl. Acad. Sci. USA 109: 4098-4103.

Heller, H., M. Schaefer and K. Schulten. 1993. Molecular dynamics simulation of a bilayer of 200 lipids in the gel and in the liquid-crystal phases J. Phys. Chem. 97: 8343-60.

Herth, W. 1985. Plasma-membrane rosettes involved in localized wall thickening during xylem vessel formation of Lepidium sativum L. Planta 164: 12-21.

Hill, J.L., M.B. Hammudi and M. Ticn. 2014. The Arabidopsis cellulose synthase complex: proposed hexamer of CESA trimers in an equimolar stoichiometry. Plant Cell 26: 4834-4842.

Joshi, C.P., S. Thammannagowda, T. Fujino, J.-Q. Gou U. Avci, C.H. Haigler, L.M. McDonnell, S.D. Mansfield, B. Menghesa, N.C. Carpita, D.M. Harris, S. DeBolt and G.F. Peter. 2011. Perturbation of wood cellulose synthesis causes pleiotropic effects in transgenic aspen. Mol. Plant 4: 331-345.

Kimura, S., W. Laosinchai, T. Itoh, X. Cui, C.R. Linder and R.M. Brown Jr. 1999. Immunogold labeling of rosette terminal cellulose-synthesizing complexes in the vascular plant *Vigna angularis*. Plant Cell 11: 2075-2085.

Kumar, M. and S. Turner. 2014. Plant cellulose synthesis: CESA proteins crossing kingdoms. Phytochem. 112: 91-99.

Lai Kee Him, J., H. Chanzy, M. Muller, J-L. Putaux, T. Imai and V. Bulone. 2002. *In vitro* versus *in vivo* cellulose microfibrils from plant primary wall synthases: structural differences. J. Biol. Chem. 277: 36931-36939.

Landrein, G., R. Lathe, M. Bringmann, C. Vouillot, A. Ivakov, A. Boudaoud, S. Persson and O. Hamant. 2013. Impaired cellulose synthase guidance leads to stem torsion and twists phyllotactic patterns in Arabidopsis. Curr. Biol. 23: 895-900.

Landrein, G. and O. Hamant. 2013. How mechanical stress controls microtubule behavior and morphogenesis in plants: history, experiments and revisited theories. Plant J. 75: 324-338.

Lee, C.M., K. Kafle, D.W. Belias, Y.B. Park, R.E. Glick, C.H. Haigler and S.H. Kim. 2015. Comprehensive analysis of cellulose content, crystallinity, and lateral packing in *Gossypium hirsutum* and *Gossypium barbadense* cotton fibers using sum frequency generation, infrared and Raman spectroscopy, and X-ray diffraction. Cellulose 22: 971-989.

Lei, L., S. Li, J. Du, L. Bashline and Y. Gu. 2013. CELLULOSE SYNTHASE INTERACTIVE3 regulates cellulose biosynthesis in both a microtubule-dependent and microtubule-independent manner in Arabidopsis. Plant Cell, 25: 4912-4923.

Lei, L., T. Zhang, R. Strasser, C.M. Lee, M. Gonneau, L. Mach, S. Vernhettes, S.H. Kim, D. Cosgrove, S. Li and Y. Gu. 2014. The jiaoyao1 mutant is an allele of korrigan1 that abolishes endoglucanase activity and affects the organization of both cellulose microfibrils and microtubules in Arabidopsis. Plant Cell 26: 2601-2616.

Li, S., L. Lei, C.R. Somerville, and Y. Gu. 2012. Cellulose synthase interactive protein 1 (CSI1) links microtubules and cellulose synthase complexes. Proc. Natl. Acad. Sci. USA 109: 185-190.

Li, S., L. Bashline, L. Lei and Y. Gu. 2014. Cellulose synthesis and its regulation. Arabidopsis Book 12: e0169.

Liu, Z., S. Persson and C. Sanchez-Rodriguez. 2015. At the border: The plasma membrane cell wall continuum. J. Expt. Bot. 66: 1553-1563.

McFarlane, H.E, A. Döring and S. Persson. 2014. The cell biology of cellulose synthesis. Annu. Rev. Plant Biol. 65: 69-94.

McNamara, J.T., J.L.W. Morgan and J. Zimmer. 2015. A molecular description of cellulose biosynthesis. Ann. Rev. Biochem. 84: 895-921.

Miart, F., T. Desprez, E. Biot, H. Morin, K. Belcram, H. Höfte, M. Gonneau and S. Vernhettes. 2014. Spatio-temporal analysis of cellulose synthesis during cell plate formation in Arabidopsis. Plant J. 77:71-84.

Morgan, J.L., J. Strumillo, and J. Zimmer. 2013. Crystallographic snapshot of cellulose synthesis and membrane translocation. Nature 493: 181-192.

Morgan, J.L., J.T. McNamara and J. Zimmer. 2014. Mechanism of activation of bacterial cellulose synthase by cyclic di-GMP. Nat. Struct. Mol. Biol. 2: 489-496.

Newman, R.H., S.J. Hill and P.J. Harris. 2013. Wide-Angle X-Ray scattering and solid-state nuclear magnetic resonance data combined to test models for cellulose microfibrils in mung bean cell walls. Plant Physiol. 163: 1558-1567.

Nishiyama, Y. 2009. Structure and properties of the cellulose microfibril. J Wood Sci 55: 241-249.

Omadjela, O., A. Narahari, J. Strumilo, H. Mélida, O. Mazur, V. Bulone and J. Zimmer. 2013. BcsA and BcsB form the catalytically active core of bacterial cellulose synthase sufficient for *in vitro* cellulose synthesis. Proc. Natl. Acad. Sci. USA 44: 17856-17861.

Paredez, A.R., C.R. Somerville and D.W. Ehrhardt. 2006. Visualization of cellulose synthase demonstrates functional association with microtubules. Science 312: 1491-1495.

Paredez, A.R., S. Persson, D.W. Ehrhardt and C.R.Somerville. 2008. Genetic evidence that cellulose synthase activity influences microtubule cortical array organization. Plant Physiol. 147: 1723-1734.

Park, Y.B. and D.J. Cosgrove. 2015. Xyloglucan and its interaction with other components of the growing cell wall. Plant Cell Physiol 56: 180-194.

Roberts, A.W., A.O. Frost, E.M. Roberts and C.H. Haigler. 2004. Roles of microtubules and cellulose microfibril assembly in the localization of secondary cell wall synthesis in developing tracheary elements. Protoplasma 224: 217-229.

Roberts, A.W., E.M. Roberts and C.H. Haigler. 2012. Moss cell walls: structure and biosynthesis. Front. Plant Sci. 3: 166.

Sampathkumar, A., A. Yan, P. Krupinski and E.M. Meyerowitz. 2014. Physical forces regulate plant development and morphogenesis. Curr. Biol. 24: R475-R483.

Sanchez-Rodriguez, D., S. Bauer, K. Hematy, F. Saxe, A.B. Ibanez, V. Vodermaier, C. Konlechner, A. Sampathkumar, M. Ruggegerg, E. Aichinger, L. Neumetzler, I. Burgert, C. Somerville, M.-T. Hauser and S. Persson. 2012. CHITINASE-LIKE1/POM-POM1 and its homology CTLS are glucan-interacting proteins important for cellulose biosynthesis in Arabidopsis. Plant Cell 24: 589-607.

Sethaphong, L., C.H. Haigler, J.D. Kubicki, J. Zimmer, D. Bonetta, S. DeBolt and Y.G. Yingling 2013. Tertiary model of a plant cellulose synthase, Proc. Natl. Acad. Sci. USA 110: 7512-7517.

Slabaugh, E., J.K. Davis, C.H. Haigler, Y.G. Yingling and J. Zimmer. 2014. Cellulose synthases: new insights from crystallography and modeling. Trends Plant Sci. 19: 99-106.

Sorek, N., H. Sorek, A. Kijac, H.J. Szemenyei, S. Bauer, K. Hematy, D.E. Wemmer and C.R. Somerville. 2014. The Arabidopsis COBRA protein facilitates cellulose crystallization at the plasma membrane. J. Biol. Chem. 289: 34911-34920.

Sugimoto, K., R.E. Williamson and G.O. Wasteneys. 2000. New techniques enable comparative analysis of microtubule orientation, wall texture, and growth rate in intact roots of Arabidopsis. Plant Physiol. 124: 1493-1506.

Szymanski, D.B. and D.J. Cosgrove. 2009. Dynamic coordination of cytoskeletal and cell wall systems during plant cell morphogenesis. Curr. Biol. 19: R800-R811.

Taylor, N. G. 2007. Identification of cellulose synthase AtCesA7 (IRX3) *in vivo* phosphorylation sites—a potential role in regulation protein degradation. Plant Mol. Biol. 64: 161-171.

Tewari, R., E. Bailes, K.A. Bunting and J.C. Coates. 2010. Armadillo-repeat protein functions: questions for little creatures. Trends Cell Biol. 20: 470-481.

Thomas, L.H., V.T. Forsyth, A. Martel, I. Grillo, C.M. Altaner and M.C. Jarvis. 2014. Structure and spacing of cellulose microfibrils in woody cell walls of dicots. Cellulose 21: 3887-3895.

Timmers, J., S. Vernhettes, T. Desprez, J.P. Vincken, R.G. Visser and L.M. Trindade. 2009. Interactions between membrane-bond cellulose synthases involved in the synthesis of the secondary cell wall. FEBS Lett. 583: 978-982.

Tsekos, I. 1999. The sites of cellulose synthesis in algae: diversity and evolution of cellulose-synthesizing enzyme complexes. J. Phycol. 35: 635-655.

Turner, S.R., N. Taylor and L. Jones. 2001. Mutations of the secondary cell wall. Plant Mol. Biol. 47: 209-219.

Tuttle, J.R., G. Nah, M.V. Duke, D.C. Alexander, X. Guan, Q. Song, Z.J. Chen, B.E. Scheffler and C.H. Haigler. 2015. Metabolomic and transcriptomic insights into how cotton fiber transitions to secondary wall synthesis, represses lignification, and prolongs elongation. BMC Genomics 16: 477.

Vain, T., E.F. Crowell, H. Timpano, E. Biot, T. Desprez, N. Mansoori, L.M. Trindade, S. Pagant, S. Robert, H. Hofte, M. Gonneau and S. Vernhettes. 2014. The cellulase KORRIGAN is part of the cellulose synthase complex. Plant Physiol. 165: 1521-32.

Wallace, I.S. and C.R. Somerville. 2015. A blueprint for cellulose biosynthesis, deposition, and regulation in plants. pp. 65-95. *In*: H. Fukuda (ed.). Plant Cell Wall Patterning and Cell Shape. John Wiley and Sons, New York.

Wang, J., J.E. Elliott and R.E. Williamson. 2008. Features of the primary cell wall CESA complex in wild type and cellulose-deficient mutants of Arabidopsis thaliana. J. Exp. Bot. 59: 2627-2637.

Worden, N., T.E. Wilkop, V.E. Esteve, R. Jeannotte, R. Lathe, S. Vernhettes, B. Weimer, G. Hicks, J. Alonso, J. Labavitch, S. Persson, D. Ehrhardt and G. Drakakaki. 2015. CESA TRAFFICKING INHIBITOR inhibits cellulose deposition and interferes with the trafficking of cellulose synthase complexes and their associated proteins KORRIGAN1 and POM2/CELLULOSE SYNTHASE INTERACTIVE PROTEIN1. Plant Physiol. 167: 381-393.

Yang, H., J. Zimmer, Y.G. Yingling and J.D. Kubicki. 2015. How cellulose elongates - A QM/MM study of the molecular mechanism of cellulose polymerization in bacterial CESA. J. Phys. Chem. B 119: 6525-6535.

Yoneda, A., T. Ito, T. Higaki, N. Kutsuna, T. Saito, T. Ishimizu, H. Osada, S. Hasezawa, M. Matsui, and T. Demura. 2010. Cobtorin target analysis reveals that pectin functions in the deposition of cellulose microfibrils in parallel with cortical microtubules. Plant J. 64: 657-667.

Zhang, T., S. Mahgsoudy-Louyeh, B. Tittmann and D.J. Cosgrove. 2014. Visualization of the nanoscale pattern of recently-deposited cellulose microfibrils and matrix materials in never-dried primary walls of onion epidermis. Cellulose 21: 853-862.

Zhao, C., U. Avci, E.H. Grant, C.H. Haigler and E.P. Beers. 2007. XND1, a member of the NAC domain family in Arabidopsis thaliana, negatively regulates lignocellulose production and programmed cell death in xylem. Plant J. 53: 425-436.

Zhong, R., W.H. Morrison III, G.D. Freshour, M.G. Hahn and Z.-H. Ye. 2003. Expression of a mutant form of cellulose synthase AtCesA7 causes dominant negative effect on cellulose biosynthesis. Plant Physiol. 132: 786-795.

Zhong, R. and Z.-H. Ye. 2015. Secondary cell walls: Biosynthesis, patterned deposition and transcriptional regulation. Plant Cell Physiol. 56: 195-214.

9

Signaling-Dependent Cytoskeletal Dynamics and Plant Cell Growth

Stefano Del Duca[1] *and Giampiero Cai*[2,*]

Introduction

In all eukaryotic cells, the cytoskeleton is an essential protein-based matrix that determines the cell's internal order. Building a cell requires necessarily an organizational support able to provide the order and thus to regulate the various cellular functions. Classically, the cytoskeleton comprises a heterogeneous group of proteins that interact with each other to form a dynamic network that adapts to the different physiological conditions of cells. The cytoskeleton receives information as decoded by the signal transduction system and conveys them in a response that might include the moving of specific cellular components, cell division or changes in cell shape.

In plant cells, the appearance of the cell wall has dictated some limitations not found in animal cells and has affected the structure of the cytoskeleton by modifying both organization and composition of cytoskeleton proteins. The cytoskeleton also takes active part in the construction of the cell wall; because the cell wall defines the shape of a plant cell, it follows that the cytoskeleton actively participates in the shaping process. Although the cell wall is a semi-rigid structure that severely limits cell motility, the cytoskeleton is not static but is a finely-tuned highly dynamic matrix that reorganizes in response to external cues. The differences that evolved in the plant cell cytoskeleton can be macroscopic (for example, the disappearance of intermediate filaments) or more difficult to appreciate (i.e. the substitution of a few amino acids in highly conserved proteins). Although the presence of the cell wall has imposed some restrictions to both cell motility and shaping, it did not affect specific molecular repertoire or processes, e.g. neither the number of genes coding for cytoskeletal proteins nor the number of isoforms or post-translational modifications to which cytoskeletal proteins are subjected.

Although the list of cytoskeletal genes and proteins in plant cells is currently lower than that of animals, the ongoing sequencing of plant genomes can progressively increase the number of available information. The progressive identification of

[1] Dipartimento di Scienze Biologiche, Geologiche e Ambientali, University of Bologna, via Irnerio 42 Bologna, Italy.
[2] Dipartimento di Scienze della Vita, University of Siena, via Mattioli 4, Siena, Italy.
[*] Corresponding author : giampiero.cai@unisi.it

cytoskeletal proteins in plants suggested the existence of proteins almost identical to those of animal cells but also of proteins that are typical of plant cells. For this reason, understanding the various functions of cytoskeletal proteins may not be simple. The use of information available for animal or fungal cells does not imply that the protein counterparts of plant cells share the same function. In the case of exclusive plant cell proteins, identification of the function can be even more complicated (Meagher and Fechheimer 2003).

Another elusive aspect of plant cells is the dynamic regulation of the cytoskeleton. This feature is critical for understanding how the cytoskeleton adapts to external environmental stimuli. The relationship between environmental stress and plant cell response is an important topic considering the plausible modification of some environmental parameters, such as temperature (Ruelland and Zachowski 2010). Correctly interpreting the external signals and information can allow a plant cell either to define new building schemes of the cell wall (therefore a new final shape) or to divide into two daughter cells. This process can lead to substantial modification of the shape of tissues or organs in plants.

In this chapter, we describe the current state-of-art of the dynamic regulation of the plant cell cytoskeleton by focusing on some specific facets. We will describe how external signals (i.e. hormones) affect the cytoskeleton structure by interacting with signal transduction mechanisms such as GTP-binding proteins and phospholipids. Next, we will describe how the signal transduction system might regulate the assembly of new actin filaments and microtubules. This preliminary step consequently affects the construction of ordered structures, such as bundles, and it is a prerequisite for the molecular dialogue and for the functions involving an interplay between actin filaments and microtubules. This review will not cover other specific aspects (such as the intracellular transport of organelles or the response to gravitropism) that are discussed in other reviews (Perbal and Driss-Ecole 2003).

Extracellular signal for the cytoskeleton

Plant cells are constantly facing a changing environment and they possess many receptors by which they sense the presence of different physical and chemical signals. Here, we will take into account the hormonal-based signaling that can induce structural changes in the plant cell cytoskeleton. Because hormonal signals work by using different types of second messengers, we focus on those based on membrane phospholipids, whose role in the regulation of the plant cytoskeleton is much more known.

Hormone-based signaling

Auxin is one of the most important regulators of plant growth. Levels of auxin are controlled by the activity of specific auxin transporters, which cycle between the cell cytoplasm and the plasma membrane; recycling of transporters depends on the integrity of actin filaments (Nick et al. 2009). In turn, auxin participates in the organization of actin filaments most likely through the activity of ROP (Rho-of-plants) proteins. Auxin can direct the organization of actin filaments (from strands to bundles) most likely on the basis of the cellular developmental program (Holweg et al. 2004). Internalization of auxin is specifically dependent on the transporter AUX1, whose levels are closely related to the organization of actin filaments. The dialogue between auxin and actin filaments is expected to require ARF-GAP (adenosine diphosphate ribosylation factor-GTPase-activating protein) (Du and Chong 2011). Actually, ARF-GAP does not only relate to the organization of actin filaments but this protein is most likely involved in the vesicular transport that mediates the insertion of the auxin effluxer, PIN (PINFORMED), into the plasma membrane (Du et al. 2011). Therefore, auxin levels are regulated both by the AUX1-based internalization and by the PIN-based efflux. The positioning of PIN in the plasma membrane is dependent on microtubules (Boutte et al. 2006) and most likely on a second group of molecules, namely brassino-steroids. The latter are capable of regulating actin filaments (Lanza et al. 2012) as well as microtubules (Knowles et al. 2004), suggesting the existence of an extensive cross-talk between auxin, brassinosteroids, actin filaments and microtubules, which regulates synergistically the hormone-based organization of the cytoskeleton. The activity of brassinosteroids on microtubules could be mediated by specific proteins such as MICROTUBULE DESTABILIZING PROTEIN40 (MDP40) (Wang et al. 2012).

Another important mechanism of hormone-based signaling is linked to ethylene, which is able to regulate many aspects of plant shaping (such as branching of trichomes). The activity of ethylene is closely linked to the ETR2 (ETHYLENE RESPONSE) receptor, which in turn interfaces with the microtubule organization (Plett et al. 2009) thereby suggesting yet again a cross-talk between hormone levels, microtubules and cell shaping. The hormone abscisic acid also participates in the organization of actin filaments and therefore in the regulation of cell shape such as in guard cells of *Commelina communis* L. (Eun et al. 2001).

ROP-based dynamics of actin filaments

ROP proteins are an important class of molecular switches involved in cell responses by performing a wide range of functions, including the organization of actin filaments in response to external signals. It is likely that the activity of hormones is achieved through the activation of ROP proteins, which in turn interface with additional proteins that direct the assembly or elongation of actin filaments. Among the latter, the Arp 2/3 complex will be discussed later. The use of specific mutants helped to clarify how ROP proteins regulate the dynamics of actin filaments. A constitutively active form of ROP helps redistributing actin filaments in the form of a more widespread network thereby leading cells to isotropic growth (Fu et al. 2002, Zhao and Ren 2006). It is therefore suggested that the switching between "on" and "off" forms of ROP determines the balance between interspersed actin bundles and more rigid actin strands. The key role of ROPs in the modulation of actin filaments was demonstrated in apical growing cells such as pollen tubes (Fu et al. 2001). In these cells, ROPs can activate two separate pathways, one leading to the accumulation of Ca^{2+} (and presumably to Ca^{2+}-dependent processes of actin filament fragmentation) and the other leading to the nucleation of new actin filaments. The two processes are mutually exclusive and only a correct balance enables the coordinated growth of pollen tubes (Hwang et al. 2005). The activity of ROPs is most likely required to focus the Ca^{2+} gradient in apical growing cells by marking the extending cell areas (Molendijk et al. 2001).

ROPs do not regulate actin filaments only but they also participate in the regulation of microtubules by different intermediate proteins such as RIC1 [CRIB (Cdc42/Rac-interactive binding) motif-containing ROP-interacting protein 1]. In leaf pavement cells of *Arabidopsis thaliana* (L.) Heynh., RIC1 regulates the organization of microtubules but at the same time it antagonizes ROP2 thereby feeding back on the assembly of microtubules and actin filaments; this leads to the typical interdigitated leaf pavement cells (Fu et al. 2005) (see also the chapter's end). As anticipated above, the ROP-based system most likely mediates the auxin signaling to microtubules. In Arabidopsis, the activity of auxin requires the auxin-binding protein ABP1 and downstream systems comprising a ROP protein, the RIC1 intermediate and katanin, a microtubule-binding protein that cut microtubules by generating free ends (Chen et al. 2014). In leaf epidermal pavement cells, auxin may also regulate its own levels by controlling the endocytosis of the auxin carrier PIN1; this regulatory activity is achieved through the ROP2-based organization of actin filaments (Nagawa et al. 2012). Collectively, the evidence suggests that hormones can interface with the ROP switches and that they could regulate microtubule dynamics through proteins modulating the polymerization status of the cytoskeleton.

The protein system linking ROP to the nucleation of new actin filaments is beginning to be outlined. There are presumably several intermediate proteins between ROP and actin filaments. One such protein is WAVE/SCAR (for WASP family verprolin homologous/suppressor of cyclic AMP receptor) while a second system is based on Arp 2/3 (Actin-related protein 2/3). The link between ROP and the WAVE – Arp 2/3 complex is represented by the guanine nucleotide exchange factor (GEF) SPK1, which is necessary to activate ROP. SPK1 is a dock homology region 2 (DHR2)-type of the Dock family of Rho guanine nucleotide exchange factors and is also associated with WAVE (Basu et al. 2008); this could help in defining a protein scaffold involving protein switches (ROP) and proteins that directly mediate the assembly of new actin filaments. Available data in the literature indicate that WAVE interacts with some specific subunits of the Arp 2/3 complex thereby triggering the nucleation of new actin filaments (Zhang et al. 2005). The dynamics of actin filaments is not only under the supervision of ROP proteins because proteins of the Rac subfamily are also involved (Bloch et al. 2005).

ROP signaling and calcium flux

ROPs mediate a variety of downstream effects, some of which require changes in the level of cytosolic calcium. The relationship between ROP and calcium is more intricate than one might assume from current data and it most likely requires the involvement of a number of factors, including NO (nitric oxide) and ROS (reactive oxygen species). In addition, ROS can also affect the dynamics of actin filaments thereby linking ROPs to the cytoskeleton. In apical growing cells, NO levels are also in close relationship with the concentration of cytosolic calcium. In particular, NO donor factors increase calcium levels while inhibitors of NO synthesis have the opposite effect. In both cases, actin filaments are clearly damaged thereby affecting all actin-dependent processes (Wang et al. 2009).

The relationships between calcium and actin filaments are relatively complex and probably bidirectional. The pollen tube is probably the cell model in which these relationships were mostly investigated. It was therefore realized that the intracellular calcium levels and the dynamics of actin filaments are mutually interactive; actually, damages to actin filaments mirror the oscillation of calcium levels probably through the activity of proteins (such as pumps and channels) that regulate the influx/efflux of calcium ions into pollen tubes (Cardenas et al. 2008). In addition, calcium ions may exert an inhibitory activity on actin filaments through calcium-binding proteins such as calmodulin (Yang et al. 2014). The mutual interplay between actin filaments and calcium is also modulated by the intracellular pH, which can trigger specific proteins like actin depolymerizing factors (ADFs); the increase in intracellular pH specifically activates the ADF-based fragmentation of actin filaments thereby generating free ends for the assembly of new actin filaments (Lovy-Wheeler et al. 2006). More information about ADFs will be discussed later.

The phospholipid transduction pathway

Despite the integration with intracellular calcium, ROPs exert a regulatory activity on actin filaments through a second type of biochemical pathway based on the cycle of phosphoinositide (PI). The use of membrane lipids is a tool to link the activity of cytosolic proteins to cell membranes by promoting the formation of anchoring points. The interaction between ROP and the PI cycle is very complex and probably mediated by proteins such as AGD1, a class 1 adenosine diphosphate ribosylation factor-GTPase-activating protein (ARF-GAP) (Yoo et al. 2012). Considering the several PI-based molecules, identifying the exact role of each PI phosphorylated-variant and of the inter-converting enzymes is not simple. This type of study is generally addressed using mutants and inhibitors of phosphorylation. These approaches revealed that the conversion from PI to PI(3)P or PI(4)P is critical in the regulation of actin filament dynamics in guard cells (Choi et al. 2008). Within the PI cycle, an important role is played by the enzyme PIP5 kinase, which generates PI(4, 5)P_2, a molecule that regulates the proper assembly of actin filaments. Both PI(4, 5)P_2 and its synthetizing enzyme are specifically located in regions characterized by high dynamics of actin filaments, such as the apical region of root hairs and pollen tubes (Kusano et al. 2008). The role of PI(4, 5)P_2 is most likely linked to proteins such as Rac5. In pollen tubes of *Nicotiana tabacum* L., PI(4, 5)P_2 seems to participate to the Rac5-based organization of actin filaments perhaps by affecting the correct distribution of Rac5 in the apical plasma membrane (Ischebeck et al. 2011).

A second lipid-based mechanism through which cells regulate the cytoskeleton activity is linked to phospholipases and their products. Phospholipase D (PLD) is a plasma membrane enzyme whose relationship with microtubules is beginning to be outlined. Activity of PLD is regulated by molecules such as n-butanol or by various stress conditions; therefore, PLD can regulate the dynamics of microtubules in the cell cortex (Dhonukshe et al. 2003). PLD activity is mediated by the synthesis of phosphatidic acid (PA), which controls the reorganization of microtubules. The mechanism by which PA might regulate microtubule dynamics is unclear but it is worthy of mention that PA can interact with MAP65-1 (Zhang et al. 2012) thus promoting the association of MAP65-1 with microtubules and the bundling of microtubules. In doing so, PA can promote an ordered organization of microtubules in the cell cortex. PLD could also interact directly with microtubules thereby functioning either as membrane anchor or as a protein capable of organizing microtubules (Andreeva et al. 2009). In this context, the precise relationship between PLD, PA and microtubules has not yet defined and should be clarified further, for example through the use of inhibitors such as N-Acylethanolamines (NAEs) (Motes et al. 2005). Although PLD might have a role in the reorganization of microtubules, its activity in the modulation of actin filaments is not comparably clear. Some reports indicate that PLD can directly bind to actin filaments and G-actin, suggesting that PLD might link actin filaments to the plasma membrane (Pleskot et al. 2010), with filamentous actin that simultaneously promotes PLD activity (Li et al. 2012). Studies in Arabidopsis suspension cells showed that PA can negatively modulate the activity of "capping proteins" (CP), whose role is to protect the depolymerizing end of actin filaments. Thus, PA might regulate the dynamics of actin filaments suggesting that PLD can act at the level of both actin filaments and microtubules (Huang et al. 2006). Since CPs also interact with PI(4, 5)P$_2$, CPs could be a critical link between the two signaling pathways based on the biochemistry of phospholipids. The molecular relationships between PLD, PA and actin are perhaps more complex because F-actin can also act as an activator of PLD; at the same time, production of PA can inhibit the CP activity then stimulating the shrinking of actin filaments and thus potentially leading to inhibition of PLD. This will generate a feedback that can be used to finely adjust the extension of cytoskeletal arrays (Pleskot et al. 2010). Thus, PLD is as candidate for the interplay between microtubules and actin filaments in the cell cortex.

Another important class of phospholipase is Phospholipase C (PLC). PLC acts in a way relatively different from PLD since it participates in the degradation of PI(4, 5)P$_2$. As mentioned earlier, PI(4, 5)P$_2$ is a powerful modulator of actin filament dynamics because it regulates the intracellular calcium levels as shown in pollen tubes (Dowd et al. 2006). In doing so, PLC would act indirectly by modulating the "universal" intermediate calcium ions. Actually, PLC could restrict the distribution of PI(4, 5)P$_2$ which in turn could lead to the production of localized calcium gradients. Other evidence suggests that cytosolic PLC can be associated with the actin

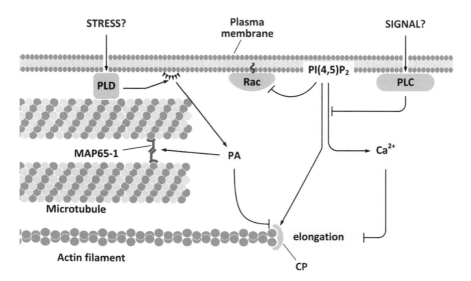

FIGURE 1   Schematic diagram of the phospholipase-based mechanism of cytoskeleton regulation. The main activities involving phospholipase D (PLD) and phospholipase C (PLC) are depicted as well as some of the main intermediate factors. PA, phosphatidic acid; CP, capping proteins. The diagram illustrates the central role of PA in modulating both microtubules and actin filaments. Even though not shown here, PLD could also bind to actin filaments and G-actin, thereby linking actin filaments to the plasma membrane; filamentous actin could also promote PLD activity. See the text for more details.

cytoskeleton; this finding increases the complexity of the system, suggesting that the mechanism linking PLC-based signals to actin is relatively multifaceted (Huang and Crain 2009). The several mechanisms of integration concerning PLD and PLC are schematically depicted in Fig. 1.

Phosphorylation

The molecular systems based on the modulation of cytosolic calcium levels and on the interconversion between different membrane phospholipids can be integrated by further mechanisms based on the activity of intracellular kinases and phosphatases. In this context, information are more limited if compared to what is known on the dynamics of calcium gradients and on the biochemistry of membrane phospholipids. Phosphorylation is involved in myriad of cellular processes, from the regulation of gene expression to metabolism to changes of cell shape. It is therefore not surprising that phosphorylation also takes part in the cytoskeletal dynamics.

Using inhibitors of phosphorylation, it was found that correct intracellular levels of phosphorylation might regulate the organization of actin filaments and microtubules in lily pollen tubes. As result of altered kinase/phosphatase activities, the cytoskeleton was damaged leading to highly disorganized growth of the cell. Hypothetically, phosphorylation can interface with other signaling processes, such as those mediated by ROP proteins (Foissner et al. 2002). This assumption is supported by evidence that mutations in genes coding for specific MAPK (mitogen-activated protein kinases) phosphatases can compromise the entire organization of microtubules thereby injuring the regular growth of cells (Naoi and Hashimoto 2004). Since MAPK kinases/phosphatases are intermediate in the signal transduction system mediated by membrane GTPases, this suggests that intracellular phosphorylation can link the ROP-based signal transduction pathway to the cytoskeleton architecture. Phosphorylation may affect specific amino acids such as tyrosine and therefore may require specific types of intracellular kinases. The reversibility of the process as shown by specific inhibitors indicates that phosphorylation of proteins involved in cytoskeleton dynamics is an effective mechanism of regulation (Zi et al. 2007).

Signaling-dependent nucleation of actin filaments

External signals (both chemical-physical and biological) may have different effects on the dynamics and organization of the cytoskeleton. One of the main effects is the synthesis of new actin filaments in specific cell areas. New actin filaments can be used for conveying cellular organelles into specific regions in order to polarize the plant cell causing its anisotropy. At present, two main systems for the nucleation of actin strands in plant cells have been characterized, one based on formins and the second based on the Arp 2/3 complex.

Formin

Formins are proteins that make bundles of unbranched actin filaments. In plants, the role of formins is not clear. The structure of plant formins is partially different from the animal counterparts and the presence and role of different types of formin should be further clarified; for example, in the moss *Physcomitrella patens*

(Hedw.) Bruch & Schimp there are at least 9 different genes coding for formins but only one is likely critical for cell development (Vidali et al. 2009). Formins usually contain two subunits that bind to the (+) end of actin filaments thus allowing the sequential addition of new actin subunits. In Arabidopsis, a formin-like protein may nucleate actin filaments in cooperation with profilin, suggesting that the assembly of new actin filaments requires the functional cooperation between different actin-binding proteins (Yi et al. 2005). Profilins are low molecular weight proteins (10–15 kDa) that bind to ATP-G-actin with 1:1 stoichiometry and form moderate affinity profilin–actin complexes. In pollen, profilin is present at levels equimolar to total actin and corresponds to 100–125 micromolar cellular concentration (Gibbon et al. 1999). As regard to the role of profilin in actin polymerization, profilin can bind to and sequester actin monomers, thereby decreasing the concentration of free actin monomers available for filament elongation. Profilins replenish the pool of ATP–actin monomers by increasing the rate of nucleotide exchange on the bound actin monomer. The profilin–ATP–actin complex interacts with the fast growing end of the actin filament and releases the ATP–actin monomer, which is added to the filament. Therefore, the elongating filament consists of ATP–actin. Along the filament, ATP is slowly hydrolyzed by the intrinsic ATPase activity of actin, which generates ADP–actin in the older part of the filament; then ADP–actin can be released slowly from the filament end by depolymerization. Moreover, profilin has a region enriched in proline and this region could interact with formin, which is the main nucleation factor for actin filaments. By its proline-enriched region, profilin drives ATP-actin to the membrane region where formin is present thereby allowing the nucleation of actin filaments (Witke 2004).

The ability of formins to nucleate new actin bundles was demonstrated in Arabidopsis pollen tubes, where the over-expression of formin 3 (AFH3) triggered the appearance of extended longitudinal actin bundles while the down-expression of AFH3 causes the disappearance of actin bundles and the interruption of intracellular transport. In this case, formin is presumably associated with the plasma membrane of pollen tubes and facilitates the synthesis of new actin bundles from the membrane (Ye et al. 2009). In Arabidopsis, a different type of formin (AFH1), whose structure is relatively different from other formins, can bind to the plasma membrane by virtue of a transmembrane domain. This formin is most likely responsible for the nucleation of new actin bundles at the apical region of pollen tubes; in doing so, formin can link the assembly of actin bundles to the process of pollen tube growth (Cheung and Wu 2004). The association between formins, cell signaling and cell growth is suggested by evidence linking specific formins to the localization of the auxin carrier, PIN. In rice cells, the protein RMD (a specific type of formin) mediates the relationships between auxin and actin by favoring the recycling of PIN in the plasma membrane (Li et al. 2014). Auxin may not be the only signal that affects the formin-dependent nucleation of actin filaments. Specific formins can also interact with the membrane phospholipid $PI(3,5)P_2$ by the PTEN domain (van Gisbergen et al. 2012). Thus, the cycle of membrane phospholipids can additionally interface with the dynamics of actin filaments by regulating the distribution of formins in the plasma membrane and then the local nucleation of new actin filaments.

Formins can probably play additional roles. Specific formins, such as the Arabidopsis FH8, are likely involved in the process of cell division and distribute differently (nuclear vs plasma membrane) in relation to the cell cycle phase (Xue et al. 2011). The formin FH5 is capable of nucleating new actin filaments in Arabidopsis (Cheung et al. 2010) but the rice homolog may also take part in the assembly of microtubule bundles. In this way, formins can interconnect the assembly of new actin bundles to the organization of microtubules and may represent a molecular link between the two cytoskeletal filaments (Zhang et al. 2011). Similar findings were observed for the FH1 Arabidopsis formin; in this case, an altered expression of formin induces alterations in the actin filament dynamics but also produces changes in the microtubule dynamics (Rosero et al. 2013). As further confirmation, the FH16 Arabidopsis formin can bind more preferentially to microtubules than to actin filaments; this formin seems unable to nucleate actin filaments but it can induce the development of microtubule bundles (Wang et al. 2013). Therefore, it seems clear that plant formins have evolved to perform different roles compared to animal formins by also molecularly linking and interfacing actin filaments to microtubules.

Arp 2/3

The Arp 2/3 complex is involved in the nucleation of actin filaments at the minus end. It is a multiprotein complex whose regulation is partially known; it possibly links the nucleation of actin filaments to the signal transduction system. The Arp 2/3 complex consists of seven subunits and their genes have been identified in Arabidopsis; consequently, the analysis of mutants provided useful information on the role of Arp 2/3 and, more specifically, of its constituent subunits in the dynamics of actin filaments (Li et al. 2003). Analysis of other organisms provided additional information. For example, in the moss *Physcomitrella patens*, the Arp3a subunit is critical in the organization of actin filaments and mutations in the encoding gene strongly affected the actin filament array (Finka et al. 2008). The ArpC1 subunit of *Physcomitrella patens* is also critical for the proper assembly of actin filaments (Harries et al. 2005). This information collectively suggests that the Arp 2/3 subunits play important roles and that mutations in each subunit may lead to profound alterations in the remodeling of the actin cytoskeleton. In Arabidopsis, the ArpC2 subunit appears critical for cellular morphogenesis and mutations in the corresponding gene can result in critical defects at the level of trichomes and cell-cell contacts. ArpC2 interacts physically with the ArpC4 subunit (El-Assall et al. 2004). Further studies in Arabidopsis also showed that the ArpC4 subunit can be the core around which the Arp 2/3 complex is assembled (Kotchoni et al. 2009). Further investigation with mutants allowed the identification of ArpC5, the smallest subunit of the Arp 2/3 complex (Mathur et al. 2003).

Despite the conserved structure of Arp 2/3, a remarkable question concerns the cellular distribution of such a complex. It is conceivable that the Arp 2/3 complex interacts with cell membranes so that it might have a role in the nucleation of actin filaments at the level of intracellular membrane-bounded organelles. Some evidence provides indications (or assumptions) about the cellular sites where the plant Arp 2/3 complex is distributed. In addition, results suggest that the binding of Arp 2/3 with membranes is dependent on its assembly condition but not on its activation status; therefore, the Arp 2/3 complex is likely assembled at the level of membranes and then activated. The cellular localization of the Arp 2/3 complex is an interesting research topic because it can affect the nucleation of actin filaments at specific cell sites where new actin arrays might be required. In BY-2 cells of tobacco, the Arp3 subunit is located on the nuclear surface or in the cell cortex; this subunit also co-localizes with the nucleation sites of new actin filaments. Arp3 also aggregates considerably in areas with polarized growth, thereby suggesting that the Arp3 subunits (and then the Arp 2/3 complex) can be distributed at sites requiring new or different actin arrays (Maisch et al. 2009).

In addition to its role in the nucleation of actin filaments, analysis of mutants revealed that Arp 2/3 could be involved in additional roles, such as the actin-dependent organization of microtubules. This finding suggests that the nucleation of new actin filaments may represent the molecular basis to correctly orient microtubules (Saedler et al. 2004a). The existing cooperation between actin filaments and microtubules might also extend beyond Arp 2/3 because proteins homologous to the ABI protein (a component of the WAVE/Scar complex) have been observed in association with microtubules. This evidence indicates that the relationship between actin and tubulin during the nucleation of actin filaments is very complex and to be further clarified (Joergens et al. 2010).

The Arp 2/3 complex is presumably regulated by a second complex, WAVE, which in turn contains several proteins (homologs of the animal counterparts WAVE/SCAR proteins, PIR121, Nap125, Abi-2, HSPC300). The specific role of these subunits can be outlined. For example, the Arabidopsis genome contains homologs to Nap and Pir and the altered pattern of actin filaments in mutants defective in Nap/Pir expression is consistent with a role for WAVE as an Arp 2/3 activator (Li et al. 2004). BRK1, which is the HSPC300 counterpart, is most likely involved in the correct assembly

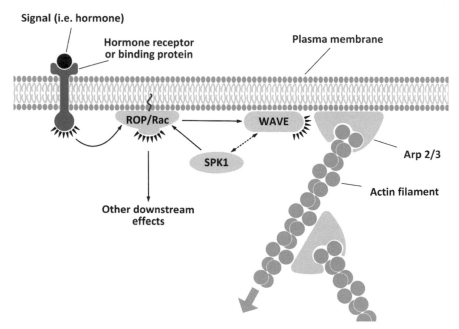

FIGURE 2 Schematic representation of the putative interactions between ROP and the Arp 2/3-based nucleation of actin filaments. Once activated by external signals, putative receptors may trigger the activation of ROP proteins. Regulation of ROP might be also dependent on proteins like SPK1, which is a guanosine exchange factor. In turn, activated ROP may interface with the WAVE/Scar complex, the latter working as an activator of the actin filament nucleator Arp 2/3. See the text for more details.

of the WAVE complex by accumulating the Scar protein. Defects in the expression of the BRK1 gene caused profound changes in the organization of actin filaments by preventing the correct assembly of the WAVE complex (Djakovic et al. 2006) and then the correct activation of the Arp 2/3 complex. Relationships between the subunits of the WAVE complex are much more difficult to infer; for example, in epidermal leaf cells of Arabidopsis, accumulation of the BRK1 subunit is dependent on Scar, suggesting that it might exist in a mutual relationship between the Scar and the BRK1 subunits. In addition, both proteins are associated with the plasma membrane and accumulate in cell sites active in growth and cell wall deposition (Dyachok et al. 2008). Therefore, the WAVE complex is important in active cell sites by probably triggering the local assembly of an actin network that mediates both growth and cell wall deposition.

As reported in animal cells, the WAVE complex could in turn be activated/regulated by proteins that link ROP to the WAVE transducer. The SRA1 homolog in Arabidopsis could play this role (Saedler et al. 2004b) by interfacing between ROP and WAVE and then linking the signal transduction system to the Arp 2/3 complex and consequently to the nucleation of actin filaments. Fig. 2 reports a scheme of proposed and hypothetical interactions existing at the level of the Arp 2/3 complex.

ROP-based organization of microtubules

The relationship between microtubules and the signal transduction system is less clear with respect to actin filaments. Evidence in the literature indicates that the effect of ROP proteins on microtubules could be relatively complex. Studies on ROP2 proteins indicated that these molecular switches could regulate both actin filaments and microtubules in an opposite way. Some evidences suggest that ROP2 can function as a positive regulator of actin filaments through the activation of the RIC4 intermediate. Therefore, when ROP2 is active, RIC4 induces the polymerization of actin filaments. At the same time, ROP2 can also function as an inhibitor of RIC1, which is in turn a positive regulator of microtubule polymerization. Therefore, when active, ROP2 could enhance the polymerization of actin filaments but decrease the polymerization of microtubules. At the same time, RIC1 is also an inhibitor of ROP2 suggesting that the entire system can be auto-regulated in such a way as to favor alternative conditions (i.e. polymerization of actin and depolymerization of microtubules or vice versa). This process is particularly characterized in the interdigitated pavement cells of the epidermis (Fu et al. 2005).

The regulation of microtubules by ROP proteins could be mediated by additional factors, such as microtubule motor proteins. Several pieces of evidence suggest that specific members of the kinesin superfamily can take part in the regulation of microtubules following activation by ROP. This is the case of the MRH2 (MORPHOGENESIS OF ROOT HAIR2) kinesin (whose activity is supposedly enhanced by ROP2), which is able to exert a positive effect on microtubules. The MRH2 kinesin is also likely capable of binding to actin filaments thereby making a direct link between actin filaments and microtubules; this suggests that actin filaments might regulate the polymerization of microtubules by using motor proteins (Yang et al. 2007). In this case, the activity of ROP on microtubules could be either direct through the MRH2 kinesin or indirect through the regulation of the actin filaments-MRH2 binding.

Another example of the role exerted by ROP concerns the motor protein kinesin 13A (KIN13A). This member of the kinesin superfamily is a negative regulator of microtubule length by playing a depolymerizing activity. KIN13A is probably activated in turn by a ROP effector, RIP3 (ROP Interacting Protein), which also binds to the microtubule surface. It is not known whether RIP3 and KIN13A form a complex on the microtubule surface but it could be an interesting aspect to analyze further. The molecular system is probably more complex than expected because other proteins, such as the plasma membrane-microtubule connector MIDD1 (microtubule depletion domain 1), are required to recruit KIN13A on microtubules and to accomplish the depolymerizing activity of KIN13A. In this case, KIN13A is required to regulate the local deposition of the cell wall (Mucha et al. 2010, Oda and Fukuda 2013).

Perspectives

Mounting evidence indicates that the cytoskeleton can interact with a scaffolding protein system located in the plasma membrane and that this protein network mediates the decoding of external cues into cytoskeleton-based responses. This decoding activity is of primary importance because it interfaces the external environment with the cell's interior. It is therefore critical to be capable of deciphering both the composition and the functioning of such a scaffolding transduction system. Efforts at proteomic, genomic and cytological level have shed some light on the fundamentals of this interplay. Nevertheless, we still miss many details on several aspects. First, we need an updated list of proteins and small molecules involved in the signal transduction interplay. Although this may be only dependent on time and technical skills, the most critical effort needs to be implemented in understanding the dynamic relationships between all molecular players involved. The formin- and PA-based communication systems are clear examples of the complexity existing between actin filaments and microtubules. It is evident that a certain regulatory system might not be exclusive for one specific cytoskeleton network and that a reciprocal regulation is clearly expected. This is not surprising because the final behavior and response of a plant cell is strictly dependent on the coordinated activity of both actin filaments and microtubules. It is therefore expected that a large research activity in the near future will be directed to deciphering how the interplay between actin filaments and microtubules is mediated by external cues and by the plasma membrane-associated transducing system.

References

Andreeva, Z., A.Y. Ho, M.M. Barthet, M. Potocky, R. Bezvoda, V. Zarsky and J. Marc. 2009. Phospholipase D family interactions with the cytoskeleton: isoform delta promotes plasma membrane anchoring of cortical microtubules. Funct. Plant Biol. 36: 600-612.

Basu, D., J. Le, T. Zakharova, E.L. Mallery and D.B. Szymanski. 2008. A SPIKE1 signaling complex controls actin-dependent cell morphogenesis through the heteromeric WAVE and ARP2/3 complexes. Proc. Natl Acad. Sci. USA 105: 4044-4049.

Bloch, D., M. Lavy, Y. Efrat, I. Efroni, K. Bracha-Drori, M. Abu-Abied, E. Sadot and S. Yalovsky. 2005. Ectopic expression of an activated RAC in Arabidopsis disrupts membrane cycling. Mol. Biol. Cell 16: 1913-1927.

Boutte, Y., M.T. Crosnier, N. Carraro, J. Traas and B. Satiat-Jeunemaitre. 2006. The plasma membrane recycling pathway and cell polarity in plants: studies on PIN proteins. J. Cell Sci. 119: 1255-1265.

Cardenas, L., A. Lovy-Wheeler, J.G. Kunkel and P.K. Hepler. 2008. Pollen tube growth oscillations and intracellular calcium levels are reversibly modulated by actin polymerization. Plant Physiol. 146: 1611-1621.

Chen, X., L. Grandont, H. Li, R. Hauschild, S. Paque, A. Abuzeineh, H. Ralcusova, E. Benkova, C. Perrot-Rechenmann and J. Friml. 2014. Inhibition of cell expansion by rapid ABP1-mediated auxin effect on microtubules. Nature 516: 90-U206.

Cheung, A.Y. and H.M. Wu. 2004. Overexpression of an Arabidopsis formin stimulates supernumerary actin cable formation from pollen tube cell membrane. Plant Cell 16: 257-269.

Cheung, A.Y., S. Niroomand, Y. Zou and H.M. Wu. 2010. A transmembrane formin nucleates subapical actin assembly and controls tip-focused growth in pollen tubes. Proc. Natl Acad. Sci. USA 107: 16390-16395.

Choi, Y., Y. Lee, B.W. Jeon, C.J. Staiger and Y. Lee. 2008. Phosphatidylinositol 3-and 4-phosphate modulate actin filament reorganization in guard cells of day flower. Plant Cell Environ. 31: 366-377.

Dhonukshe, P., A.M. Laxalt, J. Goedhart, T.W.J. Gadella and T. Munnik. 2003. Phospholipase D activation correlates with microtubule reorganization in living plant cells. Plant Cell 15: 2666-2679.

Djakovic, S., J. Dyachok, M. Burke, M.J. Frank and L.G. Smith. 2006. BRICK1/HSPC300 functions with SCAR and the ARP2/3 complex to regulate epidermal cell shape in Arabidopsis. Development 133: 1091-1100.

Dowd, P.E., S. Coursol, A.L. Skirpan, T.H. Kao and S. Gilroy. 2006. Petunia phospholipase C1 is involved in pollen tube growth. Plant Cell 18: 1438-1453.

Du, C. and K. Chong. 2011. ARF-GTPase activating protein mediates auxin influx carrier AUX1 early endosome trafficking to regulate auxin dependent plant development. Plant Signal. Behav. 6: 1644-1646.

Du, C., Y. Xu, Y. Wang and K. Chong. 2011. Adenosine diphosphate ribosylation factor-GTPase-activating protein stimulates the transport of AUX1 endosome, which relies on actin cytoskeletal organization in rice root development. J. Integr. Plant Biol. 53: 698-709.

Dyachok, J., M.R. Shao, K. Vaughn, A. Bowling, M. Facette, S. Djakovic, L. Clark and L. Smith. 2008. plasma membrane-associated SCAR complex subunits promote cortical F-actin accumulation and normal growth characteristics in Arabidopsis roots. Mol. Plant 1: 990-1006.

El-Assall, S.E., J. Le, D. Basu, E.L. Mallery and D.B. Szymanski. 2004. DISTORTED2 encodes an ARPC2 subunit of the putative Arabidopsis ARP2/3 complex. Plant J. 38: 526-538.

Eun, S.O., S.H. Bae and Y. Lee. 2001. Cortical actin filaments in guard cells respond differently to abscisic acid in wild-type and abi1-1 mutant Arabidopsis. Planta 212: 466-469.

Finka, A., Y. Saidi, P. Goloubinoff, J.M. Neuhaus, J.P. Zryd and D.G. Schaefer. 2008. The knock-out of ARP3a gene affects F-actin cytoskeleton organization altering cellular tip growth, morphology and development in moss Physcomitrella patens. Cell Motil. Cytoskel. 65: 769-784.

Foissner, I., F. Grolig and G. Obermeyer. 2002. Reversible protein phosphorylation regulates the dynamic organization of the pollen tube cytoskeleton: effects of calyculin A and okadaic acid. Protoplasma 220: 1-15.

Fu, Y., Y. Gu, Z. Zheng, G. Wasteneys and Z. Yang. 2005. Arabidopsis interdigitating cell growth requires two antagonistic pathways with opposing action on cell morphogenesis. Cell 120: 687-700.

Fu, Y., H. Li and Z.B. Yang. 2002. The ROP2 GTPase controls the formation of cortical fine F-actin and the early phase of directional cell expansion during Arabidopsis organogenesis. Plant Cell 14: 777-794.

Fu, Y., G. Wu and Z.B. Yang. 2001. Rop GTPase-dependent dynamics of tip-localized F-actin controls tip growth in pollen tubes. J. Cell Biol. 152: 1019-1032.

Gibbon, B.C., L.E. Zonia, D.R. Kovar, P.J. Hussey and C.J. Staiger. 1999. Pollen profilin function depends on interaction with proline-rich motifs. Plant Cell 11: 1603-1603.

Harries, P.A., A.H. Pan and R.S. Quatrano. 2005. Actin-related protein2/3 complex component ARPC1 is required for proper cell morphogenesis and polarized cell growth in *Physcomitrella patens*. Plant Cell 17: 2327-2339.

Holweg, C., C. SuBlin and P. Nick. 2004. Capturing *in vivo* dynamics of the actin cytoskeleton stimulated by auxin or light. Plant Cell Physiol. 45: 855-863.

Huang, C.H. and R.C. Crain. 2009. Phosphoinositide-specific phospholipase C in oat roots: association with the actin cytoskeleton. Planta 230: 925-933.

Huang, S.J., L. Gao, L. Blanchoin and C.J. Staiger. 2006. Heterodimeric capping protein from Arabidopsis is regulated by phosphatidic acid. Mol. Biol. Cell 17: 1946-1958.

Hwang, J.U., Y. Gu, Y.J. Lee and Z.B. Yang. 2005. Oscillatory ROP GTPase activation leads the oscillatory polarized growth of pollen tubes. Mol. Biol. Cell 16: 5385-5399.

Ischebeck, T., I. Stenzel, F. Hempel, X. Jin, A. Mosblech and I. Heilmann. 2011. Phosphatidylinositol-4,5-bisphosphate influences Nt-Rac5-mediated cell expansion in pollen tubes of *Nicotiana tabacum.* Plant J. 65: 453-468.

Joergens, C.I., N. Gruenewald, M. Huelskamp and J.F. Uhrig. 2010. A role for ABIL3 in plant cell morphogenesis. Plant J. 62: 925-935.

Knowles, C.L., A. Koutoulis and J.B. Reid. 2004. Microtubule orientation in the brassinosteroid mutants 1k, 1ka and 1kb of pea. J. Plant Growth Reg. 23: 146-155.

Kotchoni, S.O., T. Zakharova, E.L. Mallery, J. Le, S.E.-D. El-Assal and D.B. Szymanski. 2009. The association of the Arabidopsis Actin-Related Protein2/3 complex with cell membranes is linked to its assembly status but not its activation. Plant Physiol. 151: 2095-2109.

Kusano, H., C. Testerink, J.E. Vermeer, T. Tsuge, H. Shimada, A. Oka, T. Munnik and T. Aoyama. 2008. The Arabidopsis phosphatidylinositol phosphate 5-kinase PIP5K3 is a key regulator of root hair tip growth. Plant Cell 20: 367-380.

Lanza, M., B. Garcia-Ponce, G. Castrillo, P. Catarecha, M. Sauer, M. Rodriguez-Serrano, A. Paez-Garcia, E. Sanchez-Bermejo, T. Mohan, Y. Leo del Puerto, L. Maria Sandalio, J. Paz-Ares and A. Leyva. 2012. Role of actin cytoskeleton in brassinosteroid signaling and in its integration with the auxin response in plants. Dev. Cell 22: 1275-1285.

Li, G., W. Liang, X. Zhang, H. Ren, J. Hu, M.J. Bennett and D. Zhang. 2014. Rice actin-binding protein RMD is a key link in the auxin-actin regulatory loop that controls cell growth. Proc. Natl Acad. Sci. USA 111: 10377-10382.

Li, J., J.L. Henty-Ridilla, S. Huang, X. Wang, L. Blanchoin and C.J. Staiger. 2012. Capping protein modulates the dynamic behavior of actin filaments in response to phosphatidic acid in Arabidopsis. Plant Cell 24: 3742-3754.

Li, S.D., L. Blanchoin, Z.B. Yang and E.M. Lord. 2003. The putative Arabidopsis Arp2/3 complex controls leaf cell morphogenesis. Plant Physiol. 132: 2034-2044.

Li, Y.H., K. Sorefan, G. Hemmann and M.W. Bevan. 2004. Arabidopsis NAP and PIR regulate actin-based cell morphogenesis and multiple developmental processes. Plant Physiol. 136: 3616-3627.

Lovy-Wheeler, A., J.G. Kunkel, E.G. Allwood, P.J. Hussey and P.K. Hepler. 2006. Oscillatory increases in alkalinity anticipate growth and may regulate actin dynamics in pollen tubes of lily. Plant Cell 18: 2182-2193.

Maisch, J., J. Fiserova, L. Fischer and P. Nick. 2009. Tobacco Arp3 is localized to actin-nucleating sites *in vivo*. J. Exp. Bot. 60: 603-614.

Mathur, J., N. Mathur, V. Kirik, B. Kernebeck, B.P. Srinivas and M. Hulskamp. 2003. Arabidopsis CROOKED encodes for the smallest subunit of the ARP2/3 complex and controls cell shape by region specific fine F-actin formation. Development 130: 3137-3146.

Meagher, R.B. and M. Fechheimer. 2003. The Arabidopsis cytoskeletal genome. pp. 2:e0096. The American Society of Plant Biologists (eds.). The Arabidopsis Book.

Molendijk, A.J., F. Bischoff, C.S.V. Rajendrakumar, J. Friml, M. Braun, S. Gilroy and K. Palme. 2001. Arabidopsis thaliana Rop GTPases are localized to tips of root hairs and control polar growth. Embo J. 20: 2779-2788.

Motes, C.M., P. Pechter, C.M. Yoo, Y.S. Wang, K.D. Chapman and E.B. Blancaflor. 2005. Differential effects of two phospholipase D inhibitors, 1-butanol and N-acylethanolamine, on *in vivo* cytoskeletal organization and Arabidopsis seedling growth. Protoplasma 226: 109-123.

Mucha, E., C. Hoefle, R. H³ckelhoven and A. Berken. 2010. RIP3 and AtKinesin-13A - A novel interaction linking Rho proteins of plants to microtubules. Eur. J. Cell Biol. 89: 906-916.

Nagawa, S., T. Xu, D. Lin, P. Dhonukshe, X. Zhang, J. Friml, B. Scheres, Y. Fu and Z. Yang. 2012. ROP GTPase-dependent actin microfilaments promote PIN1 polarization by localized inhibition of clathrin-dependent endocytosis. Plos Biology 10: e1001299.

Naoi, K. and T. Hashimoto. 2004. A semidominant mutation in an Arabidopsis mitogen-activated protein kinase phosphatase-like gene compromises cortical microtubule organization. Plant Cell 16: 1841-1853.

Nick, P., M.J. Han and G. An. 2009. Auxin stimulates its own transport by shaping actin filaments. Plant Physiol. 151: 155-167.

Oda, Y. and H. Fukuda. 2013. Rho of plant GTPase signaling regulates the behavior of Arabidopsis kinesin-13A to establish secondary cell wall patterns. Plant Cell 25: 4439-4450.

Perbal, G. and D. Driss-Ecole. 2003. Mechanotransduction in gravisensing cells. Trends Plant Sci. 8: 498-504.

Pleskot, R., M. Potocky, P. Pejchar, J. Linek, R. Bezvoda, J. Martinec, O. Valentova, Z. Novotna and V. Zarsky. 2010. Mutual regulation of plant phospholipase D and the actin cytoskeleton. Plant J. 62: 494-507.

Plett, J.M., J. Mathur and S. Regan. 2009. Ethylene receptor ETR2 controls trichome branching by regulating microtubule assembly in Arabidopsis thaliana. J. Exp. Bot. 60: 3923-3933.

Rosero, A., V. Zarsky and F. Cvrckova. 2013. AtFH1 formin mutation affects actin filament and microtubule dynamics in *Arabidopsis thaliana*. J. Exp. Bot. 64: 585-597.

Ruelland, E. and A. Zachowski. 2010. How plants sense temperature. J. Exp. Bot. 69: 225-232.

Saedler, R., N. Mathur, B.P. Srinivas, B. Kernebeck, M. Hulskamp and J. Mathur. 2004a. Actin control over microtubules suggested by DISTORTED2 encoding the Arabidopsis ARPC2 subunit homolog. Plant Cell Physiol. 45: 813-822.

Saedler, R., I. Zimmermann, M. Mutondo and M. Hulskamp. 2004b. The Arabidopsis KLUNKER gene controls cell shape changes and encodes the AtSRA1 homolog. Plant Mol. Biol. 56: 775-782.

van Gisbergen, P.A., M. Li, S.Z. Wu and M. Bezanilla. 2012. Class II formin targeting to the cell cortex by binding PI(3,5)P-2 is essential for polarized growth. J. Cell Biol. 198: 235-250.

Vidali, L., P.A. van Gisbergen, C. Guerin, P. Franco, M. Li, G.M. Burkart, R.C. Augustine, L. Blanchoin and M. Bezanilla. 2009. Rapid formin-mediated actin-filament elongation is essential for polarized plant cell growth. Proc. Natl Acad. Sci. USA 106: 13341-13346.

Wang, J., Y. Zhang, J. Wu, L. Meng and H. Ren. 2013. At FH16, an Arabidopsis type II formin, binds and bundles both microfilaments and microtubules, and preferentially binds to microtubules. J. Integr. Plant Biol. 55: 1002-1015.

Wang, X., J. Zhang, M. Yuan, D.W. Ehrhardt, Z. Wang and T. Mao. 2012. Arabidopsis microtubule destabilizing protein40 is involved in brassinosteroid regulation of hypocotyl elongation. Plant Cell 24: 4012-4025.

Wang, Y., T. Chen, C. Zhang, H. Hao, P. Liu, M. Zheng, F. Baluska, J. Samaj and J. Lin. 2009. Nitric oxide modulates the influx of extracellular Ca^{2+} and actin filament organization during cell wall construction in Pinus bungeana pollen tubes. New Phytologist 182: 851-862.

Witke, W. 2004. The role of profilin complexes in cell motility and other cellular processes. Trends Cell Biol. 14: 461-469.

Xue, X.H., C.Q. Guo, F. Du, Q.L. Lu, C.M. Zhang and H.Y. Ren. 2011. AtFH8 is involved in root development under effect of low-dose Latrunculin B in dividing cells. Mol. Plant 4: 264-278.

Yang, G., P. Gao, H. Zhang, S. Huang and Z.L. Zheng. 2007. A mutation in MRH2 kinesin enhances the root hair tip growth defect caused by constitutively activated ROP2 small GTPase in Arabidopsis. Plos One 2: e1074.

Yang, X., S.S. Wang, M. Wang, Z. Qiao, C.C. Bao and W. Zhang. 2014. Arabidopsis thaliana calmodulin-like protein CML24 regulates pollen tube growth by modulating the actin cytoskeleton and controlling the cytosolic Ca^{2+} concentration. Plant Mol. Biol. 86: 225-236.

Ye, J., Y. Zheng, A. Yan, N. Chen, Z. Wang, S. Huang and Z. Yang. 2009. Arabidopsis formin3 directs the formation of actin cables and polarized growth in pollen tubes. Plant Cell 21: 3868-3884.

Yi, K.X., C.Q. Guo, D. Chen, B.B. Zhao, B. Yang and H.Y. Ren. 2005. Cloning and functional characterization of a formin-like protein (AtFH8) from Arabidopsis. Plant Physiol. 138: 1071-1082.

Yoo, C.M., L. Quan, A.E. Cannon, J. Wen and E.B. Blancaflor. 2012. AGD1, a class 1 ARF-GAP, acts in common signaling pathways with phosphoinositide metabolism and the actin cytoskeleton in controlling Arabidopsis root hair polarity. Plant J. 69: 1064-1076.

Zhang, Q., F. Lin, T. Mao, J. Nie, M. Yan, M. Yuan and W. Zhang. 2012. Phosphatidic acid regulates microtubule organization by interacting with MAP65-1 in response to salt stress in Arabidopsis. Plant Cell 24: 4555-4576.

Zhang, X.G., J. Dyachok, S. Krishnakumar, L.G. Smith and D.G. Oppenheimer. 2005. Irregular trichome branch1 in Arabidopsis encodes a plant homolog of the Actin-Related Protein2/3 complex activator Scar/Wave that regulates actin and microtubule organization. Plant Cell 17: 2314-2326.

Zhang, Z., Y. Zhang, H. Tan, Y. Wang, G. Li, W. Liang, Z. Yuan, J. Hu, H. Ren and D. Zhang. 2011. Rice morphology determinant encodes the type II formin Fh5 and regulates rice morphogenesis. Plant Cell 23: 681-700.

Zhao, H.P. and H.Y. Ren. 2006. Rop1Ps promote actin cytoskeleton dynamics and control the tip growth of lily pollen tube. Sex. Plant Reprod. 19: 83-91.

Zi, H., Y. Xiang, M. Li, T. Wang and H. Ren. 2007. Reversible protein tyrosine phosphorylation affects pollen germination and pollen tube growth via the actin cytoskeleton. Protoplasma 230: 183-191.

10

The Regulation of Plant Cell Expansion: Auxin-Induced Turgor-Driven Cell Elongation

Koji Takahashi[1,*] *and Toshinori Kinoshita*[1,2]

Introduction

The plant body, which originates from a single-celled zygote, is formed by numerous cell divisions and increases in cell size. Differentiated plant cells are of various sizes and diverse shapes, including cylindrical cells (internodal cells of *Chara* and *Nitella* and epidermal and cortical cells in the axial organs, such as stems and roots), tubular and filamentous cells (pollen tubes and root hairs), two- or three-branched trichomes, jigsaw-puzzle-shaped dicot leaf pavement cells, and kidney-shaped guard cells (Fig. 1).

These diverse plant cell shapes are dependent on the arrangement of cortical microtubule and cellulose microfibrils in the cell wall. Irreversible enlargement of plant cells is caused by water uptake into the cytoplasm and the subsequent synthesis of cell constituents and secretion of materials for the plasma membrane and cell wall. The final size of plant cells is mostly proportional to the ploidy generated by endoreduplication, which facilitates somatic nuclear polyploidization without cytokinesis, in hypocotyl cells, trichomes, and leaf pavement cells (Hülskamp et al. 1994, Melaragno et al. 1993, Gendreau et al. 1998, Kondorosi et al. 2000, Larkins et al. 2001, Sugimoto-Shirasu and Roberts 2003, Lee et al. 2009) (Fig. 2).

A cell enlargement profile can be observed along the apical–basal axis of the hypocotyls and roots. Arabidopsis hypocotyls elongate without any marked division of their epidermal and cortical cells (Gendreau et al. 1997); an epidermal cell in the hypocotyls of 36-hour-old etiolated seedlings elongates longitudinally from 50 to 500 μm during the following 36 hours, and the elongated hypocotyls exhibit 4C, 8C, and 16C ploidy levels. Thus, irreversible cell enlargement occurs after these various biophysical and biochemical steps and is affected by physiological signals, such as phytohormones and environmental stimuli. Control of plant cell volume contributes to plant

---

[1] Division of Biological Science, Graduate School of Science, Nagoya University, Chikusa, Nagoya, 464-8602, Japan.
[2] Institute of Transformative Bio-Molecules (WPI-ITbM), Nagoya University, Chikusa, Nagoya 464-8602, Japan.
[*] Corresponding author : takahashi@bio.nagoya-u.ac.jp

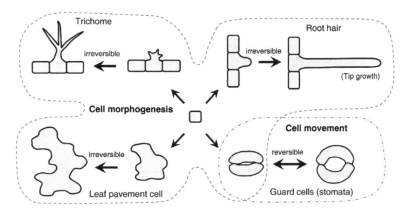

FIGURE 1 Plant cell morphogenesis and movement due to cell expansion. Plant cells exhibit a wide diversity in size and shape, even if restricted to the cells derived from epidermal cells, such as trichome, root hair, leaf pavement cells, guard cells. A trichome, meaning "hair", and a root hair are formed by irreversible elongation from a single epidermal cell of aerial and underground organs, respectively. In root hairs, only the surface of the tip region expands; this type of cell growth is known as "tip growth." Pollen tubes also elongate by tip growth. The jigsaw-puzzle shape of leaf pavement cells in most dicots is caused by an irreversible increase in the surface area of the lobe regions of leaf epidermal cells. Guard cells, which develop by asymmetric cell division, can undergo repeat expansion and shrinkage in response to environmental stimuli, which enables control of vapour and gas transport through the pore. The diverse morphologies of plant cells are determined primarily by the mechanical properties of the cell wall.

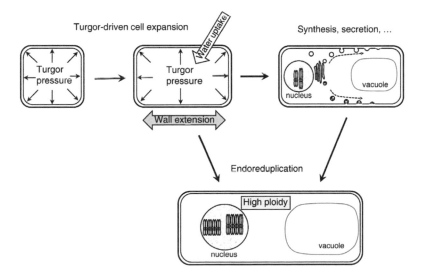

FIGURE 2 Plant cell enlargement. The enlargement of plant cells begins with turgor-driven cell expansion due to water uptake into the cytoplasm and extension of the rigid cell wall, and is followed by *de novo* synthesis and supply of cell materials. High ploidy due to endoreduplication is often correlated with cell size. Combination of these steps determines the size and shape of the plant cells.

morphogenesis and growth, as well as plant cell movement, such as that of stomatal guard cells and pulvinar motor cells (Moran 2007, Kinoshita and Hayashi 2011) (Fig. 1). The turgor-driven cycle of expansion and contraction of guard cells and pulvinar motor cells regulates the stomatal aperture and the leaf positioning, respectively. In these cases, the size of the cells is controlled in a reversible manner in response to physiological signals including light, carbon dioxide, and phytohormone ABA.

In this chapter, with regard to the various steps in plant cell enlargement, we will focus on turgor-driven cell expansion, with emphasis on the processes of water uptake and cell wall extension.

Water Absorption by Plant Cells

Changes in the volume of plant cells are caused by water transport between the symplast and apoplast through the plasma membrane. Water flow across a membrane (J_v), derived from the theory of irreversible thermodynamics, is generally expressed using the equation shown below (Kedem and Katchalsky 1958, Nobel 2009):

$$J_V = L_p \Delta \Psi_W \qquad [1]$$

where J_v is the rate of water flow across a membrane per unit membrane area (m^3 m^{-2} sec^{-1}), L_p is the hydraulic conductivity of the membrane (m sec^{-1} MPa^{-1}), and $\Delta \Psi_W$ is the water potential difference (MPa) between the two sides of the membrane. Therefore, the relationship between plant cell expansion and water transport across the plasma membrane is expressed using the following equation:

$$dV/dt = AL_p (\Delta \Psi_W) \qquad [2]$$

where dV/dt is the change in cell volume per unit time and A is the surface area of the plasma membrane. Water potential (Ψ_w) consists of solute potential (Ψ_s), pressure potential (Ψ_p), and gravity potential (Ψ_g). However, because the gravitational potential in water transport across the plasma membrane is extremely small, the water potential involved in water transport via the plasma membrane is simplified to the sum of the solute potential and pressure potential, as below:

$$\Psi_W = \Psi_S + \Psi_p \qquad [3]$$

Substituting the terms from Equation [3] into Equation [2], the following equation is obtained:

$$dV/dt = AL_p (\Delta \Psi_S + \Delta \Psi_p) = AL_p [\Delta \Psi_S + (\Psi_{p \, (outside)} - \Psi_{p \, (cell)})] \qquad [4]$$

where $\Delta \Psi_W$ is the water potential difference between intracellular and extracellular water potentials ($\Psi_{p \, (outside)} - \Psi_{p \, (cell)}$). Because extracellular pressure is equal to the atmospheric pressure, $\Psi_{p \, (outside)} = 0$. Therefore,

$$dV/dt = -AL_p (\Psi_{p \, (cell)} - \Delta \Psi_S) \qquad [5]$$

Equation [5] indicates that the cell expansion rate (dV/dt) is dependent on the hydraulic conductivity of the membrane (L_p), turgor pressure (pressure potential in a cell, $\Psi_{p\,(cell)}$), and the solute potential difference ($\Delta\Psi_s$). Water can move across the lipid bilayer of biological membranes, although water permeability is dependent on lipid composition and the osmotic water permeability coefficient (P_f), which is 10×10^{-4} to 100×10^{-4} cm sec^{-1} (Graziani and Livne 1972, Mathai et al. 2008) and defined in the equation below:

$$P_f = L_p RT/V_w \qquad [6]$$

where R is the gas constant, T is the absolute temperature, and V_w is the partial molar volume of water. Biological membranes in plant cells have 10–100-fold higher P_f values ($\sim$1,200 $\times$ 10^{-4} cm sec^{-1}) than pure lipid bilayers, suggesting that water is effectively transported through water-permeable transmembrane proteins, known as aquaporins (Maurel 1997, Tyerman et al. 1999). Aquaporins facilitate the transport of water, as well as uncharged small solutes (urea, boric acid, silicic acid, etc.) and gases (carbon dioxide), and influence cell expansion and plant development (Maurel et al. 2008). Water permeability is thought to be regulated by the amount of aquaporins in the plasma membrane and their transport activity by post-translational modification, but the direction of water transport and the driving force are determined by the solute potential gradient across the membrane ($\Delta\Psi_s$). Because solute potential is obtained by applying a minus sign to the osmotic pressure, it is determined by the concentration of osmoticum, which are the osmotic solutes responsible for osmotic pressure, including ions, amino acids, and sugars. Among these solutes, potassium is the main cationic osmoticum in plant cells (Marschner 1995) and plays an important role in

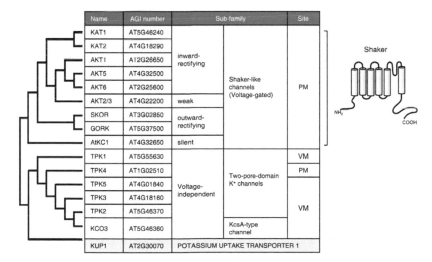

FIGURE 3 Potassium channels in Arabidopsis. Potassium transport *in planta* is mediated through potassium channels, potassium transporters, and Na$^+$(K$^+$)/H$^+$ antiporters (Wang and Wu 2013). Among them, Arabidopsis has 15 members of the superfamily of potassium channel proteins, which are divided into the voltage-gated Shaker-like channels, two-pore-domain potassium channels, and KcsA-type potassium channels. The Shaker-like potassium channels are localized to the plasma membrane (PM) and the majority of two-pore potassium channels are localized to the vacuolar membrane (VM). Shaker-like potassium channels possess six transmembrane domains and both the N- and C-terminal domains face the cytosolic space. The phylogenetic inferences from the amino acid sequences of the potassium channels were based on a Maximum Likelihood analysis.

the turgor-driven control of plant cell volume, including cell elongation and movement of stomata and pulvinar motor cells (Marrè et al. 1974, Schnabl and Raschke
1980, Zeiger 1983, Lowen and Satter 1989, Okazaki et al. 2000). In this chapter, we
will focus on the potassium transport, although the transports of other solutes are
also involved in cell expansion (Ward et al. 2009). Turgor-driven cell expansion is
facilitated by potassium transport across the plasma membrane, mainly through the
Shaker-like voltage-gated K^+ channels (Schroeder et al. 1987, Moran et al. 1988, Thiel
et al. 1996, Claussen et al. 1997). This family of K^+ channels includes the inward-rectifying channels (such as KAT1 and KAT2), which mediate K^+ uptake dependent on
hyperpolarizing membrane potentials, and the outward-rectifying channels (such as
SKOR and GORK), which mediate K^+ efflux dependent on depolarizing membrane
potentials (Hedrich 2012) (Fig. 3). In plants, the membrane potential of the plasma
membrane is generated primarily by the action of plasma membrane H^+-ATPases,
which function to extrude protons by using energy obtained from ATP hydrolysis.
In addition to the voltage-gated channels, the KUP potassium transporter family
proteins are reported to control potassium homeostasis in cell growth (Osakabe et
al. 2013).

Turgor-driven Cell-wall Extension

The cell wall, which consists of a complex network of a variety of polysaccharides—
including cellulose microfibrils, hemicelluloses, and pectins—forms a load-bearing
structure and withstands the internal hydrostatic pressure of the cell. On the other
hand, the combination of viscoelastic and plastic properties of the cell wall regulates
turgor-driven cell expansion and the anisotropy of wall extension finally determines
the size and shape of plant cells. Irreversible cell wall extension is dependent on the
plastic extensibility of the wall and the applied force, but a force below a certain
value cannot drive irreversible wall extension (Probine and Preston 1962). Thus, the
rheological behaviour of the cell wall is approximated by a Bingham fluid (Bingham
1916), for which the shear rate is linearly proportional to the shear stress beyond the
yield threshold. Lockhart (1965) proposed the relationship between irreversible cell
expansion and the turgor pressure ($\Psi_{p\,(cell)}$) as the following equation:

$$dV/dt = m\,(\Psi_{p\,(cell)} - Y)\ \text{for}\ \Psi_{p\,(cell)} > Y \qquad [7]$$

where m is the plastic extensibility and Y is the yield threshold of the cell wall. Equation
[7] holds only when the value of $\Psi_{p\,(cell)}$ is higher than that of Y. Irreversible wall extension is affected by pH and induced by wall-loosening proteins, known as expansins,
which may weaken the load-bearing structure of the cell wall under acidic conditions
(Hager et al. 1971, McQueen-Mason et al. 1992, Cosgrove 2000). Because apoplastic
pH is regulated largely by the action of plasma membrane H^+-ATPases, the resultant
acidification contributes to irreversible wall extension *in planta*. In addition to expansin, xyloglucan endotransglucosylase/hydrolases (XTH) and a substrate-specific
fungal endoglucanase have been reported to modulate irreversible plastic wall extension (Yuan et al. 2001, Park and Cosgrove 2012, Miedes et al. 2013). Measurement of
wall mechanical properties revealed that expansins and endoglucanase affect the wall
parameters, m and Y (Takahashi et al. 2006, Miedes et al. 2013).

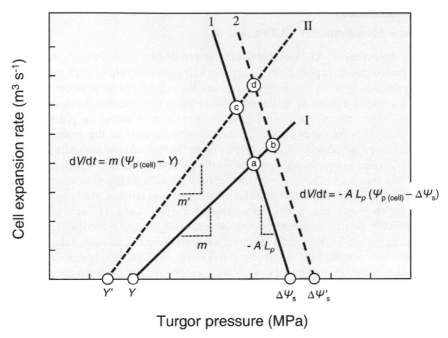

FIGURE 4 Graphical representation of the relationship between cell expansion rate and turgor pressure. The equation of wall extension ($dV/dt = m$ ($\Psi_{p\,(cell)}$ − Y)) is represented by the straight lines, "I" and "II." The slopes of these lines and their intercepts with the "Turgor pressure" axis show the wall extensibility and yield threshold values, respectively. The equation of water uptake ($dV/dt = - AL_p$ ($\Psi_{p\,(cell)}$ − $\Delta\Psi_s$)) is represented by the straight lines, "1" and "2." The slopes of these lines and their intercepts with the "Turgor pressure" axis show the "- AL_p" values and their solute potential differences, respectively. The intersections of the two straight lines representing wall extension and water uptake are shown by "a" to "d."

A graphical representation of Equations [5] and [7] aids understanding of the contribution of the two physiological processes of water uptake and irreversible wall extension to cell expansion. When the two equations are plotted on the same graph (Boyer et al. 1985, Katou and Furumoto 1986a), the cell expansion rate (dV/dt) and the turgor pressure ($\Psi_{p\,(cell)}$) are represented by the intersection of two straight lines (I and 1) (Fig. 4). If the solute potential difference increases from $\Delta\Psi_s$ to $\Delta\Psi'_s$ due to the uptake of osmoticum, the intersection point moves from "a" to "b", resulting in increases in both the cell expansion rate and turgor pressure. If the wall extensibility increases and the yield threshold decreases due to the action of expansins, the intersection point moves from "a" to "c", resulting in an increase in cell expansion rate despite a decrease in turgor pressure. As shown in Fig. 4, irreversible expansion of plant cells is modulated by the mechanical properties of the cell wall, solute potential difference, and hydraulic conductivity of the membrane. Cell-wall properties and solute potential differences are indirectly regulated by plasma membrane H⁺-ATPases via the actions of expansins and the voltage-gated inward-rectifying K⁺-channels. Therefore, the function of plasma membrane H⁺-ATPases is an important determinant of turgor-driven cell expansion. In the next section, we will discuss the structure and regulatory mechanism of plasma membrane H⁺-ATPases.

Plasma Membrane H⁺-ATPases

Plasma membrane H^+-ATPases (hereinafter referred to as "H^+-ATPases") are electrogenic proton pumps in plants, which actively extrude protons from cells by utilizing energy derived from ATP hydrolysis to generate an electrochemical proton gradient, thereby creating a potential difference (inside being negative and frequently in the range −200 to −100 mV) and a pH gradient of ~1.5 units across the plasma membrane (Sze 1985, Palmgren 2001). The membrane potential and the proton gradient established by the activity of H^+-ATPases energise multiple ion channels and various H^+-coupled symporters/antiporters on the plasma membrane, which drives the secondary transport of ions, sugars, and amino acids. In addition, proton transport through H^+-ATPases controls cytosolic pH homeostasis (around pH 7) and directly regulates apoplastic pH, leading to acidifying the cell wall (pH 5–6). These processes finally mediate diverse physiological responses, including phloem loading and unloading, xylem loading and unloading, seed germination, solute uptake in roots, stomatal movements, leaf movement, tip growth, and cell expansion (Palmgren 2001, Hager 2003, Sondergaard et al. 2004, Inoue et al. 2005). Eleven genes encode H^+-ATPases (*AHA1–11*) in Arabidopsis, and a double-knockout mutant of *AHA1* and *AHA2*, which are predominantly expressed in all plant tissues and organs, has an embryonic lethal phenotype. Therefore, H^+-ATPases play important roles in

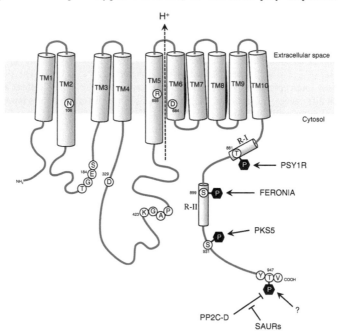

FIGURE 5 Basic structure of plasma membrane H^+-ATPases. H^+-ATPases possess 10 transmembrane domains (TM1-10) and a cytosolic C-terminal autoinhibitory domain containing the R-I and R-II regions. The C-terminal domain harbours multi-phosphorylation sites (Thr881, Ser899, Ser931, and Thr947) whose phosphorylation levels are regulated by protein kinases and protein phosphatases, such as PSY1R, FERONIA, PKS5, and PP2C-D. The numbering of the amino acid residues corresponds to Arabidopsis H^+-ATPase2 (AHA2).

the growth, development, and physiological responses to environmental stresses throughout the life cycle of plants.

Plasma membrane H^+-ATPases, a family of P-type ATPases, consist of a functional polypeptide possessing 10 transmembrane domains (TM1–TM10) and three cytosolic domains containing the N- and C-termini (Fig. 5). The relationships between the mechanism of proton transport and the structure of the H^+-ATPase have been elucidated in numerous biochemical studies, mutation analyses, and by derivation of the crystal structure of Arabidopsis H^+-ATPase2 (AHA2) (Palmgren 2001, Pedersen et al. 2007, Buch-Pedersen et al. 2009). Proton transfer across the plasma membrane occurs through the transmembrane domains via a proton acceptor/donor (Asp684 in AHA2) located in TM6. An asparagine residue (Asn106 in AHA2) in TM2 and an arginine residue (Arg655 in AHA2) in TM5 were proposed to serve as gatekeeper residues and prevent proton backflow, respectively. A lysine residue (Lys423 in AHA2) in the conserved motif KGAP, and an aspartate residue (Asp329 in AHA2), which is involved in the formation of a phosphorylated intermediate with phosphorus due to ATP hydrolysis, are located in the large cytosolic loop between TM4 and TM5, which serves as a catalytic domain. The small cytosolic loop between TM2 and TM3 functions as an actuator, which stimulates dephosphorylation of the phosphorylated intermediate; a glutamic acid residue in the loop (Glu184 in AHA2) in the conserved motif TGES is responsible for this function.

The proteolytic removal of the C-terminal fragment of H^+-ATPases enhances their catalytic activity (Palmgren et al. 1990, 1991) and two regions of the regulatory domain, R-I and R-II, have been identified in the C-terminal cytosolic domain (Axelsen et al. 1999, Pedersen et al. 2012). Thus, the C-terminal cytosolic domain as an autoinhibitory domain is thought to be involved in the regulation of catalytic activity through a conformational change due to post-translational modification. It was demonstrated that the catalytic activity of H^+-ATPase is regulated by phosphorylation of the penultimate threonine at the C-terminus and the subsequent interaction of 14-3-3 protein with the phosphorylated C-terminus (Fuglsang et al. 1999, Kinoshita and Shimazaki 1999, Svennelid et al. 1999, Maudoux et al. 2000). Physiological signals—such as light, salts, sucrose, auxin, gibberellin, and abscisic acid—modulate the phosphorylation level of the penultimate threonine within several minutes (Fuglsang et al. 1999, Kinoshita and Shimazaki 1999, Svennelid et al. 1999, Kerkeb et al. 2002, Zhang et al. 2004, Inoue et al. 2005, Niittylä et al. 2007, Chen et al. 2010, Okumura et al. 2012, Takahashi et al. 2012, Hayashi et al. 2014). The protein kinase responsible for the phosphorylation of the penultimate threonine has not been identified despite much effort, whereas it was recently demonstrated that the phosphorylated penultimate threonine of the H^+-ATPases is dephosphorylated by D-clade type 2C protein phosphatases (PP2Cs) (Spartz et al. 2014).

In addition to the penultimate threonine, the C-terminal cytosolic domain contains multiple phosphorylation sites (Fuglsang et al. 2007, 2014, Rudashevskaya et al. 2012, Haruta et al. 2014). Among them, phosphorylation of three residues was demonstrated to affect H^+-ATPase activity. The phosphorylation of a threonine residue (Thr881 in AHA2) in R-I is induced by a receptor kinase, PSY1R, and application of the peptide hormone PSY1 increased proton efflux (Fuglsang et al. 2014). Moreover, the FERONIA receptor kinase phosphorylates a serine residue (Ser899 in AHA2) in R-II, which negatively regulates proton efflux (Haruta et al. 2014). In addition, the serine residue (Ser931 in AHA2) between R-II and the C-terminus is phosphorylated by the Ser/Thr protein kinase, PKS5, which inhibits interaction of the H^+-ATPase with 14-3-3 protein, leading to a decrease in H^+-ATPase activity (Fuglsang et al. 2007). The physical interaction of chaperone J3 with PKS5 induces the activation of H^+-ATPase by repressing PKS5 activity (Yang et al. 2010). Indeed, the interaction of type 2A protein phosphatase scaffolding subunit A and RIN4 was also reported to affect H^+-ATPase activity (Fuglsang et al. 2006, Riu et al. 2009). Thus, H^+-ATPase activity is post-translationally regulated by phosphorylation of multiple sites in the C-terminal domain and/or protein–protein interactions.

The translocation of H^+-ATPases by constitutive cycling between the plasma membrane and endosomes is thought to regulate the level of H^+-ATPases in the plasma membrane (Paciorek et al. 2005) and has reported to be mediated by a Muc13-like protein, PATROL1 (Hashimoto-Sugimoto et al. 2013). Therefore, together with post-translational modification, the subcellular localization of H^+-ATPases may influence their function.

Auxin-induced Cell Expansion

The phytohormone auxin plays important roles in the various processes related to plant cell enlargement, including turgor-driven cell expansion, regulation of the expression of genes related to cell expansion, regulation of vesicle trafficking, organization of the cytoskeleton, and endoreduplication (Frías et al. 1996, Philippar et al. 1999, Paciorek et al. 2005, Ishida et al. 2010, Perrot-Rechenmann 2010, Robert et al. 2010, Xu et al. 2010). Among them, turgor-driven cell expansion in plant tissues is enhanced within 15 minutes by the application of exogenous auxin (Rayle et al. 1970). Since the 1970s, early-phase auxin-induced elongation of plant organs, such as hypocotyls and coleoptiles, has been explained by the acid-growth theory; *i.e.* the irreversible extension of the cell wall and plant tissue elongation are enhanced in acidic solution (Rayle and Cleland 1970, 1972, Hager et al. 1971). This suggests that auxin induces cell elongation by stimulating proton extrusion from cells and the subsequent acidification induces irreversible wall extension through the functions of cell-wall proteins. In fact, the wall-yielding properties, extensibility (m) and yield threshold (Y), are affected by treatment with auxin and acid *in vivo* (Nakahori et al. 1991, Mizuno et al. 1993). McQueen-Mason et al. (1992) identified a group of cell-wall proteins, expansins, which induce cell wall loosening under acidic conditions (pH 4–5). Expansins, which comprise two major groups: α- and β-expansins, may affect the load-bearing tethers, cellulose-xyloglucan junctions and arabinoxylans, and enhance turgor-driven cell expansion by loosening of the cell wall without hydrolysing the target polysaccharides (Cosgrove 2014). Analysis of the mechanical properties of cell-wall specimens treated with expansins revealed that α-expansins enhance irreversible wall extension due to an acid-induced increase in wall extensibility (m) and a downward shift in yield threshold (Y), but do not affect wall viscoelasticity (Yuan et al. 2001, Takahashi et al. 2006). Regarding water uptake, auxin appears not to alter water permeability, at least in the early phase (Ordin and Bonner 1956, Dowler and Rayle 1974, Cosgrove and Cleland 1983), although the intracellular localization and function of aquaporin were reported to be regulated by auxin (Paciorek et al. 2005, Péret et al. 2012). Auxin-induced cell expansion is dependent on potassium uptake through K^+-channels, which include the voltage-dependent, inward-rectifying KAT1 and KAT2 channels (De Boer et al. 1985, Claussen et al. 1997, Philippar et al. 2004), indicating that the solute potential difference between symplasts and apoplasts is increased by auxin treatment (Katou and Furumoto 1986b). Thus, auxin induces water uptake by increasing the solute potential difference without changing water permeability.

Changes in the growth parameters in response to auxin will be explained using the graph of the relationship of cell expansion rate vs. turgor pressure (Fig. 4). When the two equations for wall extension and water uptake represent straight lines, I and 1, in their initial states, auxin-induced changes in m and Y shift the straight line of I to II. In addition, the straight line 1 (representing water uptake) is moved to line 2 because auxin increases $\Delta\Psi_s$, but not L_p. Note that the intersection point of the two straight lines before and after auxin treatment ("a" and "d", respectively) indicates an identical turgor pressure value, which is reflected by the experimental results that turgor pressure remains stable during auxin-induced turgor-driven cell expansion (Cosgrove and Cleland 1983, Nakahori et al. 1991). That is, auxin affects water uptake and wall extension in a simultaneous and co-ordinated manner. This cooperative regulation may be governed by H⁺-ATPases.

Auxin enhances the phosphorylation level of the penultimate threonine in the C-terminus of H⁺-ATPases within several minutes following the application of auxin (Takahashi et al. 2012). This phosphorylation precedes the increase in catalytic activity of the H⁺-ATPases, as mentioned above. Auxin-induced phosphorylation of the penultimate threonine was not suppressed in the presence of the auxin antagonist, PEO-IAA or the proteasome inhibitor, MG-132, which inhibit the well-known auxin signal transduction system that operates via the E3 ubiquitin-ligase complex SCR[TIR1/AFB] and includes the auxin receptor, TRANSPORT INHIBITOR RESPONSE 1/AUXIN SIGNALING F-BOX (TIR1/AFB) protein (Dharmasiri et al. 2005, Kepinski and Leyser 2005, Mockaitis and Estelle 2008). In addition, the auxin-induced phosphorylation was not affected in a double mutant of the TIR1/AFB family proteins, *tir1-1 afb2-3* (Takahashi et al. 2012) and early-phase auxin-induced hypocotyl elongation occurs in the *tir1-1 afb1-3 afb2-3 afb3-4* quadruple mutant (Schenck et al. 2010), suggesting that auxin-triggered cell expansion occurs without involvement of TIR1/AFBs. Another candidate for the auxin receptor is auxin-binding protein1 (ABP1), because it was shown to be involved in auxin-induced stimulation of the plasma membrane current by H⁺-ATPase in the protoplasts of *Zea mays* coleoptiles (Rück et al. 1993) and in the auxin-induced swelling of protoplasts in elongating *Pisum sativum* internodes (Yamagami et al. 2004). Recent research demonstrated that clathrin-dependent endocytosis, ROP GTPase signalling via the plasma membrane-localised transmembrane kinase (TMK), and gene expression via TIR1/AFBs are associated with ABP1 (Robert et al. 2010, Xu et al. 2010, 2014, Perrot-Rechenmann 2010, Grones et al. 2015). However, the actual physiological function of ABP1 is largely enigmatic, because knockout mutants of ABP1—which is present as a single copy in *A. thaliana*—had normal phenotypes (Gao et al. 2015). Thus, elucidation of the mechanism by which auxin mediates the phosphorylation of H⁺-ATPases warrants further research.

The protein kinase responsible for the phosphorylation of the penultimate threonine has not been identified, despite strenuous efforts. On the other hand, it has been demonstrated that the phosphorylated penultimate threonine of the H⁺-ATPases is dephosphorylated by PP2Cs (Hayashi et al. 2010), especially clade D of PP2Cs

(Spartz et al. 2014). *SMALL AUXIN-UP RNA* (SAUR) genes comprise a large multi-gene family of early-auxin-responsive genes (Hagen and Guilfoyle 2002) and SAUR proteins regulate plant growth in response to physiological signals (Ren and Gray 2015). In particular, some SAUR proteins—including SAUR9, SAUR19, SAUR40, and SAUR72—inhibit clade D of PP2C through physical interaction, which results in an increase in the phosphorylation level of the penultimate threonine of H^+-ATPases (Spartz et al. 2014). These results suggest that auxin-induced H^+-ATPase phosphorylation is mediated, at least in part, through early induction of gene expression by TIR1/AFBs. Currently, the acid-growth theory is understood as follows (Fig. 6). Auxin enhances proton extrusion via the activation of plasma membrane H^+-ATPases by increasing the phosphorylation level of the penultimate threonine, which is mediated through the suppression of PP2C-D activity by SAURs. This process induces acidification of the cell wall, thereby promoting wall loosening by the action of expansins. Besides, the hyperpolarization of the plasma membrane that is generated by H^+-ATPases provides the driving force for K^+ uptake through voltage-gated inward-rectifying K^+ channels and subsequent water uptake. Thus, wall extension and water uptake, which simultaneously occur in response to auxin, leads to the cell expansion. In addition to the post-translational modification of H^+-ATPases, auxin modulates their intercellular localization by affecting endocytosis and constitutive cycling between the plasma membrane and endosomes (Hager et al. 1991, Paciorek et al. 2005, Robert et al. 2010) (Fig. 6). The regulation of H^+-ATPase activity by modulation of their subcellular localization and post-translational modification is a subject worthy of further investigation.

In this chapter, we have discussed the water uptake and wall extension in the process of irreversible plant cell expansion. Unlike the irreversibly elongating hypocotyl

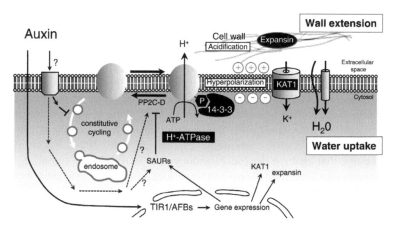

FIGURE 6   Auxin-induced turgor-driven cell expansion via H^+-ATPase activation. Auxin activates H^+-ATPases by phosphorylation of the penultimate threonine in the C-terminus and the subsequent interaction with 14-3-3 protein, leading to wall acidification and hyperpolarization of the plasma membrane. These processes induce (i) expansin-mediated wall extension and (ii) water uptake by increased solute potential differences through potassium channels, such as KAT1. Wall extension and water uptake, which simultaneously occur in response to auxin, induce cell expansion. The phosphorylation level of H^+-ATPases is enhanced by SAUR-mediated suppression of PP2C-D, whereas the upstream signalling pathway is not clear.

cells, the volume of guard cells and pulvinar motor cells reversibly changes dependent on the increase and decrease in their turgor pressure (Satter and Galston 1981, Zeiger 1983). In these cells, the change in turgor pressure controls elastic deformation rather than plastic extension of the cell wall, suggesting that these cell walls must harbour exceptionally greater elasticity compared to those of other cell types. On the other hand, the water uptake basically occurs by the same regulatory mechanism as the cells in the elongating tissues: the physiological stimuli mediate the regulation of H^+-ATPase activity, the subsequent regulation of voltage-gated potassium channels, and water uptake dependent on the resultant water potential (Inoue et al. 2005, Kinoshita and Hayashi 2011). Taken together, water uptake across the plasma membrane via activation of H^+-ATPases is prerequisite for turgor-driven cell expansion regardless of reversible or irreversible expansion. In general, the cytoplasm of the differentiated and enlarged plant cells is occupied by a large central vacuole and the vacuolar membrane shows the higher water permeability. These mean that the accumulation of potassium in the vacuole contributes to the cell expansion by the effective water uptake into the vacuole. The regulatory mechanism of potassium uptake into the vacuole has not been fully established in the process of cell expansion, although $Na^+,K^+/H^+$ antiporters, NHX1 and NHX2 localized in vacuolar membrane have been demonstrated to regulate cell expansion and stomatal movement by mediating potassium uptake into vacuoles (Bassil et al. 2011, Barragán et al. 2012).

To uncover the mechanism of turgor-driven cell expansion we should understand the complicated physical properties of the cell wall and the water relations correlated with the transport of various solutes.

Acknowledgements

Research in the authors' laboratories has been supported by Grants-in-Aid for Scientific Research from the Ministry of Education, Culture, Sports, Science, and Technology, Japan (26440140 to K.T.; 15H05956 and 15H04386 to T.K.).

References

Axelsen, K.B., K. Venema, T. Jahn, L. Baunsgaard and M.G. Palmgren. 1999. Molecular dissection of the C-terminal regulatory domain of the plant plasma membrane H^+-ATPase AHA2: Mapping of residues that when altered give rise to an activated enzyme. Biochemistry 38: 7227-7234.

Barragán, V., E.O. Leidi, Z. Andrés, L. Rubio, A.D. Luca, J.A. Fernández, B. Cubero and J.M. Pardo. 2012. Ion exchangers NHX1 and NHX2 mediate active potassium uptake into vacuoles to regulate cell turgor and stomatal function in *Arabidopsis*. Plant Cell 24: 1127-1142.

Bassil, E., H. Tajima, Y.-C. Liang, M. Ohto, K. Ushijima, R. Nakano, T. Esumi, A. Coku, M. Belmonte and E. Blumwald. 2011. The *Arabidopsis* Na^+/H^+ antiporters NHX1 and NHX2 control vacuolar pH and K^+ homeostasis to regulate growth, flower development, and reproduction. Plant Cell 23: 3482-3497.

Bingham, E.C. 1916. An investigation of the laws of plastic flow. Bulletin of the Bureau of Standards 13: 309-353.

Boyer, J.S., A.J. Cavalieri and E.-D. Schulze. 1985. Control of the rate of cell enlargement: excision, wall relaxation, and growth-induced water potentials. Planta 163: 527-543.

Buch-Pedersen, M.J., B.P. Pedersen, B. Veierskov, P. Nissen and M.G. Palmgren. 2009. Protons and how they are transported by proton pumps. Eur. J. Physiol. 457: 573-579.

Chen, Y., W. Hoehenwarter and W. Weckwerth. 2010. Comparative analysis of phytohormone-responsive phosphoproteins in *Arabidopsis thaliana* using TiO_2-phosphopeptide enrichment and mass accuracy precursor alignment. Plant J. 63: 1-17.

Claussen, M., H. Lüthen, M. Blatt and M. Böttger. 1997. Auxin-induced growth and its linkage to potassium channels. Planta 201: 227-234.

Cosgrove, D.J. 2000. Loosening of plant cell walls by expansins. Nature 407: 321-326.

Cosgrove, D.J. 2014. Plant cell growth and elongation. *In*: eLS. John Wiley & Sons, Ltd: Chichester.

Cosgrove, D.J. and R.E. Cleland. 1983. Osmotic properties of pea internodes in relation to growth and auxin action. Plant Physiol. 72: 332-338.

Dharmasiri, N., S. Dharmasiri and M. Estelle. 2005. The F-box protein TIR1 is an auxin receptor. Nature 435: 441-445.

De Boer, A.H., K. Katou, A. Mizuno, H. Kojima and H. Okamoto. 1985. The role of electrogenic xylem pumps in K^+ absorption from the xylem of *Vigna unguiculata*: the effects of auxin and fusicoccin. Plant Cell Environ. 8: 579-586.

Dowler, M.J. and D.L. Rayle. 1974. Auxin does not alter the permeability of pea segments to tritium-labeled water. Plant Physiol. 53: 229-232.

Frías, I., M.T. Caldeira, J.R. Pérez-Castiñeira, J.P. Navarro-Aviñó, F.A. Culiañez-Maciá, O. Kuppinger, H. Stransky, M. Pagés, A. Hager and R. Serrano. 1996. A major isoform of the maize plasma membrane H^+-ATPase: Characterization and induction by auxin in coleoptiles. Plant Cell 8: 1533-1544.

Fuglsang, A.T., Y. Guo, T.A. Cuin, Q. Qiu, C. Song, K.A. Kristiansen, K. Bych, A. Schulz, S. Shabala, K.S. Schumaker, M.G. Palmgren and J.-K. Zhu. 2007. *Arabidopsis* protein kinase PKS5 inhibits the plasma membrane H^+-ATPase by preventing interaction with 14-3-3 protein. Plant Cell 19: 1617-1634.

Fuglsang, A.T., A. Kristensen, T.A. Cuin, W.X. Schulze, J. Persson, K.H. Thuesen, C.K. Ytting, C.B. Oehlenschlaeger, K. Mahmood, T.E. Sondergaard, S. Shabala and M.G. Palmgren. 2014. Receptor kinase-mediated control of primary active proton pumping at the plasma membrane. Plant J. 80: 951-964.

Fuglsang, A.T., G. Tulinius, N. Cui and M.G. Palmgren. 2006. Protein phosphatase 2A scaffolding subunit A interacts with plasma membrane H^+-ATPase C-terminus in the same region as 14-3-3 protein. Physiol. Plant 128: 334-340.

Fuglsang, A.T., S. Visconti, K. Drumm, T. Jahn, A. Stensballe, B. Mattei, O.N. Jensen, P. Aducci and M.G. Palmgren. 1999. Binding of 14-3-3 protein to the plasma membrane H^+-ATPase AHA2 involves the three C-terminal residues Tyr^{946}-Thr-Val and requires phosphorylation of Thr^{947}. J. Biol. Chem. 274: 36774-36780.

Gao, Y., Y. Zhang, D. Zhang, X. Dai, M. Estelle and Y. Zhao. 2015. Auxin binding protein 1 (ABP1) is not required for either auxin signaling or *Arabidopsis* development. Proc. Natl. Acad. Sci. USA 112: 2275-2280.

Gendreau, E., H. Höfte, O. Grandjean, S. Brown and J. Traas. 1998. Phytochrome controls the number of endoreduplication cycles in the *Arabidopsis thaliana* hypocotyl. Plant J. 13: 221-230.

Gendreau, E., J. Traas, T. Desnos, O. Grandjean, M. Caboche and H. Höfte. 1997. Cellular basis of hypocotyl growth in *Arabidopsis thaliana*. Plant Physiol. 114: 295-305.

Graziani, Y. and A. Livne. 1972. Water permeability of bilayer lipid membranes: sterol-lipid interaction. J. Membrane Biol. 7: 275-284.

Grones, P., X. Chen, S. Simon, W.A. Kaufmann, R. De Rycke, T. Nodzyński, E. Zažímalová and J. Friml. 2015. Auxin-binding pocket of ABP1 is crucial for its gain-of-function cellular and developmental roles. J. Exp. Bot. 66: 5055-5065.

Hagen, G. and T. Guilfoyle. 2002. Auxin-responsive gene expression: genes, promoters and regulatory factors. Plant Mol. Biol. 49: 373-385.

Hager, A. 2003. Role of the plasma membrane H^+-ATPase in auxin-induced elongation growth: historical and new aspects. J. Plant Res. 116: 483-505.

Hager, A., G. Debus, H.-G. Edel, H. Stransky and R. Serrano. 1991. Auxin induces exocytosis and the rapid synthesis of a high-turnover pool of plasma membrane H^+-ATPase. Planta 185: 527-537.

Hager, A., H. Menzel and A. Krauss. 1971. Versuche und Hypothese zur Primärwirkung des Auxins beim Streckungswachstum. Planta 100: 47-75.

Haruta, M., G. Sabat, K. Stecker, B.B. Minkoff and M.R. Sussman. 2014. A peptide hormone and its receptor protein kinase regulate plant cell expansion. Science 343: 408-411.

Hashimoto-Sugimoto, M., T. Higaki, T. Yaeno, A. Nagami, M. Irie, M. Fujimi, M. Miyamoto, K. Akita, J. Negi, K. Shirasu, S. Hasezawa and K. Iba. 2013. A Munc13-like protein in Arabidopsis mediates H^+-ATPase translocation that is essential for stomatal responses. Nature Commun. 4: 2215.

Hayashi, Y., S. Nakamura, A. Takemiya, K. Shimazaki and T. Kinoshita. 2010. Biochemical characterization of *in vitro* phosphorylation and dephosphorylation of the plasma membrane H^+-ATPase. Plant Cell Physiol. 51: 1186-1196.

Hayashi, Y., K. Takahashi, S. Inoue and T. Kinoshita. 2014. Abscisic acid suppresses hypocotyl elongation by dephosphorylating plasma membrane H^+-ATPase in *Arabidopsis thaliana*. Plant Cell Physiol. 55: 845-853.

Hedrich, R. 2012. Ion channels in plants. Physiol. Rev. 92: 1777-1811.

Hülskamp, M., S. Miséra and G. Jürgens. 1994. Genetic dissection of trichome cell development in Arabidopsis. Cell 76: 555-566.

Inoue, S., T. Kinoshita and K. Shimazaki. 2005. Possible involvement of phototropins in leaf movement of kidney bean in response to blue light. Plant Physiol. 138: 1994-2004.

Ishida, T., S. Adachi, M. Yoshimura, K. Shimizu, M. Umeda and K. Sugimoto. 2010. Auxin modulates the transition from the mitotic cycle to the endocycle in *Arabidopsis*. Development 137: 63-71.

Katou, K. and M. Furumoto. 1986a. A mechanism of respiration-dependent water uptake in higher plants. Protoplasma 130: 80-82.

Katou, K. and M. Furumoto. 1986b. A mechanism of respiration-dependent water uptake enhanced by auxin. Protoplasma 133: 174-185.

Kedem, O. and A. Katchalsky. 1958. Thermodynamic analysis of the permeability of biological membranes to non-electrolytes. Biochim. Biophys. Acta 27: 229-246.

Kepinski, S and O. Leyser. 2005. The *Arabidopsis* F-box protein TIR1 is an auxin receptor. Nature 435: 446-451.

Kerkeb, L., K. Venema, J.P. Donaire and M.P. Rodríguez-Rosales. 2002. Enhanced H^+/ATP coupling ratio of H^+-ATPase and increased 14-3-3 protein content in plasma membrane of tomato cells upon osmotic shock. Physiol. Plant 116: 37-41.

Kinoshita, T. and Y. Hayashi. 2011. New insights into the regulation of stomatal opening by blue light and plasma membrane H^+-ATPase. Int. Rev. Cell Mol. Biol. 289: 89-115.

Kinoshita, T. and K. Shimazaki. 1999. Blue light activates the plasma membrane H^+-ATPase by phosphorylation of the C-terminus in stomatal guard cells. EMBO J. 18: 5548-5558.

Kondorosi, E., F. Roudier and E. Gendreau. 2000. Plant cell-size control: growing by ploidy? Curr. Opinion Plant Biol. 3: 488-492.

Larkins, B.A., B.P. Dilkes, R.A. Dante, C.M. Coelho, Y. Woo and Y. Liu. 2001. Investigating the hows and whys of DNA endoreduplication. J. Exp. Bot. 52: 183-192.

Lee, H.O., J.M. Davidson and R.J. Duronio. 2009. Endoreplication: polyploidy with purpose. Genes Dev. 23: 2461-2477.

Lockhart, J.A. 1965. An analysis of irreversible plant cell elongation. J. Theoret. Biol. 8: 264-275.

Lowen, C.Z. and R.L. Satter. 1989. Light-promoted changes in apoplastic K^+ activity in the *Samanea saman* pulvinus, monitored with liquid membrane microelectrodes. Planta 179: 421-427.

Marrè, E., P. Lado, F. Rasi-Caldogno, R. Colombo and M.I. De Michelis. 1974. Evidence for the coupling of proton extrusion to K^+ uptake in pea internode segments treated in fusicoccin or auxin. Plant Sci. Lett. 3: 365-379.

Marschner, H. 1995. Mineral Nutrition of Higher Plants, 2nd Edn. London: Academic Press.

Mathai, J.C., S. Tristram-Nagle, J.F. Nagle and M.L. Zeidel. 2008. Structural determinants of water permeability through the lipid membrane. J. Gen. Physiol. 131: 69-76.

Maudoux, O., H. Batoko, C. Oecking, K. Gevaert, J. Vandekerckhove, M. Boutry and P. Morsomme. 2000. A plant plasma membrane H^+-ATPase expressed in yeast is activated by phosphorylation at its penultimate residue and binding of 14-3-3 regulatory proteins in the absence of fusicoccin. J. Biol. Chem. 275: 17762-17770.

Maurel, C. 1997. Aquaporins and water permeability of plant membranes. Annu. Rev. Plant Physiol. Plant Mol. Biol. 48: 399-429.

Maurel, C., L. Verdoucq, D.-T. Luu and V. Santoni. 2008. Plant aquaporins: membrane channels with multiple integrated functions. Annu. Rev. Plant Biol. 59: 595-624.

McQueen-Mason, S., D.M. Durachko and D.J. Cosgrove. 1992. Two endogenous proteins that induce cell wall extension in plants. Plant Cell 4: 1425-1433.

Melaragno, J.E., B. Mehrotra and A.W. Coleman. 1993. Relationship between endopolyploidy and cell size in epidermal tissue of Arabidopsis. Plant Cell 5: 1661-1668.

Miedes, E., D. Suslov, F. Vandenbussche, K. Kenobi, A. Ivakov, D.V.D. Straeten, E.P. Lorences, E.J. Mellerowicz, J.-P. Verbelen and K. Vissenberg. 2013. Xyloglucan endotransglucosylase/hydrolase (XTH) overexpression affects growth and cell wall mechanics in etiolated *Arabidopsis* hypocotyls. J. Exp. Bot. 64: 2481-2497.

Mizuno, A., K. Nakahori and K. Katou. 1993. Acid-induced changes in the *in vivo* wall-yielding properties of hypocotyl sections of *Vigna unguiculata*. Physiol. Plant. 89: 693-698.

Mockaitis, K. and M. Estelle. 2008. Auxin receptors and plant development: a new signaling paradigm. Annu. Rev. Cell Dev. Biol. 24: 55-80.

Moran, N. 2007. Osmoregulation of leaf motor cells. FEBS Lett. 581: 2337-2347.

Moran, N., G. Ehrenstein, K. Iwasa, C. Mischke, C. Bare and R.L. Satter. 1988. Potassium channels in motor cells of *Samanea saman*. Plant Physiol. 88: 643-648.

Nakahori, K., K. Katou and H. Okamoto. 1991. Auxin changes both the extensibility and the yield threshold of the cell wall of *Vigna* hypocotyls. Plant Cell Physiol. 32: 121-129.

Niittylä, T., A.T. Fuglsang, M.G. Palmgren, W.B. Frommer and W.X. Schulze. 2007. Temporal analysis of sucrose-induced phosphorylation changes in plasma membrane proteins of *Arabidopsis*. Mol. Cell. Proteomics. 6: 1711-1726.

Nobel, P.S. 2009. Physicochemical and Environmental Plant Physiology 4th ed. Academic Press/Elsevier, Amsterdam.

Okazaki, Y., K. Azuma and Y. Nishizaki. 2000. A pulse of blue light induces a transient increase in activity of apoplastic K^+ in laminar pulvinus of *Phaseolus vulgaris* L. Plant Cell Physiol. 41: 230-233.

Okumura, M., S. Inoue, K. Takahashi, K. Ishizaki, T. Kohchi and T. Kinoshita. 2012. Characterization of the plasma membrane H^+-ATPase in the Liverwort *Marchantia polymorpha*. Plant Physiol. 159: 826-834.

Ordin, L. and J. Bonner. 1956. Permeability of Avena coleoptile sections to water measured by diffusion of deuterium hydroxide. Plant Physiol. 31: 53-57.

Osakabe, Y., N. Arinaga, T. Umezawa, S. Katsura, K. Nagamachi, H. Tanaka, H. Ohiraki, K. Yamada, S.-U. Seo, M. Abo, E. Yoshimura, K. Shinozaki and K. Yamaguchi-Shinozaki. 2013. Osmotic stress responses and plant growth controlled by potassium transporters in *Arabidopsis*. Plant Cell 25: 609-624.

Paciorek, T., E. Zažímalová, N. Ruthardt, J. Petrásek, Y.-D. Stierhof, J. Kleine-Vehn, D.A. Morris, N. Emans, G. Jürgens, N. Geldner and J. Friml. 2005. Auxin inhibits endocytosis and promotes its own efflux from cells. Nature 435: 1251-1256.

Palmgren, M.G. 2001. Plant plasma membrane H^+-ATPases: powerhouses for nutrient uptake. Annu. Rev. Plant Physiol. Plant Mol. Biol. 52: 817-845.

Palmgren, M.G., C. Larsson and M. Sommarin. 1990. Proteolytic activation of the plant plasma membrane H^+-ATPase by removal of a terminal segment. J. Biol. Chem. 265: 13423-13426.

Palmgren, M.G., M. Sommarin, R. Serrano and C. Larsson. 1991. Identification of an autoinhibitory domain in the C-terminal region of the plant plasma membrane H^+-ATPase. J. Biol. Chem. 266: 20470-20475.

Park, Y.B. and D.J. Cosgrove. 2012. A revised architecture of primary cell walls based on biomechanical changes induced by substrate-specific endoglucanases. Plant Physiol. 158: 1933-1943.

Pedersen, B.P., M.J. Buch-Pedersen, J.P. Morth, M.G. Palmgren and P. Nissen. 2007. Crystal structure of the plasma membrane proton pump. Nature. 450: 1111-1114.

Pedersen, C.N.S., K.B. Axelsen, J.F. Harper and M.G. Palmgren. 2012. Evolution of plant P-type ATPases. Frontiers in Plant Sci. 3: 31.

Péret, B., G. Li, J. Zhao, L.R. Band, U. Voß, O. Postaire, D.-T. Luu, O. Da Ines, I. Casimiro, M. Lucas, D.M. Wells, L. Lazzerini, P. Nacry, J.R. King, O.E. Jensen, A.R. Schäffner, C. Maurel and M.J. Bennett. 2012. Auxin regulates aquaporin function to facilitate lateral root emergence. Nature Cell Biol. 14: 991-998.

Perrot-Rechenmann, C. 2010. Cellular responses to auxin: division versus expansion. Cold Spring Harb. Perspect. Biol. 2: a001446.

Philippar, K., I. Fuchs, H. Lüthen, S. Hoth, C.S. Bauer, K. Haga, G. Thiel, K. Ljung, G. Sandberg, M. Böttger, D. Becker and R. Hedrich. 1999. Auxin-induced K^+ channel expression represents an essential step in coleoptile growth and gravitropism. Proc. Natl. Acad. Sci. USA 96: 12186-12191.

Philippar, K., N. Ivashikina, P. Ache, M. Christian, H. Lüthen, K. Palme and R. Hedrich. 2004. Auxin activates *KAT1* and *KAT2*, two K^+-channel genes expressed in seedlings of *Arabidopsis thaliana*. Plant J. 37: 815-827.

Probine, M.C. and R.D. Preston. 1962. Cell growth and the structure and mechanical properties of the wall in internodal cells of *Nitella opaca*: II. Mechanical properties of the walls. J. Exp. Bot. 13: 111-127.

Rayle, D.L. and R. Cleland. 1970. Enhancement of wall loosening and elongation by acid solutions. Plant Physiol. 46: 250-253.

Rayle, D.L. and R. Cleland. 1972. The *in-vitro* acid-growth response: relation to *in-vivo* growth responses and auxin action. Planta 104: 282-296.

Rayle, D.L., M.L. Evans and R. Hertel. 1970. Action of auxin on cell elongation. Proc. Natl. Acad. Sci. 65: 184-191.

Ren, H. and W.M. Gray. 2015. SAUR proteins as effectors of hormonal and environmental signals in plant growth. Mol. Plant 8: 1153-1164.

Riu, J., J.M. Elmore, A.T. Fuglsang, M.G. Palmgren, B.J. Staskawicz, G. Coaker. 2009. RIN4 functions with plasma membrane H⁺-ATPases to regulate stomatal apertures during pathogen attack. PLoS Biol. 7: e10000139.

Robert, S., J. Kleine-Vehn, E. Barbez, M. Sauer, T. Paciorek, P. Baster, S. Vanneste, J. Zhang, S. Simon, M. Covanová, K. Hayashi, P. Dhonukshe, Z. Yang, S.Y. Bednarek, A.M. Jones, C. Luschnig, F. Aniento, E. Zazímalová and J. Friml. 2010. ABP1 mediates auxin inhibition of clathrin-dependent endocytosis in *Arabidopsis*. Cell 143: 111-121.

Rück, A., K. Palme, M.A. Venis, R.M. Napier and H.H. Felle. 1993. Patch-clamp analysis establishes a role for an auxin binding protein in the auxin stimulation of plasma membrane current in *Zea mays* protoplasts. Plant J. 4: 41-46.

Rudashevskaya, E.L., J. Ye, O.N. Jensen, A.T. Fuglsang and M.G. Palmgren. 2012. Phosphosite mapping of P-type plasma membrane H⁺-ATPase in homologous and heterologous environments. J. Biol. Chem. 287: 4904-4913.

Satter, R.L. and A.W. Galston. 1981. Mechanisms of control of leaf movements. Annu. Rev. Plant Physiol. 32: 83-110.

Schenck, D., M. Christian, A. Jones and H. Lüthen. 2010. Rapid auxin-induced cell expansion and gene expression: a four-decade-old question revisited. Plant Physiol. 152: 1183-1185.

Schnabl, H. and K. Raschke. 1980. Potassium chloride as stomatal osmoticum in *Allium cepa* L., a species devoid of starch in guard cells. Plant Physiol. 65: 88-93.

Schroeder, J.I., K. Raschke and E. Neher. 1987. Voltage dependence of K⁺ channels in guard-cell protoplasts. Proc. Natl. Acad. Sci. USA 84: 4108-4112.

Sondergaard, T.E., A. Schulz and M.G. Palmgren. 2004. Energization of transport processes in plants. Roles of the plasma membrane H⁺-ATPase. Plant Physiol. 136: 2475-2482.

Spartz, A.K., H. Ren, M.Y. Park, K.N. Grandt, S.H. Lee, A.S. Murphy, M.R. Sussman, P.J. Overvoorde and W.M. Gray. 2014. SAUR inhibition of PP2C-D phosphatases activates plasma membrane H⁺-ATPases to promote cell expansion in *Arabidopsis*. Plant Cell 26: 2129-2142.

Sugimoto-Shirasu, K. and K. Roberts. 2003. "Big it up": endoreduplication and cell-size control in plants. Curr. Opinion Plant Biol. 6: 544-553.

Svennelid, F., A. Olsson, M. Piotrowski, M. Rosenquist, C. Ottman, C. Larsson, C. Oecking and M. Sommarin. 1999. Phosphorylation of Thr-948 at the C terminus of the plasma membrane H⁺-ATPase creates a binding site for the regulatory 14-3-3 protein. Plant Cell 11: 2379-2391.

Sze, H. 1985. H⁺-translocating ATPases: Advances using membrane vesicles. Annu. Rev. Plant Physiol. 36: 175-208.

Takahashi, K., K. Hayashi and T. Kinoshita. 2012. Auxin activates the plasma membrane H⁺-ATPase by phosphorylation during hypocotyl elongation in Arabidopsis. Plant Physiol. 159: 632-641.

Takahashi, K., S. Hirata, N. Kido and K. Katou. 2006. Wall-yielding properties of cell walls from elongating cucumber hypocotyls in relation to the action of expansin. Plant Cell Physiol. 47: 1520-1529.

Thiel, G., A. Brüdern and D. Gradmann. 1996. Small inward rectifying K⁺ channels in coleoptiles: inhibition by external Ca²⁺ and function in cell elongation. J. Membrane Biol. 150: 9-20.

Tyerman, S.D., H.J. Bohnert, C. Maurel, E. Steudle and J.A.C. Smith. 1999. Plant aquaporins: their molecular biology, biophysics and significance for plant water relations. J. Exp. Bot. 50: 1055-1071.

Wang, Y. and W.-H. Wu. 2013. Potassium transport and signalling in higher plants. Annu. Rev. Plant Biol. 64: 451-476.

Ward, J.M., P. Mäser and J.I. Schroeder. 2009. Plant ion channels: gene families, physiology, and functional genomics analyses. Annu. Rev. Physiol. 71: 59-82.

Xu, T., N. Dai, J. Chen, S. Nagawa, M. Cao, H. Li, Z. Zhou, X. Chen, R. De Rycke, H. Rakusová, W. Wang, A.M. Jones, J. Friml, S.E. Patterson, A.B. Bleecker and Z. Yang. 2014. Cell surface ABP1-TMK auxin-sensing complex activates ROP GTPase signaling. Science 343: 1025-1028.

Xu, T., M. Wen, S. Nagawa, Y. Fu, J.-G. Chen, M.-J. Wu, C. Perrot-Rechenmann, J. Friml, A.M. Jones and Z. Yang. 2010. Cell surface- and Rho GTPase-based auxin signaling controls cellular interdigitation in *Arabidopsis*. Cell 143: 99-110.

Yamagami, M., K. Haga, R.M. Napier and M. Iino. 2004. Two distinct signaling pathways participate in auxin-induced swelling of pea epidermal protoplasts. Plant Physiol. 134: 735-747.

Yang Y., Y. Qin, C. Xie, F. Zhao, J. Zhao, D. Liu, S. Chen, A.T. Fuglsang, M.G. Palmgren, K.S. Schumaker, X.W. Den and Y. Guo. 2010. The *Arabidopsis* chaperone J3 regulates the plasma membrane H+-ATPase through interaction with the PKS5 kinase. Plant Cell 22: 1313-1332.

Yuan, S., Y. Wu and D.J. Cosgrove. 2001. A fungal endoglucanase with plant cell wall extension activity. Plant Physiol. 127: 324-333.

Zeiger, E. 1983. The biology of stomatal guard cells. Annu. Rev. Plant Physiol. 34: 441-475.

Zhang, X., H. Wang, A. Takemiya, C.-P. Song, T. Kinoshita and K. Shimazaki. 2004. Inhibition of blue light-dependent H+ pumping by abscisic acid through hydrogen peroxide-induced dephosphorylation of the plasma membrane H+-ATPase in guard cell protoplasts. Plant Physiol. 136: 4150-4158.

11

How Plant Hormones and Their Interactions Affect Cell Growth

Stephen Depuydt[1,2,3,*], *Stan Van Praet*[1,2,3], *Hilde Nelissen*[2,3], *Bartel Vanholme*[2,3] *and Danny Vereecke*[4]

Introduction

The development and growth of plant organs requires orchestration between two seemingly separate phases: cell proliferation and cell expansion. These two phases are followed by cell maturation, which is mainly achieved by stiffening of the plant cell wall (Sachs 1882). The first phase of cell proliferation drives primordium growth, whereas the second phase of organ growth relies mainly on postmitotic cell expansion of the differentiating cells via either endoreduplication and/or vacuolization (den Boer and Murray 2000). Note that in plants the term 'cell expansion' refers merely to cell size increase, as the process is not necessarily accompanied by an increase in cytoplasmic content. It is also still unclear whether cell expansion or cell division initiates organ formation in plants (Sampathkumar et al. 2014).

The classical model of Heyn (1940) states that cell expansion occurs once a critical internal pressure, i.e. turgor pressure, is reached. A plant hormone, auxin, would subsequently mediate the extension of the plant cell wall. Lately, this turgor-driven cell expansion theory has however been challenged by Wei and Lintilhac (2003, 2007) who claim that the cell wall would lose stability, rather than being rearranged by a signaling cascade. Nevertheless, these two hypotheses do not differ in terms of the fundamental mechanism responsible for cell elongation: the cell wall needs to be altered in order for a plant cell to expand (Kutschera and Niklas 2013).

Indeed, plant cells are unique in the sense that they have fairly rigid cell walls that constitute their outermost layer and provide a protective physical barrier against

[1] Ghent University Global Campus, Songdomunhwa-Ro 119, Yeonsu-Gu, Incheon, 406-840 South Korea.
[2] Department of Plant Biotechnology and Bioinformatics, Ghent University, Technologiepark 927, B-9052 Ghent, Belgium.
[3] Department of Plant Systems Biology, VIB, Technologiepark 927, B-9052 Ghent, Belgium.
[4] Department of Applied Biosciences, Ghent University, Valentin Vaerwyckweg 1, B-9000 Ghent, Belgium.
[*] Corresponding author : stephen.depuydt@ghent.ac.kr

biotic and abiotic stresses. The (cumulative) strength of the cell walls of individual cells provides plants with a structural alternative for a skeleton. As such, plants can deal with gravity forces that can be considerable, for example in giant trees, or with osmotic pressure within the cells. Additionally, this enables them to penetrate compact soil or withstand stormy weather (Burton et al. 2010). Nevertheless, the function of the plant cell walls ranges far beyond a structural role. The plant cell wall is also involved in signal transduction, for instance to switch on the defense machinery following pathogen attack (e.g. Cantu et al. 2008), in cell-to-cell adhesion (e.g. Zhu et al. 1993), and in primary defense against pathogenic microorganisms (e.g. Malinovsky et al. 2014). Next to that, it is postulated that plant cell walls represent a significant source of metabolizable energy, required during the germination of many crops seeds, such as rice, wheat, maize, and barley (Knox 2008). The basic building blocks for plant cell walls are polysaccharides, but the different functions of the plant cell walls are somewhat reflected in their structure and composition: in young and growing cells, the so-called primary cell wall consists mainly of cellulose microfibrils embedded in a matrix of pectins and hemicellulose (predominantly xyloglucans in dicots and non-graminaceous monocots) (Somerville et al. 2004) and forms a tough, but extensible, extracellular matrix. In contrast, the secondary cell wall of terminally differentiated cells that stopped growing often becomes impregnated with lignin to increase wall strength (Boerjan et al. 2003), which also helps to resist compressive stresses. Thus, cell walls are highly dynamic with a great structural and compositional plasticity. At the same time, the plant cell wall plays a crucial role in directing plant growth and development, as it determines the shape and architecture of the plant body: postmitotic cell expansion is controlled by the cell wall because it can either restrict or tolerate growth of the cell (and thus the tissue and/or organ) in one or more directions. As the local flexibility of the cell wall matrix controls cell expansion, one might actually argue that plant morphogenesis is the result of differential organ growth at the level of the cell walls.

During plant growth and development, the cell proliferation and cell expansion phases and, thus, the biosynthesis and differentiation of cell walls, need to be tightly regulated in a spatiotemporal way and highly coordinated between different cells of the same organ or tissue in order to preserve organ integrity. To achieve such a complex coordination and, hence, to actually steer plant growth and development, different growth effectors have been recruited throughout plant evolution (reviewed in e.g. Depuydt and Hardtke 2011, Ubeda-Tomás et al. 2012, Wang et al. 2015). Plant hormones are major intrinsic growth regulators that can adjust plant growth depending on environmental and/or developmental cues. As such, it would seem logic that plant hormone action and cell wall expansion are directly linked. However, the major hormones that are involved in cell expansion (see below) show surprisingly little overlap in transcriptional targets with one important exception: the EXPANSINS (EXPs) and their homologs that mediate loosening of the plant cell wall (Cosgrove 2000, Nemhauser et al. 2006). Although beyond the scope of this book chapter, besides hormone signaling, several other signaling pathways are also controlling elongation *in planta*, including light signaling and circadian clock. Mounting evidence suggests that these pathways could be integrated by PHYTOCHROME INTERACTING FACTORS (PIFs), members of the basic-loop-helix family of transcriptional regulators that take a central position in photomorphogenic plant development (for a recent review, see Leivar and Monte 2014). This chapter will focus on how hormones and downstream signaling affect specifically plant cell elongation and review their independent and concerted activities. Briefly the link with cell wall remodeling will be covered, but the reader is referred to the comprehensive review by Sánchez-Rodríguez et al. (2010) for more in-depth information on this matter.

The Process of Cell Elongation

Postmitotic cell enlargement, which usually follows cell proliferation, is considered to be the most important contributor to the final size of plant cells. As mentioned, turgor pressure is crucial for cell expansion, as well as cell wall extensibility (Cosgrove 2005, Wolf et al. 2012). Although the 'acid growth theory' (also referred to as 'auxin-induced acidification') was proposed over 40 years ago, until today it is still believed to be key for cell expansion (Heyn 1931, 1940, Hager 2003). In short, plasma membrane proton pumps, specifically activated by the plant hormone auxin, would lower the pH in the apoplast, and as a result of the action of EXPs, amongst others, the cell wall would loosen, allowing turgor-driven cell enlargement (Hager 2003, Cosgrove 1998, 2000). The specific working mechanisms of EXPs remain obscure as it was shown that they do not alter covalent linkages between cell wall polysaccharides but rather they loosen non-covalent bonds in acidic conditions, temporarily releasing the cross-link between hemicellulose and cellulose fibrils (Cosgrove 2000). The auxin class of phytohormones (see below) is considered to be the major and crucial orchestrator of plant development that originated early in evolution as well (Cooke et al. 2002, Sergeeva et al. 2002) but in the past years, it has been demonstrated that *EXP* gene expression can be altered by nearly all other classes of plant hormones including brassinosteroids, gibberellins, cytokinins, abscicic acid, salicylic acid, jasmonic acid and ethylene (Cho and Kende 1997, Fleming et al. 1997, Downes and Crowell 1998, Cho and Cosgrove 2002, Sánchez-Rodríguez et al. 2010, Sun et al. 2010, Zhao et al. 2012).

Werner et al. (2001, 2003) showed that overexpression of cytokinin dehydrogenase retards shoot development but enhances root development, one of many more examples illustrating that plant hormones act in a tissue-specific way. Consequently, cell elongation and the associated cell wall regulation processes need to be studied in different organ and cell types in order to infer 'general' mechanisms and modes of action. Several model systems have significantly contributed to the understanding of cell elongation, hormones, and wall extensibility. For instance, pollen tubes show highly polarized cell expansion (reviewed in Hepler et al. 2013) and root hairs uniquely grow via cell expansion after rather complex stages of cell fate acquisition (reviewed in Mendrinna and Persson 2015). However, here the two most important model systems will be described in which the link between phytohormones and cell elongation is relatively well characterized: the hypocotyl and primary root growth in *Arabidopsis thaliana*. In Arabidopsis hypocotyls hardly any cell division occurs so almost all the cells originate from the embryo and undergo cell elongation. Additionally, in etiolated hypocotyls, synthesis, addition, and reorganization of cell wall material occur in time-separate phases (Gendreau et al. 1997). The Arabidopsis primary root is also considered to be a good model system for the action of phytohormones on cell elongation as it shows distinct zones of cell proliferation, cell elongation, and differentiation (Dolan et al. 1993, Bennett and Scheres 2010).

Different Hormones Have Been Implicated in Cell Elongation Processes

In the past years the function, biosynthesis and signaling pathways of several classes of phytohormones became relatively well understood. Whereas auxins, cytokinins, gibberellins, and brassinosteroids take a central position in the context of plant growth and development, salicylic, jasmonic and abscisic acids act more as auxillary signals in plant growth and are also involved in plant-biotic and plant-abiotic stresses (Derksen et al. 2013). Therefore, the latter three hormones will not be discussed in this chapter.

It is difficult to pinpoint the action of a single hormone to specific effects on either cell proliferation or cell enlargement for three main reasons. Firstly, hormones have highly overlapping roles in plant development. Secondly, hormones interact with each other (a few examples are given below). Thirdly, whether hormones restrict or promote growth depends on the specific cell type, the concentration and the intensity of the signal. Although this adds to the complexity of studying hormone action *in planta*, it has been generally accepted for a long time that gibberellins, brassinosteroids and auxins promote cell elongation, while ethylene restricts cell enlargement. Cytokinins on the other hand, but auxins as well, are considered to regulate cell proliferation. More recently, the strigolactones, have been indirectly implicated in cell elongation as well, mainly via cross-talk with other hormonal pathways.

The Cell Elongation-Promoting Hormones: Brassinosteroids, Gibberellins, and Auxins

Brassinosteroids

The brassinosteroids have clear effects on root growth. High levels of brassinosteroids inhibit root growth while lower levels stimulate it (Müssig and Altmann 2003). Now, it is however clear that brassinosteroids are required for cell division as well as cell elongation in root meristems. This is exemplified by the short root phenotype of several mutants in brassinosteroid signaling and biosynthesis (González-García et al. 2011, Hacham et al. 2011).

In short, brassinosteroids are detected at the membrane, by the receptor-like kinase BRASSINOSTEROID INSENSITIVE1 (BRI1), of which there are three homologs in Arabidopsis (Caño-Delgado et al. 2004, Zhou et al. 2004). Although single mutants disrupted in this receptor (such as *bri1-116*) display a short root phenotype (Li and Chory, 1997, Friedrichsen et al. 2000), this root phenotype could be rescued by *CYCLIN D3;1* (*CYCD3;1*) expression (González-García et al. 2011) suggesting that it is caused by the altered expression of several cell cycle genes, such as *CYCLIN B1;1* (*CYCB1;1*), *KIP-RELATED PROTEIN 2* (*KRP2*), and *KNOLLE (KN)*, all involved in cell division processes.

After binding of brassinosteroid hormone molecules to the BRI receptor, an association with coreceptors BRI1-ASSOCIATED RECEPTOR KINASE1 (BAK1), SOMATIC EMBRYOGENESIS RECEPTOR KINASE 1 (SERK1), and BAK1-LIKE KINASE 1 (BKK1) takes place (Wang et al. 2005, Gou et al. 2012). Interestingly, the *bak1-4/bkk1-1/serk1-8* triple mutant, blocked in brassinosteroid signal transduction (Du et al.

2012) also shows a short root phenotype. Subsequent signaling cascades, that involve multiple phosphorylation and dephosphorylation steps, activate the transcription factors BRASSINOZOLE-RESISTANT 1 (BRZ1) and BRI1-EMS SUPPRESSOR 1 (BES1), which are considered to be the main BR-related gene expression regulators (He et al. 2002, Wang et al. 2002, Yin et al. 2005). The latter seems to be a downstream target of strigolactone signaling too (Wang et al. 2013). These transcription factors are, in the absence of brassinosteroids, inactivated by phosphorylation through BRASSINOSTEROID-INSENSITIVE 2 (BIN2) activity (Kim and Wang 2010). BIN2 activity itself is inhibited by the receptor complex that is induced by binding of brassinosteroids to the BRI1 receptor. Among the target genes of BZR1 and BES1, there are a number of cell wall biosynthesis and remodeling enzymes (Sun et al. 2010, Yu et al. 2011). In addition, the levels of enzymes involved in cellulose synthesis and in linking cellulose microfibrils, were reduced in mutants impaired in brassinosteroid responses and/or signaling (reviewed in Sánchez-Rodríguez et al. 2010). Moreover, in soybean hypocotyls, it has been demonstrated that adding brassinosteroids induces the activity of xyloglucan cleaving enzymes (Oh et al. 1998), although the functional significance of this finding requires further research. Nevertheless, the general importance of brassinosteroids in elongation processes is further illustrated by the clearly shorter hypocotyls observed in brassinosteroid synthesis and response mutants, such as *bri1, constitutively photomorphogenic dwarf (cpd), de-etiolated2 (det2), dwarf1 (dwf1), brassinosteroid light and sugar1 (bls1),* and *boule1 (bul1)* (Szekeres et al. 1996, Catterou et al. 2001, Laxmi et al. 2004).

Auxins

Because auxins are important regulators of the plant cell cycle (reviewed by del Pozo and Manzano, 2014), many developmental processes in plants - from germination until senescence - depend on the general auxin status of the plant cells. Auxin has a prominent role in the acid-growth hypothesis and there is ample evidence that auxin influences the expression and activity of crucial cell wall modifiers, such as expansins, pectin methylesterases, and polygalacturonases in roots (Laskowski et al. 2006, Kumpf et al. 2013). Elegant experiments show that local application of auxin leads to demethylation of pectins at the shoot apical meristem, which causes tissue softening (Braybrook and Peaucelle 2013). Notably, several links between auxin and secondary wall formation have also been reported lately, for instance in lignification processes (Cecchetti et al. 2013, Hentrich et al. 2013).

Along the apical-basal axis in roots, auxin shows a graded distribution, which in turn will lead to clear patterns of cell proliferation and differentiation (Sabatini et al. 1999, Friml et al. 2002a, Benková et al. 2003). High levels of auxin inhibit cell proliferation e.g. in the quiescent centre (QC), and at concentrations higher than 10^{-6} M even root cell elongation is arrested (Evans et al. 1994). Lower levels of auxin are generally associated with cell proliferation in the cell division zone, substantiating classical experiments where it was shown that exogenously applied auxin stimulates cell division in cultured plant cells and plant tissues. Very low levels of auxin on the other hand result in arrest of the mitotic activity and promote cell differentiation and elongation (Grieneisen et al. 2007). Auxin import in cells is mediated by AUXIN-RESISTANT 1 (AUX1) and LIKE AUXIN-RESISTANT (LAX) carrier proteins and mutants therein show several auxin-related developmental defects (Swarup et al. 2001, Péret et al. 2012). The auxin distribution is mainly possible because of the highly specialized auxin transport system in plants. The main contributors in that system are the PIN-formed (PIN) auxin-efflux carriers (for a recent review, see Adamowski and Friml 2015). Certain PIN proteins are membrane bound, expressed specifically in the basal or apical membrane of the cells, and in specific cell types of the root, and they are continuously recycled by endocytosis. Altogether, this creates an effective and highly controlled auxin (re)distribution system called polar auxin transport (PAT) that allows shoot-originated auxin to be (re)cycled in the root. The resulting typical graded auxin distribution pattern along the root's main trajectory provides positional information and steers cell fate. Single and higher order mutants in the PIN proteins display severe growth defects, and modulation of PIN activity affects elongation rates and final cell sizes (Luschnig et al. 1998, Friml et al. 2002a, b, 2003, Benková et al. 2003, Reinhardt et al. 2003, Blilou et al. 2005, Vieten et al. 2005, Scarpella et al. 2006). The PAT system is extremely important for root development and for controlling root meristem size, and it seems to be a point of convergence for other hormones to exert their action (see below).

Tryptophan-dependent pathways and tryptophan-independent pathways both contribute to the pool of indole-3-acetic acid (IAA) *in planta* (Woodward and Bartel 2005, Zhang et al. 2008, Chandler 2009, Normanly 2010). IAA molecules are mainly synthesized in the leaves and the shoot apical meristem (SAM) although root borne auxin is important for development too. Indeed, higher order YUCCA mutants (*yuc1/ yuc4/yuc10/yuc11*), involved in tryptophan-dependent auxin biosynthesis, do not even develop a root meristem at all (Cheng et al. 2006). TRANSPORT INHIBITOR-RESISTANT 1/AUXIN SIGNALING F-BOX (TIR1/AFB) and AUXIN BINDING PROTEIN1 (ABP1) are the specific receptors of IAA (Jones et al. 1998, Dharmasiri et al. 2005, Kepinski and Leyser, 2005). In the absence of auxin, AUX/IAA proteins will heterodimerize with AUXIN RESPONSE FACTORS (ARFs) preventing transcription of auxin responsive genes (reviewed in Salehin et al. 2015). The ARFs indeed regulate the expression of target genes that have an auxin responsive element (ARE) (Boer et al 2014). Such AREs have for instance been found in the promotors of cyclins, such as CYCB; 1 and CYCA; 2 (Hu et al. 2003, Roudier et al. 2003). Upon interaction with auxin, the binding of the TIR/AFB receptors to AUX/ IAA proteins is promoted. This leads to AUX/IAA degradation via the proteasome by ubiquitination. Indeed, TIR1 and AFB1-5 contain an F-box and participate in the ubiquitination ligase complex SCF^{TIR1} (reviewed in Salehin et al. 2015). Various gain-of-function mutants in AUX/IAA genes in Arabidopsis have short hypocotyls as a consequence of cell elongation defects (Liscum and Reed 2002, Mockaitis and Estelle 2008), whereas ABP1 was shown to be involved in auxin-induced enlargement of Arabidopsis hypocotyls as well (Steffens et al. 2001).

To summarize, under high auxin concentrations, AUX/IAA degradation effectively releases ARF proteins that dimerize and regulate transcription. Under low auxin concentrations, AUX/IAA repress ARF activity and prevent them from transcribing auxin responsive genes (reviewed in Salehin et al. 2015).

Gibberellins

There are over 130 gibberellin metabolites in plants, but only a few are bioactive and their levels are tightly regulated (Yamaguchi 2008, Shani et al. 2013). Not only biosynthesis but also catabolism and metabolic deactivation will determine the pool of active gibberellins in the plant. Gibberellin can also inhibit its own biosynthesis via negative feedback mechanisms (Schomburg et al. 2003, reviewed in Hedden and Thomas 2012). Gibberellins primarily promote cell growth: for instance, exogenous treatment promotes hypocotyl elongation (Derbyshire et al. 2007) and gibberellin biosynthesis genes are associated with growing tissues in several plant systems. Additionally, gibberellin inactivation leads to clear growth defects throughout the plant kingdom (reviewed in Claeys et al. 2014). Indeed, gibberellin biosynthetic mutants (*ga1-3* and *ga3ox1/ga3ox2*) have shorter roots than wild-type plants (Ubeda-Tomás et al. 2009) and *ga1-3* and *sly1-10* F-box receptor mutants are dwarf as a result of inhibited gibberellin-induced cell growth (Dill and Sun 2001, Strader et al. 2004). However, in Arabidopsis, upregulation of gibberellin biosynthesis seems to affect both cell proliferation and cell expansion, dependent on the tissue/organ. In the proximal root meristem, gibberellins induce cell proliferation (Achard et al. 2007) and in leaves for instance, both cell proliferation and cell expansion are stimulated by gibberellins (Gonzalez et al. 2010). It has also been demonstrated that gibberellin-mediated cell expansion is required for adjacent cells to divide in the epidermis (Ubeda-Tomás et al. 2009). In the hypocotyl, gibberellins clearly link to cell expansion. The latter happens via several mechanisms, including cell wall relaxation via induced expression of EXPs and XYLOGLUCAN ENDOTRANSGLUCOSYLASES/HYDROLASES (Liu et al. 2007), but also by controlling auxin biosynthesis and transport.

Arabidopsis has three gibberellin receptors that act redundantly because only double receptor mutants show dwarfisms (Suzuki et al. 2009). Upon binding of gibberellin to these GIBBERELLIN INSENSITIVE DWARF1 (GID1) receptors, the DELLA growth inhibitors (DELLA) are destructed via an ubiquitin-proteasome pathway. These DELLA proteins are generally regarded as growth repressors since they are repressors of transcription factors that mediate gibberellin responses (Peng et al. 1997, Silverstone et al. 2001). Indeed, quadruple DELLA loss-of-function mutants show an exaggerated hypocotyl elongation (Alabadí et al. 2004). DELLAs also form important points of cross-talk with other signaling cascades, including phytohormones, but also light signaling and circadian clock regulation. For instance, DELLAs interact with PIF3 and PIF4, directly controlling cell expansion regulators (Achard et al. 2007). This interaction can be inhibited by a negative regulator of jasmonic acid signaling which likely explains the negative effect of jasmonates on growth (Yang et al. 2012). Also, gibberellins (via DELLAs) and phytochromes seem to share a common transcription module with brassinosteroids, in which DELLA degradation releases BZR1 inhibition (Bai et al. 2012). Furthermore, BZR1 also interacts with PIF4 and as such activates transcription of cell expansion genes (Bai et al. 2012). Auxin and ethylene also affect GA-mediated DELLA degradation (Achard et al. 2003, Fu and Harberd 2003). Next to that, DELLAs upregulate the CYCLIN DEPENDENT KINASE (CDK) inhibitors KRP2 and SIAMESE, thus affecting cell cycle progression (Achard et al. 2009). Finally, DELLAs directly affect auxin transport via transcriptional and posttranslational control of PIN carriers. Auxin biosynthesis on the other hand can activate gibberellin biosynthesis too (Moubayidin et al. 2010, Willige et al. 2011).

Ethylene, the Growth-Inhibiting Hormone

Ethylene regulates many processes *in planta*, but its growth inhibitory role became clearly evident with the so-called triple-response, i.e. upon treatment of etiolated seedlings with ethylene, exaggerated apical hooks, shorter roots, induction of stem swelling, and inhibition of elongation growth were observed (Guzmán and Ecker 1990). In the dark, ethylene inhibits Arabidopsis hypocotyl elongation, whereas, contradictory, ethylene and its precursor 1-aminocyclopropane-1-carboxylate (ACC) increase hypocotyl elongation in light-grown Arabidopsis seedlings (Smalle et al. 1997, Liang et al. 2012). In root meristems, ethylene inhibits cell elongation (Tanimoto et al. 1995, Pitts et al. 1998, Růžička et al. 2007). Indeed, mutants with increased levels of ethylene (e.g. *ctr1-1* and *etol-2*) display shorter roots (Ortega-Martínez et al. 2007, Thomann et al. 2009).

The ethylene signaling cascade is more complex than initially put forward and involves many regulatory circuits (reviewed in Yang et al. 2015). In short, five putative endoplasmic reticulum-bound receptors are involved: the His-Kinase two-component regulators ETHYLENE RESPONSE 1 (ETR1), ETR2, ETHYLENE RESPONSE SENSOR 1 (ERS1), ERS2, and ETHYLENE INSENSITIVE 4 (EIN4) (Bleecker et al. 1998, Hua et al. 1995, Hua and Meyerowitz 1998, Sakai et al. 1998). Upon binding with ethylene, CONSTITUTIVE TRIPLE RESPONSE 1 (CTR1), a serine/threonine protein kinase that acts downstream of the ethylene receptors is inactivated (Kieber et al. 1993, Zhong and Chang 2012). In the absence of ethylene, CTR1 is active and represses downstream components, among which EIN2. Upon ethylene binding, EIN2 is derepressed which in turn allows transcriptional changes via EIN3, which will activate ERF1 and ETHYLENE INSENSITIVE3 LIKE (EIL) transcription factors. The *ein2* mutant is defective in all ethylene responses, demonstrating its critical importance in ethylene signaling. Furthermore, in the absence of ethylene, specific F-box proteins trigger proteasome-mediated degradation of EIN2 and EIN3/EIL (Guo and Ecker 2003, Gagne et al. 2004, Konishi and Yanagisawa 2008, Qiao et al. 2009).

Inhibition of cell expansion through ethylene takes place because of complex interactions with both auxin biosynthesis and auxin transport. Stepanova et al. (2005, 2008) demonstrated that auxin biosynthesis as well as basipetal auxin transport is stimulated by ethylene in root tips, setting the right conditions for inhibiting cell elongation via altering auxin disposibility (Luschnig et al. 1998, Růžička et al. 2007, Swarup et al. 2007, Negi et al. 2008, Robles et al. 2013). The increased auxin levels in their turn enhance ethylene synthesis, creating a positive feedback loop (Swarup et al. 2007). Also interactions and cross-talk between ethylene and gibberellins have been noted: for instance, DELLA double mutants are less responsive to ACC than wild type, and gibberellin can at least partly revert the ethylene mediated inhibition of root growth (Achard et al. 2003). Likewise, it was shown that ethylene production in Arabidopsis seedlings can stimulate brassinosteroid production, highlighting an important cross-talk between ethylene and brassinosteroids (Woeste et al. 1999).

Cytokinins and Strigolactones Indirectly Affect Elongation

Cytokinins

Classically, cytokinins have been described as controlling proliferation in the shoot and differentiation in the root by, among others, inducing *CYCD3;1* expression (Riou-Khamlichi et al. 1999, Ferreira and Kieber 2005, Dello Ioio et al. 2007, 2008). In short, there are three membrane-bound histidine kinase receptors: ARABIDOPSIS HISTIDINE KINASE 2 (AHK2), AHK3, and AHK4 (Hwang and Sheen 2001, Inoue et al. 2001), that after binding to cytokinin, induce a phosphorylation cascade involving ARABIDOPSIS HISTIDINE PHOSPHOTRANSFER PROTEINs (AHPs) that ultimately will phosphorylate ARABIDOPSIS RESPONSE REGULATORs (ARRs) in the nucleus. The so-called B-type ARRs are positive transcriptional regulators and the A-type are repressors that likely act through cooperation with other transcription factors (To et al. 2004, Argueso et al. 2010). Specific members of these AHP and ARR protein families (e.g. ARR1, ARR12, and AHP6) represent important points of interplay between cytokinins and auxins. It is that specific interplay that will ultimately determine root (meristem) size. In protoxylem development for instance, high auxin levels upregulate cytokinin biosynthesis but at the same time *AHP6* expression is upregulated, repressing cytokinin signaling (De Rybel et al. 2014). For establishing the promotion of cell differentiation in the transition zone of the root meristem, cytokinin acts via dampening the PAT: ARR1 and ARR12 control an AUX/IAA gene, SHORT HYPOCOTYL 2 (SHY2), that inhibits *PIN* gene expression (Dello Ioio et al. 2007, 2008) and in turn promotes differentiation and elongation of cells in the root meristem. At early stages of root meristem development, *ARR1* expression in turn is controlled by a specific DELLA factor, REPRESSOR OF GA1-3 (RGA), providing a clear link with gibberellin signaling (Moubayidin et al. 2010). Also brassinosteroid signaling, via BREVIX RADIX (BRX) is involved in this process: the BRX transcriptional coregulator (Mouchel et al. 2004) is a target of the auxin response factor MONOPTEROS (MP) that can induce *PIN3* expression in complex feedback loops with cytokinin signaling via SHY2 (Scacchi et al. 2010). Even more, cytokinins inhibit cell elongation also by interplaying with ethylene signaling via ETR1 and EIN2 (Růžička et al. 2009, Kushwah et al. 2011).

Strigolactones

Strigolactones are the newest addition to the hormone classes *in planta*. Their hormonal actions were originally attributed to inhibitory effects on the outgrowth of axillary buds (Gomez-Roldan et al. 2008, Umehara et al. 2008), but nowadays multiple roles in plant growth and development have been described. At least two key components participate in perception of strigolactones leading to an ubiquitin-mediated protein degradation via an SCF mediated signaling pathway: MAX2, an F-box protein, and D14, an α/β hydrolase, in Arabidopsis (Stirnberg et al. 2007, Hamiaux et al. 2012). Among the downstream targets that have been described, there is BRANCHED 1/TEOSINTE BRANCHED1 LIKE1 (BRC1) (Braun et al. 2012) and, interestingly, BES1, the positive regulator of the brassinosteroid signaling pathway (Wang et al. 2013), pointing toward cross-talk with brassinosteroids.

Only very recent data shed light on possible effects of strigolactones on cell growth. In pea for instance, it has been demonstrated that strigolactones affect stem elongation by stimulating cell division in the epidermal internode cells (de Saint Germain et al. 2013). During germination and growth of rice, however, strigolactones were found to negatively regulate cell division in the mesocotyl in darkness (Hu et al. 2010, 2014). Nonetheless, both reports clearly indicate that strigolactones did not interfere with cell elongation processes. Conversely, it was shown that strigolactones inhibit hypocotyl elongation in Arabidopsis (Tsuchiya et al. 2010), which is dependent on several components of light signaling pathways, with the transcription factor LONG HYPOCOTYL 5 (HY5) as a possible integrator of the strigolactone and light signaling pathways (Tsuchiya et al. 2010, Jia et al. 2014). Lauressergues et al. (2015) showed a reduced shoot elongation in strigolactone-deficient *Medicago truncatula* mutants and demonstrated that this was not caused by reduced resource allocation as in strigolactone-related dwarf mutants from other species (Zou et al. 2006, Kohlen et al. 2012), but they have yet to examine this at the cellular level.

In root systems, strigolactones regulate lateral root development (Ruyter-Spira et al. 2011, Kapulnik et al. 2011) and they increase cell numbers in the primary root meristem and transition zones probably through inhibitory effects on auxin efflux carriers (Ruyter-Spira et al. 2011, Koren et al. 2013). Strigolactones may indeed dampen auxin transport, as demonstrated by PIN1 depletion from the membrane in Arabidopsis stems (Shinohara et al. 2013). In root meristems however, although PIN1 was reported to be less abundant after long-term treatment with synthetic strigolactones (Ruyter-Spira et al. 2011), there is no direct evidence of this so far (Shinohara et al. 2013). Nonetheless, PIN localization and polarization seems to be affected by strigolactone treatment or altered transport (Pandya-Kumar et al. 2014, Sasse et al. 2015). Furthermore, auxin can enhance strigolactone biosynthesis (Hayward et al. 2009) and strigolactones in turn can inhibit auxin biosynthesis (Sang et al. 2014). It was also recorded that strigolactones have a positive effect on root hair growth, which is, after cell fate acquisition, solely dependent on cell elongation. This effect however requires ethylene synthesis, and root hair elongation by strigolactones and ethylene seems to be relying on a common regulatory pathway, with ethylene being epistatic (Kapulnik et al. 2011). Furthermore, auxin signaling contributed to the strigolactone mediated root hair response, but was not absolutely required (Kapulnik et al. 2011). The crosstalk between strigolactones and auxin seems to be key for every developmental aspect regulated by this new class of hormones, and future research will undoubtedly yield interesting points of convergence and divergence with other hormonal pathways.

In terms of mechanisms of strigolactone-mediated cell expansion, more research is also required, but an interesting link has been uncovered between *EXP* genes and karrikinolide, a compound in smoke that structurally resembles strigolactones and even relies in part on the strigolactone signaling pathway (Jain et al. 2008, Scaffidi et al. 2014).

Cross-talk in Elongation: Influencing Biosynthesis, Transport, and Signal Transduction

An overview of the main players in the hormonal cross-talk for elongating cells in the hypocotyl and the proximal root meristem is presented (Fig. 1). Cross-talk *sensu strictu* occurs when two pathways are not independent. It can be positive (synergistic or additive) and it can be negative.

One important thing to note is that all hormonal interactions are context-specific and tissue-dependent, which greatly adds to the complexity of understanding hormonal networks and their actions. Figure. 1 shows clearly that there are numerous points of cross-talk, sometimes involving more than three hormones, again adding to the complexity of studying these interaction. It is also clear that the levels and/ or signaling pathways of one hormone affect biosynthesis (or rather homeostasis) of

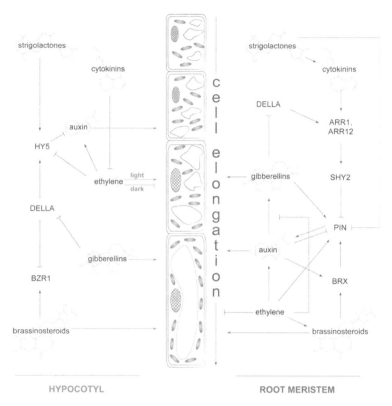

FIGURE 1   Scheme presenting the main players in some of the cross-talk mechanisms that regulate hormone-mediated cell expansion in the hypocotyl (right) and the root meristem (left). For abbreviations, see text.

other hormones, which in turn leads to altered signaling events that then again have repercussions on cell responses. Additionally, apart from influencing biosynthesis of other hormones, cross-talk can also affect transport, with the polar auxin transport machinery playing a very crucial role in this. Next to that, one hormone clearly influences signaling components of other hormones. Since the signaling machineries for the different classes of hormones show great similarities, this is very plausible. For instance, auxin, strigolactones, ethylene, gibberellins and jasmonic acids all rely on the ubiquitin dependent proteasome system for hormone perception, de-repression of hormone signaling pathways, degradation of hormone specific transcription factors, and/or regulation of hormone biosynthesis (Santner and Estelle 2010). Furthermore, downstream targets of a specific hormone pathway can be important regulators and integrators of other hormonal processes too. In that respect, DELLAs for instance are involved in at least four hormonal pathways during hypocotyl growth, but also the most recently discovered class of phytohormones, the strigolactones, seem to share transcriptional regulators with other pathways. Altogether, this leads to the conclusion that the action of a specific hormone on plant cell elongation cannot be isolated from other hormones, and that this in fact is one of the main mechanisms, partly explaining the pleiotropy of hormone-mediated processes *in planta* as well.

Acknowledgments

The authors thank Martine De Cock for help with the manuscript and Christian Hardtke for critically reading the manuscript. S.D. was indebted to the Research Foundation Flanders for a post-doctoral fellowship.

References

Achard, P., A. Gusti, S. Cheminant, M. Alioua, S. Dhondt, F. Coppens, G.T.S. Beemster, and P. Genschik. 2009. Gibberellin signaling controls cell proliferation in Arabidopsis. Curr. Biol. 19: 1188-1193.

Achard, P., L. Liao, C. Jiang, T. Desnos, J. Bartlett, X. Fu and N.P. Harberd. 2007. DELLAs contribute to plant photomorphogenesis. Plant Physiol. 143: 1163-1172.

Achard, P., W.H. Vriezen, D. Van Der Straeten and N.P. Harberd. 2003. Ethylene regulates Arabidopsis development via the modulation of DELLA protein growth repressor function. Plant Cell 15: 2816-2825.

Adamowski, M. and J. Friml. 2015. PIN-dependent auxin transport: action, regulation, and evolution. Plant Cell 27: 20-32.

Alabadí, D., J. Gil, M.A. Blázquez and J.L. García-Martínez. 2004. Gibberellins repress photomorphogenesis in darkness. Plant Physiol. 134: 1050-1057.

Argueso, C.T., T. Raines and J.J. Kieber. 2010. Cytokinin signaling and transcriptional networks. Curr. Opin. Plant Biol. 13: 533-539.

Bai, M.-Y., J.-X. Shang, E. Oh, M. Fan, Y. Bai, R. Zentella, T.-p. Sun and Z.-Y. Wang. 2012. Brassinosteroid, gibberellin and phytochrome impinge on a common transcription module in Arabidopsis. Nat. Cell Biol. 14: 810-817.

Benková, E., M. Michniewicz, M. Sauer, T. Teichmann, D. Seifertová, G. Jürgens and J. Friml. 2003. Local, efflux-dependent auxin gradients as a common module for plant organ formation. Cell 115: 591-602.

Bennett, T. and B. Scheres. 2010. Root development - two meristems for the price of one? pp. 67-102 *In*: M.C.P. Timmermans (ed.). Plant Development (Current Topics in Developmental Biology, Vol. 91). Elsevier, Amsterdam.

Bleecker, A.B., J.J. Esch, A.E. Hall, F.I. Rodríguez and B.M. Binder. 1998. The ethylene-receptor family from Arabidopsis: structure and function. Philos. Trans. R. Soc. Lond. B-Biol. Sci. 353: 1405-1412.

Blilou, I., J. Xu, M. Wildwater, V. Willemsen, I. Paponov, J. Friml, R. Heidstra, M. Aida, K. Palme and B. Scheres. 2005. The PIN auxin efflux facilitator network controls growth and patterning in Arabidopsis roots. Nature 433: 39-44.

Boer, D.R., A. Freire-Rios, W.A.M. van den Berg, T. Saaki, I.W. Manfield, S. Kepinski, I. López-Vidrieo, J.M. Franco-Zorrilla, S.C. de Vries, R. Solano, D. Weijers and M. Coll. 2014. Structural basis for DNA binding specificity by the auxin-dependent ARF transcription factors. Cell 156: 577-589.

Boerjan, W., J. Ralph and M. Baucher. 2003. Lignin biosynthesis. Annu. Rev. Plant Biol. 54: 519-546.

Braun, N., A. de Saint Germain, J.-P. Pillot, S. Boutet-Mercey, M. Dalmais, I. Antoniadi, X. Li, A. Maia-Grondard, C. Le Signor, N. Bouteiller, D. Luo, A. Bendahmane, C. Turnbull and C. Rameau. 2012. The pea TCP transcription factor PsBRC1 acts downstream of strigolactones to control shoot branching. Plant Physiol. 158: 225-238.

Braybrook, S.A. and A. Peaucelle. 2013. Mechano-chemical aspects of organ formation in *Arabidopsis thaliana*: the relationship between auxin and pectin. PLoS ONE 8: e57813.

Burton, R.A., M.J. Gidley and G.B. Fincher. 2010. Heterogeneity in the chemistry, structure and function of plant cell walls. Nat. Chem. Biol. 6: 724-732.

Cantu, D., A.R. Vicente, J.M. Labavitch, A.B. Bennett and A.L.T. Powell. 2008. Strangers in the matrix: plant cell walls and pathogen susceptibility. Trends Plant Sci. 13: 610-617.

Caño-Delgado, A., Y. Yin, C. Yu, D. Vafeados, S. Mora-García, J.-C. Cheng, K.H. Nam, J. Li and J. Chory. 2004. BRL1 and BRL3 are novel brassinosteroid receptors that function in vascular differentiation in Arabidopsis. Development 131: 5341-5351.

Catterou, M., F. Dubois, H. Schaller, L. Aubanelle, B. Vilcot, B.S. Sangwan-Norreel and R.S. Sangwan. 2001. Brassinosteroids, microtubule and cell elongation in *Arabidopsis thaliana*. II. Effects of brassinosteroids on microtubule and cell elongation in the *bul1* mutant. Planta 212: 673-683.

Cecchetti, V., M.M. Altamura, P. Brunetti, V. Petrocelli, G. Falasca, K. Ljung, P. Costantino and M. Cardarelli. 2013. Auxin controls Arabidopsis anther dehiscence by regulating endothecium lignification and jasmonic acid biosynthesis. Plant J. 74: 411-422.

Chandler, J.W. 2009. Local auxin production: a small contribution to a big field. BioEssays 31: 60-70.

Cheng, Y., X. Dai and Y. Zhao. 2006. Auxin biosynthesis by the YUCCA flavin monooxygenases controls the formation of floral organs and vascular tissues in Arabidopsis. Genes Dev. 20: 1790-1799.

Cho, H.-T. and D.J. Cosgrove. 2002. Regulation of root hair initiation and expansin gene expression in Arabidopsis. Plant Cell 14: 3237-3253.

Cho, H.-T. and H. Kende. 1997. Expression of expansin genes is correlated with growth in deepwater rice. Plant Cell 9: 1661-1671.

Claeys, H., S. De Bodt and D. Inzé. 2014. Gibberellins and DELLAs: central nodes in growth regulatory networks. Trends Plant Sci. 19: 231-239.

Cooke, T.J., D. Poli, A.E. Sztein and J.D. Cohen. 2002. Evolutionary patterns in auxin action. Plant Mol Biol. 49: 319-338.

Cosgrove, D.J. 1998. Molecular regulation of plant cell wall extensibility. Gravit. Space Biol. Bull. 11: 61-70.

Cosgrove, D.J. 2000. Loosening of plant cell walls by expansins. Nature 407: 321-326.

Cosgrove, D.J. 2005. Growth of the plant cell wall. Nat. Rev. Mol. Cell Biol. 6: 850-861.

De Rybel, B., M. Adibi, A.S. Breda, J.R. Wendrich, M.E. Smit, O. Novák, N. Yamaguchi, S. Yoshida, G. Van Isterdael, J. Palovaara, B. Nijsse, M.V. Boekschoten, G. Hooiveld, T. Beeckman, D. Wagner, K. Ljung, C. Fleck and D. Weijers. 2014. Integration of growth and patterning during vascular tissue formation in Arabidopsis. Science 345: 125-215.

de Saint Germain, A., S. Bonhomme, F.-D. Boyer and C. Rameau. 2013. Novel insights into strigolactone distribution and signalling. Curr. Opin. Plant Biol. 16: 583-589.

del Pozo, J.C. and C. Manzano. 2014. Auxin and the ubiquitin pathway. Two players-one target: the cell cycle in action. J. Exp. Bot. 65: 2617-2632.

Dello Ioio, R., K. Nakamura, L. Moubayidin, S. Perilli, M. Taniguchi, M.T. Morita, T. Aoyama, P. Costantino and S. Sabatini. 2008. A genetic framework for the control of cell division and differentiation in the root meristem. Science 322: 1380-1384.

Dello Ioio, R., F. Scaglia Linhares, E. Scacchi, E. Casamitjana-Martinez, R. Heidstra, P. Costantino and S. Sabatini. 2007. Cytokinins determine Arabidopsis root-meristem size by controlling cell differentiation. Curr. Biol. 17: 678-682.

den Boer, B.G.W. and J.A.H. Murray. 2000. Control of plant growth and development through manipulation of cell-cycle genes. Curr. Opin. Biotechnol. 11: 138-145.

Depuydt, S. and C.S. Hardtke. 2011. Hormone signalling crosstalk in plant growth regulation. Curr. Biol. 21: R365-R373.

Derbyshire, P., M.C. McCann and K. Roberts. 2007. Restricted cell elongation in Arabidopsis hypocotyls is associated with a reduced average pectin esterification level. BMC Plant Biol. 7: 31.

Derksen, H., C. Rampitsch and F. Daayf. 2013. Signaling cross-talk in plant disease resistance. Plant Sci. 207: 79-87.

Dharmasiri, N., S. Dharmasiri and M. Estelle. 2005. The F-box protein TIR1 is an auxin receptor. Nature 435: 441-445.

Dill, A. and T.-p. Sun. 2001. Synergistic derepression of gibberellin signaling by removing RGA and GAI function in *Arabidopsis thaliana*. Genetics 159: 777-785.

Dolan, L., K. Janmaat, V. Willemsen, P. Linstead, S. Poethig, K. Roberts and B. Scheres. 1993. Cellular organisation of the *Arabidopsis thaliana* root. Development 119: 71-84.

Downes, B.P. and D.N. Crowell. 1998. Cytokinin regulates the expression of a soybean ß-expansin gene by a post-transcriptional mechanism. Plant Mol. Biol. 37: 437-444.

Du, J., H. Yin, S. Zhang, Z. Wei, B. Zhao, J. Zhang, X. Gou, H. Lin and J. Li. 2012. Somatic embryogenesis receptor kinases control root development mainly via brassinosteroid-independent actions in *Arabidopsis thaliana*. J. Integr. Plant Biol. 54: 388-399.

Evans, M.L., H. Ishikawa and M.A. Estelle. 1994. Responses of Arabidopsis roots to auxin studied with high temporal resolution: Comparison of wild type and auxin-response mutants. Planta 194: 215-222.

Ferreira, F.J. and J.J. Kieber. 2005. Cytokinin signaling. Curr. Opin. Plant Biol. 8: 518-525.

Fleming, A.J., S. McQueen-Mason, T. Mandel and C. Kuhlemeier. 1997. Induction of leaf primordia by the cell wall protein expansin. Science 276: 1415-1418.

Friedrichsen, D.M., C.A.P. Joazeiro, J. Li, T. Hunter and J. Chory. 2000. Brassinosteroid-insensitive-1 is a ubiquitously expressed leucine-rich repeat receptor serine/threonine kinase. Plant Physiol. 123: 1247-1255.

Friml, J., E. Benková, I. Blilou, J. Wisniewska, T. Hamann, K. Ljung, S. Woody, G. Sandberg, B. Scheres, G. Jürgens and K. Palme. 2002a. AtPIN4 mediates sink-driven auxin gradients and root patterning in Arabidopsis. Cell 108: 661-673.

Friml, J., J. Wiśniewska, E. Benková, K. Mendgen and K. Palme. 2002b. Lateral relocation of auxin efflux regulator PIN3 mediates tropism in Arabidopsis. Nature 415: 806-809.

Friml, J., A. Vieten, M. Sauer, D. Weijers, H. Schwarz, T. Hamann, R. Offringa and G. Jürgens. 2003. Efflux-dependent auxin gradients establish the apical--basal axis of Arabidopsis. Nature 426: 147-153.

Fu, X. and N.P. Harberd. 2003. Auxin promotes Arabidopsis root growth by modulating gibberellin response. Nature 421: 740-743.

Gagne, J.M., J. Smalle, D.J. Gingerich, J.M. Walker, S.-D. Yoo, S. Yanagisawa and R.D. Vierstra. 2004. Arabidopsis EIN3-binding F-box 1 and 2 form ubiquitin-protein ligases that repress ethylene action and promote growth by directing EIN3 degradation. Proc. Natl. Acad. Sci. USA 101: 6803-6808.

Gendreau, E., J. Traas, T. Desnos, O. Grandjean, M. Caboche and H. Höfte. 1997. Cellular basis of hypocotyl growth in *Arabidopsis thaliana*. Plant Physiol. 114: 295-305.

Gomez-Roldan, V., S. Fermas, P.B. Brewer, V. Puech-Pagès, E.A. Dun, J.-P. Pillot, F. Letisse, R. Matusova, S. Danoun, J.-C. Portais, H. Bouwmeester, G. Bécard, C.A. Beveridge, C. Rameau and S.F. Rochange. 2008. Strigolactone inhibition of shoot branching. Nature 455: 189-194.

Gonzalez, N., S. De Bodt, R. Sulpice, Y. Jikumaru, E. Chae, S. Dhondt, T. Van Daele, L. De Milde, D. Weigel, Y. Kamiya, M. Stitt, G.T.S. Beemster and D. Inzé. 2010. Increased leaf size: different means to an end. Plant Physiol. 153: 1261-1279.

González-García, M.-P., J. Vilarrasa-Blasi, M. Zhiponova, F. Divol, S. Mora-García, E. Russinova and A.I. Caño-Delgado. 2011. Brassinosteroids control meristem size by promoting cell cycle progression in Arabidopsis roots. Development 138: 849-859.

Gou, X., H. Yin, K. He, J. Du, J. Yi, S. Xu, H. Lin, S.D. Clouse and J. Li. 2012. Genetic evidence for an indispensable role of somatic embryogenesis receptor kinases in brassinosteroid signaling. PLoS Genet. 8: e1002452.

Grieneisen, V.A., J. Xu, A.F.M. Marée, P. Hogeweg and B. Scheres. 2007. Auxin transport is sufficient to generate a maximum and gradient guiding root growth. Nature 449: 1008-1013.

Guo, H. and J.R. Ecker. 2003. Plant responses to ethylene gas are mediated by SCF[EBF1/EBF2]-dependent proteolysis of EIN3 transcription factor. Cell 115: 667-677.

Guzmán, P. and J.R. Ecker. 1990. Exploiting the triple response of Arabidopsis to identify ethylene-related mutants. Plant Cell 2: 513-523.

Hacham, Y., N. Holland, C. Butterfield, S. Ubeda-Tomas, M.J. Bennett, J. Chory and S. Savaldi-Goldstein. 2011. Brassinosteroid perception in the epidermis controls root meristem size. Development 138: 839-848.

Hager, A. 2003. Role of the plasma membrane H^+-ATPase in auxin-induced elongation growth: historical and new aspects. J. Plant Res. 116: 483-505.

Hamiaux, C., R.S.M. Drummond, B.J. Janssen, S.E. Ledger, J.M. Cooney, R.D. Newcomb and K.C. Snowden. 2012. DAD2 is an a/ß hydrolase likely to be involved in the perception of the plant branching hormone, strigolactone. Curr. Biol. 22: 2032-2036.

Hayward, A., P. Stirnberg, C. Beveridge and O. Leyser. 2009. Interactions between auxin and strigolactone in shoot branching control. Plant Physiol. 151: 400-412.

He, J.-X., J.M. Gendron, Y. Yang, J. Li and Z.-Y. Wang. 2002. The GSK3-like kinase BIN2 phosphorylates and destabilizes BZR1, a positive regulator of the brassinosteroid signaling pathway in Arabidopsis. Proc. Natl. Acad. Sci. USA 99: 10185-10190.

Hedden, P. and S.G. Thomas. 2012. Gibberellin biosynthesis and its regulation. Biochem. J. 444: 11-25.

Hentrich, M., B. Sánchez-Parra, M.-M. Pérez Alonso, V. Carrasco Loba, L. Carrillo, J. Vicente-Carbajosa, J. Medina and S. Pollmann. 2013. *YUCCA8* and *YUCCA9* over-expression reveals a link between auxin signaling and lignification through the induction of ethylene biosynthesis. Plant Signal. Behav. 8: e26363.

Hepler, P.K., C.M. Rounds and L.J. Winship. 2013. Control of cell wall extensibility during pollen tube growth. Mol. Plant 6: 998-1017.

Heyn, A.N.J. 1931. Der Mechanismus der Zellstreckung. Rec. Trav. Bot. Neerl. 28: 113-244.

Heyn, A.N.J. 1940. The physiology of cell elongation. Bot. Rev. 6: 515-574.

Hu, Y., Q. Xie and N.-H. Chua. 2003. The Arabidopsis auxin-inducible gene *ARGOS* controls lateral organ size. Plant Cell 15: 1951-1961.

Hu, Z., T. Yamauchi, J. Yang, Y. Jikumaru, T. Tsuchida-Mayama, H. Ichikawa, I. Takamure, Y. Nagamura, N. Tsutsumi, S. Yamaguchi, J. Kyozuka and M. Nakazono. 2014. Strigolactone and cytokinin act antagonistically in regulating rice mesocotyl elongation in darkness. Plant Cell Physiol. 55: 30-41.

Hu, Z., H. Yan, J. Yang, S. Yamaguchi, M. Maekawa, I. Takamure, N. Tsutsumi, J. Kyozuka and M. Nakazono. 2010. Strigolactones negatively regulate mesocotyl elongation in rice during germination and growth in darkness. Plant Cell Physiol. 51: 1136-1142.

Hua, J. and E.M. Meyerowitz. 1998. Ethylene responses are negatively regulated by a receptor gene family in *Arabidopsis thaliana*. Cell 94: 261-271.

Hua, J., C. Chang, Q. Sun and E.M. Meyerowitz. 1995. Ethylene insensitivity conferred by Arabidopsis *ERS* gene. Science 269: 1712-1714.

Hwang, I. and J. Sheen. 2001. Two-component circuitry in Arabidopsis cytokinin signal transduction. Nature 413: 383-389.

Inoue, T., M. Higuchi, Y. Hashimoto, M. Seki, M. Kobayashi, T. Kato, S. Tabata, K. Shinozaki and T. Kakimoto. 2001. Identification of CRE1 as a cytokinin receptor from Arabidopsis. Nature 409: 1060-1063.

Jain, N., G.D. Ascough and J. Van Staden. 2008. A smoke-derived butenolide alleviates HgCl₂ and ZnCl₂ inhibition of water uptake during germination and subsequent growth of tomato – possible involvement of aquaporins. J. Plant Physiol. 165: 1422-1427.

Jia, K.-P., Q. Luo, S.-B. He, X.-D. Lu and H.-Q. Yang. 2014. Strigolactone-regulated hypocotyl elongation is dependent on cryptochrome and phytochrome signaling pathways in Arabidopsis. Mol. Plant 7: 528-540.

Jones, A.M., K.-H. Im, M.A. Savka, M.-J. Wu, N.G. DeWitt, R. Shillito and A.N. Binns. 1998. Auxin-dependent cell expansion mediated by overexpressed auxin-binding protein 1. Science 282: 1114-1117.

Kapulnik, Y., P.-M. Delaux, N. Resnick, E. Mayzlish-Gati, S. Wininger, C. Bhattacharya, N. Séjalon-Delmas, J.-P. Combier, G. Bécard, E. Belausov, T. Beeckman, E. Dor, J. Hershenhorn and H. Koltai. 2011. Strigolactones affect lateral root formation and root-hair elongation in Arabidopsis. Planta 233: 209-216.

Kepinski, S. and O. Leyser. 2005. The Arabidopsis F-box protein TIR1 is an auxin receptor. Nature 435: 446-451.

Kieber, J.J., M. Rothenberg, G. Roman, K.A. Feldmann and J.R. Ecker. 1993. *CTR1*, a negative regulator of the ethylene response pathway in Arabidopsis, encodes a member of the Raf family of protein kinases. Cell 72: 427-441.

Kim, T.-W. and Z.-Y. Wang. 2010. Brassinosteroid signal transduction from receptor kinases to transcription factors. Annu. Rev. Plant Biol. 61: 681-704.

Knox, J.P. 2008. Revealing the structural and functional diversity of plant cell walls. Curr. Opin. Plant Biol. 11: 308-313.

Kohlen, W., T. Charnikhova, M. Lammers, T. Pollina, P. Tóth, I. Haider, M.J. Pozo, R.A. de Maagd, C. Ruyter-Spira, H.J. Bouwmeester and J.A. López-Ráez. 2012. The tomato *CAROTENOID CLEAVAGE DIOXYGENASE8* (*SlCCD8*) regulates rhizosphere signaling, plant architecture and affects reproductive development through strigolactone biosynthesis. New Phytol. 196: 535-547.

Konishi, M. and S. Yanagisawa. 2008. Two different mechanisms control ethylene sensitivity in Arabidopsis via the regulation of *EBF2* expression. Plant Signal. Behav. 3: 749-751.

Koren, D., N. Resnick, E. Mayzlish Gati, E. Belausov, S. Weininger, Y. Kapulnik and H. Koltai. 2013. Strigolactone signaling in the endodermis is sufficient to restore root responses and involves SHORT HYPOCOTYL 2 (SHY2) activity. New Phytol. 198: 866-874.

Kumpf, R.P., C.-L. Shi, A. Larrieu, I.M. Stø, M.A. Butenko, B. Péret, E.S. Riiser, M.J. Bennett and R.B. Aalen. 2013. Floral organ abscission peptide IDA and its HAE/ HSL2 receptors control cell separation during lateral root emergence. Proc. Natl. Acad. Sci. USA 110: 5235-5240.

Kushwah, S., A.M. Jones and A. Laxmi. 2011. Cytokinin interplay with ethylene, auxin, and glucose signaling controls Arabidopsis seedling root directional growth. Plant Physiol. 156: 1851-1866.

Kutschera, U. and K.J. Niklas. 2013. Cell division and turgor-driven stem elongation in juvenile plants: a synthesis. Plant Sci. 207: 45-56.

Laskowski, M., S. Biller, K. Stanley, T. Kajstura and R. Prusty. 2006. Expression profiling of auxin-treated *Arabidopsis* roots: toward a molecular analysis of lateral root emergence. Plant Cell Physiol. 47: 788-792.

Lauressergues, D., O. André, J. Peng, J. Wen, R. Chen, P. Ratet, M. Tadege, K.S. Mysore and S.F. Rochange. 2015. Strigolactones contribute to shoot elongation and to the formation of leaf margin serrations in *Medicago truncatula* R108. J. Exp. Bot. 66: 1237-1244.

Laxmi, A., L.K. Paul, J.L. Peters and J.P. Khurana. 2004. Arabidopsis constitutive photomorphogenic mutant, *bls1*, displays altered brassinosteroid response and sugar sensitivity. Plant Mol. Biol. 56: 185-201.

Leivar, P. and E. Monte. 2014. PIFs: systems integrators in plant development. Plant Cell 26: 56-78.

Li, J. and J. Chory. 1997. A putative leucine-rich repeat receptor kinase involved in brassinosteroid signal transduction. Cell 90: 929-938.

Liang, X., H. Wang, L. Mao, Y. Hu, T. Dong, Y. Zhang, X. Wang and Y. Bi. 2012. Involvement of COP1 in ethylene- and light-regulated hypocotyl elongation. Planta 236: 1791-1802.

Liscum, E. and J.W. Reed. 2002. Genetics of Aux/IAA and ARF action in plant growth and development. Plant Mol. Biol. 49: 387-400.

Liu, Y., D. Liu, H. Zhang, H. Gao, X. Guo, D. Wang, X. Zhang and A. Zhang. 2007. The α- and β-expansin and xyloglucan endotransglucosylase/hydrolase gene families of wheat: molecular cloning, gene expression, and EST data mining. Genomics 90: 516-529.

Luschnig, C., R.A. Gaxiola, P. Grisafi and G.R. Fink. 1998. EIR1, a root-specific protein involved in auxin transport, is required for gravitropism in *Arabidopsis thaliana*. Genes Dev. 12: 2175-2187.

Malinovsky, F.G., J.U. Fangel and W.G.T. Willats. 2014. The role of the cell wall in plant immunity. Front. Plant Sci. 5: 178.

Mendrinna, A. and S. Persson. 2015. Root hair growth: it's a one way street. F1000 Prime Rep. 7: 23.

Mockaitis, K. and M. Estelle. 2008. Auxin receptors and plant development: a new signaling paradigm. Annu. Rev. Cell Dev. Biol. 24: 55-80.

Moubayidin, L., S. Perilli, R. Dello Ioio, R. Di Mambro, P. Costantino and S. Sabatini. 2010. The rate of cell differentiation controls the Arabidopsis root meristem growth phase. Curr. Biol. 20: 1138-1143.

Mouchel, C.F., G.C. Briggs and C.S. Hardtke. 2004. Natural genetic variation in Arabidopsis identifies *BREVIS RADIX*, a novel regulator of cell proliferation and elongation in the root. Genes Dev. 18: 700-714.

Müssig, C. and T. Altmann. 2003. Genomic brassinosteroid effects. J. Plant Growth Regul. 22: 313-324.

Negi, S., M.G. Ivanchenko and G.K. Muday. 2008. Ethylene regulates lateral root formation and auxin transport in *Arabidopsis thaliana*. Plant J. 55: 175-187.

Nemhauser, J.L., F. Hong and J. Chory. 2006. Different plant hormones regulate similar processes through largely nonoverlapping transcriptional responses. Cell 126: 467-475.

Normanly, J. 2010. Approaching cellular and molecular resolution of auxin biosynthesis and metabolism. Cold Spring Harb. Perspect. Biol. 2: a001594.

Oh, M.-H., W.G. Romanow, R.C. Smith, E. Zamski, J. Sasse and S.D. Clouse. 1998. Soybean *BRU1* encodes a functional xyloglucan endotransglycosylase that is highly expressed in inner epicotyl tissues during brassinosteroid-promoted elongation. Plant Cell Physiol. 39: 124-130.

Ortega-Martínez, O., M. Pernas, R.J. Carol and L. Dolan. 2007. Ethylene modulates stem cell division in the *Arabidopsis thaliana* root. Science 317: 507-510.

Pandya-Kumar, N., R. Shema, M. Kumar, E. Mayzlish-Gati, D. Levy, H. Zemach, E. Belausov, S. Wininger, M. Abu-Abied, Y. Kapulnik and H. Koltai. 2014. Strigolactone analog GR24 triggers changes in PIN2 polarity, vesicle trafficking and actin filament architecture. New Phytol. 202: 1184-1196.

Peng, J., P. Carol, D.E. Richards, K.E. King, R.J. Cowling, G.P. Murphy and N.P. Harberd. 1997. The Arabidopsis *GAI* gene defines a signaling pathway that negatively regulates gibberellin responses. Genes Dev. 11: 3194-3205.

Péret, B., K. Swarup, A. Ferguson, M. Seth, Y. Yang, S. Dhondt, N. James, I. Casimiro, P. Perry, A. Syed, H. Yang, J. Reemmer, E. Venison, C. Howells, M.A. Perez-Amador, J. Yun, J. Alonso, G.T.S. Beemster, L. Laplaze, A. Murphy, M.J. Bennett, E. Nielsen and R. Swarup. 2012. *AUX/LAX* genes encode a family of auxin influx transporters that perform distinct functions during Arabidopsis development. Plant Cell 24: 2874-2885.

Pitts, R.J., A. Cernac and M. Estelle. 1998. Auxin and ethylene promote root hair elongation in Arabidopsis. Plant J. 16: 553-560.

Qiao, H., K.N. Chang, J. Yazaki and J.R. Ecker. 2009. Interplay between ethylene, ETP1/ETP2 F-box proteins, and degradation of EIN2 triggers ethylene responses in Arabidopsis. Genes Dev. 23: 512-521.

Reinhardt, D., E.-R. Pesce, P. Stieger, T. Mandel, K. Baltensperger, M. Bennett, J. Traas, J. Friml and C. Kuhlemeier. 2003. Regulation of phyllotaxis by polar auxin transport. Nature 426: 255-260.

Riou-Khamlichi, C., R. Huntley, A. Jacqmard and J.A.H. Murray. 1999. Cytokinin activation of *Arabidopsis* cell division through a D-type cyclin. Science 283: 1541-1544.

Robles, L., A. Stepanova and J. Alonso. 2013. Molecular mechanisms of ethylene–auxin interaction. Mol. Plant 6: 1734-1737.

Roudier, F., E. Fedorova, M. Lebris, P. Lecomte, J. Györgyey, D. Vaubert, G. Horvath, P. Abad, A. Kondorosi and E. Kondorosi. 2003. The *Medicago* species A2-type cyclin is auxin regulated and involved in meristem formation but dispensable for endoreduplication-associated developmental programs. Plant Physiol. 131: 1091-1103.

Růžička, K., K. Ljung, S. Vanneste, R. Podhorská, T. Beeckman, J. Friml and E. Benková. 2007. Ethylene regulates root growth through effects on auxin biosynthesis and transport-dependent auxin distribution. Plant Cell 19: 2197-2212.

Růžička, K., M. Šimášková, J. Duclercq, J. Petrášek, E. Zažímalová, S. Simon, J. Friml, M.C.E. Van Montagu and E. Benková. 2009. Cytokinin regulates root meristem activity via modulation of the polar auxin transport. Proc. Natl. Acad. Sci. USA 106: 4284-4289.

Ruyter-Spira, C., W. Kohlen, T. Charnikhova, A. van Zeijl, L. van Bezouwen, N. de Ruijter, C. Cardoso, J.A. Lopez-Raez, R. Matusova, R. Bours, F. Verstappen and H. Bouwmeester. 2011. Physiological effects of the synthetic strigolactone analog GR24 on root system architecture in Arabidopsis: another belowground role for strigolactones? Plant Physiol. 155: 721-734.

Sabatini, S., D. Beis, H. Wolkenfelt, J. Murfett, T. Guilfoyle, J. Malamy, P. Benfey, O. Leyser, N. Bechtold, P. Weisbeek and B. Scheres. 1999. An auxin-dependent distal organizer of pattern and polarity in the Arabidopsis root. Cell 99: 463-472.

Sachs, J. 1882. Vorlesungen über Pflanzen-Physiologie. Wilhelm Engelmann, Leipzig, Germany.

Sakai, H., J. Hua, Q.G. Chen, C. Chang, L.J. Medrano, A.B. Bleecker and E.M. Meyerowitz. 1998. *ETR2* is an *ETR1*-like gene involved in ethylene signaling in Arabidopsis. Proc. Natl. Acad. Sci. USA 95: 5812-5817.

Salehin, M., R. Bagchi and M. Estelle. 2015. SCF*TIR1/AFB*-based auxin perception: mechanism and role in plant growth and development. Plant Cell 27: 9-19.

Sampathkumar, A., A. Yan, P. Krupinski and E.M. Meyerowitz. 2014. Physical forces regulate plant development and morphogenesis. Curr. Biol. 24: R475-R483.

Sánchez-Rodríguez, C., I. Rubio-Somoza, R. Sibout and S. Persson. 2010. Phytohormones and the cell wall in *Arabidopsis* during seedling growth. Trends Plant Sci. 15: 291-301.

Sang, D., D. Chen, G. Liu, Y. Liang, L. Huang, X. Meng, J. Chu, X. Sun, G. Dong, Y. Yuan, Q. Qian, J. Li and Y. Wang. 2014. Strigolactones regulate rice tiller angle by attenuating shoot gravitropism through inhibiting auxin biosynthesis. Proc. Natl. Acad. Sci. USA 111: 11199-11204.

Santner, A. and M. Estelle. 2010. The ubiquitin-proteasome system regulates plant hormone signaling. Plant J. 61: 1029-1040.

Sasse, J., S. Simon, C. Gübeli, G.-W. Liu, X. Cheng, J. Friml, H. Bouwmeester, E. Martinoia and L. Borghi. 2015. Asymmetric localizations of the ABC transporter PaPDR1 trace paths of directional strigolactone transport. Curr. Biol. 25: 647-655.

Scacchi, E., P. Salinas, B. Gujas, L. Santuari, N. Krogan, L. Ragni, T. Berleth and C.S. Hardtke. 2010. Spatio-temporal sequence of cross-regulatory events in root meristem growth. Proc. Natl. Acad. Sci. USA 107: 22734-22739.

Scaffidi, A., M.T. Waters, Y.K. Sun, B.W. Skelton, K.W. Dixon, E.L. Ghisalberti, G.R. Flematti and S.M. Smith. 2014. Strigolactone hormones and their stereoisomers signal through two related receptor proteins to induce different physiological responses in Arabidopsis. Plant Physiol. 165: 1221-1232.

Scarpella, E., D. Marcos, J. Friml and T. Berleth. 2006. Control of leaf vascular patterning by polar auxin transport. Genes Dev. 20: 1015-1027.

Schomburg, F.M., C.M. Bizzell, D.J. Lee, J.A.D. Zeevaart and R.M. Amasino. 2003. Overexpression of a novel class of gibberellin 2-oxidases decreases gibberellin levels and creates dwarf plants. Plant Cell 15: 151-163.

Sergeeva, E., A. Liaimer and B. Bergman. 2002. Evidence for production of the phytohormone indole-3-acetic acid by cyanobacteria. Planta 215: 229-238.

Shani, E., R. Weinstain, Y. Zhang, C. Castillejo, E. Kaiserli, J. Chory, R.Y. Tsien and M. Estelle. 2013. Gibberellins accumulate in the elongating endodermal cells of Arabidopsis root. Proc. Natl. Acad. Sci. USA 110: 4834-4839.

Shinohara, N., C. Taylor and O. Leyser. 2013. Strigolactone can promote or inhibit shoot branching by triggering rapid depletion of the auxin efflux protein PIN1 from the plasma membrane. PLoS Biol. 11: e1001474.

Silverstone, A.L., H.-S. Jung, A. Dill, H. Kawaide, Y. Kamiya and T.-p. Sun. 2001. Repressing a repressor: Gibberellin-induced rapid reduction of the RGA protein in Arabidopsis. Plant Cell 13: 1555-1565.

Smalle, J., M. Haegman, J. Kurepa, M. Van Montagu and D. Van Der Straeten. 1997. Ethylene can stimulate Arabidopsis hypocotyl elongation in the light. Proc. Natl. Acad. Sci. USA 94: 2756-2761.

Somerville, C., S. Bauer, G. Brininstool, M. Facette, T. Hamann, J. Milne, E. Osborne, A. Paredez, S. Persson, T. Raab, S. Vorwerk and H. Youngs. 2004. Toward a systems approach to understanding plant cell walls. Science 306: 2206-2211.

Steffens, B., C. Feckler, K. Palme, M. Christian, M. Böttger and H. Lüthen. 2001. The auxin signal for protoplast swelling is perceived by extracellular ABP1. Plant J. 27: 591-599.

Stepanova, A.N., J.M. Hoyt, A.A. Hamilton and J.M. Alonso. 2005. A link between ethylene and auxin uncovered by the characterization of two root-specific ethylene-insensitive mutants in Arabidopsis. Plant Cell 17: 2230-2242.

Stepanova, A.N., J. Robertson-Hoyt, J. Yun, L.M. Benavente, D.-Y. Xie, J. Doležal, A. Schlereth, G. Jürgens and J.M. Alonso. 2008. *TAA1*-mediated auxin biosynthesis is essential for hormone crosstalk and plant development. Cell 133: 177-191.

Stirnberg, P., I.J. Furner and H.M.O. Leyser. 2007. MAX2 participates in an SCF complex which acts locally at the node to suppress shoot branching. Plant J. 50: 80-94.

Strader, L.C., S. Ritchie, J.D. Soule, K.M. McGinnis and C.M. Steber. 2004. Recessive-interfering mutations in the gibberellin signaling gene *SLEEPY1* are rescued by overexpression of its homologue, *SNEEZY*. Proc. Natl. Acad. Sci. USA 101: 12771-12776.

Sun, Y., X.-Y. Fan, D.-M. Cao, W. Tang, K. He, J.-Y. Zhu, J.-X. He, M.-Y. Bai, S. Zhu, E. Oh, S. Patil, T.-W. Kim, H. Ji, W.H. Wong, S.Y. Rhee and Z.-Y. Wang. 2010. Integration of brassinosteroid signal transduction with the transcription network for plant growth regulation in *Arabidopsis*. Dev. Cell 19: 765-777.

Suzuki, H., S.-H. Park, K. Okubo, J. Kitamura, M. Ueguchi-Tanaka, S. Iuchi, E. Katoh, M. Kobayashi, I. Yamaguchi, M. Matsuoka, T. Asami and M. Nakajima. 2009. Differential expression and affinities of Arabidopsis gibberellin receptors can explain variation in phenotypes of multiple knock-out mutants. Plant J. 60: 48-55.

Swarup, R., J. Friml, A. Marchant, K. Ljung, G. Sandberg, K. Palme and M. Bennett. 2001. Localization of the auxin permease AUX1 suggests two functionally distinct hormone transport pathways operate in the Arabidopsis root apex. Genes Dev. 15: 2648-2653.

Swarup, R., P. Perry, D. Hagenbeek, D. Van Der Straeten, G.T.S. Beemster, G. Sandberg, R. Bhalerao, K. Ljung and M.J. Bennett. 2007. Ethylene upregulates auxin biosynthesis in *Arabidopsis* seedlings to enhance inhibition of root cell elongation. Plant Cell 19: 2186-2196.

Szekeres, M., K. Németh, Z. Koncz-Kálmán, J. Mathur, A. Kauschmann, T. Altmann, G.P. Rédei, F. Nagy, J. Schell and C. Koncz. 1996. Brassinosteroids rescue the deficiency of CYP90, a cytochrome P450, controlling cell elongation and de-etiolation in Arabidopsis. Cell 85: 171-182.

Tanimoto, M., K. Roberts and L. Dolan. 1995. Ethylene is a positive regulator of root hair development in *Arabidopsis thaliana*. Plant J. 8: 943-948.

Thomann, A., E. Lechner, M. Hansen, E. Dumbliauskas, Y. Parmentier, J. Kieber, B. Scheres and P. Genschik. 2009. Arabidopsis *CULLIN3* genes regulate primary root growth and patterning by ethylene-dependent and -independent mechanisms. PLoS Genet. 5: e1000328.

To, J.P.C., G. Haberer, F.J. Ferreira, J. Deruère, M.G. Mason, G.E. Schaller, J.M. Alonso, J.R. Ecker and J.J. Kieber. 2004. Type-A Arabidopsis response regulators are partially redundant negative regulators of cytokinin signaling. Plant Cell 16: 658-671.

Tsuchiya, Y., D. Vidaurre, S. Toh, A. Hanada, E. Nambara, Y. Kamiya, S. Yamaguchi and P. McCourt. 2010. A small-molecule screen identifies new functions for the plant hormone strigolactone. Nat. Chem. Biol. 6: 741-749.

Ubeda-Tomás, S., G.T.S. Beemster and M.J. Bennett. 2012. Hormonal regulation of root growth: integrating local activities into global behaviour. Trends Plant Sci. 17: 326-331.

Ubeda-Tomás, S., F. Federici, I. Casimiro, G.T.S. Beemster, R. Bhalerao, R. Swarup, P. Doerner, J. Haseloff and M.J. Bennett. 2009. Gibberellin signalling in the endodermis controls Arabidopsis root meristem size. Curr. Biol. 19: 1194-1199.

Umehara, M., A. Hanada, S. Yoshida, K. Akiyama, T. Arite, N. Takeda-Kamiya, H. Magome, Y. Kamiya, K. Shirasu, K. Yoneyama, J. Kyozuka and S. Yamaguchi. 2008. Inhibition of shoot branching by new terpenoid plant hormones. Nature 455: 195-200.

Vieten, A., S. Vanneste, J. Wiśniewska, E. Benková, R. Benjamins, T. Beeckman, C. Luschnig and J. Friml. 2005. Functional redundancy of PIN proteins is accompanied by auxin-dependent cross-regulation of PIN expression. Development 132: 4521-4531.

Wang, C., Y. Liu, S.-S. Li and G.-Z. Han. 2015. Insights into the origin and evolution of the plant hormone signaling machinery. Plant Physiol. 167: 872-886.

Wang, X., M.B. Goshe, E.J. Soderblom, B.S. Phinney, J.A. Kuchar, J. Li, T. Asami, S. Yoshida, S.C. Huber and S.D. Clouse. 2005. Identification and functional analysis of *in vivo* phosphorylation sites of the Arabidopsis BRASSINOSTEROID-INSENSITIVE1 receptor kinase. Plant Cell 17: 1685-1703.

Wang, Y., S. Sun, W. Zhu, K. Jia, H. Yang and X. Wang. 2013. Strigolactone/MAX2-induced degradation of brassinosteroid transcriptional effector BES1 regulates shoot branching. Dev. Cell 27: 681-688.

Wang, Z.-Y., T. Nakano, J. Gendron, J. He, M. Chen, D. Vafeados, Y. Yang, S. Fujioka, S. Yoshida, T. Asami and J. Chory. 2002. Nuclear-localized BZR1 mediates brassinosteroid-induced growth and feedback suppression of brassinosteroid biosynthesis. Dev. Cell 2: 505-513.

Wei, C. and P.M. Lintilhac. 2007. Loss of stability: a new look at the physics of cell wall behavior during plant cell growth. Plant Physiol. 145: 763-772.

Wei, C. and P.M. Lintilhac. 2003. Loss of stability—a new model for stress relaxation in plant cell walls. J. Theor. Biol. 224: 305-312.

Werner, T., V. Motyka, M. Strnad and T. Schmülling. 2001. Regulation of plant growth by cytokinin. Proc. Natl. Acad. Sci. USA 98: 10487-10492.

Werner, T., V. Motyka, V. Laucou, R. Smets, H. Van Onckelen and T. Schmülling. 2003. Cytokinin-deficient transgenic Arabidopsis plants show multiple developmental alterations indicating opposite functions of cytokinins in the regulation of shoot and root meristem activity. Plant Cell 15: 2532-2550.

Willige, B.C., E. Isono, R. Richter, M. Zourelidou and C. Schwechheimer. 2011. Gibberellin regulates PIN-FORMED abundance and is required for auxin transport-dependent growth and development in *Arabidopsis thaliana*. Plant Cell 23: 2184-2195.

Woeste, K.E., C. Ye and J.J. Kieber. 1999. Two Arabidopsis mutants that overproduce ethylene are affected in the posttranscriptional regulation of 1-aminocyclopropane-1-carboxylic acid synthase. Plant Physiol. 119: 521-529.

Wolf, S., K. Hématy and H. Höfte. 2012. Growth control and cell wall signaling in plants. Annu. Rev. Plant Biol. 63: 381-407.

Woodward, A.W. and B. Bartel. 2005. Auxin: regulation, action, and interaction. Ann. Bot. 95: 707-735.

Yamaguchi, S. 2008. Gibberellin metabolism and its regulation. Annu. Rev. Plant Biol. 59: 225-251.

Yang, C., X. Lu, B. Ma, S.-Y. Chen and J.-S. Zhang. 2015. Ethylene signaling in rice and *Arabidopsis*: conserved and diverged aspects. Mol. Plant 8: 495-505.

Yang, D.-L., J. Yao, C.-S. Mei, X.-H. Tong, L.-J. Zeng, Q. Li, L.-T. Xiao, T.-p. Sun, J. Li, X.-W. Deng, C.M. Lee, M.F. Thomashow, Y. Yang, Z. He and S.Y. He. 2012. Plant hormone jasmonate prioritizes defense over growth by interfering with gibberellin signaling cascade. Proc. Natl. Acad. Sci. USA 109: E1192-E1200.

Yin, Y., D. Vafeados, Y. Tao, S. Yoshida, T. Asami and J. Chory. 2005. A new class of transcription factors mediates brassinosteroid-regulated gene expression in Arabidopsis. Cell 120: 249-259.

Yu, X., L. Li, J. Zola, M. Aluru, H. Ye, A. Foudree, H. Guo, S. Anderson, S. Aluru, P. Liu, S. Rodermel and Y. Yin. 2011. A brassinosteroid transcriptional network revealed by genome-wide identification of BES1 target genes in *Arabidopsis thaliana*. Plant J. 65: 634-646.

Zhang, R., B. Wang, J. Ouyang, J. Li and Y. Wang. 2008. *Arabidopsis* indole synthase, a homolog of tryptophan synthase alpha, is an enzyme involved in the Trp-independent indole-containing metabolite biosynthesis. J. Integr. Plant Biol. 50: 1070-1077.

Zhao, M., Y. Han, Y. Feng, F. Li and W. Wang. 2012. Expansins are involved in cell growth mediated by abscisic acid and indole-3-acetic acid under drought stress in wheat. Plant Cell Rep. 31: 671-685.

Zhong, S. and C. Chang. 2012. Ethylene signalling: the CTR1 protein kinase. pp. 147-168. *In*: M.T. McManus (ed.). The Plant Hormone Ethylene (Annual Plant Reviews, Vol. 44). Wiley-Blackwell, Oxford.

Zhou, A., H. Wang, J.C. Walker and J. Li. 2004. BRL1, a leucine-rich repeat receptor-like protein kinase, is functionally redundant with BRI1 in regulating Arabidopsis brassinosteroid signaling. Plant J. 40: 399-409.

Zhu, J.-K., J. Shi, U. Singh, S.E. Wyatt, R.A. Bressan, P.M. Hasegawa and N.C. Carpita. 1993. Enrichment of vitronectin- and fibronectin-like proteins in NaCl-adapted plant cells and evidence for their involvement in plasma membrane--cell wall adhesion. Plant J. 3: 637-646.

Zou, J., S. Zhang, W. Zhang, G. Li, Z. Chen, W. Zhai, X. Zhao, X. Pan, Q. Xie and L. Zhu. 2006. The rice *HIGH-TILLERING DWARF1* encoding an ortholog of Arabidopsis MAX3 is required for negative regulation of the outgrowth of axillary buds. Plant J. 48: 687-696.

Plant Cell Differentiation

12

Cellular Dynamics of the Primary Shoot and Root Meristem

Lam Dai Vu[1,2,3,4] and Ive De Smet[1,2,]*

Introduction

Plants continue to grow during their whole life, and this spectacular growth is largely due to their continuously active stem cells and meristems. In extreme cases, this yields giant redwood trees that are hundreds of years old with active meristems or the trembling giant, a clonal colony of a single male quaking aspen. The necessary primary root and shoot meristems are established early in development – during embryogenesis – and remain active throughout the plant's life. In addition, some meristems are established *de novo* and post-embryonically, such as during lateral root initiation, and these also contribute to the final plant size and shape.

In this chapter, the focus is on the cellular dynamics of the primary shoot and root meristems following their establishment and the various gradients that are present in the meristems. We will highlight the key stages from the birth of a cell over its differentiated stage to – if this occurs – its death, and flag some – but surely not all – key regulatory mechanisms and associated core components involved in regulating and maintaining this. We will mainly focus on data obtained from *Arabidopsis thaliana* (Arabidopsis) research, and occasionally flag important observations from other plant species.

Early Stages of Embryogenesis Preceding the Establishment of Meristems

As in any plant developmental process, embryogenesis is largely driven by (oriented) cell divisions (Yoshida et al. 2014). In Arabidopsis, cell divisions occur in a highly ordered sequence which defines the cell patterning in early embryogenesis

[1] Department of Plant Systems Biology, VIB, B-9052 Ghent, Belgium.
[2] Department of Plant Biotechnology and Genetics, Ghent University, B-9052 Ghent, Belgium.
[3] Department of Medical Protein Research, VIB, B-9000 Ghent, Belgium.
[4] Department of Biochemistry, Ghent University, B-9000 Ghent, Belgium.
[*] Corresponding author : ivsme@psb.vib-ugent.be

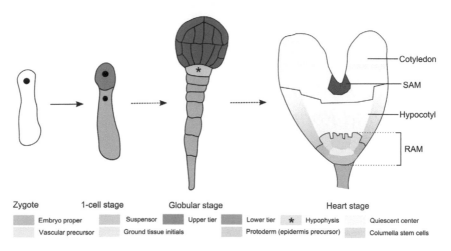

FIGURE 1   Four key stages of Arabidopsis embryogenesis. Dashed arrows represent several transition stages in between.

(Mansfield and Briarty 1991). After fertilization, the zygote loses the polarity that is observed in the egg cell and the nucleus moves to the center of the cell (Ueda et al. 2011, Faure et al. 2002). Before the zygote elongates, the polarity is re-established and subsequently an asymmetric cell division generates a smaller apical cell and a larger basal cell, which mark the apical–basal axis (Mayer et al. 1993, De Smet et al. 2010) (Fig. 1).

Recently, a 4D map following cell division planes improved our understanding of the geometric framework underlying Arabidopsis early embryogenesis (Yoshida et al. 2014). This 4D map visualizes and demonstrates an auxin-mediated derivation from a default division rule that defines the division plane as the smallest wall area going through the center of the cell, thus allowing asymmetric cell division. These initial, coordinated processes that set up the embryo pattern lie at the basis of primary shoot and root meristem establishment. Several components that play a role in these processes have been identified but precise mechanisms are largely unknown (Lukowitz et al. 2004, Breuninger et al. 2008, Bayer et al. 2009, Jeong et al. 2011, Rademacher et al. 2011, 2012, Ueda et al. 2011, Waki et al. 2011, Yoshida et al. 2014). However, this is outside the scope of this chapter and we therefore refer to some comprehensive reviews (De Smet et al. 2010, Wendrich and Weijers 2013).

Separation of Shoot and Root Domain and Initiation of the First Stem Cells During Embryogenesis

At the globular stage, a pair of closely related leucine-rich repeat receptor-like kinases (LRR-RLK), RECEPTOR-LIKE PROTEIN KINASE1 (RPK1) and TOADSTOOL2 (TOAD2), are required for the maintenance of outer and inner cell separation (Nodine et al. 2007). At this stage, the upper and lower tier of the inner cells begin to express different gene sets. The lower tier divides into the ground tissue precursors and vascular stem cell initials (procambium). This division is marked by the expression of

MONOPTEROS (*MP*) which is important for the specification of the uppermost suspensor cell, called the hypophysis, and cells that form the embryonic root (Berleth and
Jürgens 1993, Hardtke and Berleth 1998, Weijers et al. 2006) (Fig. 1). Irregular vascular development is observed in *mp* mutants, which is caused by defects in polar auxin
transport. This is linked to a reduced expression of *PIN-FORMED 1* (*PIN1*), a major
auxin efflux carrier (Friml et al. 2003). The MP-promoted PIN1-dependent auxin
transport to the hypophysis will activate the auxin response governed by AUXIN
RESPONSE FACTORs (ARFs) and AUX/IAA transcriptional repressors, such as
ARF9 and IAA10, which are required for hypophysis specification and prevent transformation to embryo identity (Rademacher et al. 2012). Interestingly, the signalling
output for hypophysis specification is not directly derived from the primary auxin
response but rather adjacent cell-to-cell signalling. This is mediated by the transient
interaction of the AUX/IAA protein IAA12/BODENLOS (BDL) with MP (Weijers
et al. 2006). Two downstream targets of MP, *TARGET OF MONOPTEROS 5* (*TMO5*)
and *TMO7*, which encode two basic helix-loop-helix (bHLH) transcription factors, are
expressed in the lower tier. TMO7 moves from its transcription zone to the hypophysis, where it is necessary to establish a root (Schlereth et al. 2010), while TMO5 has
been shown to activate cytokinin biosynthesis (De Rybel et al. 2013, 2014). A mutually inhibitory feedback loop between auxin and cytokinin has been reported to determine post-embryonic vascular pattern in the primary root (Bishopp et al. 2011).

MP activity also controls the expression of *PLT1* and *PLT2* from the AP2-domain
PLETHORA (*PLT*) family, but not directly (Aida et al. 2004, Schlereth et al. 2010).
PLT1 and *PLT2* expression is restricted in the lower tier cells at the octant stage
and later in the lens-shaped quiescent center (QC) progenitor cell, following the
asymmetric division of the hypophysis (Aida et al. 2004). Two other *PLT* members,
PLT3 and *PLT4/BABY BOOM* (*BBM*), start to be expressed from the heart stage and
accumulate in provascular cells and the lens-shaped cell (Galinha et al. 2007). A
severe rootless phenotype was observed in *plt1 plt2 plt3 bbm* quadruple mutants.
Furthermore, TOPLESS (TPL) which is recruited as a co-repressor of several AUX/
IAA proteins during embryonic root development (Szemenyei et al. 2008), directly
represses *PLT1* and *PLT2* expression (Smith and Long 2010). Loss of *TPL* function
leads to misexpression of *PLT1* and *PLT2* in the apical domain of the embryo, resulting in formation of double root seedlings (Smith and Long 2010). This implicates the
central role of *PLT* genes in root apical meristem (RAM) specification and maintenance under control of auxin signalling.

The class III homeodomain-leucine zipper (*HD-ZIP III*) transcription factor
genes *PHABULOSA* (*PHB*), *PHAVOLUTA* (*PHV*), *REVOLUTA* (*REV*) and *CORONA*
(*CNA*)/*ATHB15* (Smith and Long 2010) occupy an expression domain in the upper tier
cells at the globular stage under control of microRNA miR165/166 family members
(Emery et al. 2003, Mallory et al. 2004). They perform several overlapping, distinct,
and antagonistic functions that are important for apical embryo patterning, shoot
apical meristem (SAM) initiation and vascular development (Prigge et al. 2005).
HD-ZIP III and *PLT* genes act in an antagonistic manner in controlling apical and
basal embryonic cell fate (Smith and Long 2010). Expression of the auxin efflux carrier gene *PIN4*, which is dependent on PLT1/PLT2 activity, is lost in cells ectopically
expressing *REV*. In addition, gain-of-function mutations of *HD-ZIP III* genes were
shown to repress the double root phenotype in a *tpl* background. Furthermore, when
HD-ZIP III genes are ectopically expressed in the embryonic basal tier, seedlings

exhibit a second shoot pole instead of the root pole. This implicates the important role of HD-ZIP III genes in promotion of SAM and antagonism in RAM development (Smith and Long 2010).

The formation and development of the SAM is marked by *WUSCHEL* (*WUS*) which is initiated in the upper tier during the globular stage of embryogenesis (Mayer et al. 1998). *WUS* expression in the embryo is controlled by class III HD-ZIP factors. Simultaneous loss-of-function mutation of the class II *HD-ZIP* (*HD-ZIP II*) genes *ATHB2* and *HAT3*, expressed in the early embryo under the direct regulation of REV, reduced the expression of *WUS* and hence, reducing SAM activity (Turchi et al. 2013). Therefore, these *HD-ZIP II* genes may play an important role in downstream regulation of SAM development (Brandt et al. 2012, Reinhart et al. 2013).

Maintaining a Developmental Gradient in the Shoot Apical Meristem

In this section, key aspects associated with the shoot apical meristem will be discussed, and some gradients that exist will be pointed out (Fig. 2A-B). Stem cell proliferation is tightly controlled by the intercellular communication between the organization center (OC) and the central zone (CZ) of the SAM, coordinated by a negative feedback loop, consisting of WUS and CLAVATA3 (CLV3) as the central regulators (Mayer et al. 1998, Fletcher et al. 1999, Schoof et al. 2000) (Fig. 2B). WUS activates stem cell fate in a non-cell autonomous manner, correlated to its migration from the expression zone in the OC to the CZ (Yadav et al. 2011, Daum et al. 2014). The mobility of WUS after expression is highly regulated and promoted by the WUS homeodomain which mediates WUS homodimerization, while being attenuated by a non-conserved sequence between the homeodomain and the WUS-box to prevent WUS from spreading into the peripheral cell and to restrict stem cell induction in the CZ (Daum et al. 2014). Mutations leading to immobilization of WUS in the OC or cell-specific degradation of WUS in the CZ result in loss of stem cells, confirming functional relevance of WUS migration to the CZ. Upon WUS signalling, stem cells in the CZ secrete the peptide CLV3 which represses *WUS* expression in the OC via several proteins, including the receptor kinase CLV1, the receptor protein CLV2 and the pseudokinase CORYNE (CRN) (Fletcher et al. 1999, Jeong et al. 1999, Brand et al. 2000, Schoof et al. 2000, Müller et al. 2008). In maize, the stem cell-restrictive signal is further transmitted by COMPACT PLANT2, an α-subunit of a heterotrimeric GTP binding protein (Bommert et al. 2013).

The SAM activity and cytokinin signalling pathways are linked by the repressive activity of WUS towards *ARABIDOPSIS RESPONSE REGULATOR* (*ARR*) genes, such as *ARR5, ARR6, ARR7* and *ARR15,* which participate in a negative cytokinin regulatory feedback loop (Leibfried et al. 2005) (Fig. 2B). Conversely, plants overexpressing *ARR7* have lower *WUS* RNA level without exhibiting any obvious defects in SAM. However, mutations producing constitutively active ARR7 result in aberrant SAM formation, similar to the phenotype observed in *wus* mutants. In rice, loss of LONELY GUY (LOG) function cause severe defects in cytokinin biosynthesis and premature termination of the SAM (Kurakawa et al. 2007). In Arabidopsis, the function of *LOG* genes is more redundant, however, multiple loss-of-function mutations in *LOGs* result in partially reduced SAM activity. Additionally, *LOG4* promoter

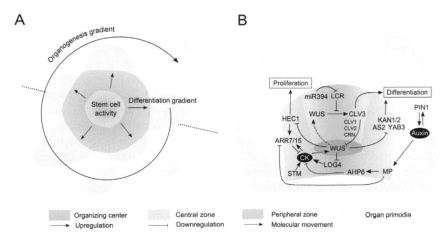

FIGURE 2 Developmental regulation in the shoot apical meristem. (A) The main developmental gradients in the shoot apical meristem. The dotted line represents a perpendicular plane illustrated in (B). (B) Key molecular regulators mediate communication between OC, CZ and PZ, CK, cytokinin.

and cytokinin promoter activity is drastically reduced in *clv3* mutants. The coupled antagonistic effects of CLV3 and cytokinin are suggested to control *WUS* domain dynamics, and there is a negative feedback loop between *WUS* function and cytokinin biosynthesis (Chickarmane et al. 2012).

SHOOTMERISTEMLESS (STM), a *KNOTTED-1 (KN1)* HOMEOBOX (KNOX) family member in Arabidopsis, is expressed throughout the SAM and suppresses cell differentiation by inhibiting the activity of differentiation factor ASYMMETRIC LEAVES 1 (AS1) and downregulating the biosynthesis of the differentiation-promoting hormone gibberellic acid (GA) (Byrne et al. 2000, 2002, Hay et al. 2002). STM also promotes *WUS* expression by increasing biosynthesis of cytokinin (Hay et al. 2002, Jasinski et al. 2005) (Fig. 2B). Mutants with a reduced cytokinin level in a constitutive GA signalling background are detrimental for SAM function (Jasinski et al. 2005).

MicroRNAs also play a key role in stem cell maintenance in SAM. MiR394, a mobile signal acting downstream of WUS, targets *LEAF CURLING RESPONSIVENESS (LCR)*, encoding an F-Box protein that is suggested to target proteins involved in WUS function (Knauer et al. 2013) (Fig. 2B). In contrast to miR394, the accumulation of miR165/miR166 members leads to loss of stem cell competence in the SAM (Zhou et al. 2007). The sequestration of miR165/miR166 by the ARGONAUTE (AGO) protein AGO10/ZWILLE is essential to prevent the degradation of miR165/miR166 targets, e.g. *HD-ZIP III* genes, which are expressed in the SAM to maintain it (Zhu et al. 2011, Zhang and Zhang 2012).

The bHLH transcription factor HECATE1 (HEC1) was found to regulate SAM function by promoting stem cell proliferation by uncoupling the WUS-CLV3 feedback loop (Schuster et al. 2014) (Fig. 2B). Its expression level is repressed by WUS and this signal is important for stem cell niche integrity. Enhancing HEC1 activity in the CZ results in a dramatic SAM expansion, while enhanced expression in the OC leads to meristem termination. In addition, HEC1 cell-autonomously activates the expression of *ARR7* and *ARR15*. Furthermore, cell-to-cell movement was observed for ARR7, implying a communication pathway between the OC and CZ and the peripheral zone (PZ) (Schuster et al. 2014). The PZ surrounding the CZ is composed of more rapidly dividing cells that will later differentiate to form lateral organs (Reddy et al. 2004). Non-cell autonomous activity of the WUS-CLV3 pathway has an important function in controlling cell division in the PZ, and thus, defining the boundary between CZ and PZ (Laufs et al. 1998, Reddy and Meyerowitz 2005). Elevated WUS levels or reduction in CLV3 activity induces the expansion of the CZ and at the same time, increase the cell division rate in the PZ, indicatively due to re-specification of PZ cells (Reddy and Meyerowitz 2005, Yadav et al. 2010). Moreover, WUS-mediated transcriptional repression of differentiation-promoting transcription factor genes such as *KANADI1* (*KAN1*), and *KAN2, ASYMETRIC LEAVES2* (*AS2*) and *YABBI3* (*YAB3*) prevents premature differentiation of stem cell progenitors (Yadav et al. 2013).

The formation of lateral organs is initiated when the differentiated cells moves from the PZ towards the outer zone of the SAM. The precise arrangement of plant organs, called phyllotaxis, is largely dependent on auxin efflux and signalling (Reinhardt et al. 2003, Traas 2012). The major efflux carrier *PIN1* is highly expressed in the summit cell layer of the CZ. Loss-of-function in *PIN1* inhibits auxin transport to the meristem surface and creates a naked inflorescence stem. However the phenotype is rescued by external addition of auxin. Hence, a network of cell-cell interactions and PIN networks is sufficient to generate proper auxin distribution that produces a robust phyllotactic patterning in the shoot apex (Reinhardt et al. 2003, Heisler et al. 2005, De Reuille et al. 2006).

Although the auxin flux was shown to be preferentially directed towards the SAM center, the low local (in the SAM center) expression of the E3 ligase *TRANSPORT INHIBITOR RESPONSE 1* (*TIR1*) restricts degradation of AUX/IAA proteins to control the transcriptional auxin output (De Reuille et al. 2006; Vernoux et al. 2011). In general, AUX/IAA networks that acquire differential auxin sensitivity in the SAM may help to generate a robust signalling output by buffering the fluctuation in signalling input (Vernoux et al. 2011).

MP plays a central role at the crossroad between auxin and cytokinin signalling pathways and governs the activity of a set of genes that controls SAM activity and formation of lateral organs (Chandler and Werr 2015) (Fig. 2B). The expression of the cytokinin signalling inhibitor *ARABIDOPSIS HISTIDINE PHOSPHOTRANSFER PROTEIN 6* (*AHP6*) in the SAM is lost in a *mp* background (Besnard et al. 2014a). The synergistic action of auxin and cytokinin during organ initiation defines the spatio-temporal distribution of AHP6 which is required for the precise rhythmicity of organogenesis (Besnard et al. 2014a, 2014b). Auxin, mediated through MP, negatively regulates the cytokinin-activated *ARR7* and *ARR15*, demonstrating the ability of such transcription factors that are able to integrate auxin and cytokinin signals (Zhao et al. 2010). On the other hand, *MP* regulates the transcription factor *LEAFY* (*LFY*), which promotes auxin signalling to initiate floral primordia (Weigel et al. 1992, Yamaguchi et al. 2013).

Once the primordium is produced, it is separated from the SAM by a boundary region with decreased cell division rate (Breuil-Broyer et al. 2004). Brassinosteroid homeostasis is essential for proper organ boundary formation (Bell et al. 2012, Gendron et al. 2012), and auxin has been shown to stimulate brassinosteroid biosynthesis and also act synergistically on a largely overlapping transcriptome (Vert et al. 2005, Chung et al. 2011). The expression of *CUP-SHAPED COTYLEDON1* (*CUC1*) and *CUC2*, which are upregulated by auxin, is regulated by brassinosteroids through the transcriptional repressor BRASSINAZOLE-RESISTANT 1 (BZR1) in the boundary region (Gendron et al. 2012). A reduced local presence of BZR1 is necessary for normal *CUC* expression to maintain the boundary formation. Low brassinosteroid signalling is secured by the boundary-specific transcription factor LATERAL ORGAN BOUNDARIES (LOB). The *lob* mutations lead to organ fusion and ectopic expression of LOB has been shown to reduce brassinosteroid signalling (Bell et al. 2012). However, increased brassinosteroid concentration promotes the expression of *LOB*. This may establish a feedback-loop to stabilize the boundary domain (Bell et al. 2012).

In general, an intricate network of post-translationally modified peptides, receptor kinases, hormones and transcription factors maintains gradients in the shoot apical meristem with respect to the stem cell niche and the formation of new lateral shoot organs (Fig. 2B).

Maintaining a Developmental Gradient in the Root Apical Meristem

In this section, we will explore key developmental gradients in the root apical meristem and highlight a number of key players (Fig. 3). A developmental gradient marks specific zones in the Arabidopsis root (Fig. 3A). The quiescent center (QC)

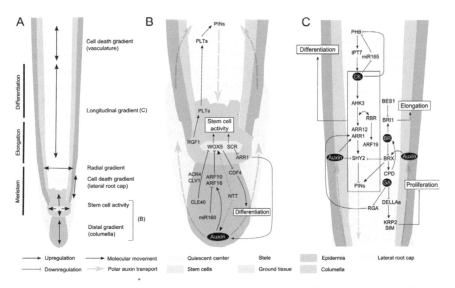

FIGURE 3 Developmental regulation in the root apical meristem. (A) Developmental gradients in the root apical meristem. Developmental zones are indicated. (B) Molecular networks that control communication between the columella and the QC and the regulation of stem cell activity in the meristem. Dotted line, required for proper signalling, yet neither upregulation or downregulation is known. (C) Longitudinal developmental gradient controls cell elongation and differentiation above the stem cell properties. CK, cytokinin; BR, brassinosteroids; GA; gibberellic acid.

is surrounded by the stem cells and non-cell autonomously maintains their activity. Different stem cells generate different daughter cells after an asymmetric division. These daughter cells will give rise to various cell types and will undergo proliferative cell divisions before they elongate and differentiate.

A distal stemness gradient below the quiescent center

The QC, the equivalent of the OC in the SAM, is marked by the activity of the transcription factor *WUSCHEL-RELATED HOMEOBOX 5* (*WOX5*), a homologue of *WUS* (Sarkar et al. 2007). Like the WUS-CLV3 signalling module in the SAM, the expression domain of *WOX5* is tightly controlled by the CLAVATA3/EMBRYO SURROUNDING-RELATED (ESR) 40 (CLE40) peptide, which is expressed in differentiated columella cells and promotes distal stem cell differentiation by suppressing the stem cell promoting activity of *WOX5* (Stahl et al. 2009). The activity of CLE40 is mediated by the RLKs ARABIDOPSIS CRINKLY4 (ACR4) and CLV1 which also transmits CLV3 signals in the SAM (Stahl et al. 2013, Ogawa et al. 2008). Loss-of-function mutations in *CLE40, ACR4* or *CLV1* give raise to additional columella stem cell layers distal to the QC (De Smet et al. 2008, Stahl et al. 2009, 2013) (Fig. 3B).

Similar to WUS, WOX5 exhibits mobility, moving from the QC to the columella stem cell and directly binds on the cis-regulatory elements and represses the expression of the differentiation-promoting transcription factor *CYCLING DOF FACTOR 4* (*CDF4*) (Pi et al. 2015) (Fig. 3B). The repression of *CDF4* transcription is mediated by TPL and the HISTON DEACETYLASE 19 (HDA19), which form a complex with WOX5 as co-repressors. The complex catalyses the histone H3 deacetylation, resulting in *CDF4* transcriptional inhibition in the columella stem cells and the QC, thus maintaining QC activity and preventing the loss of pluripotent stem cells (Pi et al. 2015).

Another important pathway controlling distal stem cell differentiation in the root cap includes the transcriptional repressor IAA17/AXR3 and the auxin response factors ARF10 and ARF16 (Ding and Friml 2010) (Fig. 3B). Increased auxin levels and *ARF16* activity result in reduction of *WOX5* expression and promote distal stem cell differentiation. Interestingly, WOX5 positively regulates auxin production in the QC and this activity is counterbalanced by IAA17-dependent repression of auxin signalling (Tian et al. 2014). On the other hand, overexpression of miR160 or loss-of-function double mutants *arf10 arf16* display excessive cell division in the root distal region, loss of columella cell identity and agravitropic growth (Ding and Friml 2010). It is suggested that the root developmental program is connected to the auxin signalling network by the regulation of ARF subsets by miR160 (Wang et al. 2005, Ding and Friml 2010) (Fig. 3B).

Recently, loss-of-function mutations of *NO TRANSMITTING TRACT* (*NTT*) and two closely related paralogs *WIP DOMAIN PROTEIN 4* (*WIP4*) and *WIP5* display the same rootless phenotype as *mp* mutants (Crawford et al. 2015). *MP* expression in the hypophysis promotes root meristem initiation by activating *NTT, WIP4* and *WIP5.* Further *NTT, WIP4* and *WIP5* are expressed continuously in the root meristem during mature root development. Treatment with auxin rescues the rootless phenotype in *ntt wip4 wip5* mutants but results in a disorganized root meristem and misexpression of the RAM marker QC45. Thus, *NTT, WIP4* and *WIP5* are required for appropriate auxin signalling in the RAM and development of the QC and distal root meristem (Crawford et al. 2015) (Fig. 3B).

A longitudinal developmental gradient in the root

A longitudinal developmental gradient is generated between three distinct zones in the root (Fig. 3A): the apical meristem, the basal elongation zone (where cell expansion occurs before entering differentiation) and the differentiation zone (Ishikawa and Evans 1995, Verbelen et al. 2006). This zonation is maintained by the control of the balance between cell division and cell differentiation (Fig. 3C). First, the maintenance of stem cell niche function is ensured by local auxin biosynthesis and polar transport which produce an auxin maximum in the root tip (Blilou et al. 2005, Scacchi et al. 2010). This generates a signalling output largely depending on the transcriptional domain consisting of *PLT* genes which represent expression levels correlated to the auxin concentration maxima in the stem cell niche (Aida et al. 2004). Additionally, PLTs are required for the transcription of *PINs*, thus ensuring a feedback loop to stabilize the auxin maximum in the distal root tip (Blilou et al. 2005). Noticeably, the activity of *PLTs* has also been shown to be dosage-dependent, demonstrated by the promotion of stem cell identity at high protein levels and the tendency to endorse towards cell differentiation at lower protein concentration (Galinha et al. 2007). In contrast to the rapid effect of auxin on the zonation, the *PLT* expression domain only expands significantly and creates a protein gradient through growth dilution and cell-to-cell movement after prolonged treatment with IAA, which indicates that the PLT protein gradient is not a rapid readout of the auxin gradient (Mähönen et al. 2014). Therefore, the transient interplay between auxin and *PLT* enables both the fast responses to environmental tropic stimuli and stable zonation dynamics necessary for coordinated cell differentiation.

The transcription factor SCR inhibits cytokinin signalling in the QC by directly repressing the expression of the differentiation-promoting cytokinin responsive *ARR1*, thereby modulating auxin production in the root tip and restricting meristem size (Dello Ioio et al. 2008, Moubayidin et al. 2013). Moreover, by controlling auxin production in the QC, SCR non-cell autonomously regulates *ARR1* activity in the transition zone and thus controls differentiation rate (Moubayidin et al. 2013) (Fig. 3B).

Interestingly, auxin-independent pathways may also play a crucial role in regulating and/or maintaining the longitudinal gradient. One example is a group of tyrosine-sulfated peptides known as ROOT MERISTEM GROWTH FACTORs (RGFs) (Matsuzaki et al. 2010). One post-translationally modified peptide, RGF1, acts through the tyrosylprotein sulfotransferase (TPST) and restores the meristematic activity that is lost in the *tpst* mutation (Matsuzaki et al. 2010, Zhou et al. 2010). Treatment of seedlings with synthetic RGF1 or overexpression of *RGF1* results in enlarged meristems. In addition, *rgf1 rgf2 rgf3* triple mutants exhibit a short root phenotype with a decreased number of meristematic cells. Furthermore, RGF1 positively regulates the expression of *PLT* genes on the posttranscriptional level, defining the *PLT* expression domain without altering the pattern of PLT transcripts (Matsuzaki et al. 2010) (Fig. 3B).

In contrast to the cell proliferation promoting effects in the SAM, cytokinin mainly antagonizes auxin as a differentiation promoter in the RAM (Dello Ioio et al. 2007) (Fig. 3C). At the transition zone before cells enter differentiation cytokinins are perceived by the ARABIDOPSIS HISTIDINE KINASE3 (AHK3), transferring the signal via phosphorelay to the nucleus and leading to the activation of two downstream primary type B cytokinin response factors ARR1 and ARR12 (Dello Ioio et al. 2007, Nishimura et al. 2004, Higuchi et al. 2004), which in turn activate the auxin signalling repressor gene *SHORT HYPOCOTYL2* (*SHY2*) (Tian et al. 2002). SHY2 promotes the redistribution of auxin by repressing the transcription of *PIN* genes, thus locally initiating cell differentiation (Dello Ioio et al. 2008). Conversely, the degradation of SHY2 mediated by auxin is important to sustain proper activity of PINs and cell division.

Similar to its homologue in animals, RETINOBLASTOMA-RELATED (RBR) protein is a key regulator of cell differentiation in plant cells (Perilli et al. 2013) (Fig. 3C). In the root apex, RBR interacts with the AHK3/ARR12 pathway, and both synergistically regulate the expression of *ARF19*, which is involved in promoting cell differentiation. The activation of *ARF19* by the RBR-AHK3/ARR12 module occurs mostly at the transition zone, thus defining the boundary between the cell division and differentiation and thereby determining meristem size, independently from the AHK3–ARR12–SHY2 pathway (Perilli et al. 2013). However, how RBR and ARR12 interact to regulate *ARF19* expression is still elusive.

Cell differentiation can also be directly promoted through activation of cytokinin biosynthesis genes. An example is *IPT7* which is activated by the HD-ZIPIII transcription factor PHABULOSA (PHB), thus promoting cytokinin biosynthesis in meristem vasculature (Dello Ioio et al. 2012). Cytokinin is then transported to the proximal transition zone to promote differentiation. Interestingly, the feedback signal from increased cytokinin levels represses the expression of both *PHB* and miR165, a negative regulator of *PHB*. Therefore, the dynamics of this feedback loop may provide a robust boundary between dividing and differentiating cells upon cytokinin fluctuations (Dello Ioio et al. 2012).

GA responses in the root endodermis have been shown to be of functional importance for root growth, upregulating cell proliferation by mediating the degradation of the root growth repressor DELLA proteins in a subset of meristem cells in the endodermis (Ubeda-Tomás et al. 2008, 2009, Achard et al. 2009). DELLAs impair cell production by enhancing the transcription of cell cycle inhibitor genes *KIP-RELATED PROTEIN2* (*KRP2*) and *SIAMESE* (*SIM*) (Achard et al. 2009) (Fig. 3C). Another DELLA protein, REPRESSOR OF GA (RGA) represses ARR1 during the early stage of meristem development, thus upregulating polar auxin transport by PINs and cell division by keeping SHY2 levels low (Moubayidin et al. 2010) (Fig. 3C). By promoting DELLA degradation and the expansion of endodermal dividing cells in both meristematic and elongation zone, GA responses indirectly increase cell proliferation of other root tissues, thus accelerating root growth (Ubeda-Tomás et al. 2009). This is in contrast to the shoot case, where GA was shown to inhibit SAM activity, reflecting the distinct features in hormone regulatory pathways leading to a different cellular organization in shoot and root (Hay et al. 2002, Jasinski et al. 2005, Ubeda-Tomás et al. 2009).

Brassinosteroids play an important role in the maintenance of root meristem homeostasis. The brassinosteroid-insensitive loss-of-function *brassinosteroid receptor1* (*bri1*) mutant displays a reduced meristem size, resulting from the decreased mitotic activity in the RAM (González-García et al. 2011). Treatment with brassinolide or using a brassinosteroid-signalling enhanced mutation, such as loss-of-function *bri1-ems-suppressor1* (*bes1*) results in early differentiation of meristematic cells, thus also negatively affecting the meristem size and impairing root growth (González-García et al. 2011) (Fig. 3C).

Polar auxin transport appears as a major regulatory linkage between auxin and brassinosteroids (Fig. 3C). The interaction is mediated by *BREVIS RADIX* (*BRX*) of which the promoter contains predicted binding sites for both auxin- and brassinosteroid-controlled transcription factors (Mouchel et al. 2006). In fact, *BRX* expression is strongly induced by auxin treatment and mildly repressed by treatment with brassinolide. In a *brx* background, the expression of *CONSTITUTIVE PHOTOMORPHOGENESIS AND DWARF* (*CPD*), encoding a rate-limiting enzyme for brassinosteroid biosynthesis, is also significantly down regulated, leading to brassinosteroid deficiency (Mouchel et al. 2006). Treatment of short-rooted *brx* seedlings with brassinolide results in recovery of normal expression of auxin transporter genes, which is defective in *brx* (e.g. *PIN3* expression is greatly perturbed), and enhancement of root growth. *BRX* is therefore a feedback mediator between brassinosteroid levels and auxin signalling. This entails a spatio-temporal cross regulatory sequences during root growth, whereas *BRX* competes with the *AUX/IAA* gene *SHY2* in the interaction with the auxin responsive factor *MP*, resulting in the down regulation *SHY2* in the root of young seedlings. However, induction of *SHY2* expression by cytokinin in the transition zone will eventually enable SHY2 to dominate in ARF regulation, leading to a decrease of *BRX* expression (Scacchi et al. 2010).

A radial developmental gradient in the root

While the longitudinal signalling networks regulate the root primary growth, at the same time a second radial developmental gradient controls and maintains the formation of radially concentric circles of various tissues and the root vascular system, consisting of conductive tissues, xylem and phloem, which are responsible for the transport of water, minerals and nutrients (Fig. 3A).

Xylem cell lineages arise early from the vascular initials, while periclinal asymmetric divisions give raise to the phloem in the upper part of the meristem. Loss of function of the histidine kinase and cytokinin receptor gene *AHK4* shows defects in periclinal divisions, and causes cells within the vasculature to differentiate into protoxylem (Mähönen et al. 2000, Inoue et al. 2001). The absence of the negative regulator of cytokinin signalling and auxin-reducible *AHP6*, expressed in the protoxylem and adjacent cells, can partially suppress the *ahk4* phenotype (Bishopp et al. 2011, Mähönen et al. 2006a, b). The negative regulation of the cytokinin responsive targets provides a positive feedback loop on cytokinin signalling which promotes auxin transport to the xylem axis (Bishopp et al. 2011). The auxin-MP-TMO5/LHW pathway promotes both cytokinin biosynthesis and the expression of *AHP6* in xylem precursor cells (De Rybel et al. 2014, Ohashi-Ito et al. 2014). Cytokinins move from their synthesis site to the surrounding procambial cells where they promote periclinal division, thus keeping the CK levels in xylem precursor cells low and impairing periclinal division to occur there.

The primary xylem cell fate is regulated by radial signalling underlying the activity of SHORTROOT (SHR)-SCARECROW (SCR) complex (Helariutta et al. 2000, Carlsbecker et al. 2010). SHR moves to the endodermis after being produced in the stele and interacts with SCR to activate the expression of *miR165* and *miR166*. The *miR165/166* then move back to the stele where they target *HD-ZIP III* transcripts for degradation in a dosage-dependent manner (Nakajima et al. 2001, Carlsbecker et al. 2010). Loss of all HD-ZIP III activity inhibits differentiation and leads to the complete absence of xylem and increase numbers of procambial cells (Carlsbecker et al. 2010). Further, the miR165/166-resistant *phb-7d* mutant shows an expansion of *PHB* transcript throughout as well as outside the stele, suggesting the restriction of the *HD-ZIP III* expression domain by miR165/166 (Carlsbecker et al. 2010). Further, it was shown that the repression of *PHB* by miR165/166 is not only important for xylem specification, but also for ground tissue patterning and differentiation of the pericycle (Miyashima et al. 2011).

A cell death gradient

Programmed cell death (PCD) is a crucial component in plant growth and survival, controlling cell disposal in a selective manner (Pennell and Lamb 1997). In the Arabidopsis root, PCD can be observed in the development of the root cap which protects the fragile stem cells in the root tip (Fendrych et al. 2014). Since new columella and lateral root cap (LRC) cells are constantly generated by distal root stem cells, outermost root cap cells are removed in order to restrict the root cap size and its position at the very end of the root tip (Barlow 2002). In Arabidopsis, this process includes LRC cell death at the edge of the elongation zone and the shedding of these dying cells (Del Campillo et al. 2004, Fendrych et al. 2014). PCD events in the Arabidopsis LRC that result in rapid and direct cell death, such as an abrupt drop of cytoplasmic pH, disintegration and permeabilization of the plasma membrane, collapse of the vacuole, release of vacuolar hydrolytic enzymes, occur sequentially (Fendrych et al. 2014). The initiation of LRC PCD is marked by the expression of the senescence-associated *BIFUNCTIONAL NUCLEASE 1* (*BFN1*) and the aspartyl protease gene *PASPA3* and depends on the NAC domain transcription factor SOMBREBRO (SMB) (Fendrych et al. 2014). The *smb* LRC cells showed a prolonged division activity and delayed differentiation, thus fail to mature and detach from the root tip (Bennett et al. 2010).

In the vascular system, PCD is a crucial step in the final stage of the formation of xylem. Mature xylem is composed of conductive cells called tracheary elements (TEs). One xylem-specific developmental feature is the formation and thickening of the secondary cell wall, which is tightly coupled with PCD (Schuetz et al. 2012, Escamez and Tuominen 2014). TE cell death results in the formation of reinforced, empty vessels that are capable of the transport of water and several solutes (Bollhöner et al. 2012, Schuetz et al. 2012). In contrast to the LRC case, the first observed morphological changes in TE PCD are the expansion of the vacuole and the rupture of the tonoplast, leading to the acidification of the cytoplasm. This is followed by the rapid destruction of the protoplasmic content and TE cell walls are modified to become resistant to final hydrolysis (Bollhöner et al. 2012, Obara et al. 2001). There is evidence that hormone signalling has been shown to participate in inducing xylem PCD, including brassinosteroid, ethylene and polyamine signalling pathways (Kubo et al. 2005, Pesquet and Tuominen 2011, Muñiz et al. 2008). Additionally, loss-of-function mutations in the spermine synthase gene *ACAULIS5* (*ACL5*) lead to defects in the secondary cell wall structure and premature PCD initiation (Muñiz et al. 2008). The expression of *XYLEM CYSTEINE PROTEASE1* (*XCP1*) and *XCP2*, which are involved in PCD and micro-autolysis of cellular aggregates during xylogenesis in the Arabidopsis root, is directly controlled by *VASCULAR-RELATED NAC-DOMAIN6* (*VND6*) and *VND7* (Avci et al. 2008, Ohashi-Ito et al. 2010, Yamaguchi et al. 2011). *xcp1 xcp2* double mutants have been shown to display incompletely degraded cellular content within the TEs. Additionally, the expression of several lipases and DNA and RNA hydrolytic enzymes is activated by VND6 and VND7, including the nuclease *BFN1* which is also active during LRC PCD (Farage-Barhom et al. 2008, Ito and Fukuda 2002). Further, loss of function of the only xylem-specific metacaspase AtMC9 causes defects in TE post-mortem autolysis (Bollhöner et al. 2013). While *XCP2* appears to be an AtMC9 target *in vitro*, analysis of *atmc9 xcp1 xcp2* triple mutants suggest that AtMC9 acts in a XCP1 and XCP2-independent manner and may regulate another protease cascade.

Conclusion

As described above, primary meristems are extremely dynamic and cells go through several stages from birth to differentiation to death. All of these stages are tightly regulated, and underlying this are networks of transcription factors that are under the control of phytohormones, post-translationally modified peptides, environment, etc. In future, it will be important to unravel how all these components interact to establish fairly distinct developmental gradient boundaries.

References

Achard, P., A. Gusti, S. Cheminant, M. Alioua, S. Dhondt, F. Coppens, G.T.S. Beemster and P. Genschik. 2009. Gibberellin signaling controls cell proliferation rate in Arabidopsis. Curr. Biol. 19: 1188-1193.

Aida, M., D. Beis, R. Heidstra, V. Willemsen, I. Blilou, C. Galinha, L. Nussaume, Y.S. Noh, R. Amasino and B. Scheres. 2004. The *PLETHORA* genes mediate patterning of the Arabidopsis root stem cell niche. Cell 119: 119-120.

Avci, U., H.E. Petzold, I.O. Ismail, E.P. Beers and C.H. Haigler. 2008. Cysteine proteases XCP1 and XCP2 aid micro-autolysis within the intact central vacuole during xylogenesis in Arabidopsis roots. Plant J. 56: 303-315.

Barlow, P.W. 2002. The root cap: cell dynamics, cell differentiation and cap function. J. Plant Growth Regul. 21: 261-286.

Bayer, M., T. Nawy, C. Giglione, M. Galli, T. Meinnel and W. Lukowitz. 2009. Paternal control of embryonic patterning in *Arabidopsis thaliana*. Science 323: 1485-1488.

Bell, E.M., W. Lin, A.Y. Husbands, L. Yu, V. Jaganatha, B. Jablonska, A. Mangeon, M.M. Neff, T. Girke and P.S. Springer. 2012. Arabidopsis LATERAL ORGAN BOUNDARIES negatively regulates brassinosteroid accumulation to limit growth in organ boundaries. Proc. Nat. Acad. Sci. USA. 109: 21146-21151.

Bennett, T., A. van den Toorn, G.F. Sanchez-Perez, A. Campilho, V. Willemsen, B. Snel and B. Scheres. 2010. SOMBRERO, BEARSKIN1, and BEARSKIN2 regulate root cap maturation in Arabidopsis. Plant Cell 22: 640-654.

Berleth, T. and G. Jürgens. 1993. The role of the monopteros gene in organising the basal body region of the Arabidopsis embryos. Trends Genet 9: 299.

Besnard, F., Y. Refahi, V. Morin, B. Marteaux, G. Brunoud, P. Chambrier, F. Rozier, V. Mirabet, J. Legrand, S. Lainé, E. Thévenon, E. Farcot, C. Cellier, P. Das, A. Bishopp, R. Dumas, F. Parcy, Y. Helariutta, A. Boudaoud, C. Godin, J. Traas, Y. Guédon and T. Vernoux. 2014a. Cytokinin signalling inhibitory fields provide robustness to phyllotaxis. Nature. 505: 417-421.

Besnard, F., F. Rozier and T. Vernoux. 2014b. The AHP6 cytokinin signaling inhibitor mediates an auxin-cytokinin crosstalk that regulates the timing of organ initiation at the shoot apical meristem. Plant Signal. Behav. 9: e28788.

Bishopp, A., H. Help, S. El-Showk, D. Weijers, B. Scheres, J. Friml, E. Benková, A.P. Mähönen and Y. Helariutta. 2011. A mutually inhibitory interaction between auxin and cytokinin specifies vascular pattern in roots. Curr. Biol. 21: 917-926.

Blilou, I., J. Xu, M. Wildwater, V. Willemsen, I. Paponov, J. Friml, R. Heidstra, M. Aida, K. Palme and B. Scheres. 2005. The PIN auxin efflux facilitator network controls growth and patterning in Arabidopsis roots. Nature 433: 39-44.

Bollhöner, B., J. Prestele and H. Tuominen. 2012. Xylem cell death: Emerging understanding of regulation and function. J. Exp. Bot. 63: 1081-1094.

Bollhöner, B., B. Zhang, S. Stael, N. Denancé, K. Overmyer, D. Goffner, F. Van Breusegem and H. Tuominen. 2013. Post mortem function of AtMC9 in xylem vessel elements. New Phytol. 200: 498-510.

Bommert, P., B. Il Je, A. Goldshmidt and D. Jackson. 2013. The maize Gα gene *COMPACT PLANT2* functions in CLAVATA signalling to control shoot meristem size. Nature 502: 555-558.

Brand, U., J.C. Fletcher, M. Hobe, E.M. Meyerowitz and R. Simon. 2000. Dependence of stem cell fate in Arabidopsis on a feedback loop regulated by CLV3 activity. Science 289: 617-619.

Brandt, R., M. Salla-Martret, J. Bou-Torrent, T. Musielak, M. Stahl, C. Lanz, F. Ott, M. Schmid, T. Greb, M. Schwarz, S.B. Choi, M.K. Barton, B.J. Reinhart, T. Liu, M. Quint, J.C. Palauqui, J.F. Martínez-García and S. Wenkel. 2012. Genome-wide binding-site analysis of REVOLUTA reveals a link between leaf patterning and light-mediated growth responses. Plant J. 72: 31-42.

Breuil-Broyer, S., P. Morel, J. De Almeida-Engler, V. Coustham, I. Negrutiu and C. Trehin. 2004. High-resolution boundary analysis during *Arabidopsis thaliana* flower development. Plant J. 38: 182-192.

Breuninger, H., E. Rikirsch, M. Hermann, M. Ueda and T. Laux. 2008. Differential expression of *WOX* genes mediates apical-basal axis formation in the Arabidopsis embryo. Dev. Cell 14: 867-876.

Byrne, M.E., R. Barley, M. Curtis, J.M. Arroyo, M. Dunham, a Hudson and R. a Martienssen. 2000. Asymmetric leaves1 mediates leaf patterning and stem cell function in Arabidopsis. Nature 408: 967-971.

Byrne, M.E., J. Simorowski and R.A Martienssen. 2002. *ASYMMETRIC LEAVES1* reveals *knox* gene redundancy in Arabidopsis. Development 129: 1957-1965.

Carlsbecker, A., J.-Y. Lee, C.J. Roberts, J. Dettmer, S. Lehesranta, J. Zhou, O. Lindgren, M. A. Moreno-Risueno, A. Vatén, S. Thitamadee, A. Campilho, J. Sebastian, J.L. Bowman, Y. Helariutta and P.N. Benfey. 2010. Cell signalling by microRNA165/6 directs gene dose-dependent root cell fate. Nature 465: 316-321.

Chandler, J.W. and W. Werr. 2015. Cytokinin-auxin crosstalk in cell type specification. Trends Plant Sci. 20: 291-300.

Chickarmane, V.S., S.P. Gordon, P.T. Tarr, M.G. Heisler and E.M. Meyerowitz. 2012. Cytokinin signaling as a positional cue for patterning the apical-basal axis of the growing Arabidopsis shoot meristem. Proc. Nat. Acad. Sci. USA. 109: 4002-4007.

Chung, Y., P.M. Maharjan, O. Lee, S. Fujioka, S. Jang, B. Kim, S. Takatsuto, M. Tsujimoto, H. Kim, S. Cho, T. Park, H. Cho, I. Hwang and S. Choe. 2011. Auxin stimulates *DWARF4* expression and brassinosteroid biosynthesis in Arabidopsis. Plant J. 66: 564-578.

Crawford, B.C.W., J. Sewell, G. Golembeski, C. Roshan, J.A. Long and M.F. Yanofsky. 2015. Genetic control of distal stem cell fate within root and embryonic meristems. Science 347: 655-659.

Daum, G., A. Medzihradszky, T. Suzaki and J.U. Lohmann. 2014. A mechanistic framework for noncell autonomous stem cell induction in Arabidopsis. Proc. Nat. Acad. Sci. USA. 111: 14619-14624.

Del Campillo, E., A. Abdel-Aziz, D. Crawford and S.E. Patterson. 2004. Root cap specific expression of an endo-beta-1,4-D-glucanase (cellulase): a new marker to study root development in Arabidopsis. Plant Mol. Biol. 56: 309-323.

Dello Ioio, R., C. Galinha, A.G. Fletcher, S.P. Grigg, A. Molnar, V. Willemsen, B. Scheres, S. Sabatini, D. Baulcombe, P.K. Maini and M. Tsiantis. 2012. A PHABULOSA/cytokinin feedback loop controls root growth in Arabidopsis. Curr. Biol. 22: 1699-1704.

Dello Ioio, R., F.S. Linhares, E. Scacchi, E. Casamitjana-Martinez, R. Heidstra, P. Costantino and S. Sabatini. 2007. Cytokinins determine Arabidopsis root-meristem size by controlling cell differentiation. Curr. Biol. 17: 678-682.

Dello Ioio, R., K. Nakamura, L. Moubayidin, S. Perilli, M. Taniguchi, M.T. Morita, T. Aoyama, P. Costantino and S. Sabatini. 2008. A genetic framework for the control of cell division and differentiation in the root meristem. Science 322: 1380-1384.

De Reuille, P.B., I. Bohn-Courseau, K. Ljung, H. Morin, N. Carraro, C. Godin and J. Traas. 2006. Computer simulations reveal properties of the cell-cell signaling network at the shoot apex in Arabidopsis. Proc. Nat. Acad. Sci. USA. 103: 1627-1632.

De Rybel, B., M. Adibi, A. S. Breda, J.R. Wendrich, M.E. Smit, O. Novak, N. Yamaguchi, S. Yoshida, G. Van Isterdael, J. Palovaara, B. Nijsse, M. V. Boekschoten, G. Hooiveld, T. Beeckman, D. Wagner, K. Ljung, C. Fleck and D. Weijers. 2014. Integration of growth and patterning during vascular tissue formation in Arabidopsis. Science. 345: 1255215.

De Rybel, B., B. Möller, S. Yoshida, I. Grabowicz, P. Barbier de Reuille, S. Boeren, R.S. Smith, J.W. Borst and D. Weijers. 2013. A bHLH complex controls embryonic vascular tissue establishment and indeterminate growth in Arabidopsis. Dev. Cell. 24: 426-437.

De Smet, I., S. Lau, U. Mayer and G. Jürgens. 2010. Embryogenesis - The humble beginnings of plant life. Plant J. 61: 959-970.

De Smet, I., V. Vassileva, B. De Rybel, M.P. Levesque, W. Grunewald, D. Van Damme, G. Van Noorden, M. Naudts, G. Van Isterdael, R. De Clercq, J.Y. Wang, N. Meuli, S. Vanneste, P. Hilson, G. Jürgens, G.C. Ingram, D. Inzé, P.N. Benfey and T. Beeckman. 2008. Receptor-like kinase ACR4 restricts formative cell divisions in the Arabidopsis root. Science 322: 594-597.

Ding, Z. and J. Friml. 2010. Auxin regulates distal stem cell differentiation in Arabidopsis roots. Proc. Nat. Acad. Sci. USA. 107: 12046-12051.

Emery, J.F., S.K. Floyd, J. Alvarez, Y. Eshed, N.P. Hawker, A. Izhaki, S.F. Baum and J.L. Bowman. 2003. Radial patterning of Arabidopsis shoots by class III *HD-ZIP* and *KANADI* genes. Curr. Biol. 13: 1768-1774.

Escamez, S. and H. Tuominen. 2014. Programmes of cell death and autolysis in tracheary elements: when a suicidal cell arranges its own corpse removal. J. Exp. Bot. 65: 1313-1321.

Farage-Barhom, S., S. Burd, L. Sonego, R. Perl-Treves and A. Lers. 2008. Expression analysis of the *BFN1* nuclease gene promoter during senescence, abscission, and programmed cell death-related processes. J. Exp. Bot. 59: 3247-3258.

Faure, J.E., N. Rotman, P. Fortuné and C. Dumas. 2002. Fertilization in *Arabidopsis thaliana* wild type: Developmental stages and time course. Plant J. 30: 481-488.

Fendrych, M., T. Van Hautegem, M. Van Durme, Y. Olvera-Carrillo, M. Huysmans, M. Karimi, S. Lippens, C.J. Guérin, M. Krebs, K. Schumacher and M.K. Nowack. 2014. Programmed cell death controlled by ANAC033/SOMBRERO determines root cap organ size in Arabidopsis. Curr. Biol. 24: 931-940.

Fletcher, J.C., U. Brand, M.P. Running, R. Simon and E.M. Meyerowitz. 1999. Signaling of cell fate decisions by CLAVATA3 in Arabidopsis shoot meristems. Science 283: 1911-1914.

Friml, J., A. Vieten, M. Sauer, D. Weijers, H. Schwarz, T. Hamann, R. Offringa and G. Jürgens. 2003. Efflux-dependent auxin gradients establish the apical-basal axis of Arabidopsis. Nature 426: 147-153.

Galinha, C., H. Hofhuis, M. Luijten, V. Willemsen, I. Blilou, R. Heidstra and B. Scheres. 2007. PLETHORA proteins as dose-dependent master regulators of Arabidopsis root development. Nature 449: 1053-1057.

Gendron, J.M., J.-S. Liu, M. Fan, M.-Y. Bai, S. Wenkel, P.S. Springer, M.K. Barton and Z.-Y. Wang. 2012. Brassinosteroids regulate organ boundary formation in the shoot apical meristem of Arabidopsis. Proc. Nat. Acad. Sci. USA. 109: 21152-21157.

González-García, M.P., J. Vilarrasa-Blasi, M. Zhiponova, F. Divol, S. Mora-García, E. Russinova and A.I. Caño-Delgado. 2011. Brassinosteroids control meristem size by promoting cell cycle progression in Arabidopsis roots. Development 138: 849-859.

Hardtke, C.S. and T. Berleth. 1998. The Arabidopsis gene *MONOPTEROS* encodes a transcription factor mediating embryo axis formation and vascular development. EMBO J. 17: 1405-1411.

Hay, A., H. Kaur, A. Phillips, P. Hedden, S. Hake and M. Tsiantis. 2002. The gibberellin pathway mediates KNOTTED1-type homeobox function in plants with different body plans. Curr. Biol. 12: 1557-1565.

Heisler, M.G., C. Ohno, P. Das, P. Sieber, G. V. Reddy, J. A. Long and E.M. Meyerowitz. 2005. Patterns of auxin transport and gene expression during primordium development revealed by live imaging of the Arabidopsis inflorescence meristem. Curr. Biol. 15: 1899-1911.

Helariutta, Y., H. Fukaki, J. Wysocka-Diller, K. Nakajima, J. Jung, G. Sena, M.T. Hauser and P.N. Benfey. 2000. The *SHORT-ROOT* gene controls radial patterning of the Arabidopsis root through radial signaling. Cell 101: 555-567.

Higuchi, M., M.S. Pischke, A.P. Mähönen, K. Miyawaki, Y. Hashimoto, M. Seki, M. Kobayashi, K. Shinozaki, T. Kato, S. Tabata, Y. Helariutta, M.R. Sussman and T. Kakimoto. 2004. *In planta* functions of the Arabidopsis cytokinin receptor family. Proc. Nat. Acad. Sci. USA. 101: 8821-8826.

Inoue, T., M. Higuchi, Y. Hashimoto, M. Seki, M. Kobayashi, T. Kato, S. Tabata, K. Shinozaki and T. Kakimoto. 2001. Identification of CRE1 as a cytokinin receptor from Arabidopsis. Nature 409: 1060-1063.

Ishikawa, H. and M.L. Evans. 1995. Specialized zones of development in roots. Plant Physiol. 109: 725-727.

Ito, J. and H. Fukuda. 2002. ZEN1 is a key enzyme in the degradation of nuclear DNA during programmed cell death of tracheary elements. Plant Cell 14: 3201-3211.

Jasinski, S., P. Piazza, J. Craft, A. Hay, L. Woolley, I. Rieu, A. Phillips, P. Hedden and M. Tsiantis. 2005. KNOX action in Arabidopsis is mediated by coordinate regulation of cytokinin and gibberellin activities. Curr. Biol. 15: 1560-1565.

Jeong, S., T.M. Palmer and W. Lukowitz. 2011. The RWP-RK factor GROUNDED promotes embryonic polarity by facilitating YODA MAP kinase signaling. Curr. Biol. 21: 1268-1276.

Jeong, S., A.E. Trotochaud and S.E. Clark. 1999. The Arabidopsi*s* *CLAVATA2* gene encodes a receptor-like protein required for the stability of the CLAVATA1 receptor-like kinase. Plant Cell 11: 1925-1934.

Knauer, S., A.L. Holt, I. Rubio-Somoza, E.J. Tucker, A. Hinze, M. Pisch, M. Javelle, M.C. Timmermans, M.R. Tucker and T. Laux. 2013. A Protodermal miR394 signal defines a region of stem cell competence in the Arabidopsis shoot meristem. Dev. Cell 24: 125-132.

Kubo, M., M. Udagawa, N. Nishikubo, G. Horiguchi, M. Yamaguchi, J. Ito, T. Mimura, H. Fukuda and T. Demura. 2005. Transcription switches for protoxylem and metaxylem vessel formation. Genes Dev. 19: 1855-1860.

Kurakawa, T., N. Ueda, M. Maekawa, K. Kobayashi, M. Kojima, Y. Nagato, H. Sakakibara and J. Kyozuka. 2007. Direct control of shoot meristem activity by a cytokinin-activating enzyme. Nature. 445: 652-655.

Laufs, P., O. Grandjean, C. Jonak, K. Kiêu and J. Traas. 1998. Cellular parameters of the shoot apical meristem in Arabidopsis. Plant Cell 10: 1375-1390.

Leibfried, A., J.P.C. To, W. Busch, S. Stehling, A. Kehle, M. Demar, J.J. Kieber and J.U. Lohmann. 2005. WUSCHEL controls meristem function by direct regulation of cytokinin-inducible response regulators. Nature 438: 1172-1175.

Lukowitz, W., A. Roeder, D. Parmenter and C. Somerville. 2004. A MAPKK kinase gene regulates extra-embryonic cell fate in Arabidopsis. Cell 116: 109-119.

Mähönen, A.P., A. Bishopp, M. Higuchi, K.M. Nieminen, K. Kinoshita, K. Törmäkangas, Y. Ikeda, A. Oka, T. Kakimoto and Y. Helariutta. 2006a. Cytokinin signaling and its inhibitor AHP6 regulate cell fate during vascular development. Science 311: 94-98.

Mähönen, A.P., M. Higuchi, K. Törmäkangas, K. Miyawaki, M.S. Pischke, M.R. Sussman, Y. Helariutta and T. Kakimoto. 2006b. Cytokinins regulate a bidirectional phospho-relay network in Arabidopsis. Curr. Biol. 16: 1116-1122.

Mähönen, A.P., M. Bonke, L. Kauppinen, M. Riikonen, P.N. Benfey and Y. Helariutta. 2000. A novel two-component hybrid molecule regulates vascular morphogenesis of the Arabidopsis root. Genes Dev. 14: 2938-2943.

Mähönen, A.P., K.T. Tusscher, R. Siligato, O. Smetana, S. Díaz-Triviño, J. Salojärvi, G. Wachsman, K. Prasad, R. Heidstra and B. Scheres. 2014. PLETHORA gradient formation mechanism separates auxin responses. Nature 515: 125-129.

Mallory, A.C., B.J. Reinhart, M.W. Jones-Rhoades, G. Tang, P.D. Zamore, M.K. Barton and D.P. Bartel. 2004. MicroRNA control of *PHABULOSA* in leaf development: importance of pairing to the microRNA 5' region. EMBO J. 23: 3356-3364.

Mansfield, S.G. and L.G. Briarty. 1991. Early embryogenesis in *Arabidopsis thaliana*. II. The developing embryo. Can. J. Bot. 69: 461-476.

Matsuzaki, Y., M. Ogawa-Ohnishi, A. Mori and Y. Matsubayashi. 2010. Secreted peptide signals required for maintenance of root stem cell niche in Arabidopsis. Science 329: 1065-1067.

Mayer, K.F.X., H. Schoof, A. Haecker, M. Lenhard, G. Jürgens and T. Laux. 1998. Role of WUSCHEL in regulating stem cell fate in the Arabidopsis shoot meristem. Cell 95: 805-815.

Mayer, U., G. Büttner and G. Jürgens. 1993. Apical-basal pattern formation in the Arabidopsis embryo : studies on the role of the gnom gene. Development 1: 149-162.

Miyashima, S., S. Koi, T. Hashimoto and K. Nakajima. 2011. Non-cell-autonomous microRNA165 acts in a dose-dependent manner to regulate multiple differentiation status in the Arabidopsis root. Development 138: 2303-2313.

Moubayidin, L., R. Di Mambro, R. Sozzani, E. Pacifici, E. Salvi, I. Terpstra, D. Bao, A. van Dijken, R. Dello Ioio, S. Perilli, K. Ljung, P.N. Benfey, R. Heidstra, P. Costantino and S. Sabatini. 2013. Spatial coordination between stem cell activity and cell differentiation in the root meristem. Dev Cell. 26: 405-15.

Moubayidin, L., S. Perilli, R. Dello Ioio, R. Di Mambro, P. Costantino and S. Sabatini. 2010. The rate of cell differentiation controls the Arabidopsis root meristem growth phase. Curr. Biol. 20: 1138-1143.

Mouchel, C.F., K.S. Osmont and C.S. Hardtke. 2006. BRX mediates feedback between brassinosteroid levels and auxin signalling in root growth. Nature 443: 458-461.

Müller, R., A. Bleckmann and R. Simon. 2008. The receptor kinase CORYNE of Arabidopsis transmits the stem cell-limiting signal CLAVATA3 independently of CLAVATA1. Plant Cell 20: 934-946.

Muñiz, L., E.G. Minguet, S.K. Singh, E. Pesquet, F. Vera-Sirera, C.L. Moreau-Courtois, J. Carbonell, M. a Blázquez and H. Tuominen. 2008. ACAULIS5 controls Arabidopsis xylem specification through the prevention of premature cell death. Development 135: 2573-2582.

Nakajima, K., G. Sena, T. Nawy and P.N. Benfey. 2001. Intercellular movement of the putative transcription factor SHR in root patterning. Nature 413: 307-311.

Nishimura, C., Y. Ohashi, S. Sato, T. Kato, S. Tabata and C. Ueguchi. 2004. Histidine kinase homologs that act as cytokinin receptors possess overlapping functions in the regulation of shoot and root growth in Arabidopsis. Plant Cell 16: 1365-1377.

Nodine, M.D., R. Yadegari and F.E. Tax. 2007. RPK1 and TOAD2 are two receptor-like kinases redundantly required for arabidopsis embryonic pattern formation. Dev. Cell 12: 943-956.

Obara, K., H. Kuriyama and H. Fukuda. 2001. Direct evidence of active and rapid nuclear degradation triggered by vacuole rupture during programmed cell death in Zinnia. Plant Physiol. 125: 615-626.

Ogawa, M., H. Shinohara, Y. Sakagami and Y. Matsubayashi. 2008. Arabidopsis CLV3 peptide directly binds CLV1 ectodomain. Science 319: 294.

Ohashi-Ito, K., Y. Oda and H. Fukuda. 2010. Arabidopsis VASCULAR-RELATED NAC-DOMAIN6 directly regulates the genes that govern programmed cell death and secondary wall formation during xylem differentiation. Plant Cell 22: 3461-3473.

Ohashi-Ito, K., M. Saegusa, K. Iwamoto, Y. Oda, H. Katayama, M. Kojima, H. Sakakibara and H. Fukuda. 2014. A bHLH complex activates vascular cell division via cytokinin action in root apical meristem. Curr. Biol. 24: 2053-2058.

Pennell, R. and C. Lamb. 1997. Programmed cell death in plants. Plant Cell 9: 1157-1168.

Perilli, S., J.M. Perez-perez, R. Di Mambro, L. Peris, S. Díaz-triviño, M. Del Bianco, E. Pierdonati, L. Moubayidin, A. Cruz-Ramírez, P. Costantino, B. Scheres, S. Sabatini, R. Di Mambro, C.L. Peris and M. Del Bianco. 2013. RETINOBLASTOMA-RELATED protein stimulates cell differentiation in the Arabidopsis root meristem by interacting with cytokinin signaling. Plant Cell 25: 4469-4478.

Pesquet, E. and H. Tuominen. 2011. Ethylene stimulates tracheary element differentiation in Zinnia elegans cell cultures. New Phytol. 190: 138-149.

Pi, L., E. Aichinger, E. van der Graaff, C.I. Llavata-Peris, D. Weijers, L. Hennig, E. Groot and T. Laux. 2015. Organizer-derived WOX5 signal maintains root columella stem cells through chromatin-mediated repression of *CDF4* expression. Dev. Cell 33: 576-588.

Prigge, M.J., D. Otsuga, J.M. Alonso, J.R. Ecker, G.N. Drews and S.E. Clark. 2005. Class III homeodomain-leucine zipper gene family members have overlapping, antagonistic and distinct roles in Arabidopsis development. Plant Cell 17: 61-76.

Rademacher, E.H., A.S. Lokerse, A. Schlereth, C.I. Llavata-Peris, M. Bayer, M. Kientz, A. FreireRios, J.W. Borst, W. Lukowitz, G. Jürgens and D. Weijers. 2012. Different auxin response machineries control distinct cell fates in the early plant embryo. Dev. Cell 22: 211-222.

Rademacher, E.H., B. Möller, A.S. Lokerse, C.I. Llavata-Peris, W. Van Den Berg and D. Weijers. 2011. A cellular expression map of the Arabidopsis *AUXIN RESPONSE FACTOR* gene family. Plant J. 68: 597-606.

Reddy, G.V., M.G. Heisler, D.W. Ehrhardt and E.M. Meyerowitz. 2004. Real-time lineage analysis reveals oriented cell divisions associated with morphogenesis at the shoot apex of *Arabidopsis thaliana*. Development 131: 4225-4237.

Reddy, G.V. and E.M. Meyerowitz. 2005. Stem cell homeostasis and growth dynamics can be uncoupled in the Arabidopsis shoot apex. Science 310: 663-668.

Reinhardt, D., E.-R. Pesce, P. Stieger, T. Mandel, K. Baltensperger, M. Bennett, J. Traas, J. Friml and C. Kuhlemeier. 2003. Regulation of phyllotaxis by polar auxin transport. Nature 426: 255-260.

Reinhart, B.J., T. Liu, N.R. Newell, E. Magnani, T. Huang, R. Kerstetter, S. Michaels and M.K. Barton. 2013. Establishing a framework for the Ad/abaxial regulatory network of Arabidopsis: ascertaining targets of class III homeodomain leucine zipper and KANADI regulation. Plant Cell 25: 3228-3249.

Sarkar, A.K., M. Luijten, S. Miyashima, M. Lenhard, T. Hashimoto, K. Nakajima, B. Scheres, R. Heidstra and T. Laux. 2007. Conserved factors regulate signalling in *Arabidopsis thaliana* shoot and root stem cell organizers. Nature 446: 811-814.

Scacchi, E., P. Salinas, B. Gujas, L. Santuari, N. Krogan, L. Ragni, T. Berleth and C.S. Hardtke. 2010. Spatio-temporal sequence of cross-regulatory events in root meristem growth. Proc. Nat. Acad. Sci. USA. 107: 22734-22739.

Schlereth, A., B. Möller, W. Liu, M. Kientz, J. Flipse, E.H. Rademacher, M. Schmid, G. Jürgens and D. Weijers. 2010. MONOPTEROS controls embryonic root initiation by regulating a mobile transcription factor. Nature 464: 913-916.

Schoof, H., M. Lenhard, a Haecker, K.F. Mayer, G. Jürgens and T. Laux. 2000. The stem cell population of Arabidopsis shoot meristems in maintained by a regulatory loop between the *CLAVATA* and *WUSCHEL* genes. Cell 100: 635-644.

Schuetz, M., R. Smith and B. Ellis. 2012. Xylem tissue specification, patterning, and differentiation mechanisms. J. Exp. Bot. 64: 11-31.

Schuster, C., C. Gaillochet, A. Medzihradszky, W. Busch, G. Daum, M. Krebs, A. Kehle and J.U. Lohmann. 2014. A regulatory framework for shoot stem cell control integrating metabolic, transcriptional and phytohormone signals. Dev. Cell 28: 438-449.

Smith, Z.R. and J.A. Long. 2010. Control of Arabidopsis apical-basal embryo polarity by antagonistic transcription factors. Nature 464: 423-426.

Stahl, Y., S. Grabowski, A. Bleckmann, R. Kühnemuth, S. Weidtkamp-Peters, K.G. Pinto, G.K. Kirschner, J.B. Schmid, R.H. Wink, A. Hülsewede, S. Felekyan, C.A.M. Seidel and R. Simon. 2013. Moderation of Arabidopsis root stemness by CLAVATA1 and ARABIDOPSIS CRINKLY4 receptor kinase complexes. Curr. Biol. 23: 362-371.

Stahl, Y., R.H. Wink, G.C. Ingram and R. Simon. 2009. A signaling module controlling the stem cell niche in Arabidopsis root meristems. Curr. Biol. 19: 909-914.

Szemenyei, H., M. Hannon and J.A. Long. 2008. TOPLESS mediates auxin-dependent transcriptional repression during Arabidopsis embryogenesis. Science. 319: 1384-1386.

Tian, H., K. Wabnik, T. Niu, H. Li, Q. Yu, S. Pollmann, S. Vanneste, W. Govaerts, J. Rolčík, M. Geisler, J. Friml and Z. Ding. 2014. WOX5-IAA17 feedback circuit-mediated cellular auxin response is crucial for the patterning of root stem cell niches in Arabidopsis. Mol. Plant. 7: 277-289.

Tian, Q., N.J. Uhlir and J.W. Reed. 2002. Arabidopsis SHY2/IAA3 inhibits auxin-regulated gene expression. Plant Cell 14: 301-319.

Traas, J. 2012. Phyllotaxis. Development 140: 249-253.

Turchi, L., M. Carabelli, V. Ruzza, M. Possenti, M. Sassi, A. Peñalosa, G. Sessa, S. Salvi, V. Forte, G. Morelli and I. Ruberti. 2013. Arabidopsis HD-ZIP II transcription factors control apical embryo development and meristem function. Development. 140: 2118-2129.

Ubeda-Tomás, S., F. Federici, I. Casimiro, G.T.S. Beemster, R. Bhalerao, R. Swarup, P. Doerner, J. Haseloff and M.J. Bennett. 2009. Gibberellin signaling in the endodermis controls Arabidopsis root meristem size. Curr. Biol. 19: 1194-1199.

Ubeda-Tomás, S., R. Swarup, J. Coates, K. Swarup, L. Laplaze, G.T.S. Beemster, P. Hedden, R. Bhalerao and M.J. Bennett. 2008. Root growth in Arabidopsis requires gibberellin/DELLA signalling in the endodermis. Nat. Cell. Biol. 10: 625-628.

Ueda, M., Z. Zhang and T. Laux. 2011. Transcriptional activation of Arabidopsis axis patterning genes *WOX8/9* links zygote polarity to embryo development. Dev. Cell 20: 264-270.

Verbelen, J.-P., T. De Cnodder, J. Le, K. Vissenberg and F. Baluska. 2006. The root apex of *Arabidopsis thaliana* consists of four distinct zones of growth activities: meristematic zone, transition zone, fast elongation zone and growth terminating zone. Plant Signal Behav. 1: 296-304.

Vernoux, T., G. Brunoud, E. Farcot, V. Morin, H. Van den Daele, J. Legrand, M. Oliva, P. Das, A. Larrieu, D. Wells, Y. Guédon, L. Armitage, F. Picard, S. Guyomarc'h, C. Cellier, G. Parry, R. Koumproglou, J.H. Doonan, M. Estelle, C. Godin, S. Kepinski, M. Bennett, L. De Veylder and J. Traas. 2011. The auxin signalling network translates dynamic input into robust patterning at the shoot apex. Mol. Syst. Biol. 7: 508.

Vert, G., J.L. Nemhauser, N. Geldner, F. Hong and J. Chory. 2005. Molecular mechanisms of steroid hormone signaling in plants. Annu. Rev. Cell Dev. Biol. 21: 177-201.

Waki, T., T. Hiki, R. Watanabe, T. Hashimoto and K. Nakajima. 2011. The Arabidopsis RWP-RK protein RKD4 triggers gene expression and pattern formation in early embryogenesis. Curr. Biol. 21: 1277-1281.

Wang, J.-W., L.-J. Wang, Y.-B. Mao, W.-J. Cai, H.-W. Xue and X.-Y. Chen. 2005. Control of root cap formation by MicroRNA-targeted auxin response factors in Arabidopsis. Plant Cell 17: 2204-2216.

Weigel, D., J. Alvarez, D.R. Smyth, M.F. Yanofsky and E.M. Meyerowitz. 1992. LEAFY controls floral meristem identity in Arabidopsis. Cell 69: 843-859.

Weijers, D., A. Schlereth, J.S. Ehrismann, G. Schwank, M. Kientz and G. Jürgens. 2006. Auxin triggers transient local signaling for cell specification in Arabidopsis embryogenesis. Dev. Cell 10: 265-270.

Wendrich, J.R. and D. Weijers. 2013. The Arabidopsis embryo as a miniature morphogenesis model. New Phytol. 199: 14-25.

Yadav, R.K., M. Perales, J. Gruel, T. Girke, H. Jönsson and G.V. Reddy. 2011. WUSCHEL protein movement mediates stem cell homeostasis in the Arabidopsis shoot apex. Genes Dev. 25: 2025-2030.

Yadav, R.K., M. Perales, J. Gruel, C. Ohno, M. Heisler, T. Girke, H. Jönsson and G.V. Reddy. 2013. Plant stem cell maintenance involves direct transcriptional repression of differentiation program. Mol. Syst. Biol. 9: 654.

Yadav, R.K., M. Tavakkoli and G.V. Reddy. 2010. WUSCHEL mediates stem cell homeostasis by regulating stem cell number and patterns of cell division and differentiation of stem cell progenitors. Development 137: 3581-3589.

Yamaguchi, M., N. Mitsuda, M. Ohtani, M. Ohme-Takagi, K. Kato and T. Demura. 2011. VASCULAR-RELATED NAC-DOMAIN 7 directly regulates the expression of a broad range of genes for xylem vessel formation. Plant J. 66: 579-590.

Yamaguchi, N., M.F. Wu, C.M. Winter, M.C. Berns, S. Nole-Wilson, A. Yamaguchi, G. Coupland, B. a. Krizek and D. Wagner. 2013. A molecular framework for auxin-mediated initiation of flower primordia. Dev. Cell 24: 271-282.

Yoshida, S., P. Barbier de Reuille, B. Lane, G.W. Bassel, P. Prusinkiewicz, R.S. Smith and D. Weijers. 2014. Genetic control of plant development by overriding a geometric division rule. Dev. Cell 29: 75-87.

Zhang, Z. and X. Zhang. 2012. Argonautes compete for miR165/166 to regulate shoot apical meristem development. Curr. Opin. Plant Biol. 15: 652-658.

Zhao, Z., S.U. Andersen, K. Ljung, K. Dolezal, A. Miotk, S.J. Schultheiss and J.U. Lohmann. 2010. Hormonal control of the shoot stem-cell niche. Nature 465: 1089-1092.

Zhou, G., M. Kubo, R. Zhong, T. Demura and Z.-H. Ye. 2007. Overexpression of miR165 affects apical meristem formation, organ polarity establishment and vascular development in Arabidopsis. Plant Cell Physiol. 48: 391-404.

Zhou, W., L. Wei, J. Xu, Q. Zhai, H. Jiang, R. Chen, Q. Chen, J. Sun, J. Chu, L. Zhu, C.-M. Liu and C. Li. 2010. Arabidopsis tyrosylprotein sulfotransferase acts in the auxin/PLETHORA pathway in regulating postembryonic maintenance of the root stem cell niche. Plant Cell 22: 3692-3709.

Zhu, H., F. Hu, R. Wang, X. Zhou, S. Sze, L.W. Liou, A. Barefoot, M. Dickman and X. Zhang. 2011. Arabidopsis Argonaute10 specifically sequesters miR166/165 to regulate shoot apical meristem development. Cell 145: 242-256.

13

The Cell Cycle in Nodulation

*Jeremy D. Murray**

Introduction

The most obvious outcome of the legume-rhizobia symbiosis is the formation of a new lateral root organ called the nodule. Nodule organogenesis involves cell divisions in the cortex that give rise to large, highly endore duplicated cells, which house the nitrogen-fixing rhizobia. Some legumes form indeterminate nodules that feature a developmental gradient, ranging from tiny meristematic cells to large nitrogen-fixing cells. The extensive manipulation of the cell cycle required across different nodule zones present it as an excellent model to study cell cycle regulation in plants. In addition to its importance in organogenesis, recent findings point to a role for the cell cycle in the rhizobial root-hair infection process. This is unexpected since root hair cells undergoing infection form trans-cellular structures called infection threads, but do not divide. During both organogenesis and infection cell cycle changes are accompanied by alterations in hormone signalling, especially cytokinin and auxin.

Nodule Development Comprises Infection and Organogenesis

Nodulation begins with a molecular signal exchange between the host and symbiont; the host produces flavonoids which induce the production of lipo-chitooligosaccharides called Nod factors. Perception of Nod factors by the host induces two developmental programmes, rhizobial infection and nodule organogenesis (Oldroyd et al. 2010). Infection involves the entrapment of attached rhizobia within a curl that forms at the root hair tip, and the subsequent formation of a transcellular tubular invagination of the plasma membrane and cell wall called the infection thread. The infection thread extends from the infection site within the initial curl through the root hair cell and into underlying cortical cells, forming a conduit for passage of the rhizobia into the inner cell layers. As the infection thread is growing, beneath it cell divisions in the cortex and pericycle and endodermis begin to occur, forming the nodule primordial and vascular strands, respectively (Xiao et al. 2014). As the infection thread reaches the cells of the primordia, it branches and extends into many cells where it

Department of Cell & Developmental Biology, John Innes Centre, Norwich Research Park, Norwich, NR4 7UH, UK.
* Corresponding author : Email: jeremy.murray@jic.ac.uk

releases its rhizobial contents through endocytosis forming organelle-like structures containing one or a few bacteria bounded by host membrane called symbiosomes where the nitrogen fixation takes place. In this manner the nodule cells gradually become filled with symbiosomes and at the same time undergo several rounds of endoreduplication while expanding greatly in size.

In legumes that form determinate nodules such as soybean and *Lotus japonicus*, once the cell division and infection phases are complete the nodule cells differentiate into large nitrogen fixing cells which are regularly interspersed with non-infected cells and the nodules do not have any obvious zonation and the nodule is spherical in shape. In legume species that form elongated indeterminate nodules, the nodules develop an apical meristem that consists of tiny dividing cells that are devoid of rhizobia and adjacent to the meristem they maintain an active infection zone containing cells that are forming infection threads from which bacteria are released. Within this zone cells are enlarging and undergoing endoreduplication. The infection zone is bounded proximally by the interzone, where cell divisions and endocycling cease, which is followed by the nitrogen fixation zone containing large cells filled with symbiosomes. Therefore cells originating in the meristem are destined to become infected and differentiate into nitrogen-fixing cells, spending some time in each of the developmental zones. Recently, Roux et al. (2014) have used the laser-capture microdissection technique to sample each of the nodule zones for transcript profiling providing a useful window into the cell cycle changes that occur across these zones.

Nodule Ontogeny

Different root cell layers are involved in nodule formation. The process has been most well studied in *Medicago truncatula* (Timmers et al. 1999, Xiao et al. 2014). In Medicago, infection of a root hair is accompanied by anticlinal cell divisions occurring in the pericycle and the cortical cell layers of the root. The initiation of nodules, as is the case with lateral roots, only occurs opposite protoxylem poles, hemmed in by ethylene produced in the region of the phloem (Heidstra et al. 1997, Penmetsa et al. 2003). The area of cell division in the pericycle expands to include the endodermis and together these tissues give rise to vascular tissue that grows and branches into several peripheral vascular strands, continuous with the root stele, which extend towards the nodule apex. The meristem, which appears around two days after the initial cell divisions, is formed from the third layer of cortical cells (Xiao et al. 2014). The growth of the meristem produces cells that will later become infected and develop into nitrogen fixing cells, giving the nodule its elongated appearance. Cortical cells in layers 4 and 5 also divide, and become infected at the primordia stage, even as the meristem is developing, leading to formation of the first nitrogen fixing cells at the base of the nodule (Xiao et al. 2014). Infection threads, while most apparent in the infection zone, can also be seen at a low frequency within the nitrogen fixing zone. Understanding the genetic controls underlying these events is one of the great emerging challenges in nodulation research. While nodulation researchers have progressed greatly on understanding Nod factor signalling there has been limited progress in understanding the downstream developmental events. For instance it was recently discovered that the transcription factor NF-YA1 is required for the formation and/or maintenance of the nodule meristem, but its targets remain unknown (Xiao et al. 2014).

The Cell Cycle in Plants

A simplified model of plant cell cycle control (reviewed by Scofield et al. 2014) is provided in Figure 1. Entry into the cell cycle is dependent on Cyclin Dependent Kinase A (CDKA) complexed with D-type Cyclins, which phosphorylate the transcriptional repressor Retinoblastoma-Related protein (RBR; Boniotti and Gutierrez 2001). This CDKA/CyclinD phosphorylation of RBR inactivates it, releasing the transcription factor E2F/DEL1, which de-represses *CELL CYCLE SWITCH 52a* (*CCS52a*) leading to endocycling (Cebolla et al. 1999, Vinardell et al. 2003, Vlieghe et al. 2005, Lammens et al. 2008). In opposition to endocycling are the negative regulators OMISSION OF SECOND DIVISION (OSD1) which inhibits the ANAPHASE PROMOTING COMPLEX/CENTROSOME (APC/C; Cromer et al. 2012), and HIGH PLOIDY TWO (HPY2) which acts downstream of PLETHORA proteins (Ishida et al. 2009). Another endocycle repressor, the trihelix transcription factor, GT2–LIKE 1 (GTL1), was shown in Arabidopsis to directly repress the transcription of *CCS52* to stop endocycle progression (Breuer et al. 2012). The APC/C is required for both mitosis and endocycling, consequence of its activation by CDC20 or CCS52A, respectively (Kevei et al. 2011). Other major players include the Kip-related proteins (KRPs) that block the G1-S transition through inhibition of CDK activity and thereby suppressing both endocycling and mitosis through inhibition of D-type cyclins (Nakai

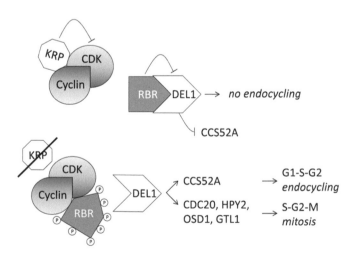

FIGURE 1   A simplified model of the plant cell cycle. Binding of RBR to the transcriptional activator DEL1 represses it, preventing G1-S progression. This occurs when the cyclin-dependent kinase inhibitor KRP binds to the Cyclin-CDK complex which inhibits CDK. In the absence of KRP the CDK-Cyclin complex is active, leading to the phosphorylation of RBR and the release of DEL1 which activates G1-S progression the outcome of which is determined by differential activation of the APC/C by either CCS52A to promote endocycling, or CDC20 to promote mitosis . HPY2, OSD1 and GTL1 repress endocycling to promote mitosis.

et al. 2006). These cell cycle components are conserved in plants, and some of these components have been shown to have roles in nodulation.

Expression of Cell Cycle Regulators Across Indeterminate Nodule Zones

As mentioned above, indeterminate nodules, such as found in the model legume *Medicago truncatula*, recapitulate all the early stages of nodule development, having zones of cell division, cell differentiation/endore duplication (accompanied by infection), and mature cells, present in a single organ. The recent study by Roux et al. (2014) used laser capture microdissection coupled with genome-wide transcriptomics to characterize gene expression across these zones. In Medicago nodules, endocycling is thought to occur mainly in the infection zone which comprises several layers of cells sandwiched between the rapidly dividing cells of the meristem and the differentiating cells of the interzone which are preparing for the onset of N-fixation. This model is recapitulated by the pattern of gene expression of the key regulators of the cell cycle (Fig. 2). *E2F/DEL1* has apical expression ending in the proximal infection zone, serving to repress the endocycle via CCS52A and preserve the mitotic state while *CCS52A* is required to activate the APC/C to promote exit from the cell cycle and entry into endocycling. Knockdown of *CCS52A* resulted in decreased

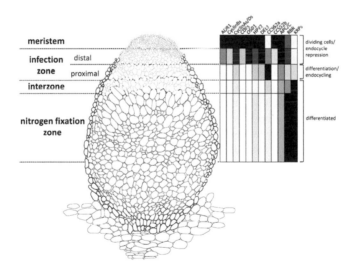

FIGURE 2 Expression of cell cycle components across *Medicago truncatula* nodule zones. Right side: shading reflects relative transcript levels of each gene or group of genes across nodule zones (black indicates maximum expression for the gene). Data extracted from Roux et al. (2014). *AUR1* (Medtr3g110405), *OSD1* (Medtr7g092570), *HYP2* (Medtr2g078640), *DEL1* (Medtr4g106540), *CCS52a* (Medtr4g102510), *CCS52b* (Medtr3g067940), *RBR* (Medtr8g040560), *KRPs*: Medtr7g057920, Medtr5g017600, and Medtr1g072860. *CDKs*: Medtr4g007750 (*CDKB1*), Medtr1g075610 (*CDKB2*), and Medtr2g032060 (*CDKA*). *APC/C*: Medtr5g082890, Medtr7g021880, Medtr4g082265, Medtr4g082150, Medtr4g090990, Medtr5g091020, Medtr8g058380, Medtr2g019850, Medtr8g021140, Medtr1g103750, Medtr8g077200, Medtr7g005900, Medtr4g007410. The cyclins are listed in Figure legend 1).

ploidy levels and cell size in nodules (Vinardell et al. 2003). *CCS52A* is therefore an excellent functional marker for endore duplication; it shows a peak in expression in the proximal infection zone, with significant expression in the distal infection zone and the interzone. The closest Medicago homologue of *GTL1*, Medtr1g098900, is expressed mostly in the meristem with some expression in the distal infection zones, and therefore may have a role in restricting the expression of *CCS52A* (not shown; Roux et al. 2014). Interestingly, *CCS52b,* a homologue of *CCS52a,* is expressed mainly in the nodule meristem (Roux et al. 2014). This is consistent with *CCS52b* expression being restricted to the G2-M phase, and inability of the encoded protein to bind to the yeast APC, in contrast to CCS52a (Tarayre et al. 2004). The expression of the closest *M. truncatula CDC20* homologue, Medtr5g008010, is also restricted to the meristem and distal infection zone, but is absent in the proximal infection zone where CCS52A expression peaks reflecting the complementary function of these proteins in activation of the APC/C.

In yeast and animals CDC25 activates CDKs by dephosphorylating them. A Medicago *CDC25* homologue (Medtr7g102310) is constitutively expressed at high levels in all nodule tissues (not shown; Roux et al. 2014), consistent with early reports that its expression is not correlated with cell proliferation (Landrieu et al. 2004). *RBR* is expressed throughout the nodule, but its expression is lowest in the proximal infection zone, which may serve to promote endocycling in this zone (Fig. 2). The significance of *RBR* expression in the N-fixation zone is unclear, but may be related to its functions outside the cell cycle (Kuwabara and Gruissem 2014). Three *KRPs* are expressed in nodules, mainly in the interzone and nitrogen fixation zone, consistent with their role of blocking CDK-mediated entry into S-phase (Fig. 2). These regulatory proteins provide a counter-gradient to the cyclins and CDKs, presumably preventing activation of the cell cycle in the basal nodule zones. Interestingly, studies of Arabidopsis roots have shown that *KRP* expression is strongest in the pericycle cells opposite the protophloem poles and that ectopic over expression of *KRP2* inhibits lateral root formation which is normally restricted to protoxylem poles (Himanen et al. 2002). It is therefore possible a similar mechanism operates to constrain nodule formation to protoxylem poles, potentially under the control of ethylene. A further role for the KRPs in nodule development is also possible. As mentioned above, a major difference between roots and legume nodules is that the vascular bundles of roots are centrally positioned, while in nodules the vascular bundles that initiate from the root stele branch form several strands around the nodule periphery. In addition to the above-proposed role in controlling the sites of nodule initiation, the expression of KRPs at the nodule base may also have a role in preventing the development of central vasculature in the nodule, through the prevention of cell divisions at the nodule base.

Over twenty cyclins belonging to A, B and D classes are expressed at appreciable levels in nodules, mainly in the nodule apex (Figs. 2 and 3). The B-type Cyclins, which are required for G2-M transitions, are most highly expressed in the meristem with some expression in the distal infection zone and have very little expression in the basal nodule zones. The apparent overlap between B-type cyclins and *CCS52A* in the distal infection zone suggests that this zone contains a mixed population of cells that are either dividing or differentiating. The near absence of these cyclins in the proximal infection zone where endore duplication takes place is probably significant as these B Cyclins have been shown to repress the endocycle in yeast (Wuarin et al. 2002). The absence of these cyclins and the strongly reduced expression of the endocycle respressors *OSD1* and *HYP2* in the proximal infection zone further indicate that cell division has stopped and has given over to endore duplication (Fig. 2). D-class cyclins, which are limiting for the G1-S phase transitions (Menges et al. 2006), and A-type cyclins which are required for nodule meristem formation or maintenance (Roudier et al. 2003) and act to repress endocycles (Imai et al. 2006), had a similar expression profile to that of the B-type cyclins. However, unlike the B-type cyclins which all showed a clear apical-basal gradient of expression, several A- and D-type cyclins have expression that is as high or higher in the distal infection zone as in the meristem, and several were also expressed, albeit at lower levels, in the nitrogen fixation zone. The genes encoding the various subunits of the APC/C, which has a central role in both cell cycle progression in mitosis and endocycling, are expressed throughout the nodule, with highest expression in the meristem region (Fig. 2). Intriguingly, knockdown of *APC6/CDC16* , a subunit of the APC/C, results in increased numbers of nodules which have an extended infection zone (Kuppusamy et al. 2009). The authors speculate that the extended infection zone in this mutant is a consequence of impeded endore duplication that would result from inhibition of the APC/C, such as is seen when *CCS52A* expression is diminished (Vinardell et al. 2003). This finding also suggests a role for the APC/C in nodule initiation; one possibility is that lowered activity of the APC/C enhances the accumulation of mitotic cyclins, lowering the threshold for the initial cell divisions in the root cortex and pericycle. A further clue to this effect may be the apparent link between the APC/C and hormone signalling; over expression of the APC10 subunit caused

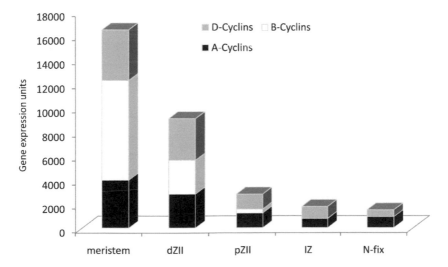

FIGURE 3 Transcript levels (normalized reads) for A-, B- and D-class cyclins in different nodule zones. (Data from Roux et al. 2104). dZII, distal infection zone; pZII, proximal infection zone; IZ, interzone; N-fix, nitrogen fixation zone. A Cyclins: Medtr3g102530, Medtr1g018680, Medtr3g102520, Medtr1g011470, Medtr3g088415, Medtr2g102520, Medtr3g102510, Medtr3g107970, and Medtr2g102550. B Cyclins: Medtr7g089080, Medtr5g088980, Medtr8g074000, and Medtr5g023790. D Cyclins: Medtr4g094942, Medtr5g035360, Medtr3g100710, Medtr5g032550, Medtr3g102310, Medtr1g107535, Medtr2g089490, and Medtr8g063120.

phenotypes reminiscent of ethylene sensitivity (thickened hypocotyls and a short thick root; Lindsay et al. 2011). Ethylene strongly inhibits nodulation through direct effects on Nod factor signalling (Penmetsa et al. 2003, Oldroyd et al. 2001), so it may be instructive to test whether altered ethylene sensitivity is responsible for increased nodulation in *APC6*-knockdown roots.

Auxin's Role in Cell Cycle Regulation

Auxin's ubiquitous involvement in cell developmental programmes promises a central role for it in nodulation, a role that has only begun to be explored. While few functional genetic studies have been carried out, a few key insights have been made. Auxin markers, such as the synthetic promoter consisting of a series of *Auxin Response Elements*, and reporters for auxin-responsive genes such as *GH3*, indicate that increases in auxin signalling occur at the site of primordium formation (Mathesius et al. 1998, Pacios-Bras et al. 2003, Suzaki et al. 2012, 2013, Breakspear et al. 2014). Analysis of mutants for flavonoid production and ethylene-sensing has demonstrated a positive correlation of root auxin levels and nodulation (Wasson et al. 2006, Prayitno et al. 2006). With all these studies it has become clear that the exact choice of reporter is important; early auxin responsive genes such as *GH3*, *SAUR*, *ARF*, and *AUX/IAA* belong to large families and it has become apparent that each family member may be under different regulation. For example, many early

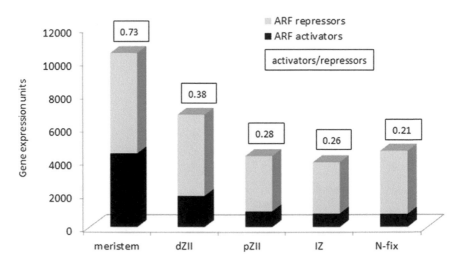

FIGURE 4 Transcript levels (normalized reads) for all putative ARF activators and ARF repressors across nodule zones. Data from Roux et al. 2014. Ratios for transcripts levels (activators/repressors) shown in boxes. dZII, distal infection zone; pZII, proximal infection zone; IZ, interzone; N-fix, nitrogen fixation zone. ARF repressors (putative): Medtr4g021580, Medtr2g005240, Medtr4g058930, Medtr2g094570, Medtr1g094960, Medtr1g064430, Medtr3g073420, Medtr8g027440, Medtr8g100050, Medtr2g014770, Medtr4g088210, Medtr4g060460, and Medtr7g062540; ARF activators (putative): Medtr2g043250, Medtr8g101360, Medtr4g124900, Medtr1g024025, Medtr2g018690, Medtr8g079492, Medtr5g076270, and Medtr3g064050.

auxin-response genes are expressed in nodules, but only a specific subset is specifically upregulated in cells undergoing infection by rhizobia (Breakspear et al. 2014).

With this in mind, genome-wide transcriptomic studies can provide a more complete picture of auxin signalling during nodulation. Auxin Response Factors (ARFs) can be divided into two types, activators and repressors of transcription. A survey of the entire family of Medicago ARFs across the nodule shows that ARF activators and repressors are both expressed in all the nodule zones but display a continuous gradient of expression, with the highest expression in the meristem (Figure 4; Roux et al. 2014). A second observation that can be made is that the ratio of transcripts of ARF activators to that of ARF repressors decreases across from the apex to the nodule base (Figure 4). Based on the current model of auxin signalling, ARF repressors compete with ARF activators at target promoters. The decreasing ratio of ARF activators to ARF repressors across nodule zones further supports the idea that auxin responses are highest in the nodule meristem and lowest in the nitrogen fixation zone. While specialized roles for individual ARFs cannot be ruled out, these data suggest some degree of redundancy of function, at least for the ARF repressors. Overall, this pattern of gene expression is consistent with the evidence suggesting that higher auxin responses, as mediated through TIR1-AUX/IAA-ARF complexes, promote mitosis and inhibit endore duplication (Ishida et al. 2010).

Cytokinin's Role in Regulation of the Cell Cycle in Nodulation

Responses to cytokinins in plants are activated via cytokinin receptors that activate a two-component response system that leads to increased expression of two types of transcription factors that have been classified as negative (Type A) and positive (Type B) response regulators, designated RRAs and RRBs respectively (Heyl et al. 2013). Cytokinin's play a central role in nodulation; the earliest divisions in nodule formation are dependent on cytokinin signalling; phenotypes of loss-of-function and gain-of-function mutants in the major root cytokinin receptor demonstrate that cytokinin responses are both necessary and sufficient for nodule formation, respectively (Gonzalez-Rizzo et al. 2006, Tirichine et al. 2007, Murray et al. 2007, Plet et al. 2011, Held et al. 2014). In *L. japonicus* nodules can be induced by cytokinin treatments in absence of rhizobia (Heckmann et al. 2011). Several genes belonging to both classes of response regulators are expressed in nodules including several B-type response regulators with close homology to *ARR2, ARR10* and *ARR12* which have been implicated in cytokinin responsiveness in *A. thaliana* (Mason et al. 2005, Yokoyama et al. 2007, Hill et al. 2013). These genes are expressed most strongly in the meristem and N-fixation zone and lowest in the interzone, however, in contrast to the ARFs, their expression does not vary greatly across the nodule zones (not shown). The most highly expressed cytokinin receptor, *CRE1*, which is required for timely nodule initiation (Murray et al. 2007), is expressed most highly in the meristem and interzone (Roux et al. 2014). The role of cytokinin signalling in mature nodules is unknown, but a role in cellular proliferation and differentiation has been proposed (Plet et al. 2011). Some genes acting downstream of cytokinin signalling in nodulation have been identified, including the response regulator *RRA4* (previously *ARR4*), the GRAS transcription factor *NSP2*, and a bHLH transcription factor (Ariel et al. 2012) but the links with cell cycle activation, if any, are not yet understood. In

Arabidopsis evidence has been found for cytokinin-mediated cell cycle activation through D-type cyclins (Riou-Khamlichi et al. 1999), and in tobacco, cytokinins were shown to activate CDKA1 (Zhang et al. 1996), the orthologue of which is most highly expressed in the nodule meristem (Fig. 2, Roux et al. 2014). Recent evidence revealed that exogenously supplied flavonoids can rescue nodulation in the *cre1* cytokinin receptor mutant, which is associated with restoration of auxin transport control (Ng et al. 2015). Therefore one major role for cytokinin signalling may be to regulate auxin transport, potentially via flavonoids.

The Role of the Cell Cycle in Rhizobial Infection

While the role of the cell cycle in nodule formation is self-evident, its role in rhizobial infection of root hairs is not. The epidermal cells that develop infection threads do not divide, but instead form a trans-cellular tube, bounded by cell wall and membrane that initiates from near the root hair tip and extends to the base of the epidermal cell, forming a tunnel through the cell (Fig. 5). The process then repeats itself in the underlying cells, forming a connected series of these tubes which serve as a conduit for the rhizobia into the inner cell layers of the root and the rapidly dividing cells of

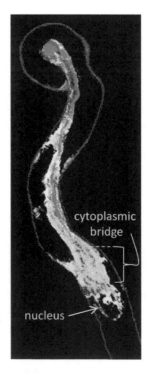

FIGURE 5   A rhizobial infection thread traversing a root hair. A confocal image of a *Medicago truncatula* root-hair cell being colonized by cyan-fluorescent protein-tagged *Sinorhizobium meliloti* 2011. The transgenic root is expressing a GFP-HDEL construct (mgfp4-ER) which highlights the endoplasmic reticulum (green), and makes the cytoplasmic bridge visible. Image credit Joëlle Fournier.

TABLE 1 Cell Cycle-related genes induced by rhizobia or Nod factors.

Gene symbol	Gene model	Treatment				
		1 dpi	3 dpi	5 dpi	skl	Nod Factors
ABAP1	Medtr1g012360	1.16	2.06	**3.42***	**5.46***	1.58
APC6	Medtr8g058380	1.17	1.23	1.21	1.47	**1.72***
AUR1	Medtr3g110405	1.02	1.85	3.17	**3.88***	1.14
CDKA;1	Medtr2g032060	1.01	1.12	1.20	**1.27***	1.27
CDKB2;2	Medtr1g075610	1.27	1.87	2.26	**4.06***	0.60
CDKC;1	Medtr3g040810	1.10	0.62	0.68	0.75	**0.67***
CDKD1;1	Medtr6g080470	1.03	0.88	1.06	1.03	**1.26***
CycA3;1	Medtr3g102530	1.61	2.80	**5.98***	5.26	3.41
CycC1;2	Medtr7g055650	1.10	0.81	0.86	0.94	**1.32***
CycD1;1	Medtr8g063120	1.19	1.50	1.40	**3.18ᴬ**	**3.18***
CycD1;2	Medtr5g035360	0.97	1.46	1.20	1.71	**2.01***
CycD4;1	Medtr3g100710	0.92	1.16	1.72	1.87	**2.39***
ELC	Medtr4g075250	1.01	1.14	**0.86***	0.79	**0.86***
KRP3	Medtr7g057920	1.00	1.07	0.91	0.75	**0.79***
OSD1	Medtr7g092570	0.94	2.94	**5.23***	**14.99***	0.89
SMT2	Medtr3g114780	1.51	1.46	**2.45***	**8.26***	1.31

* Significantly different than control treatment (see Breakspear et al. 2014).

the nodule primordia. The first clue of the possible involvement of the cell cycle to be noticed by nodulation researchers came from subcellular structures that formed prior to formation of the infection thread in cortical cells. The cells were seen to develop an enlarged nucleus which would move to the centre of a wide cytoplasmic bridge that formed anticlinally across the cells, predicting the future path of the growing infection thread (Dart 1974). Brewin (1991) pointed out that this structure closely resembles a phragmosome, the trans-cellular structure which is visible in highly vacuolated cells prior to their division which predicts the site of cell plate formation (Lloyd 1991, Venverloo and Libbenga 1987). Later studies using cell cycle markers established cell cycle involvement at the molecular level (Yang et al. 1994). Recently a group of cell cycle regulators were found to have altered expression during root hair infection and/or after treatment by Nod factors (Table 1; Breakspear et al. 2014). This included D- and A-type cyclins, the latter having been linked to the S phase and the G2/M transition (Roudier et al. 2000), suggesting mitotic-associated events might be involved. One of the induced cyclins clusters separately from the A, B and D type cyclins and has homology to C-type cyclins in animals. The C-type cyclins are not well studied, but existing evidence suggests that they are involved in the G0-G1 transition (Ren and Rollins 2004), with specialized roles in DNA repair in cells such as postmitotic neurons which are otherwise fixed in G0 (Tomashevski et al. 2010). Such a cyclin, if it is involved, could play a central role in initiating the cell cycle-related events that occur during infection. Also noted were changes in the expression of regulators of the endocycle, *ELC*, *SMT2* and *OSD1* (Table 1), and many genes involved in DNA replication, and an increase in the levels of *APC6*, which together indicate that the endocycle is being supressed, possibly following a round of DNA synthesis (Breakspear et al. 2014). Also induced was *ARMADILLO BTB PROTEIN*

1 (ABAP1) whose expression is negatively associated with cell proliferation (Table 1; Masuda et al. 2008).

Although the purpose of cell cycle induction during infection is far from clear, one possible explanation is that the cell division programme provided an established strategy for trans-cellular deposition of cell wall material that could be co-opted for infection thread formation. Prior to cytokinesis, phragmosome formation is associated with the appearance of markers that position the site of subsequent cell wall growth (Lloyd and Buschmann 2007). In the case of infection thread formation in root hairs and cortical cells, cell wall deposition is restricted to a tube that initiates at the distal tip of root hair cell or distal wall of the cortical cell, and in the root hair

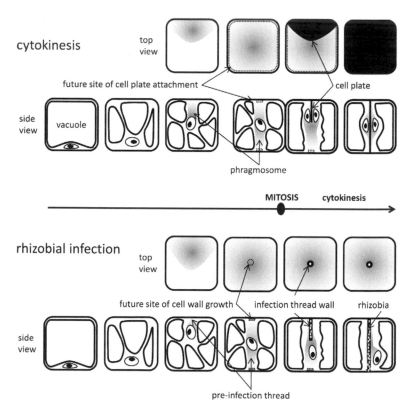

FIGURE 6 Events occurring during rhizobial infection and cytokinesis. The model for cytokinesis was adapted from Venverloo and Libbenga (1987), updated to feature polar formation of the cell plate. In each case the nucleus which is initially appressed to the cell wall in vacuolated cells becomes enlarged and moves to a central position which is accompanied by vacuolar fragmentation. The phragmosome, a microtubule-rich cytoplasmic bridge, forms, dividing the cell in two halves. In cytokinesis, specific proteins mark the future site where the nascent cell wall will form; similar markers may also designate sites for infection thread initiation. Following mitosis the phragmoplast initiates cell plate formation from one side of the cell. During rhizobial infection, the tubular ingrowth of the infection thread initiates from the outer cell wall, extending through the cytoplasmic bridge to eventually merge with the inner cell wall. As it grows the infection thread becomes colonized by rhizobia.

its growth through the cell is accompanied by the movement of the nucleus (Fournier 2008). Similarly, during cytokinesis the formation of cell wall is typically polarized, always forming from one side of the cell (Cutler and Ehrhardt 2002). In both cases the new cell wall spans the cell connecting opposing cell walls (Fig. 6).

One clue to the mechanisms underlying these events is the induction of *Aurora Kinase 1* during infection (Table 1), which suggests that some mitotic machinery may be recruited during the formation of pre-infection structures. Interestingly, Arabidopsis mutants in AUR1/AUR2 kinases have randomly oriented cell plates (Van Damme et al. 2011), hinting at a potential role for AURORA kinases in orientation of the pre-infection thread, which forms anticlinally. Whether and which components of the cell cycle are required for infection awaits future studies, but it is notable that cells in the infection zone can become infected while actively dividing, demonstrating that the processes are not exclusive (Voroshilova et al. 2009). Regardless of the details, the phenomenon appears to be of widespread importance in plants. It has been noted that pre-infection threads resemble the analogous pre-penetration structures that form in advance of arbuscular mycorrhizal infection, a symbiosis that occurs in over 80% of land plants (Parniske 2008), and so a better understanding of these structures promises insights into the fundamentals of endosymbiont accommodation in general. Perhaps importantly, cytokininesis can be induced by wounding/contact (Sinnott and Bloch 1940). It is possible that rhizobial attachment or formation of the mycorrhizal hyphopodium might provide a similar stimulus perhaps providing an evolutionary link between the cell cycle and endosymbiotic events.

Acknowledgements

The author thanks Silvia Costa for proofreading and comments on the manuscript and Mathias Brault and Florian Frugier (IPS2/Gif sur Yvette, France) and Frédéric Debellé (LIPM/Castanet-Tolosan, France) for the annotated list of cytokinin response regulators and Sonali Roy for the nodule artwork and Joëlle Fournier for the image of the infection thread.

References

Ariel, F., M. Brault-Hernandez, C. Laffont, E. Huault, M. Brault, J. Plet, M. Moison, S. Blanchet, J.L.Ichanté, M. Chabaud, S. Carrere, M.Crespi, R.L.Chan and F.Frugier. 2012. Two direct targets of cytokinin signaling regulate symbiotic nodulation in *Medicago truncatula*. Plant Cell 24: 3838-3852.

Boniotti, M.B. and C. Gutierrez. 2001. A cell-cycle-regulated kinase activity phosphorylates plant retinoblastoma protein and contains, in Arabidopsis, a CDKA/cyclin D complex. Plant J. 28: 341-350.

Breakspear, A., C. Liu, S. Roy, N. Stacey, C. Rogers, M. Trick, G. Morieri, K.S. Mysore, J. Wen, G.E.D. Oldroyd, J.A. Downie and J.D. Murray. 2014. The root hair 'infectome' of *Medicago truncatula* uncovers changes in cell cycle genes and reveals a requirement for auxin signalling in rhizobial infection. Plant Cell 26: 4680-4701.

Breuer, C., K. Morohashi, A. Kawamura, N. Takahashi, T. Ishida, U. Umeda, E. Grotewold and K. Sugimoto. 2012. Transcriptional repression of the APC/C activator CCS52A1 promotes active termination of cell growth. EMBO J. 31: 4488-4501.

Brewin, N. 1991. Development of the legume root nodule. Annual Review of Cell Biology 7: 191-226.

Cebolla, A., J.M. Vinardell, E. Kiss, B. Oláh, F. Roudier, A. Kondorosi and E. Kondorosi. 1999. The mitotic inhibitor ccs52 is required for endoreduplication and ploidy-dependent cell enlargement in plants. EMBO J. 18: 4476-4484.

Cromer, L., J. Heyman, S. Touati, H. Harashima, E. Araou, C. Girard, C. Horlow, K. Wassmann, A. Schnittger, L. De Veylder and R. Mercier. 2012. *OSD1* promotes meiotic progression via APC/C inhibition and forms a regulatory network with TDM and CYCA1;2/TAM. PLoS Genet.8:e1002865. doi: 10.1371/journal.pgen.1002865.

Cutler, S.R. and D.W. Ehrhardt. 2002. Polarized cytokinesis in vacuolate cells of Arabidopsis. Proc Natl Acad Sci USA. 99: 2812-2817.

Dart, P.J. 1974. The infection process. pp. 381-429. *In*: A. Quispel (ed.). The Biology of Nitrogen Fixation. North-Holland Publishing Co, Amsterdam.

Fournier, J., A.C. Timmers, B.J. Sieberer, A. Jauneau, M. Chabaud and D.G. Barker. 2008. Mechanism of infection thread elongation in root hairs of *Medicago truncatula* and dynamic interplay with associated rhizobial colonization. Plant Physiol. 148: 1985-1995.

Gonzalez-Rizzo, S., M. Crespi and F. Frugier. 2006 The *Medicago truncatula* CRE1 cytokinin receptor regulates lateral root development and early symbiotic interaction with *Sinorhizobium meliloti*. Plant Cell 18: 2680-2693.

Heckmann, A.B., N. Sandal, A. Bek, L.H. Madsen, A. Jurkiewicz, M.W. Nielsen, L.Tirichine and J. Stougaard. 2011. Cytokinin induction of root nodule primordia in *Lotus japonicus* is regulated by a mechanism operating in the root cortex. Mol. Plant Microbe Interact. 24: 1385-1395.

Heidstra, R., W.C. Yang, Y. Yalcin, S. Peck, A.M. Emons, A. van Kammen and T. Bisseling. 1997. Ethylene provides positional information on cortical cell division but is not involved in Nod factor-induced root hair tip growth in Rhizobium-legume interaction. Development 124:1781-1787.

Held, M., H. Hou, M. Miri, C. Huynh, L. Ross, M.S. Hossain, S. Sato, S. Tabata, J. Perry, T.L. Wang and K. Szczyglowski. 2014. *Lotus japonicus* cytokinin receptors work partially redundantly to mediate nodule formation. Plant Cell 26: 678-694.

Heyl A, M. Brault, F. Frugier, A. Kuderova, A.C. Lindner, V. Motyka, A.M. Rashotte, K.V. Schwartzenberg, R. Vankova and G.E. Schaller. 2013. Nomenclature for members of the two-component signaling pathway of plants. Plant Physiol. 161: 1063-1065.

Hill, K., D.E. Mathews, H.J. Kim, I.H. Street, S.L. Wildes, Y.H. Chiang, M.G. Mason, J.M. Alonso, J.R. Ecker, J.J. Kieber and G.E. Schaller. 2013. Functional characterization of type-B response regulators in the Arabidopsis cytokinin response. Plant Physiol. 162: 212-224.

Himanen, K., E. Boucheron, S. Vanneste, J. de Almeida Engler, D. Inzé and T. Beeckman. 2002. Auxin-mediated cell cycle activation during early lateral root initiation. Plant Cell 14: 2339-2351.

Imai, K.K., Y. Ohashi, T. Tsuge, Y. Yoshizumi, M. Matsui, A. Oka and T. Aoyama. 2006. The A-type cyclin CYCA2;3 is a key regulator of ploidy levels in Arabidopsis endoreduplication. Plant Cell 18: 382-396.

Ishida, T., S. Adachi, M. Yoshimura, K. Shimizu, M. Umeda and K. Sugimoto. 2010. Auxin modulates the transition from the mitotic cycle to the endocycle in Arabidopsis. Development 137: 63-71.

Ishida, T., S. Fujiwara, K. Miura, N. Stacey, M. Yoshimura, K. Schneider, S. Adachi, K. Minamisawa, M. Umeda and K. Sugimoto. 2009. SUMO E3 ligase HIGH PLOIDY2 regulates endocycle onset and meristem maintenance in Arabidopsis. Plant Cell 21: 2284-2297.

Kevei, Z., M. Baloban, O. Da Ines, H. Tiricz, A. Kroll, K. Regulski, P. Mergaert and E. Kondorosi. 2011. Conserved CDC20 cell cycle functions are carried out by two of the five isoforms in *Arabidopsis thaliana*. PLoS One.6 :e20618. doi: 10.1371/journal.pone.0020618.

Kuppusamy, K.T., S. Ivashuta, B. Bucciarelli, C.P. Vance, J.S. Gantt and K.A. Vandenbosch. 2009. Knockdown of CELL DIVISION CYCLE16 reveals an inverse relationship between lateral root and nodule numbers and a link to auxin in *Medicago truncatula*. Plant Physiol. 151: 1155-1166.

Kuwabara, A. and W. Gruissem. 2014. Arabidopsis RETINOBLASTOMA-RELATED and Polycomb group proteins: cooperation during plant cell differentiation and development. J Exp Bot. 65: 2667-2676.

Lammens, T., V. Boudolf, L. Kheibarshekan, L.P. Zalmas, T. Gaamouche, S.Maes, M. Vanstraelen, E. Kondorosi, N.B. La Thangue, W. Govaerts, D. Inzé and L. De Veylder. 2008. A typical E2F activity restrains APC/CCCS52A2 function obligatory for endocycle onset. Proc. Natl. Acad. Sci. USA. 105: 14721-14726.

Landrieu, I., S. Hassan, M. Sauty, F. Dewitte, J.M. Wieruszeski, D. Inzé, D.L. De Veylder and G. Lippens. 2004. Characterization of the *Arabidopsis thaliana* Arath;CDC25 dual-specificity tyrosine phosphatase. Biochem. Biophys. Res. Commun. 322: 734-749.

Lindsay, D.L., P.C. Bonham-Smith, S. Postnikoff, G.R. Gray and T.A. Harkness. 2011. A role for the anaphase promoting complex in hormone regulation. Planta 233: 1223-1235.

Lloyd, C. and H. Buschmann. 2007. Plant division: remembering where to build the wall. Curr Biol.17: R1053-1055.

Lloyd, C.W. 1991. Cytoskeletal elements of the phragmosome establish the division plane in vacuolated higher plant cells. pp. 245-257. *In*: C.W. Lloyd (ed.). The Cytoskeletal Basis of Plant Growth and Form. Academic Press, San Diego.

Mason, M.G., D.E. Mathews, D.A. Argyros, B.B. Maxwell, J.J. Kieber, J.M. Alonso, J.R. Ecker and G.E. Schaller. 2005. Multiple type-B response regulators mediate cytokinin signal transduction in Arabidopsis. Plant Cell 17: 3007-3018.

Masuda, H.P., L.M. Cabral, L. De Veylder, M. Tanurdzic, J. de Almeida Engler, G. Geelen, D. Inzé, D, R.A. Martienssen, P.C. Ferreira and A.S. Hemerly. 2008. ABAP1 is a novel plant Armadillo BTB protein involved in DNA replication and transcription. EMBO J. 27: 2746-2756.

Mathesius, U., H.R. Schlaman, H.P. Spaink, C. Of Sautter, B.G. Rolfe and M.A. Djordjevic. 1998. Auxin transport inhibition precedes root nodule formation in white clover roots and is regulated by flavonoids and derivatives of chitin oligosaccharides. Plant J. 14: 23-34.

Menges, M., A.K. Samland, S. Planchais and J.A.H. Murray. 2006. The D-Type Cyclin CYCD3;1 is limiting for the G1-to-S-Phase transition in Arabidopsis. Plant Cell 18: 893-906.

Murray, J.D., B.J. Karas, S. Sato, S. Tabata, L. Amyot, L. and K. Szczyglowski. 2007. A cytokinin perception mutant colonized by Rhizobium in the absence of nodule organogenesis. Science 315: 101-104.

Nakai, T., K. Kato, A. Shinmyo and M. Sekine. 2006. Arabidopsis KRPs have distinct inhibitory activity toward cyclin D2-associated kinases, including plant-specific B-type cyclin-dependent kinase. FEBS Lett. 580: 336-340.

Ng, J.L., S. Hassan, T.T. Truong, C.H. Hocart, C. Laffont, F. Frugier and U. Mathesius. 2015. Flavonoids and auxin transport inhibitors rescue symbiotic nodulation in the *Medicago truncatula* cytokinin perception mutant *cre1*. Plant Cell. Aug 7. pii: tpc.15.00231.

Oldroyd, G.E., E.M. Engstrom and S.R. Long. 2001. Ethylene inhibits the Nod factor signal transduction pathway of *Medicago truncatula*. Plant Cell 13: 1835-1849.

Oldroyd, G.E., J.D. Murray, P. S. Poole and J.A. Downie. 2010. The rules of engagement in the legume-rhizobial symbiosis. *Annual Review of Genetics*. 45: 119-144.

Pacios-Bras, C., H.R. Schlaman, K. Boot, P. Admiraal, J.M. Langerak, J. Stougaard and H.P. Spaink. 2003. Auxin distribution in *Lotus japonicus* during root nodule development. Plant Mol Biol.52: 1169-1180.

Parniske, M. 2008. Arbuscular mycorrhiza: the mother of plant root endosymbiosis. Nat. Rev. Microbiol. 6: 763-775.

Penmetsa, R.V., J.A. Frugoli, L.S. Smith, S.R. Long and D.R. Cook. 2003. Dual genetic pathways controlling nodule number in *Medicago truncatula*. Plant Physiol. 131: 998-1008.

Plet, J., A. Wasson, F. Ariel, C. Le Signor, D. Baker, U. Mathesius, M. Crespi, and F. Frugier. 2011. MtCRE1-dependent cytokinin signaling integrates bacterial and plant cues to coordinate symbiotic nodule organogenesis in *Medicago truncatula*. Plant J. 65: 622-633.

Prayitno, J., B.G. Rolfe and U. Mathesius. 2006. The ethylene-insensitive *sickle* mutant of *Medicago truncatula* shows altered auxin transport regulation during nodulation. Plant Physiol. 142: 168-180.

Ren, S. and B.J. Rollins. 2004. Cyclin C/cdk3 promotes Rb-dependent G0 exit. Cell 117: 239-251.

Riou-Khamlichi, C., R. Huntley, A. Jacqmard and J.A. Murray. 1999. Cytokinin activation of Arabidopsis cell division through a D-type cyclin. Science 283: 1541-1544.

Roudier, F., E. Fedorova, J. Györgyey, A. Feher, S.Brown, A. Kondorosi and E. Kondorosi. 2000. Cell cycle function of a *Medicago sativa* A2-type cyclin interacting with a PSTAIRE-type cyclin-dependent kinase and a retinoblastoma protein. Plant J. 23: 73-83.

Roudier, F., E. Fedorova, M. Lebris, P. Lecomte, J. Györgyey, D. Vaubert, G. Horvath, P. Abad, A. Kondorosi and E. Kondorosi. 2003. The *Medicago* species A2-type cyclin is auxin regulated and involved in meristem formation but dispensable for endoreduplication-associated developmental programs. Plant Physiol. 131: 1091-1103.

Roux, B., N. Rodde, M.F. Jardinaud, T. Timmers, L. Sauviac, I. Cottret, S. Carrère, E. Sallet, E. Courcelle, S. Moreau, F. Debellé, D. Capela, F. de Carvalho-Niebel, J. Gouzy, C. Bruand and P. Gamas. 2014. An integrated analysis of plant and bacterial gene expression in symbiotic root nodules using laser-capture microdissection coupled to RNA sequencing. Plant J. 77: 817-837.

Scofield, S., A. Jones and J.A. Murray. 2014. The plant cell cycle in context. J. Exp. Bot. 65: 2557-2562.

Sinnott, E.W. and R. Bloch. 1940. Cytoplasmic behavior during division of vacuolate plant cells. Proc. Natl. Acad. Sci. USA. 26: 223-227.

Suzaki, T., M. Ito and M. Kawaguchi. 2013. Induction of localized auxin response during spontaneous nodule development in *Lotus japonicus*.Plant Signal. Behav. 8: e23359. doi: 10.4161/psb.23359.

Suzaki, T., K. Yano, M. Ito, Y. Umehara, N. Suganuma and M. Kawaguchi. 2012. Positive and negative regulation of cortical cell division during root nodule development in *Lotus japonicus* is accompanied by auxin response. Development 139: 3997-4006. doi: 10.1242/dev.084079.

Tarayre, S., J.M. Vinardell, A. Cebolla, A. Kondorosi and E. Kondorosi. 2004. Two classes of the CDh1-type activators of the anaphase-promoting complex in plants: novel functional domains and distinct regulation. Plant Cell 16: 422-434.

Timmers, A.C, M.C. Auriac and G. Truchet. 1999. Refined analysis of early symbiotic steps of the Rhizobium-*Medicago* interaction in relationship with microtubular cytoskeleton rearrangements. Development. 126: 3617-3628.

Tirichine, L., N. Sandal, L.H. Madsen, S. Radutoiu, A.S. Albrektsen, S. Sato, E. Asamizu, S. Tabata and J. Stougaard. 2007. A gain-of-function mutation in a cytokinin receptor triggers spontaneous root nodule organogenesis. Science 315: 104-107.

Tomashevski, A., D.R. Webster, P. Grammas, M. Gorospe and I.I. Kruman. 2010. Cyclin-C-dependent cell-cycle entry is required for activation of non-homologous end joining DNA repair in postmitotic neurons. Cell Death Differ. 17: 1189-1198.

Van Damme, D., B. De Rybel, G. Gudesblat, D. Demidov, W. Grunewald, I. De Smet, A. Houben, T. Beeckman and E. Russinova. 2011. *Arabidopsis* α Aurora kinases function in formative cell division plane orientation. Plant Cell 23: 4013-4024.

Venverloo, C. and K. Libbenga. 1987. Regulation of the plane of cell-division in vacuolated cells. 1. The function of nuclear positioning and phragmosome formation. J. Plant Phys. 131: 267-284.

Vinardell, J.M., E. Fedorova, A. Cebolla, Z. Kevei, G. Horvath, Z. Kelemen, S. Tarayre, F. Roudier, P, Mergaer, A. Kondorosi and E. Kondorosi. 2003. Endoreduplication mediated by the anaphase-promoting complex activator CCS52A is required for symbiotic cell differentiation in *Medicago truncatula* nodules. Plant Cell 15: 2093-2105.

Vlieghe, K., V. Boudolf, G.T. Beemster, S. Maes, Z. Magyar, A. Atanassova, J. de Almeida Engler, R. De Groodt, D. Inzé and L. De Veylder. 2005. The DP-E2F-like gene DEL1 controls the endocycle in *Arabidopsis thaliana*. Curr Biol. 15: 59-63.

Voroshilova, V.A., K.N. Demchenko, N.J. Brewin, A.Y. Borisov and I.A. Tikhonovich. 2009. Initiation of a legume nodule with an indeterminate meristem involves proliferating host cells that harbour infection threads. New Phytol. 181: 913-923.

Wasson, A.P., F.I. Pellerone and U. Mathesius. 2006. Silencing the flavonoid pathway in *Medicago truncatula* inhibits root nodule formation and prevents auxin transport regulation by rhizobia. Plant Cell 18: 1617-1629.

Wuarin, J., V. Buck, P. Nurse and J.B.A. Millar. 2002. Stable association of mitotic Cyclin B/Cdc2 to replication origins prevents endoreduplication. Cell 111 : 419-431

Xiao, T.T., S. Schilderink, S. Moling, E.E. Deinum, E. Kondorosi, H. Franssen, O. Kulikova, A. Niebel and T. Bisseling. 2014. Fate map of *Medicago truncatula* root nodules. Development 141: 3517-3528.

Yang, W.C, C. de Blank, I. Meskiene, H. Hirt, J. Bakker, A. van Kammen, H. Franssen and T. Bisseling. 1994. Rhizobium nod factors reactivate the cell cycle during infection and nodule primordium formation, but the cycle is only completed in primordium formation. Plant Cell 6: 1415-1426.

Yokoyama, A., T. Yamashino, Y. Amano, Y. Tajima, A. Imamura, H. Sakakibara and T. Mizuno 2007. Type-B ARR transcription factors, ARR10 and ARR12, are implicated in cytokinin-mediated regulation of protoxylem differentiation in roots of *Arabidopsis thaliana*. Plant Cell Physiol. 48: 84-96.

Zhang, K., D.S. Letham and P.C. John. 1996. Cytokinin controls the cell cycle at mitosis by stimulating the tyrosine dephosphorylation and activation of p34cdc2-like H1 histone kinase. Planta 200: 2-12.

14

Cellular and Molecular Features of the Procambium and Cambium in Plant Vascular Tissue Development

Xin-Qiang He and Li-Jia Qu*

Introduction

The procambium and cambium are plant vascular meristems at two meristematic stages. These tissues produce the vascular tissues and form the vascular system in the plant body. Vascular tissues, which are composed of several types of xylem and phloem cells, connect the shoot organs with the roots, provide physical strength to plant bodies, and enable efficient long-distance transport between organs. The xylem is the main tissue for transporting water and solute minerals, whereas the phloem is the route for transporting photosynthates and various signaling molecules (Lucas et al. 2013).

The procambium appears during embryogenesis. Procambium cells, which are also known as provascular tissue cells, are embryonic cells located at the site of the future vascular cylinder during the globular stage (Miyashima et al. 2013, De Rybel et al. 2014). During postembryonic development, the procambium appears in the shoot and root apexes as one of the apical meristems. It produces the primary phloem and primary xylem during primary growth, the type of growth that originates from the shoot apical meristem (SAM) and root apical meristem (RAM). The cambium is a lateral meristem that produces the secondary phloem and secondary xylem in the secondary growth of plants, especially in woody plants. Secondary growth is defined as the type of growth originating from two lateral meristems; the vascular cambium and the phellogen (cork cambium) (Aloni 2013). Thus, the procambium and cambium represent sequential developmental stages of the same vascular meristem during plant growth (Larson 1976).

During the last decade, there has been significant progress in understanding vascular development due to the advances in genomics and other molecular tools. Most studies have focused on the model plants *Arabidopsis* and *Populus*. Several excellent reviews have summarized recent research progress on aspects of vascular

State Key Laboratory of Protein and Plant Gene Research, Peking-Tsinghua Center for Life Sciences, School of Life Sciences, Peking University, Beijing 100871, P. R. China.
* Corresponding author : qulj@pku.edu.cn

development, such as cell-to-cell communication in vascular pattern formation (Ye 2002, Lehesranta et al. 2010, Jouannet et al. 2015), the roles of phytohormones during vascular development (Aloni 2013, Ursache et al. 2013), secondary growth (Du and Groover 2010, Agusti and Greb 2013, Ragni and Hardtke 2014), regulation and maintenance of vascular stem cells, and xylem cell differentiation (Cano-Delgado et al. 2010, Risopatron et al. 2010, Miyashima et al. 2013, Furuta et al. 2014a). However, compared with apical meristems, much less is known about vascular meristems. In this chapter, we present updated information and concepts for vascular meristems, both the procambium and cambium, and their roles in plant vascular development. In particular, we focus on the cellular and molecular regulation of the procambium and cambium in the development of vascular tissues during postembryonic growth.

Cellular Aspects of the Procambium and Cambium

During postembryonic development, the procambium appears in shoot and root apexes and gives rise to all types of vascular cells (Fig. 1A, B). Similar to other meristematic cells, procambial cells have dense peripheral cytoplasm with numerous

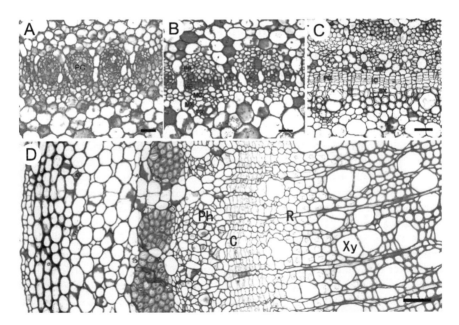

FIGURE 1 The procambium, cambium, primary and secondary vascular tissues in poplar stem. A: The procambial cells (PC) in poplar shoot at about 1 mm below apex, note the irregular cell divisions and the differentiating protophloem cells on the cortical side of the strand. B: The primary tissues in poplar shoot at about 4 mm below apex, note the protophloem (PP), metaphloem (MP), metacambium (MC), metaxylem (MX) and protoxylem (PX). C: The fascicular cambium and interfascicular cambium inpoplar shoot at about 60mm below apex, note the fascicular cambium (FC), interfascicular cambium (IC), secondary phloem (SP) and secondary xylem (SX). D: Secondary vascular tissues in poplar stem at about 1000 mm below apex, note the cambium (C), secondary phloem (Ph), secondary xylem (Xy) and ray parenchyma cells (R). Bars in A and B: 20 μm, bars in C and D: 50 μm.

ribosomes, rough endoplasmic reticulum (RER), and active dictyosomes. These narrow cells form continuous strands and traces. The strands develop acropetally in undifferentiated tissue of the residual meristem and connect with the existing procambium below (Larson 1975, 1976).

The cambium has two kinds of highly vacuolated cells; the rather isodiametric ray initials, and the elongated fusiform initials (Fig. 1D). Fusiform initials produce tracheary elements, sieve elements, and fibers, and ray initials produce the radial ray system. Active cambial cells have a dense peripheral cytoplasm with numerous ribosomes, mitochondria, RER, and active dictyosomes. The nuclei in fusiform and ray cells differ markedly in their size, shape, and possibly even in number (Iqbal and Ghouse 1990, Iqbal 1995a). In ray cells, the nucleus is more or less globular with one or two nucleoli, while in fusiform cells, the nucleus is ovoid or elongated with several aligned nucleoli and it increases in length and volume with age (Iqbal 1995a). The presence of a large vacuole is one of the important characteristics of active cambial cells. Many plasmodesmata are present in the tangential walls of ray cells, and play roles in the transmission of messenger molecules and the translocation of nutrients between the phloem and xylem (Barnett 1995)

Development of the Procambium and Primary Vascular Tissues

In the poplar shoot, procambial strands initiate acropetally through the residual meristem beneath the apex. During this process, the potentially meristematic residual tissue is stimulated to undergo cell division (Fig. 1A) (Larson 1975). The procambial strands develop in turn into mediolaterally organized vascular bundles with centripetal xylem and centrifugal phloem cells. The cells differentiating centrifugally from the originating center of the protoxylem pole are considered to be the protoxylem. These cells undergo periclinal divisions to produce the first recognizable radially aligned cells within a procambial strand, and these cells are considered to be precursors of the metacambium that follows. The metacambium is a more advanced meristematic stage of the continuum in which additional periclinal divisions are interposed within the initiating layer to unite it into a tangentially continuous series of radially aligned cells that extend across a bundle. Consequently, the metacambium shows a more definitive radial and tangential alignment and narrower radial cell diameters, compared with the metaxylem cells that differentiate centrifugally from derivatives of the metacambium (Fig. 1B) (Larson 1994). During the differentiation of xylem and phloem cells, some of the cells between the xylem and phloem retain their meristematic state and form fascicular cambial cells. However, in maize, rice, and other grasses, all procambial cells differentiate into xylem and phloem cells, and no meristematic cells remain in primary vascular tissues.

Regulation of Xylem Cell Differentiation from Procambial Cells

Xylem cells include tracheary elements (vessels and tracheids), xylem fiber cells, and xylem parenchyma cells. These cells differentiate from procambial or cambial cells. Recent comprehensive analyses of gene expression and function have revealed the key genes involved in the differentiation process into xylem cells from procambial cells (Kubo et al. 2005, Zhong et al. 2006, Ohashi-Ito et al. 2010, Yamaguchi et al. 2011, Kondo et al. 2014).

Cellular events of xylem cell differentiation

Xylem cell differentiation involves xylem initial cell division, cell expansion, secondary cell wall formation, and programmed cell death (PCD) (Han et al. 2012). Secondary cell wall deposition is regulated temporally and spatially in the various types of xylem cells. The key features of secondary cell well deposition are the annular and spiral patterns in protoxylem vessels, the reticulate and pitted patterns in metaxylem vessels, and the smeared pattern in xylem fibers. Cortical microtubule regulate the spatial pattern of the secondary cell wall by orientating cellulose deposition (Oda et al. 2010). Some microtubule associated proteins (MAPs) have been identified as proteins that regulate secondary wall pattern formation. AtMAP70 family proteins were shown to be involved in the formation of the secondary wall boundary (Pesquet et al. 2010). The plant-specific microtubule-binding protein MIDD1/RIP3 was shown to promote microtubule de-polymerization at the site where the secondary wall pit area would later form, resulting in a secondary wall depletion domain (Oda et al. 2010). The proteins ROPGEF4 and ROPGAP3 were shown to mediate local activation of the plant Rho GTPase ROP11. The activated ROP11 recruited MIDD1 to induce local disassembly of cortical microtubule (Oda and Fukuda 2012b). Cortical microtubule conversely eliminated active ROP11 from the plasma membrane via MIDD1. The mutual inhibitory interaction between active ROP domains and cortical microtubule was shown to establish the distinct pattern of secondary cell walls (Oda and Fukuda 2012a).

During xylem development, PCD occurs in tracheary elements and xylem fiber cells. This results in the removal of cell contents for the formation of dead cells with a thick secondary wall (Bollhoner et al. 2012). The PCD during tracheary element differentiation has long been recognized as an example of developmental PCD in plants (Fukuda 2004, Turner et al. 2007). The PCD process includes cell death signal induction, accumulation of autolytic enzymes in the vacuole, vacuole swelling and collapse, and degradation of cell contents before the formation of mature tracheary elements (Fukuda 2000). It was suggested that the signals for xylem cell death are produced early during xylem differentiation, and that cell death is prevented by inhibitors and by the storage of hydrolytic enzymes in the vacuole (Bollhoner et al. 2012). The death of xylem tracheary elements is defined as a vacuolar type of cell death (Kuriyama and Fukuda 2002, Van Doorn et al. 2011). Vacuolar membrane breakdown is the crucial event in tracheary element PCD. Bursting of the central vacuole triggers autolytic hydrolysis of the cell contents, leading to cell death (Bollhoner et al. 2012).

Microarray analyses of gene expression have revealed the simultaneous expression of many genes involved in both secondary wall formation and PCD (Demura et al. 2002, Milioni et al. 2002, Pesquet et al. 2005, Kubo et al. 2005, Ohashi-Ito et al. 2010). A transcriptional regulation system composed of transcription factors such as VND6 and a TERE-cis sequence regulates the simultaneous expression of genes related to both secondary wall formation and PCD in tracheary elements (Ohashi-Ito et al. 2010). These findings indicate that the master genes inducing tracheary element differentiation initiate PCD in some cells by activating PCD-related genes through binding to the TERE sequence in their promoters (Lucas et al. 2013).

Execution of PCD in developing tracheary elements involves the expression and vacuolar accumulation of several hydrolytic enzymes, such as the cysteine proteases XCP1 and XCP2 (Avci et al. 2008, Funk et al. 2002, Zhao et al. 2000), Zn^{2+}-dependent nuclease ZEN1 (Ito and Fukuda 2002), and RNases (Lehmann et al. 2001). Ca^{2+}-dependent DNases were also detected in secondary xylem cells and their activities were closely correlated with secondary xylem development (Chen et al. 2012). ATMC9 encodes a caspase-like protein, which does not function as a caspase, but as an arginine/lysine-specific cysteine protease (Vercammen et al. 2004). ATMC9 was shown to be specifically expressed in differentiating vessels but not in fully differentiated vessels (Ohashi-Ito et al. 2010), suggesting that it played a role in PCD. The application of caspase inhibitors significantly delayed the time of tracheary element formation and inhibited DNA breakdown and the appearance of TUNEL-positive nuclei in a *Zinnia* xylogenic cell culture (Twumasi et al. 2010). The protease responsible for the developing xylem-related caspase-3-like activity was purified and identified as a 20S proteasome (Han et al. 2012). In another study, treatment with a caspase-3 inhibitor Ac-DEVD-CHO caused a vein defect in *Arabidopsis* cotyledons and the proteasome inhibitor clasto-lactacystin β-lactone delayed tracheary element PCD in VND6-induced *Arabidopsis* xylogenic culture. These findings strongly suggested that the proteasome is involved in PCD during tracheary element differentiation (Han et al. 2012). Similarly, the 26S proteasome inhibitors lactacystin and MG132 delayed or blocked the differentiation of suspension-cultured tracheary elements (Endo et al. 2001, Woffenden et al. 1998). Autophagy has been implicated to function in tracheary element PCD (Weir et al. 2005). A small GTP-binding protein RabG3b was found to play a positive role in PCD during tracheary element differentiation by activating autophagy (Kwon et al. 2010).

The PCD of xylem fibers is less characterized than that of xylem tracheary elements, because it progresses very slowly. Microarray analyses revealed that many genes encoding previously uncharacterized transcription factors, as well as genes associated with ethylene, sphingolipids, light signaling, and autophagy-related factors are expressed preferentially during xylem fiber development (Courtois-Moreau et al. 2009). Further comparisons of genes related to PCD between xylem fibers and tracheary elements may shed light on the intricate process of plant PCD (Lucas et al. 2013).

Intercellular signaling pathways regulating xylem differentiation

A tracheary element differentiation inhibitory factor (TDIF), identified as the CLE-family peptide CLE41/44, was found to function as a signaling molecule that both inhibited xylem cell differentiation from procambial cells and promoted procambial cell proliferation (Ito et al. 2006, Hirakawa et al. 2008, 2010). Its receptor TDR/PXY (TDIF RECEPTOR/PHLOEM INTERCALATED WITH XYLEM) belongs to the class XI LRR-RLK (LEUCINE-RICH REPEAT RECEPTOR-LIKE KINASE) family (Hirakawa et al. 2008). CLE41/44 peptides are expressed specifically in the phloem, while TDR is expressed preferentially in the procambium (Hirakawa et al. 2008, Fisher and Turner 2007). Defects in TDR or CLE41 led to the exhaustion of procambial cells between the phloem and xylem, resulting in the formation of xylem vessels adjacent to phloem cells in hypocotyls (Hirakawa et al. 2008, 2010, Fisher and Turner 2007). The TDIF signal activates the expression of *WOX4*, a member of the WUSCHEL-related HOMEOBOX (WOX) gene family, in procambial and cambial cells (Ji et al. 2010, Hirakawa et al. 2010, Suer et al. 2011). WOX4 is required for the TDIF-dependent enhancement of procambial cell proliferation but not for the TDIF-dependent suppression of xylem differentiation (Hirakawa et al. 2010). WOX4 and WOX14 were shown to act downstream of the PXY receptor kinase to regulate plant vascular proliferation (Etchells et al. 2013). Ethylene/ERF signaling was reported as another pathway that regulates procambial/cambial cell division. It was suggested to be parallel to the CLE41-TDR/PXY pathway. TDR/PXY signaling represses the ethylene/ERF pathway under normal circumstances (Etchells et al. 2012). Recently, plant glycogen synthase kinase 3 proteins (GSK3s) were identified as crucial downstream components of the TDIF signaling pathway that suppresses xylem differentiation from procambial cells (Kondo et al. 2014). TDR interacts with GSK3s at the plasma membrane and activates GSK3s, promoting xylem differentiation. Thus, there are two pathways downstream of TDIF-TDR signaling in procambial cells. In one pathway, TDIF–TDR increases the transcription of *WOX4*, leading to the promotion of procambial cell proliferation. In the other pathway, TDIF–TDR activates GSK3s, leading to the promotion of xylem differentiation by inactivating the transcription factor BES1, a direct substrate of GSK3s in brassinosteroid signaling (Kondo et al. 2014) (Fig. 2).

Cytokinins are key regulators of xylem development (Mähönen et al. 2000, 2006, Matsumoto-Kitano et al. 2008, Bishopp et al. 2011). The crosstalk between CLE peptides and cytokinin signaling was shown to regulate xylem differentiation (Kondo et al. 2011). CLE9 and CLE10, which encode the same CLE peptide, are preferentially expressed in vascular cells of roots. Treatment with some CLE peptides including CLE9/CLE10 inhibited the formation of protoxylem vessels but not metaxylem

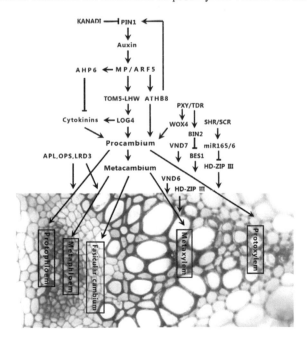

FIGURE 2 Regulation of the procambium and primary vascular development in the shoot, and a cross section of Arabidopsis stem showing the structure of the primary vascular tissues.

vessels in *Arabidopsis* roots (Kondo et al. 2011). Microarray analysis revealed that the CLE9/CLE10 peptide specifically reduces the expression of genes encoding type-A ARRs (ARABIDOPSIS RESPONSE REGULATORs, which are negative regulators of cytokinin signaling), especially ARR5 and ARR6 (Kondo et al. 2011). The CLE9/CLE10 peptide was shown to activate cytokinin signaling by repressing ARR5 and ARR6, resulting in inhibition of protoxylem vessel formation (Kondo et al. 2011).

For cell-to-cell communication, plant cells send signaling molecules via symplastic and/or apoplastic pathways. The GRAS-family transcription factor SHORT ROOT (SHR) is a signal that selectively moves through plasmodesmata. SHR proteins move from the stele to the endodermis to induce another GRAS-family transcription factor SCARECROW (SCR). Then, SHR together with SCR up-regulate the expression of target genes including miR165/6 genes (Levesque et al. 2006, Cui et al. 2007, Gallagher and Benfey 2009). The mature miR165/6 moves back from the endodermis to the pericycle and protoxylem vessel poles in the stele, probably through plasmodesmata. miR165/6 degrades the transcripts of *PHABULOSA* (*PHB*) and other members of its gene family (class III homeodomain-leucine zipper: *HD-Zip III*) (Carlsbecker et al. 2010) (Fig. 2). High levels of these transcripts in xylem precursors specify central metaxylem vessels, while low levels specify peripheral protoxylem vessels. This reciprocal signaling between the inner vascular tissues and the surrounding cell layers allows the response domain to be confined to one of two tissue compartments.

ACAULIS 5 (*ACL5*) encodes a thermospermine synthase and is expressed specifically in vessel elements and an early stage of development (Kakehi et al. 2008, Muñiz et al. 2008). The loss-of function of *ACAULIS5* (ACL5) caused excessive differentiation of xylem cells (Muñiz et al. 2008). Exogenously applied thermospermine suppressed xylem vessel differentiation in *Arabidopsis* plants and a *Zinnia* xylogenic culture (Kakehi et al. 2010). A genetic analysis of *acl5* identified a suppressor of the *acl5* phenotype, *sac51*, whose gene encodes a basic helix–loop–helix (bHLH) transcription factor (Imai et al. 2006). It was suggested that ACL5 may control xylem specification through preventing premature cell death (Muñiz et al. 2008, Vera-Sirera et al. 2010). Thermospermine was shown to regulate the translational activity of SAC51 mRNA, resulting in the suppression of xylem development (Imai et al. 2006). SAC51-mediated thermospermine signaling was shown to limit the auxin signaling that promotes xylem differentiation (Yoshimoto et al. 2012, Lucas et al. 2013).

Transcriptional regulation of xylem cell differentiation

Members of the *HD-ZIP III* subfamily function positively in xylem specification. In *phb, phv, rev, can,* and *athb8* mutants of *HD-ZIP III* subfamily members in *Arabidopsis*, procambial cells failed to differentiate into xylem cells, but proliferated actively to produce many procambium cells. However, every quadruple loss-of-function mutant of the five *HD-ZIP III* genes exhibited ectopic xylem formation in the roots (Carlsbecker et al. 2010). In contrast, a gain-of-function mutant of PHB showed ectopic metaxylem vessel formation (Carlsbecker et al. 2010), and overproduction of ATHB8 promoted xylem differentiation (Baima et al. 1995, 2001). However, the regulation of xylem differentiation by HD-ZIP III members is more complicated. MiR165/166, which degrades HD-ZIP III transcripts, promotes protoxylem vessel differentiation in roots. Therefore, high levels of HD-ZIP III members were proposed to inhibit protoxylem vessel formation but promote metaxylem vessel formation.

Exogenously applied brassinosteroids (BRs) promoted the expression of *HD-ZIP III* genes (Ohashi-Ito and Fukuda 2003) and xylem cell differentiation (Yamamoto et al. 1997). Therefore, BRs may promote xylem differentiation, at least partly, by activating genes. HD-Zip III and KANADI transcription factors were also reported to regulate cambium cell differentiation; KANADI may function by inhibiting auxin transport, while HD-Zip III was shown to promote xylem differentiation (Ilegems et al. 2010, Robischon et al. 2011) (Fig. 2).

The VNDs and NSTs are a subgroup of NAM/ATAF/CUC (NAC)-domain proteins that function as master transcription factors that induce xylem cell differentiation by their ectopic expression (Zhong and Ye 2007, Demura and Fukuda 2007). VND6 and VND7 initiate metaxylem and protoxylem vessel differentiation, respectively (Kubo et al. 2005). Similarly, SND1/NST3 and NST1 induce xylem fiber differentiation (Zhong et al. 2006, Mitsuda et al. 2005, 2007). However, a single loss-of-function mutant of each gene showed no morphological defects. This suggested that other family members may have redundant functions to induce xylem differentiation, although none of them induced xylem cell differentiation when overexpressed (Kubo et al. 2005).

Expressions of *VND7* and two genes encoding two AS2/LBD-domain-containing proteins (ASL20/LBD18 and ASL19/LBD30) form a positive feedback loop to amplify *VND7* expression (Soyano et al. 2008). This rapid amplification of the master transcription factors may drive xylem cell differentiation promptly and irreversibly. VND7 activity was also shown to be regulated at the protein level by its proteasome-mediated degradation, and the NAC-domain transcription repressor VND-INTERACTING2 (VNI2) was shown to repress VND activity via protein–protein interaction (Yamaguchi et al. 2010). VNI2 is unstable because it contains a PEST proteolysis target motif in the C-terminal region. This may allow VND7 to exert its function promptly when required.

Xylem cells have characteristic secondary cell walls. VND6, VND7, and SND1 induce a hierarchical gene expression network to regulate secondary wall formation (Zhong et al. 2008, Yamaguchi et al. 2010, Ohashi-Ito et al. 2010). These master transcription factors induce, probably directly, the expression of genes encoding other transcription factors such as MYB46, MYB83, and MYB103. MYB46 and MYB83 which regulate redundant biosynthetic pathways for all three major secondary wall components: cellulose, lignin, and xylan (Zhong and Ye 2007, McCarthy et al. 2009). NACs and 10 MYBs also act downstream of SND1 (Zhong et al. 2008, Zhou et al. 2009). Of them, MYB58, MYB63 and MYB85, which might be targets of MYB46 and/or MYB83, specifically upregulate genes related to the lignin biosynthetic pathway (Zhou et al. 2009). Thus, SND1/NST3, VND6, and VND7 function as a master regulator switch at the top of the hierarchy, to upregulate transcription factors such as MYBs. In turn, the MYBs act as the second and third regulators to upregulate the expression of genes encoding enzymes catalyzing secondary wall thickening during specific xylem cell differentiation (Lucas et al. 2013).

VND6 and VND7 also directly upregulate the expression of genes encoding enzymes such as XCP1 and CESA4, which are ranked lowest in the gene expression hierarchy, as well as genes for transcription factors such as MYB46, which are ranked higher (Ohashi-Ito et al. 2010, Yamaguchi et al. 2011). Similarly, genes for enzymes such as 4CL1 are direct targets of SND1 (Zhong et al. 2008, McCarthy et al. 2009). These results indicate that the master regulators control a sophisticated transcriptional regulation network that operates as a hierarchy. Tracheary elements and xylem fibers have different characteristics in terms of cell wall structure and PCD. Consistent with these characteristics, VND6 but not SND1 induces the expression of genes related to rapid PCD such as *XCP1* and *XCPs*, while SND1 but not VND6 preferentially upregulates genes related to lignin monomer synthesis such as *PAL1, 4CL3*, and *CCoAOMT* (Ohashi-Ito et al. 2010). An 11-bp cis-element known as the tracheary-element-regulating cis-element (TERE), which is found in the upstream sequences of many genes expressed in xylem vessel cells, was shown to be responsible for xylem vessel cell-specific expression (Pyo et al. 2007). VND6 binds the TERE sequence and activates the TERE-containing promoter *in planta*. VND6 was unable to bind to a mutated promoter with substitutions in the TERE sequence (Ohashi-Ito et al. 2010). VND7 also binds TERE (Yamaguchi et al. 2011). These results demonstrate that TERE is a target sequence of VND6. In contrast, SND1 specifically binds to a 19-bp sequence known as the SECONDARY WALL NAC BINDING ELEMENT (SNBE), thereby activating its target genes (Zhong et al. 2010, Lucas et al. 2013).

Phloem Cell Differentiation from Procambial Cells

The phloem consists of sieve elements (SEs), companion cells (CCs), phloem fibers, and parenchyma cells. SEs form the main conductive tissue of the phloem, and CCs are specialized phloem parenchyma cells that are tightly connected to SEs, supporting them and loading and unloading nutrients. Both SEs and CCs originate from phloem precursor cells in the procambium in primary vascular tissue (Iqbal 1995b).

Cellular aspects of phloem cell differentiation from the procambium

During primary phloem development, all phloem cells undergo dramatic changes in their morphology. For example, SEs undergo extensive degradation of their organelles as they mature. The nucleus, vacuoles, RER, and Golgi are degraded, but not in the same way as in TEs in the xylem. The SEs remain alive, as they retain a plasma membrane and some smooth endoplasmic reticulum, plastids, and mitochondria. The residual endoplasmic reticulum is localized near the plasmodesmata that connect the SEs to their neighboring CCs (Iqbal 1995b, van Bel 2003, Wu and Zheng 2003, Truernit et al. 2008). These cellular reorganizations are orchestrated by the genetically redundant NAC domain–containing transcription factors, NAC45 and NAC86 (Furuta et al. 2014b). Thus, sieve element differentiation is a special autolysis process.

Another key characteristic of SE differentiation is callose deposition around their plasmodesmata. When the SEs mature, these callose deposits and the middle lamella in these regions of the cell wall are removed, forming a sieve plate with enlarged pores (Lucas et al. 1993). The lateral cell walls of SEs also develop lateral sieve areas that are specialized areas of plasmodesmata-derived pores. The phloem-specific callose synthase genes *CALS7* and *CALS3* are involved in depositing plasmodesmata-callose during this developmental process (Vatén et al. 2011, Xie et al. 2011).

The survival and differentiation of SEs depends on a close association with their neighboring CCs, a specialized type of parenchyma cell. The cytoplasm of the CC is unusually dense, partly because of large numbers of plastids, mitochondria, and free ribosomes (Cronshaw 1981). The CCs are connected to their adjacent SEs by numerous branched plasmodesmata. Through these connections, the enucleate SEs are supplied with energy, assimilates, and macromolecular compounds such as proteins and RNA (Lough and Lucas 2006). Several phloem markers, Phloem Differentiation 1–5 (PD1–PD5), were identified by screening the Versailles collection of gene-trap mutants for plant lines expressing the uidA reporter gene in immature vascular tissues (Bauby et al. 2007). PD1–4 were shown to be restricted to protophloem cells, as determined by their cell shape, while *PD5* was shown to be expressed in both protophloem and metaphloem cells. The specific gene expression patterns of the phloem cell types were revealed by transcriptome studies (Lee et al. 2006, Brady et al. 2011).

Molecular regulation of phloem differentiation from procambial cells

Whereas there has been substantial progress understanding the regulation of xylem development, less is known about the specific regulatory factors involved in phloem development. Currently, only a few factors are known to be specific to phloem development (Fig. 2).

ALTERED PHLOEM DEVELOPMENT (APL) is a MYB coil-coil transcription factor essential for the proper differentiation of both SEs and CCs (Bonke et al. 2003). It is expressed both in SEs and CCs and has been shown to be nuclear-localized. APL is responsible for the later stages of phloem development, rather than establishment of the phloem cell lineage. Additionally, APL contributes to the spatial limitation of xylem differentiation (Miyashima et al. 2013).

OCTOPUS (OPS) encodes a membrane-associated protein that localizes to the apical end of phloem SE cells (Benschop et al. 2007, Nagawa et al. 2006). It is expressed in the phloem and procambium initials near the QC in the root, and it is required for phloem continuity during phloem development (Nagawa et al. 2006, Truernit et al. 2012).

ARABIDOPSIS LATERAL ROOT DEVELOPMENT 3 (LRD3) is another important regulator of early phloem development. It controls the transport function of the phloem. The *LRD3* gene encodes a LIM-domain protein of unknown function, and it is specifically expressed in CCs. It has been suggested that LRD3 may have a non-cell-autonomous role in early phloem development (Ingram et al. 2011).

Cambium and Secondary Vascular Tissue Development

Cambium initiation and activity is the key feature of plant secondary growth and secondary vascular tissue development. In the shoot, the cambium initiates from the procambium and parenchyma cells that regain the capacity to divide and form a layer of meristematic cells between the primary vascular bundles. The part of the cambium cylinder that develops within the vascular bundles is the fascicular cambium, while the part that develops from the dedifferentiating parenchyma cells between the vascular bundles is the interfascicular cambium. The fascicular cambium and the interfascicular cambium form a continuous ring of meristematic tissue (Fig. 1C). In the root, part of the cambium forms from the procambial cells within the vascular bundle, whereas pericycle cells give rise to the vascular cambium next to the xylem poles.

The activity of the cambium produces secondary vascular tissues; the secondary xylem (consisting of xylem vessels, xylem fibers, and parenchyma cells), and the secondary phloem (consisting of sieve elements, companion cells, fibers, and parenchyma cells) (Fig. 1D). In addition to the axial vascular tissue cells, secondary vascular tissue has a radial system of rays arranged longitudinally relative to the long axis of the stem or root. The vascular rays, consisting mostly of parenchyma cells, serve to transport substances including photosynthates and water across the stem or root between the secondary xylem and phloem.

When a cambial cell divides, one of the new cells retains the cambial meristematic identity, while the other undergoes a committed fate transition to a xylem or phloem mother cell. The mother cell is able to divide several times, giving rise to multiple cells that are required for secondary vascular tissues (Lachaud et al. 1999). The dividing cells, the cambial stem cells, and the mother cells produced through their activity are located in the middle of the cambial zone. The expanding xylem cells and the xylem cells that form secondary cell walls are located at the internal side of the cambial zone. The expanding phloem cells that will gradually differentiate into mature phloem are located on the peripheral side (Fig. 1D).

The radial component of the secondary vascular tissues is the vascular rays (Fig. 1D), which serve as radial transport pathways between the xylem and phloem. The rays are induced and controlled by radial movement signals, and they are shaped by axial signal flows (Lev-Yadun and Aloni 1995). Ethylene moves centrifugally through the cambium from the differentiating tracheary elements towards the phloem, and is the major hormonal signal promoting ray initiation and regulating these rays in the cambium (Lev-Yadun and Aloni 1995, Aloni et al. 2013).

Regulation of Cambium and Secondary Vascular Development

Hormonal regulation of cambium and secondary vascular development

The role of plant hormones in cambial development has been studied for several decades. Phytohormones such as auxin, cytokinin, gibberellin, ethylene, jasmonate, and BRs have been shown to be important signals in the induction and maintenance of the cambium and in the development of secondary vascular tissue (Ursache et al. 2013).

Auxin plays an important role in cambium activity and secondary xylem cell differentiation. In developing woody stems, there is a strong correlation between the local concentration of auxin and the initiation of secondary vascular differentiation from the cambium (Uggla et al. 1996, Tuominen et al. 1997). Several auxin transporter genes belonging to the AUX1-like influx and PIN1-like efflux families are expressed in the cambium, and their products possibly regulate the distribution of auxin in the cambium (Schrader et al. 2003). Downregulated auxin responsiveness in transgenic poplar trees led to reduced periclinal and anticlinal cell divisions in the secondary xylem (Nilsson et al. 2008). Also, the auxin responsiveness of the cambium decreases during endogenous dormancy, a state in which the plant meristem cannot respond to growth-stimulating signals (Baba et al. 2011). Thus, active auxin transport is required for cambium activity and secondary vascular development.

Cytokinin is another key hormone that regulates cambium activity. An *Arabidopsis* quadruple mutant lacking four key cytokinin biosynthetic enzymes failed to form a cambium, but showed accelerated root elongation (Matsumoto-Kitano et al. 2008). External application of cytokinins or genetic engineering of the cytokinin biosynthesis or signaling pathways resulted in increased secondary growth (Mähönen et al. 2006, Matsumoto-Kitano et al. 2008). Transgenic trees with decreased cytokinin signaling showed reduced cell division in the cambium (Nieminen et al. 2008).

Gibberellins have been shown to stimulate cell division in the cambium and to induce fiber formation (Aloni 2001). The action of gibberellins is very important for the early stages of wood formation, including cell elongation (Israelsson et al. 2005). Elevated gibberellin concentrations in stems of transgenic plants led to more rapid growth and elongated xylem fibers (Eriksson et al. 2000). Gibberellins, as a mobile trigger, are transported from shoots to the cambial zone of hypocotyls to trigger xylem expansion (Ragni et al. 2011).

Ethylene and jasmonate also involved in regulating cambial activity. Ethylene stimulates the growth of the cambium in trees and is a key regulator of reaction wood formation in response to leaning (Love et al. 2009). In a *Zinnia* cell culture system, application of ethylene biosynthesis inhibitors prevented the differentiation of tracheary elements (Pesquet and Tuominen 2011). Application of jasmonate and an elevated level of jasmonate signaling stimulated the interfascicular cambial activity in the *Arabidopsis* inflorescence stem (Sehr et al. 2010).

Regulation of peptides and receptor-like kinases in cambium and secondary vascular development

In *Populus*, the orthologs for CLE, PXY and WOX4 are expressed in the vascular cambium, and their expression patterns within the cambial zone resemble those in *Arabidopsis* (Schrader et al. 2003). Thus, the CLE–PXY–WOX4 signaling pathway may also function in the cambium during the secondary vascular development, in a similar way that it functions in procambium and primary vascular development as described above.

Two receptor-like kinases, RUL1 and MOL1, were also shown to be involved in *Arabidopsis* cambial development (Agusti et al. 2011). The *mol1* mutant showed a substantial increase in cambial activity with enhanced formation of secondary vascular tissue in the fascicular and interfascicular regions, while *rul1* mutants showed reduced cambial activity in interfascicular regions. These results suggested that these genes act as regulators of secondary vascular tissue formation, with MOL1 functioning as a repressor and RUL1 as an activator (Agusti and Greb 2013).

Transcriptional regulation of cambium activity and secondary vascular development

A network of transcription factors regulates the formation of the vascular cambium in a defined position and the patterning of secondary vascular tissues (Fig. 3). The hormonal regulation of secondary vascular development relies on the expression of various transcription factors that regulate different aspects of cell division and differentiation (Groover and Robischon 2006, Nagawa et al. 2006, Demura and Fukuda 2007, Persson et al. 2007).

Class III homeodomain-leucine zipper (HD ZIP III) transcription factors play an important role in regulating secondary growth. A *Populus* ortholog of *REV*, *popREV-OLUTA (PRE)*, is expressed in woody stems of *Populus* (Schrader et al. 2003). This shows that PRE is expressed during secondary growth and may play a crucial role in

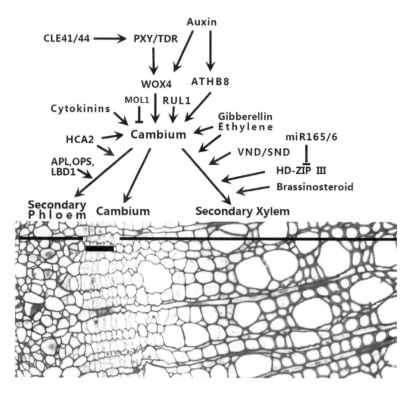

FIGURE 3 Regulation of the cambium and secondary vascular development, and a cross section of poplar stem showing the secondary vascular tissues.

initiating the cambium and in forming secondary vascular tissues. Further research showed that *HD ZIP III* genes play roles in regulating cambium activity during secondary growth (Robischon et al. 2011, Zhu et al. 2013).

Transcription factors also integrate the inputs of hormone pathways. For instance, the class *III HD-ZIP* gene *INTERFASCICULAR FIBERLESS/REVOLUTA* (*IFL1/ REV*) is essential for the normal differentiation of interfascicular fibers and the secondary xylem in the inflorescence stem (Zhong and Ye 2007), and possibly functions by promoting polar auxin transport. Indeed, *ifl1* mutants showed dramatically reduced polar auxin flows in both the inflorescence stems and the hypocotyl. Surprisingly, however, the *ifl1* mutants did not have aberrant root or hypocotyl phenotypes (Zhong and Ye 2001). The same is true for another regulator of the interfascicular cambium, the DOF class transcription factor HIGH CAMBIAL ACTIVITY 2/DOF5.6, whose gain-of-function triggered the extensive development of the interfascicular cambium to form a continuous cambium cylinder (Guo et al. 2009).

Other Class III HD-ZIP factors have been shown to play roles in xylem formation and cambium establishment in the root (Carlsbecker et al. 2010). This activity of Class III HD-ZIP factors is opposed by redundantly acting KANADI-type transcription factors, whose loss-of-function quadruple mutant produced more cells in the cambium and showed precocious secondary growth in the hypocotyl (Ilegems et al. 2010). The modeling of transcriptional networks has become possible recently, due to the development of next-generation sequencing-based technologies that can comprehensively catalog gene expression and transcription factor-binding genome-wide. This technique has been applied for network-based research and in research on poplar secondary growth and wood formation (Liu et al. 2015a). A chromatin immunoprecipitation sequencing (ChIP-seq) resource consisting of genome-wide binding regions and associated putative target genes for four *Populus* homeodomain transcription factors that are expressed during secondary growth and wood formation were reported recently (Liu et al. 2015b). These techniques and resources have provided new tools to investigate the regulatory network controlling cambium and secondary vascular development.

Conclusions and Perspectives

Our knowledge of vascular development has expanded rapidly in recent years. Vascular cells differentiate at predicted positions and times to form a specific vascular pattern, and a complex mechanism controls vascular meristem activity and vascular development. Intercellular signaling pathways such as TDIF-TDR-WOX4 and TDIF-TDR-GSK3s-BES1 regulate procambium/cambium proliferation and xylem differentiation, and hormonal and transcriptional regulators control xylem and phloem development. However, the molecular mechanisms underlying the initiation and maintenance of vascular meristems, including the procambium and cambium, are still largely unknown. Although several regulators of the procambium/cambium and vascular development have been identified, little is known about the networks of the various regulators during the developmental process. Furthermore, most of the research on procambium and vascular development has been conducted in herbaceous species such as *Zinnia* and *Arabidopsis*. More research is required to reveal the specific features of the cambium and secondary vascular tissues in trees.

References

Agusti, J., R. Lichtenberger, M. Schwarz, L. Nehlin and T. Greb. 2011. Characterization of transcriptome remodeling during cambium formation identifies MOL1 and RUL1 as opposing regulators of secondary growth. PLoS Genet.7: e1001312.

Agusti, J. and T. Greb. 2013. Going with the wind: adaptive dynamics of plant secondary meristems. Mech. Dev. 130: 34-44.

Aloni, R. 2001. Foliar and axial aspects of vascular differentiation: hypotheses and evidence. J. Plant Growth Regul. 20: 22-34.

Aloni, R. 2013. The role of hormones in controlling vascular differentiation. *In:* J. Fromm (ed.). Cellular Aspects of Wood Formation, Plant Cell Monographs 20, Springer-Verlag. Berlin. Heidelberg.

Avci, U., H.E. Petzold, I.O. Ismail, E.P. Beers and C.H. Haigler. 2008. Cysteine proteases XCP1 and XCP2 aid micro-autolysis within the intact central vacuole during xylogenesis in Arabidopsis roots. Plant J. 56: 303-315.

Baba, K., A. Karlberg, J. Schmidt, J. Schrader, T.R. Hvidsten, L. Bako and R.P. Bhalerao. 2011. Activity–dormancy transition in the cambial meristem involves stage specific modulation of auxin response in hybrid aspen. Proc. Natl. Acad. Sci. USA. 108: 3418-3423.

Baima, S., F. Nobili, G. Sessa, S. Lucchetti, I. Ruberti and G. Morelli. 1995. The expression of the athb-8 homeobox gene is restricted to provascular cells in *Arabidopsis thaliana*. Development 121: 4171-4182.

Baima, S., M. Possenti, A. Matteucci, E. Wisman and M.M. Altamura, I. Ruberti and G. Morelli. 2001. The Arabidopsis ATHB8 HD-ZIP protein acts as a differentiation-promoting transcription factor of the vascular meristems. Plant Physiol. 126: 643-55.

Barnett, J.R. 1995. Ultrastructural factors affecting xylem differentiation. pp. 107-130. *In:* M. Iqbal (ed.). The Cambial Derivatives. Encyclopedia of Plant Anatomy, Gebruder Borntraeger, Berlin.

Bauby, H., F. Divol, E. Truernit, O. Grandjean and J.C. Palauqui. 2007. Protophloem differentiation in early *Arabidopsis thaliana* development. Plant Cell Physiol. 48: 97-109.

Benschop, J.J., S. Mohammed, M. O'Flaherty, A.J.R. Heck, M. Slijper and F.L.H. Menke. 2007. Quantitative phosphoproteomics of early elicitor signaling in Arabidopsis. Mol. Cell Proteomics. 6: 1198-1214.

Bishopp, A., H. Help, S. El-Showk, D. Weijers, B. Scheres, J. Friml, E. Benková, A.P. Mähönen and Y. Helariutta. 2011. A mutually inhibitory interaction between auxin and cytokinin specifies vascular pattern in roots. Curr. Biol. 21: 917-926.

Bishopp, A., S. Lehesranta, A. Vatén, H. Help, S. El-Showk, B. Scheres, K. Helariutta, A.P. Mähönen, H. Sakakibara and Y. Helariutta. 2011. Phloem-transported cytokinin regulates polar auxin transport and maintains vascular pattern in the root meristem. Curr. Biol. 21: 927-932.

Bollhoner, B., J. Prestele and H. Tuominen. 2012. Xylem cell death: emerging understanding of regulation and function. J. Exp. Bot. 63: 1081-1094.

Bonke, M., S. Thitamadee, A.P. Mahonen, M.T. Hauser and Y. Helariutta. 2003. APL regulates vascular tissue identity in Arabidopsis. Nature. 426: 181-186.

Brady, S.M., L. Zhang, M. Megraw, N.J. Martinez, E. Jiang, C.S. Yi, W. Liu, A. Zeng, M. Taylor-Teeples, D. Kim, S. Ahnert, U. Ohler, D. Ware, A.J. Walhout and P.N. Benfey. 2011. A stele-enriched gene regulatory network in the Arabidopsis root. Mol. Syst. Biol. 7: 459.

Cano-Delgado, A., J.Y. Lee and T. Demura. 2010. Regulatory mechanisms for specification and patterning of plant vascular tissues. Annu. Rev. Cell Dev. Biol. 26: 605-637.

Carlsbecker, A., J.Y. Lee, C.J. Roberts, J. Dettmer, S. Lehesranta, J. Zhou, O. Lindgren, M.A. Moreno-Risueno, A. Vatén, S. Thitamadee, A. Campilho, J. Sebastian, J. L. Bowman, Y. Helariutta and P.N. Benfey. 2010. Cell signaling by microRNA165/6 directs gene dose-dependent root cell fate. Nature. 465: 316-321.

Chen, H.M., Y. Pang, J. Zeng, Q. Ding, S.Y. Yin, C. Liu, M.Z. Lu, K.M. Cui and X.Q. He. 2012. The Ca2+-dependent DNases are involved in secondary xylem development in *Eucommia ulmoides*. J. Integr. Plant Biol. 54: 456-470.

Courtois-Moreau, C.L., E. Pesquet, A. Sjodin, L. Muniz, B. Bollhoner, M. Kaneda, L. Samuels, S. Jansson and H. Tuominen. 2009. A unique program for cell death in xylem fibers of Populus stem. Plant J. 58: 260-274.

Cronshaw, J. 1981. Phloem structure and function. Annu. Rev. Plant Physiol. Plant Mol. Biol. 32: 465-484.

Cui, H., MP. Levesque, T. Vernoux, J.W. Jung, A.J. Paquette, K.L. Gallagher, J.Y. Wang, I. Blilou, B. Scheres and P.N. Benfey. 2007. An evolutionarily conserved mechanism delimiting SHR movement defines a single layer of endodermis in plants. Science 316: 421-425.

De Rybel, B., AS. Breda and D. Weijers. 2014. Prenatal plumbing - vascular tissue formation in the plant embryo. Physiol. Plant. 151: 126-133.

Demura, T., G. Tashiro, G. Horiguchi, N. Kishimoto, M. Kubo, N. Matsuoka, A. Minami, M. Nagata-Hiwatashi, K. Nakamura, Y. Okamura, N. Sassa, S. Suzuki, J. Yazaki, S. Kikuchi and H. Fukuda. 2002. Visualization by comprehensive microarray analysis of gene expression programs during trans-differentiation of mesophyll cells into xylem cells. Proc. Natl. Acad. Sci. USA. 99: 15794-15799.

Demura, T. and H. Fukuda. 2007. Transcriptional regulation in wood formation. Trends Plant Sci. 12: 64-70.

Du, J. and A. Groover. 2010. Transcriptional regulation of secondary growth and wood formation. J. Integr. Plant Biol. 52: 17-27.

Endo, S., T. Demura and H. Fukuda. 2001. Inhibition of proteasome activity by the TED4 protein in extracellular space: A novel mechanism for protection of living cells from injury caused by dying cells. Plant Cell Physiol. 42: 9-19.

Eriksson, M.E., M. Israelsson, O. Olsson and T. Moritz. 2000. Increased gibberellin biosynthesis in transgenic trees promotes growth, biomass production and xylem fiber length. Nature Biotech. 18: 784-788.

Etchells, J.P., C.M. Provost, L. Mishra and S.R. Turner. 2013. WOX4 andWOX14 act downstream of the PXY receptor kinase to regulate plant vascular proliferation independently of any role in vascular organisation. Development. 140: 2224-2234.

Etchells, J.P., C.M. Provost and S.R. Turner. 2012. Plant vascular cell division is maintained by an interaction between PXY and ethylene signaling. PLoS Genet. 8: e1002997.

Fisher, K. and S. Turner. 2007. PXY, a receptor-like kinase essential for maintaining polarity during plant vascular-tissue development. Curr. Biol. 17: 1061-1066.

Fukuda, H. 2000. Programmed cell death of tracheary elements as a paradigm in plants. Plant Mol. Biol. 44: 245-253

Fukuda, H. 2004. Signals that control plant vascular cell differentiation. Nat. Rev. Mol. Cell Biol. 5: 379-391.

Funk, V., B. Kositsup, C. Zhao and E.P. Beers. 2002. The Arabidopsis xylem peptidase XCP1 is a tracheary element vacuolar protein that may be a papain ortholog. Plant Physiol. 128: 84-94.

Furuta, K.M., E. Hellmann and Y. Helariutta. 2014a. Molecular control of cell specification and cell differentiation during procambial development. Annu. Rev. Plant Biol. 65: 607-638.

Furuta, K.M., S.R. Yadav, S. Lehesranta, I. Belevich, S. Miyashima, J.Heo, A. Vatén, O. Lindgren, B.D. Rybel, G. V. Isterdael, P. Somervuo, R. Lichtenberger, R. Rocha, S. Thitamadee, S. Tähtiharju, P. Auvinen, T. Beeckman, E. Jokitalo and Y. Helariutta. 2014b. Arabidopsis NAC45/86 direct sieve element morphogenesis culminating in enucleation. Science. 345: 933-937.

Gallagher, K. and P.N. Benfey. 2009. Both the conserved GRAS domain and nuclear localization are required for SHORT-ROOT movement. Plant J. 57: 785-797.

Groover, A. and M. Robischon. 2006. Developmental mechanisms regulating secondary growth in woody plants. Curr. Opin. Plant Biol. 9: 55-58.

Guo, Y. G. Qin, H. Gu and L.J. Qu. 2009. HCA2, a Dof transcription factor gene, regulates interfascicular cambium formation and vascular tissue development in Arabidopsis. Plant Cell 21: 3518-3534.

Han, J.J., W. Lin, Y. Oda, K.M. Cui, H. Fukuda and X.Q. He. 2012. The proteasome is responsible for caspase-3-like activity during xylem development. Plant J. 72: 129-141.

Hirakawa, Y., Y. Kondo and H. Fukuda. 2010. Regulation of vascular development by CLE peptide-receptor systems. J. Integr. Plant Biol. 52: 8-16.

Hirakawa, Y., H. Shinohara, Y. Kondo, A. Inoue, I. Nakanomyo, M. Ogawa, S. Sawa, K. Ohashi-Ito, Y. Matsubayashi and H. Fukuda. 2008. Non-cell-autonomous control of vascular stem cell fate by a CLE peptide/receptor system. Proc. Natl. Acad. Sci. USA. 105: 15208-15213.

Ilegems, M., V. Douet, M. Meylan-Bettex, M. Uyttewaal, L. Brand, J.L. Bowman and P. A. Stieger. 2010. Interplay of auxin, KANADI and class III HD-ZIP transcription factors in vascular tissue formation. Development 137: 975-984.

Imai, A., Y. Hanzawa, M. Komura, K.T. Yamamoto, Y. Komeda and T. Takahashi. 2006. The dwarf phenotype of the Arabidopsis *acl5* mutant is suppressed by a mutation in an upstream ORF of a bHLH gene. Development 133: 3575-3585.

Ingram, P., J. Dettmer, Y. Helariutta and J.E. Malamy. 2011. Arabidopsis lateral root development 3 is essential for early phloem development and function, and hence for normal root system development. Plant J. 68: 455-467.

Iqbal, M., and A.K.M. Ghouse. 1990. Cambial concept and organization. pp. 1-36. *In*: M. Iqbal (ed.). The Vascular Cambium. Research Studies Press. Taunton, UK.

Iqbal, M. 1995a. Structure and behavior of vascular cambium and the mechanism and control of cambial growth. pp. 1-67. *In*: M. Iqbal (ed.). The Cambial Derivatives. Encyclopedia of Plant Anatomy. Gebruder Borntraeger, Berlin.

Iqbal, M. 1995b. Ultrastructural differentiation of sieve elements. pp. 241-270. *In*: M. Iqbal (ed.). The Cambial Derivatives. Encyclopedia of Plant Anatomy. Gebruder Borntraeger, Berlin.

Israelsson, M., B. Sundberg and T. Moritz. 2005. Tissue-specific localization of gibberellins and expression of gibberellin-biosynthetic and signaling genes in wood-forming tissues in aspen. Plant J. 44: 494-504.

Ito, J. and H. Fukuda. 2002. ZEN1 is a key enzyme in the degradation of nuclear DNA during programmed cell death of tracheary elements. Plant Cell. 14: 3201-3211.

Ito, Y., I. Nakanomyo, H. Motose, K. Iwamoto, S. Sawa, N. Dohmae and H. Fukuda. 2006. Dodeca-CLE peptides as suppressors of plant stem cell differentiation. Science. 313: 842-845.

Ji, J., J. Strable, R. Shimizu, D. Koenig, N. Sinha and M.J. Scanlon. 2010. WOX4 promotes procambial development. Plant Physiol. 152: 1346-1356.

Jouannet, V., K. Brackmann and T. Greb. 2015. (Pro)cambium formation and proliferation: two sides of the same coin? Curr. Opin. Plant Biol. 23: 54-60.

Kakehi, J., Y. Kuwashiro, H. Motose, K. Igarashi and T. Takahashi. 2010. Norspermine substitutes for thermospermine in the control of stem elongation in *Arabidopsis thaliana*. FEBS Lett. 584: 3042-3046.

Kakehi, J.I., Y. Kuwashiro, Y. Niitsu and T. Takahashi. 2008. Thermospermine is required for stem elongation in *Arabidopsis thaliana*. Plant Cell Physiol. 49: 1342-1349.

Kondo, Y., Y. Hirakawa and H. Fukuda. 2011. CLE peptides can negatively regulate protoxylem vessel formation via cytokinin signaling. Plant Cell Physiol. 52: 37-48.

Kondo, Y., T. Ito, H. Nakagami, Y. Hirakawa, M. Saito, T. Tamaki, K. Shirasu and H. Fukuda. 2014. Plant GSK3 proteins regulate xylem cell differentiation downstream of TDIF–TDR signaling. Nature Commun. 5: 3504.

Kubo, M., M. Udagawa, N. Nishikubo, G. Horiguchi, M. Yamaguchi, J. Ito, T. Mimura, H. Fukuda and T. Demura. 2005. Transcription switches for protoxylem and metaxylem vessel formation. Genes Dev. 19: 1855-1860.

Kuriyama, H. and H. Fukuda. 2002. Developmental programmed cell death in plants. Curr. Opin. Plant Biol. 5: 568-573.

Kwon, S.I., H.J. Cho, J.H. Jung, K. Yoshimoto, K. Shirasu and O.K. Park. 2010. The RabGTPase RabG3b functions in autophagy and contributes to tracheary element differentiation in Arabidopsis. Plant J. 64: 151-164.

Lachaud, S., A.M. Catesson and J.L. Bonnemain. 1999. Structure and functions of the vascular cambium. Comptesrendus de l'Academie des sciences Serie III Sciences de la vie. 322: 633-650.

Larson, P.R. 1975. Development and organization of the primary vascular system in Populus deltoids according to phyllotaxy. Am. J. Bot. 62: 1084-1099.

Larson, P.R. 1976. Procambium vs. cambium and protoxylem vs. metaxylem in *Populus deltoids* seedlings. Am. J. Bot. 63: 1332-1348.

Larson, P.R. 1994. The vascular cambium: development and structure. Berlin, Springer Verlag.

Lee, J.Y., J. Colinas, J.Y. Wang, D. Mace, U. Ohler and P.N. Benfey. 2006. Transcriptional and posttranscriptional regulation of transcription factor expression in Arabidopsis roots. Proc. Natl. Acad. Sci. USA. 103: 6055-6060.

Lehesranta, S.J., R. Lichtenberger and Y. Helariutta. 2010. Cell-to-cell communication in vascular morphogenesis. Curr. Opin. Plant Biol. 13: 59-65.

Lehmann, K., B. Hause, D. Altmann and M. Kock. 2001. Tomato ribonuclease LX with the functional endoplasmic reticulum retention motifHDEFis expressed during programmed cell death processes, including xylem differentiation, germination, and senescence. Plant Physiol. 127: 436-449.

Lev-Yadun, S. and R.Aloni. 1995. Differentiation of the ray system in woody plants. Bot. Rev. 61: 45-88.

Levesque, M.P., T. Vernoux, W. Busch, H. Cui, J.Y. Wang, I. Blilou, H. Hassan, K. Nakajima, N. Matsumoto, J.U. Lohmann, B. Scheres and P.N. Benfey. 2006. Whole-genome analysis of the SHORT-ROOT developmental pathway in Arabidopsis. PLoS Biol. 4: e143.

Liu, L., M. Zinkgraf, E. Patzold, E. Beers, V. Filkov and A. Groover. 2015a. The Populus ARBORKNOX1 homeodomain transcription factor regulates woody growth through binding to evolutionarily conserved target genes of diverse function. New Phytol. 205: 682-694.

Liu, L., T. Ramsay, M. Zinkgraf, D. Sundell, N.B. Street, V. Filkov and A. Groover. 2015b. A resource for characterizing genome-wide binding and putative target genes of transcription factors expressed during secondary growth and wood formation in Populus. Plant J. 82: 887-898.

Lough, T.J. and W.J. Lucas. 2006. Integrative plant biology: role of phloem long-distance macromolecular trafficking. Annu .Rev. Plant Biol. 57: 203-232.

Love, J., S. Bjorklund, J. Vahala, M. Hertzberg, J. Kangasjarvi and B. Sundberg. 2009. Ethylene is an endogenous stimulator of cell division in the cambial meristem of Populus. Proc. Nat. Acad. Sci. USA. 106: 5984-5989.

Lucas, W.J., B. Ding and C. van der Schoot. 1993. Plasmodesmata and the supracellular nature of plants. New Phytol. 125: 435-476.

Lucas, W.J., A. Groover, R. Lichtenberger, K. Furuta, S.R. Yadav, Y. Helariutta, X.Q. He, H. Fukuda, J. Kang, S.M. Brady, J.W. Patrick, J. Sperry, A. Yoshida, A.F. Lopez-Millan, M.A. Grusak and P. Kachroo. 2013. The plant vascular system: evolution, development and functions. J. Integr. Plant Biol. 55: 294-388.

Mähönen, A.P., A. Bishopp, M. Higuchi, K.M. Nieminen, K. Kinoshita, K. Törmäkangas, Y. Ikeda, A. Oka, T. Kakimoto and Y. Helariutta. 2006. Cytokinin signaling and its inhibitor AHP6 regulate cell fate during vascular development. Science 311: 94-98.

Mähönen, A.P., M. Bonke, L. Kauppinen, M. Riikonen, P.N. Benfey and Y. Helariutta. 2000. A novel two component hybrid molecule regulates vascular morphogenesis of the Arabidopsis root. Genes Dev. 14: 2938-2943.

Matsumoto-Kitano, M., T. Kusumoto, P. Tarkowski, K. Kinoshita Tsujimura, K. Vaclavikova, K. Miyawaki and T. Kakimoto. 2008. Cytokinins are central regulators of cambial activity. Proc. Natl. Acad. Sci. USA. 105: 20027-20031.

McCarthy, R.L., R. Zhong and Z.H. Ye. 2009. MYB83 is a direct target of SND1 and acts redundantly with MYB46 in the regulation of secondary cell wall biosynthesis in Arabidopsis. Plant Cell Physiol. 50: 1950-1964.

Milioni, D., P.E. Sado, N.J .Stacey, K. Roberts and M.C. McCann. 2002. Early gene expression associated with the commitment and differentiation of a plant tracheary element is revealed by cDNA-amplified fragment length polymorphism analysis. Plant Cell. 14: 2813-2824.

Mitsuda, N., M. Seki, K. Shinozaki and M. Ohme-Takagi. 2005. The NAC transcription factors NST1 and NST2 of Arabidopsis regulate secondary wall thickenings and are required for anther dehiscence. Plant Cell. 17: 2993-3006.

Mitsuda, N., A. Iwas, H. Yamamoto, M. Yoshida, M. Seki, K. Shinozaki and M. Ohme-Takagi. 2007. NAC transcription factors, NST1 and NST3, are key regulators of the formation of secondary walls in woody tissues of Arabidopsis. Plant Cell. 19: 270-280.

Miyashima, S., J. Sebastian, Y. Lee and Y. Helariutta. 2013. Stem cell function during plant vascular development. EMBO J. 32: 178-193.

Muñiz, L., E.G. Minguet, S.K. Singh, E. Pesquet, F. Vera-Sirera, C.L. Moreau-Courtois, J. Carbonell, M.A. Blázquez and H. Tuominen. 2008. ACAULIS5 controls Arabidopsis xylem specification through the prevention of premature cell death. Development. 135: 2573-2582.

Nagawa, S., S. Sawa, S. Sato, T. Kato, S. Tabata and H. Fukuda. 2006. Gene trapping in Arabidopsis reveals genes involved in vascular development. Plant Cell Physiol. 47: 1394-1405.

Nieminen, K., J. Immanen, M. Laxell, L. Kauppinen, P. Tarkowski, K. Dolezal, S. Tähtiharju, A. Elo, M. Decourteix, K. Ljung, R. Bhalerao, K. Keinonen, V.A. Albert and Y Helariutta. 2008. Cytokinin signaling regulates cambial development in Poplar. Proc. Nat. Acad. Sci. USA. 105: 20032-20037.

Nilsson, J., A. Karlberg, H. Antti, M. Lopez-Vernaza, E. Mellerowicz, C. Perrot-Rechenmann, G. Sandberg and R.P. Bhalerao. 2008. Dissecting the molecular basis of the regulation of wood formation by auxin in hybrid aspen. Plant Cell 20: 843-855.

Oda, Y., Y. Iida, Y. Kondo and H. Fukuda. 2010. Wood cell-wall structure requires local 2D-microtubule disassembly by a novel plasma membrane-anchored protein. Curr. Biol. 20: 1197-1202.

Oda, Y. and H. Fukuda. 2012a. Initiation of cell wall pattern by a Rho- and microtubule-driven symmetry breaking. Science. 337: 1333-1336.

Oda, Y. and H. Fukuda. 2012b. Secondary cell wall patterning during xylem differentiation. Curr. Opin. Plant Biol. 15:38-44.

Ohashi-Ito, K. and H. Fukuda. 2003. HD-zip III homeobox genes that include a novel member, ZeHB-13 (Zinnia)/ATHB-15 (Arabidopsis), are involved in procambium and xylem cell differentiation. Plant Cell Physiol. 44: 1350-1358.

Ohashi-Ito, K., Y. Oda and H. Fukuda. 2010. Arabidopsis VASCULAR-RELATED NAC-DOMAIN6 directly regulates the genes that govern programmed cell death and secondary wall formation during xylem differentiation. Plant Cell 22: 3461-3473.

Persson, S., K.H. Caffall, G. Freshour, M.T. Hilley, S. Bauer, P. Poindexter, M.G. Hahn, D. Mohnen and C. Somerville. 2007. The Arabidopsis irregular xylem8 mutant is deficient in glucuronoxylan and homogalacturonan, which are essential for secondary cell wall integrity. Plant Cell 19: 237-255.

Pesquet, E., P. Ranocha, S. Legay, C. Digonnet, O. Barbier, M. Pichon and D. Goffner. 2005. Novel markers of xylogenesis in zinnia are differentially regulated by auxin and cytokinin. Plant Physiol. 139: 1821-1839.

Pesquet, E., A.V. Korolev, G. Calder and C.W. Lloyd. 2010. The microtubule associated protein AtMAP70-5 regulates secondary wall patterning in Arabidopsis wood cells. Curr. Biol. 20: 744-749.

Pesquet, E. and H. Tuominen. 2011. Ethylene stimulates tracheary element differentiation in *Zinnia elegans* cell cultures. New Phytol. 190: 138-149.

Pyo, H., T. Demura and H. Fukuda. 2007. TERE; a novel cis-element responsible for a coordinated expression of genes related to programmed cell death and secondary wall formation during differentiation of tracheary elements. Plant J. 51: 955-965.

Ragni, L. and C.S. Hardtke. 2014. Small but thick enough - the Arabidopsis hypocotyl as a model to study secondary growth. Physiol Plant. 151: 164-171.

Ragni, L., K. Nieminen, D. Pacheco-Villalobos, R. Sibout, C. Schwechheimer and C.S. Hardtke. 2011. Mobile gibberellin directly stimulates Arabidopsis hypocotyl xylem expansion. Plant Cell. 23: 1322-1336.

Robischon, M., J. Du, E. Miura and A. Groover. 2011. The Populus Class III HD ZIP, popREVOLUTA, influences cambium initiation and patterning of woody stems. Plant Physiol. 155: 1214-1225.

Risopatron, J.P.M., Y. Sun and B.J. Jones. 2010. The vascular cambium: molecular control of cellular structure. Protoplasma 247:145-161.

Schrader, J., K. Baba, S.T. May, K. Palme, M. Bennett, R.P. Bhalerao and G. Sandberg. 2003. Polar auxin transport in the wood-forming tissues of hybrid aspen is under simultaneous control of developmental and environmental signals. Proc. Natl. Acad. Sci. USA. 100: 10096-10101.

Sehr, E.M., J. Agusti, R. Lehner, E.E. Farmer, M. Schwarz and T. Greb. 2010. Analysis of secondary growth in the Arabidopsis shoot reveals a positive role of jasmonate signaling in cambium formation. Plant J. 63: 811-822.

Soyano, T., S. Thitamadee, Y. Machida, and N.H. Chua. 2008. ASYMMETRIC LEAVES2-LIKE19/ LATERAL ORGAN BOUNDARIES DOMAIN30 and ASL20/LBD18 regulate tracheary element differentiation in Arabidopsis. Plant Cell. 20: 3359-3373.

Suer, S., J. Agusti, P. Sanchez, M. Schwarz and T. Greb. 2011. WOX4 imparts auxin responsiveness to cambium cells in Arabidopsis. Plant Cell 23: 3247-3259.

Truernit, E., H. Bauby, K. Belcram, J. Barthelemy and J.C. Palauqui. 2012. OCTOPUS, a polarly localized membrane-associated protein, regulates phloem differentiation entry in *Arabidopsis thaliana*. Development 139: 1306-1315.

Truernit, E., H. Bauby, B. Dubreucq, O. Grandjean, J. Runions, J. Barthélémy and J.C. Palauqui. 2008. High-resolution whole-mount imaging of three-dimensional tissue organization and gene expression enables the study of phloem development and structure in Arabidopsis. Plant Cell 20: 1494-1503.

Tuominen, H., L. Puech, S. Fink and B. Sundberg. 1997. A radial concentration gradient of indole-3-acetic acid is related to secondary xylem development in hybrid aspen. Plant Physiol. 115: 577-585.

Turgeon, R. and S. Wolf. 2009. Phloem transport: cellular pathways and molecular trafficking. Annu. Rev. Plant Biol. 60: 207-221.

Turner, S., P. Gallois and D. Brown. 2007. Tracheary element differentiation. Annu. Rev. Plant Biol. 58: 407-433.

Twumasi, P., E.T. Iakimova, T. Qian, W. van Ieperen, J.H .Schel, A.M. Emons, O. Kooten and E.J. Woltering. 2010. Caspase inhibitors affect the kinetics and dimensions of tracheary elements in xylogenic Zinnia (*Zinnia elegans*) cell cultures. BMC Plant Biol. 10: 162.

Uggla, C., T. Moritz, G. Sandberg and B. Sundberg. 1996. Auxin as a positional signal in pattern formation in plants. Proc. Natl. Acad. Sci. USA. 93: 9282-9286.

Ursache, R., K. Nieminen and Y. Helariutta. 2013. Genetic and hormonal regulation of cambial development. Physiol Plant. 147: 36-45.

Van Bel, A.J.E. 2003. The phloem, a miracle of ingenuity. Plant Cell Environ. 26: 125-149.

Van Doorn, W., E. Beers, J. Dangl, V. Franklin-Tong, P. Gallois, I. Hara-Nishimura, A. Jones, M. Kawai-Yamada, E. Lam and J. Mundy. 2011. Morphological classification of plant cell deaths. Cell Death Differ. 18: 1241-1246.

Vatén, A., J. Dettmer, S. Wu, Y.D. Stierhof, S. Miyashima, S.R. Yadav, C.J. Roberts, A. Campilho, V. Bulone, R. Lichtenberger, S. Lehesranta, A.P. Mähönen, J.Y. Kim, E. Jokitalo, N. Sauer, B. Scheres, K. Nakajima, A. Carlsbecker, K.L. Gallagher and Y. Helariutta. 2011. Callose biosynthesis regulates symplastic trafficking during root development. Dev. Cell. 21: 1144-1155.

Vercammen, D., B. van de Cotte, G. De Jaeger, D. Eeckhout, P. Casteels, K. Vandepoele, I. Vandenberghe, J.V. Beeumen, D. Inzé and F.V. Breusegem. 2004. Type II metacaspases ATMC4 and ATMC9 of *Arabidopsis thaliana* cleave substrates after arginine and lysine. J. Biol. Chem. 279: 45329-45336.

Vera-Sirera, F., E.G. Minguet, S.K. Singh, K. Ljung, H.C.L. Tuominen, M.A. Blazquez and J. Carbonell. 2010. Role of polyamines in plant vascular development. Plant Physiol. Biochem. 48: 534-539.

Weir, I.E., R. Maddumage, A.C. Allan and I.B. Ferguson. 2005. Flow cytometric analysis of tracheary element differentiation in *Zinnia elegans* cells. Cytometry A. 68: 81-91.

Woffenden, B.J., T.B. Freeman and E.P. Beers. 1998. Proteasome inhibitors prevent tracheary element differentiation in Zinnia mesophyll cell cultures. Plant Physiol. 118: 419-430.

Wu, H. and X.F. Zheng. 2003. Ultrastructural studies on the sieve elements in root protophloem of *Arabidopsis thaliana*. Acta Bot. Sin. 45: 322-330.

Xie, B., X. Wang, M. Zhu, Z. Zhang and Z. Hong. 2011. CALS7 encodes a callose synthase responsible for callose deposition in the phloem. Plant J. 65: 1-14.

Yamaguchi, M., N. Mitsuda, M. Ohtani, M. Ohme-Takagi, K. Kato and T. Demura. 2011. VASCULARRELATED NAC-DOMAIN7 directly regulates the expression of a broad range of genes for xylem vessel formation. Plant J. 66: 579-590.

Yamaguchi, M., M. Ohtani, N. Mitsuda, M. Kubo, M. Ohme-Takagi, H. Fukuda and T. Demura. 2010. VND-INTERACTING2, a NAC domain transcription factor, negatively regulates xylem vessel formation in Arabidopsis. Plant Cell 22: 1249-1263.

Yamamoto, R., T. Demura and H. Fukuda. 1997. Brassinosteroids induce entry into the final stage of tracheary element differentiation in cultured Zinnia cells. Plant Cell Physiol. 38: 980-983.

Ye, Z.H. 2002. Vascular tissue differentiation and pattern formation in plants. Annu. Rev. Plant Biol. 53: 183-202.

Yoshimoto, K., Y. Noutoshi, K. Hayashi, K. Shirasu, T. Takahashi and H. Motose. 2012. A chemical biology approach reveals an opposite action between thermospermine and auxin in xylem development in *Arabidopsis thaliana*. Plant Cell Physiol. 53: 635-645.

Zhao, C., B.J. Johnson, B. Kositsup and E.P. Beers. 2000. Exploiting secondary growth in Arabidopsis. Construction of xylem and bark cDNA libraries and cloning of three xylem endopeptidases. Plant Physiol. 123: 1185-1196.

Zhong, R., T. Demura and Z.H. Ye. 2006. SND1, a NAC domain transcription factor, is a key regulator of secondary wall synthesis in fibers of Arabidopsis. Plant Cell 18: 3158-3170.

Zhong, R., C. Lee, J. Zhou, R.L. McCarthy and Z.H. Ye. 2008. A battery of transcription factors involved in the regulation of secondary cell wall biosynthesis in Arabidopsis. Plant Cell 20: 2763-2782.

Zhong, R., C. Lee and Z.H. Ye. 2010. Global analysis of direct targets of secondary wall NAC master switches in Arabidopsis. Mol. Plant 3: 1087-1103.

Zhong, R. and Z.H. Ye. 2007. Regulation of cell wall biosynthesis. Curr Opin Plant Biol. 10: 564-572.

Zhong, R. and Z.H. Ye. 2001. Alteration of auxin polar transport in the Arabidopsis *ifl1* mutants. Plant Physiol. 126: 549-563.

Zhou, J., C. Lee, R. Zhong and Z.H. Ye. 2009. MYB58 and MYB63 are transcriptional activators of the lignin biosynthetic pathway during secondary cell wall formation in Arabidopsis. Plant Cell. 21: 248-266.

Zhu, Y., D. Song, J. Sun, X. Wang and L. Li. 2013. PtrHB7, a class III HD-Zip gene, plays a critical role in regulation of vascular cambium differentiation in Populus. Mol. Plant. 6: 1331-1343.

15

Asymmetric Cell Division in the Zygote of Flowering Plants: The Continuing Polarized Event of Embryo Sac Development

Arturo Lòpez-Villalobos*, Ana Angela
Lòpez-Quiròz and Edward C. Yeung

Introduction

Plants are multicellular organisms which lack cell migration because plant cells are cemented together by pecto-cellulosic cell walls. During pattern formation of plants, which is the creation of structures from the generation of complex arrangements of cell fates in space and time, orientated cell divisions are of paramount importance in the absence of cell migration (Petricka et al. 2009, Willemsen and Scheres 2004). In such static conditions, when a new cell wall is formed between the two daughter cells, the relative positions of these cells are permanently defined. Moreover, plant cell division is commonly accompanied by differential deposition of cell determinants, such as transcription factors, organelles, signaling molecules, microRNAs, and ligand–receptor pairs. These cell determinants enable the generation of separate cell identities with potential differences in the function of daughter cells (Kajala et al. 2014). Hence, the location of the cell division plane is crucial to generate separate cell fates and, in general, to regulate plant tissue organization and organ architecture (Petricka et al. 2009, Willemsen and Scheres 2004).

In cell division plane-mediated cell determination, cell polarity is a major force that brings forth non-homogeneous distributions of cell fate determinants (Grieneisen et al. 2013, Yang and Lavagi 2012). A subsequent asymmetric cell division occurs to generate daughter cells with distinct cell fates. Asymmetric division is an intricate and dynamic process, since it involves the formation and specification of the location of many cell division related structures, especially the preprophase band. The generation of these structures requires the function of molecular players which respond to hormonal and physical stimuli (Rasmussen et al. 2011a,b, Paciorek and Bergmann 2010, Menke and Scheres 2009). In flowering plants, the asymmetric division of the zygote results in daughter cells with distinct cell fates. The smaller terminal cell

Department of Biological Sciences University of Calgary 2500 University Drive NW Calgary, Alberta, Canada T2N 1N4.
* Corresponding author : alopezvi@ucalgary.ca

gives rise to the embryo proper and the basal cell gives rise to a short-lived embryonic organ, the suspensor. Consequently, the positioning of the cell division plane at the time of zygotic asymmetric division is critical for promoting normal development of a plant (Musielak and Bayer 2014, Zhang and Laux 2011).

A fundamental question that plant biologists have been trying to answer for many years is how the zygote is organized cytologically and genetically to meet the conditions for the first asymmetric division. The zygote is housed within a fertilized ovule, more specifically within the embryo sac. Hence, it is not a surprise that many regulatory mechanisms and hormonal clues that regulate both ovule and embryo sac formation can play a role in zygote activities. Moreover, it is striking to discover that plants undertake biological strategies to mould ovular tissues from early stages of development with the aim of regulating future events, such as fertilization and the subsequent asymmetric division of the zygote. Polarized growth is one of such key strategies as its effect commences from early ovule development and continues until the formation of an embryo sac; a structure with a distinct organization and polarized distribution of fate determinants within it (Ceccato et al. 2013, Pagnussat et al. 2007, Balasubramanian and Schneitz 2000, 2002). During the course of embryo sac formation, special molecular networks or "cytoplasm domains" as major forces to dictate the distribution of cell determinants have been proposed (Martin et al. 2013b). The cytoplasm domains encompass molecular factors, notably transcription factors, chemical signaling molecules and cytoskeleton elements, which are grouped spatially and temporally as specific systems within the cytoplasm. The cytoplasm domains are therefore distinct as they are constituted by specific components, which respond to a particular factor, primarily a hormone or an oxidation gradient (Martin et al. 2013a, Bencivenga et al. 2011, Drews and Koltunow 2011, Pagnussat et al. 2009). Thus, through coordinated actions of cytoskeleton and physiological signals, certain fate determinant molecules are specifically transported and localized to distinctive regions of the embryo sac. This can confer cell specification as the 'syncyntial' nuclei are mobile inside the embryo sac and they can settle within a cytoplasm domain to become the major recipients of the fate determinant molecules (Martin et al. 2013a, Pagnussat et al. 2007). This enables the generation of cell-distinct gene programs resulting in the formation of an egg cell, synergids, antipodals, and the central cell (Martin et al. 2013a,b, Koszegi et al. 2011). Such enriched cell fate determinant regions can also provide positional clues that are important to regulate fertilization of the egg cell and the subsequent development and division of the zygote (Rasmussen et al. 2011a,b). Therefore, the polarized embryo sac is an indispensable structure allowing for zygote formation and its first asymmetric division (Jeong et al. 2011a, b). Given the relevance of embryo sac in zygote formation and development, in order to gain a proper understanding of the first asymmetric division of the zygote, the development of the embryo sac is first detailed.

In recent years, calcium (Ca^{2+}) has emerged as a key regulator of the fertilization process and plays a critical role in promoting polarized growth of the zygote. Therefore, upon the analysis of available evidence, we propose a Ca^{2+}-dependent mechanism as the core system responsible for the regulation of zygote elongation and the subsequent asymmetric division. Such a mechanism plays a critical role in the positioning of the cell division plane, since it coordinates the formation of cell division related structures, notably the preprophase band. It is also envisaged that the Ca^{2+}-dependent mechanism is necessary in recognizing essential clues deposited earlier in the egg cell and the central cell to guide fertilization and subsequent changes in the zygote. Despite being identified as a general mechanism, new avenues are suggested to decipher more components and molecular interactions embedded in this mechanism in order to better understand the molecular basis of zygote asymmetric division in plants.

Cytoplasm Domains are the Biological Motors that Establish Polarity and Specification of the Embryo Sac

Once the ovule primordium concludes its patterning along the distal-proximal axis, further morphogenetic changes generate two very distinct and highly differentiated structures, i.e. the mature ovule and the embryo sac (Cucinotta et al. 2014, Bencivenga et al. 2012). These morphogenetic events are intimately regulated by molecular networks or cytoplasm domains (Martin et al. 2013b). These domains can superimpose and interact upon each other to dictate cell fate determination within an embryo sac (Martin et al. 2013b, Pagnussat et al. 2007, 2009, see Fig. 1). How these cytoplasm domains are constituted and more importantly, in what manner they function and interact with each other to regulate cell specification and polarization of organelles inside the embryo sac, are detailed below.

Before presenting the mechanistic effects of cytoplasm domains on embryo sac development, it is important to emphasize that plants possess diversified embryo sac patterns (Yadegari and Drews 2004). The existence of different patterns indicates that the constitution and the mode of action of cytoplasm domains may change substantially according to the pattern of embryo sac formation. Here, we focus on the *Polygonum*-type, which is the most common pattern of ovule development among flowering plants. This pattern is also noted in *Arabidopsis thaliana* in which much information about molecular regulation of embryo sac development is available (Drews and Koltunow 2011, Endress 2011, Yang et al. 2010). Embryo sac development has been extensively described in many reviews, focusing on two key phases: megasporogenesis and megagametogenesis (Bencivenga et al. 2011, Drews and Koltunow 2011, Yang et al. 2010). During megasporogenesis, a hypodermal cell from the rudimentary nucellus of the ovule primordium differentiates to form an initial of the germline, called the archesporial cell. Subsequently, the archesporial cell enlarges and differentiates directly into the megaspore mother cell (MMC), also termed female meiocyte or megasporocyte. The MMC enters into meiosis giving rise to four haploid megaspores. Later, three of them degenerate and the megaspore located closest to the chalazal end survives. The remaining megaspore, also known as the functional megaspore (FM), enters megagametogenesis giving rise to a mature embryo sac. Initially, the FM with a single nucleus undergoes two rounds of syncytial mitosis to produce a four-nucleate cell with two nuclei at each pole. In the third mitosis, an eight-nucleate cell is formed, and is followed by the formation of phragmoplasts and cell plates between sister and non-sister nuclei. Lastly the cell wall is formed ending the process of megagametogenesis. During cellularization, one nucleus from each pole (polar nucleus) migrates to the center of the developing embryo sac, fuse together and localize in the central cell. Hence, a mature embryo sac is a seven-celled structure consisting of three antipodals, one central cell, two synergids, and an egg cell (Endress 2011, Drews and Koltunow 2011, Yang et al.

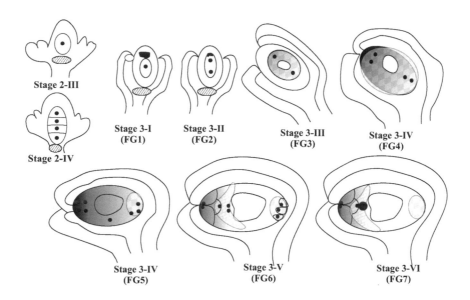

Stage 2-III

Stage 2-IV

Stage 3-I
(FG1)

Stage 3-II
(FG2)

Stage 3-III
(FG3)

Stage 3-IV
(FG4)

Stage 3-IV
(FG5)

Stage 3-V
(FG6)

Stage 3-VI
(FG7)

FIGURE 1   Positioning of cytoplasm domains during embryo sac development. Key components of the cytoplasm domains which confer polarity are depicted in this figure. The cytoplasm domains are molecular networks encompassing transcription factors, signaling molecules and cytoskeleton. These cytoplasm domains can superimpose upon each other to deposit cell specification information into cellular recipients. The syncytial nuclei are the major recipients when they are positioned in specific cytoplasm domains during motion at early stages of embryo sac development. It is envisaged that there are regions within the embryo sac with enriched components from the domains, which can confer a determined cell fate. The cytokinin cytoplasm domain is located in the chalazal region and is mainly responsible for the specification of the antipodals. In the cytokinin cytoplasm domain, polarity is primary conferred by cytokinin receptors, including the ARABIDOPSIS HISTIDINE KINASES (AHK) receptors (as circles with diagonal discontinued lines) and the CYTOKININ INDEPENDENT1 (CKI1) receptors (as diamond-shape patterned ovals). The auxin cytoplasm domain is positioned polarly in the micropylar region, to specify primarily the synergid and the egg cell fates. The auxin cytoplasm domain is exerted by the action of an auxin maximum region (depicted as black irregular shaped symbols at Stage 3-I and 3-II) and an auxin gradient (a black to grey-colored gradient inside the embryo sac with highest intensity at the micropylar end). Finally, the ROS domain (irregular shaped brick-patterned symbols) appears after cellularization of the embryo sac and specifies the central cell fate. The stages of ovule development are named according to the nomenclature of Schneitz et al. (1995, 1998) and the stages of female gametophyte are designated according to Christensen et al. (1997) cited by Sundaresan and Alandete-Saez (2010).

2010). The egg cell is also the female gamete. The formation of the highly specialized embryo sac is a result of cell specification mechanisms regulated by cytoplasm domains during megasporogenesis and megagametogenesis (Martin et al. 2013a, Bencivenga et al. 2011, Drews and Koltunow 2011, Pagnussat et al. 2009).

The cytokinin-dependent cytoplasm domain (CCD) acts early in embryo sac polarization

The cytokinin-dependent cytoplasm domain (CCD) is the first network that functions to regulate the formation and polarization of the embryo sac since the effects of the CCD go back to stage 1-II before the establishment of MMC fate (Lieber et al. 2011). In this morphogenetic event, the CCD bases its function on the activation of a transcriptional cascade dependent on the action of the SPOROCYTELESS/NOZZLE (SPL/NZZ) transcription factor (Cheng et al. 2013, Drews and Koltunow 2011, Lieber et al. 2011). By still unknown mechanisms, the CCD drives the expression of this gene to the distal location of the ovule primordium. Likewise, although the transcriptional factors that regulate *SPL* activation have not been discerned, cytokinins are essential to *SPL* transcription. The importance of cytokinin in *SPL* transcription was demonstrated by induced *SPL* expression in ovules upon benzyladenine treatments (Bencivenga et al. 2012). The *SPL* pathway includes the *NZZ/SPL, WUSCHEL (WUS), WINDHOSE1 (WH1), WH2, TORNADO2 (TRN2)* and *CINCINNATA (CIN)-like TCP (CIN-like TCPs)* genes which are highly expressed in the MMC and mutations in such genes generally converge on failure in MMC formation, and consequently, embryo sac lethality (Wei et al. 2015, Cheng et al. 2013, Lieber et al. 2011). The hierarchic positions of the components within this pathway has been demonstrated by reverse genetics which defined the SPL transcription factor as the most upstream component which immediately regulates the expression of WUS a typical homeodomain transcription factor responsive to cytokinins as well. Subsequently in this route, WUS is required to activate the *WH1-2* genes which encode small peptides that are needed together with the tetraspanin-type protein TRN2 for MMC function (Lieber et al. 2011). In the other branch of the pathway, SLP also exerts its function as a transcriptional repressor by using its EAR motif (ERF-associated amphiphilic repression) to recruit the known transcriptional co-repressor TOPLESS/TOPLESS-RELATED (TPL/TPRs) to suppress activities of CIN-like TCP transcription factors, which otherwise cause abortion of ovules in similar fashion to the abrogated *SLP* expression observed in null mutants (Wei et al. 2015).

In the subsequent developmental stage during the specification of the functional megaspore (FM), the role of CCD in the polarized growth becomes more evident. Specifically CCD promotes the creation of a cytokinin pole located in the chalazal region of the ovule which encompasses an apparatus of biosynthetic and perception factors that regulate cytokinin signaling (Cheng et al. 2013, Rabiger and Drews 2013). Indeed, gene expression studies have confirmed the specific localization of three cytokinin receptors [ARABIDOPSIS HISTIDINE KINASES 2, 3, and 4 (AHK2, AHK3 and AHK4/CRE1/WOL)] at the chalazal end of the ovule at the stage 2III, when the outer integument has initiated but meiosis of the MMC has not yet occurred (Cheng et al. 2013). The establishment of the cytokinin pole is consolidated by creating a supply of cytokinins to the chalazal region starting at the stage 2III as well. The accumulation of cytokinin at the chalaza is induced by cell-cell communications between the sporophyte and the nascent embryo sac, which results in signals to the MMC and surrounding sporophytic tissues to switch on the expression of the gene *IPT1*, which encodes isopentenyl transferase, the rate-limiting enzyme of cytokinin biosynthesis (Cheng et al. 2013, Bencivenga et al. 2012). However, in order that cytokinins to exert their effects on polarization of the nascent embryo sac, other downstream factors of the cytokinin two-component signaling (TCS) pathway, the histidine phosphotransfer proteins (AHPs) and the response regulators (ARRs), are prompted to function simultaneously with the cytokinin receptors (Bencivenga et al. 2012, Higuchi et al. 2004). Again, gene tracking expression studies confirmed this interaction of molecular factors by revealing an overlapping in the pattern of expression of genes encoding the cytokinin receptors with the genes of AHPs and ARRs at the chalaza from stages 2II (or 2-III) to 3-II. It is important to note that the FM is not only specified but also undergoes a first round of mitotic divisions. Besides cell polarization, recent evidence has demonstrated that the cytokinin pole also has a major role in the specification of the FM, as gene mutations of one of its components as showed in triple *ahk2 ahk3 ahk3* mutants result in the absence of FM development. These are unequivocal evidence indicating that the elevated cytokinin signaling in chalazal tissues conveys positional information for the FM specification. However, such a signal is also essential in the preceding event, i.e. the selection of the FM in the tetrad. By unknown mechanisms, the megaspore localized in the chalazal end of the ovule survives whilst the remaining three undergo programed cell death (Cheng et al. 2013, Drews and Koltunow 2011). Again, this finding strengthens the notion that the cytokinin pole functions as a positional clue to determine cell fate. However, so far it is not known which genes are involved in FM specification that are regulated by the CCD. This question needs to be addressed in the near future. A logical avenue would be to explore a potential link between CCD and the *SWITCH1* (*SWI1*)/ *DYAD* gene which encodes a protein involved in chromatid cohesion establishment during meiosis as well as in setting cell polarity in the MMC (Mercier et al. 2001, Motamayor et al. 2000). These future works would be particularly revealing because the *SWI1* gene also modulates the arrangement of the cytoskeleton which directs the polarization of organelles in the MMC to induce cell polarity. The polarly arranged organelles can consequently affect the selection of the FM in the tetrad by means of cell specification (Motamayor et al. 2000, Siddiqi et al. 2000). Moreover, it will be interesting to determine if the CCD regulates the expression of putative effectors of the sporophyte RNA (sRNA) silencing pathways, specifically *ARGONAUTE5* (*AGO5*) and *AGO9*, which have an important function in specification of FM and initiation of gametogenesis (Tucker et al. 2012, Olmedo-Monfil et al. 2010).

After specification of the FM, the cytokinin pole continues to drive the polarization of the embryo sac for further development. Cytokinin biosynthesis continues and is driven by IPT1 activity, which is highly expressed in the entire developing embryo sac (Cheng et al. 2013, Rabiger and Drews 2013, Benciveng et al. 2012). Downstream in the two-component system (TCS) pathway, cytokinin reception is carried out by complementing the function of AHK receptors and the CYTOKININ INDEPENDENT1 (CKI1) receptor. The latter is known to be involved in cytokinin perception, although its cytokinin-binding domain has not been identified (Glover et al. 2008, Higuchi et al. 2004, Yamada 2001). Even though the function of both types of receptor overlaps at stages 3I-II, the genes encoding the AHK receptors are still important in maintaining the polarity of the embryo sac since they are expressed at the chalazal region until stage 3-II. Later, cytokinin-mediated cell polarity is restricted to late embryo sac development, as the expression of the CKI1 gene is restricted only to the chalazal region after post-cellularization (stages 3-V to 3-VI). Beside polarization, CKI1 seems to play a role in cell specification after post-cellularization when CKI1 expression is circumscribed to the three antipodals and the central cell. The CKI1 mediated specification is consistent with the *cki1* mutant phenotype, which exhibits improper positioning of the antipodal nuclei, unfused polar nuclei and degeneration of the central cell (Cheng et al. 2013, Rabiger and Drews, 2013, Bencivenga et al. 2012, Hejatko et al. 2003). The mechanisms by which the CCD regulates both specification and polarization of embryo sac after

FM specification are unclear, however, a few reports provide some insights on the control of the CCD on key molecular players of gamete development. For example, in the *eostre* mutant, the CCD may regulate the function of the transcriptional complex BEL1-LIKE HOMEODOMAIN 1- KNOTTED1-LIKE HOMEOBOX 3- OVATE FAMILY PROTEIN 5 (BLH1-KNAT3-AtOFP5) to affect synergid cell fate by means of controlling its positioning within the embryo sac at the syncytial stage (Pagnussat et al. 2007). For operating this complex, BHL1 heteromerizes with the class II knox KNAT3 transcription factor and subsequently with AtOFP5, a protein which activates the BLH1-KNAT3 heterodimers. AtOFP5 also associates with the microtubules (Pagnussat et al. 2007, Hackbusch et al. 2005) and can regulate subcellular localization of BHL1 and KNAT3 homeodomain proteins, creating a positional control mechanism for embryo sac development. The *eostre* phenotype is caused by misexpression of BLH1 protein in the ovule, which does not occur in the wild type (Pagnussat et al. 2007). This abnormal *BLH1* expression may have prompted mislocalization of the whole transcriptional complex and affects consequently the specification of the synergids. Such assumption is based on the abnormal positioning of one of the nuclei in the central region of the embryo sac, which is closer to the chalazal pole in the *eostre* mutant, instead of having two nuclei characteristically located side by side at the micropylar pole in wild type at the stage 3IV. Cytokinin is known to repress the expression of KNAT3 (Truernit et al. 2006), therefore in the *eostre* mutant, the nucleus which localizes closer to the cytokinin-chalazal pole, may be defective in the function of the BLH1-KNAT3-AtOFP5 complex. Cytokinins may have inactivated the transcriptional complex via repression of *KNAT3*, resulting in the specification of two egg cells instead of one after one round of syncytial mitosis. Whether this assumption is correct, more research work is needed in the near future.

A more direct effect of the CCD on embryo sac development has been reported on the function of CKI1-dependent TCS pathway to promote embryo sac cellularization. The downstream factor of the CKI1 receptor is the transcription factor MYB DOMAIN PROTEIN 119 (MYB119). The mechanism undelaying this process is completely unknown but the role of these two factors in embryo sac development was demonstrated through genetic studies. Either single or double *cki1 myb119* were defective in the formation of cell walls resulting in uncellularized embryo sacs with supernumerary nuclei (Rabiger and Drews 2013). In a similar manner to previous processes, the cytokinin pole generated by a chalaza-enriched expression of the cytokinin CKI1 receptor which also affects chalazal cell identity, and in this case, logically through the function of MYB119. Mutant analysis studies also demonstrated the functional dependence of MYB119 on the perception activity of CKI1, but in this case MYB119 functions as an antagonistic molecular factor to restrict chalazal cell fate. Consequently, embryo sacs of defective plants in MYB119 experienced an expansion of chalazal cell identity at the expense of micropylar cell identity as demonstrated by tracing gametic cell marker nuclei (Rabiger and Drews 2013).

The evidence presented so far indicates that the CCD plays an essential role in polarized growth and cell specification of the embryo sac. Nevertheless, it is obvious that this network is not sufficient to build all structures and direct the whole collection of processes constituting the embryo sac. Therefore, in order to shape a completely functional embryo sac, the CCD needs to interact with other networks, including the auxin cytoplasm domain (Pagnussat et al. 2009) as well as the reactive oxygen species domain (Martin et al. 2013b), to achieve such a biological goal.

The auxin cytoplasm domain generates a gradient from the micropylar pole to regulate cell fate but not polarity of the embryo sac

Since the early studies in embryo sac development, auxin has been singled out as a major factor responsible for cell specification. To function, an auxin pole is needed in the micropylar region of the embryo sac in order to establish a gradient along this structure. Precisely, this gradient has been proposed as the major force for driving cell specification in which a specific cell fate is acquired through the nucleus, depending on its positioning along this auxin gradient. Auxin mediated cell specification occurs logically before embryo sac cellularization (Sundaresan and Alandete-Saez 2010, Pagnussat et al. 2009). Like cytokinin, auxin promotes the formation of a network to exert its cell fate role. The functioning of this network is complex since it requires a coordinate function of receptors, expression of transcription factors, and signaling molecules that can interact with the cytoskeleton. Thus, the understanding of an auxin-regulated cell fate is challenging. In order to comprehend the mechanistic action of the auxin pole, one needs to have a concise knowledge on the formation and maintenance of the auxin pole in the micropylar region during the course of embryo sac development (Bencivenga et al. 2012, Yang et al. 2010).

A vast body of evidence has indicated that a 'sequential source mechanism' is responsible for the formation of the gametophytic auxin pole. Such type of mechanism implicates that the auxin pole is continuously evolving during the course of embryo sac development, after being initially established. Thus, the establishment of the initial auxin pole is paramount and occurs immediately after the formation of the ovule primordia, through the expression of the auxin efflux PIN-FORMED 1 (PIN1) in the nucellus (Pagnussat et al. 2009). As a result of the sporophytic expression of PIN1, an auxin maximum arises after driving the sporophytically synthesized auxin to the nucellar tissue. This high concentration of auxin subsequently affects the development of the embryo sac, as demonstrated by genetic studies. In effect, ovules of the pin1-5 mutants exhibit arrest of embryo sacs at the the mono- and/or bi-nucleate stages (stage 3-I to 3-III) causing female sterility. The molecular mechanism that regulates this event has been recently unraveled and it involves components of auxin signaling as well as some factors of the CDD (Ceccato et al. 2013, Bencivenga et al. 2012). In this mechanism, PIN1 drives the TAA1-synthesized auxin to form an auxin maximum at the nucellus. However, two transcription factors, BELL 1 (BEL1) and SPL, are required for the expression and correct positioning of PIN1 in the nucellar tissue allowing the generation of a proper localized auxin maximum. The biological innovation of this mechanism relies on a cytokinin activation of BEL1 and SPL within an auxin environment, mediated by WUS, a typical cytokinin responsive transcription factor. In keeping directional polarity, BEL1 represses *WUS* in the chalaza to avoid the formation of an extra auxin pole in the chalaza and this is mediated by the activity of PIN (Bencivenga et al. 2012, Brambilla et al. 2007). In later stages of development, the auxin pole shifts from the sporophyte to the gametophyte, and more importantly, the mode of an auxin maximum formation changes from a polar efflux to localised auxin synthesis. In the latter, the YUCCA biosynthetic genes are responsible for supplying auxin to the embryo sac at the micropylar region as indicated by the auxin-dependent PLEIOTROPIC DRUG RESISTANCE 5 (*DR5*) activity and *in situ* RNA hybridizations studies (Lituiev et al. 2013, Pagnussat et al. 2009). Interestingly, the *YUC* expression occurs first in the micropylar region outside of the embryo sac at the stage 3I but later localizes to the micropylar region inside the embryo sac from stage 3III through later stages (Pagnussat et al. 2009).

Once a consistent auxin pole is established, an auxin gradient is built inside the embryo sac with a maximum at the micropylar pole and a minimum at the chalazal pole. Another independent pathway, likely involving the CCD-dependent BLH1-KNAT3-AtOFP5 transcriptional complex, positions the nuclei along the micropylar-chalazal axis of the syncytial embryo sac (Pagnussat et al. 2007, Hackbusch et al. 2005). The superimposing of the auxin gradient onto the nuclear positioning would result in exposing each nucleus to different punctual concentrations of auxin. Each nucleus would then take on a different cell fate according to the intensity of auxin signal perceived. Under this model, synergid cell fate would be determined by the highest concentration of auxin while the antipodals by the lowest. The egg and the central cell fates correspond to intermediate levels of auxin (Sundaresan and Alandete-Saez 2010, Pagnussat et al. 2007). This model of cell fate determination also conveys the partitioning of the syncytial embryo sac into different cytoplasm domains. In line with this idea, each specific cytoplasm domain can function as a positional clue as they are constituted by the distinct nature of signaling cascades of transcription factors associated with the cytoskeleton framework. Such distribution of cytoplasm networks allows positional responses only to specific small molecules, such as hormones, in determined regions of the embryo sac, by restricting transcription and translation of mRNAs. More importantly, the cytoplasm domains would maintain specific hormone gradients by regulating the site of synthesis, transport and degradation. If auxin synthesis and transport were not localized by the cytoplasm domain, small molecules like auxin would diffuse rapidly throughout the embryo sac without conferring a gradient-dependent specification (Sundaresan and Alandete-Saez 2010, Martin et al. 2013a,b, Pagnussat et al. 2007). This hypothesis has already been demonstrated by Pagnussat et al. (2009) by disrupting the auxin gradient through the overexpression of the auxin biosynthetic gene *YUC1* in the embryo sac. As expected, the generated embryo sacs displayed high concentration of auxin, which in turn induced a shift to a synergid cell fate in most of the cells within the embryo sac, including those normally specified as egg cell, central cell and antipodals. However, these elevated auxin concentrations could not uniformly specify all the cells for synergid cell fate. These findings suggest that other co-existing cytoplasm domains, likely the CCD, may antagonistically function with the auxin domain to regulate cell fate inside of the embryo sac.

In summary, the auxin cytoplasm domain performs an important role in determining micropylar cell fate in embryo sac development but not for the polarization of the embryo sac by means of affecting nucleus positioning. Consequently, other cytoplasm domains, may exert a more important function in determining embryo sac polarity in its preparation for asymmetric cell division. In the next section, the ROS domain (Martin et al. 2013a,b) is presented and its role on regulating embryo sac polarity as well as on completing the function of the auxin domain in determination of the central cell fate is discussed.

The reactive oxygen species domain affects central cell fate and polarity of the embryo sac

During embryo sac development, reactive oxygen species (ROS) are produced mainly from mitochondria in the forms of superoxide and peroxide. Interestingly, ROS and, in general, mitochondrial metabolism exerts an important function in regulating cell fate and embryo sac polarity (Martin et al. 2013a,b, 2014, Wu et al. 2013, Leon et al. 2007). An important question is how the plant exploits mitochondrial metabolism, specifically ROS molecules, as a way to control embryo sac development. Logical answers can be found through the studies of mitochondrial metabolic mutants. These studies indicate the possibility that some developmental regulators require alteration of metabolic processes to perform their roles (Leon et al. 2007). However, recent evidence has attributed ROS compounds a role as signaling molecules, which may affect more effectively in time and space gamete development without altering the overall functioning of the embryo sac (Martin et al. 2013b). This function is also in line with the specific pattern of mitochondria distribution and, consequently, the oxidative bursts that occur within the embryo sac throughout development. Elegant experiments with florescent probes showed indeed that mitochondria, although detected in the whole embryo sac, are specifically concentrated in a large number in the central zone of the developing embryo sac and in the central cell after cellularization (Martin et al. 2013b, 2014, Kägi et al. 2010). Such population of mitochondria is visualized as 'mitochondrial clouds' (Martin et al. 2014) which are evident as physical representations of the oxidative burst or the ROS cytoplasm domain. Giving its localization, it is possible that the ROS domain inherently confers central cell fate attributes to the nucleus located in the central zone. The truthfulness of this assumption was demonstrated through the *oiwa* mutant which is defective in the MSD1 enzyme, a mitochondrial Mn-superoxide dismutase that alleviates oxidative toxicity by the dismutation of superoxide to molecular oxygen and peroxide (Martin et al. 2013b, Kliebenstein et al. 1998). In wild type embryo sacs, the expression of the *MSD1* gene is confined to the egg apparatus, notably in the egg cell, at the time of cellularization to mitigate oxidative burst and keep egg cell identity. Therefore, mutation in the *MSD1* gene could categorically demonstrate the function of the oxidative burst by specifying central cell fate in egg cells. As expected in this mutant, the increased oxidative burst in the egg cell apparatus upon deficiency of MSD1 induced a shift on cell fate of the egg cell and synergids to a central cell identity, confirming in this manner the role of the ROS domain in central cell specification (Martin et al. 2013b). Furthermore, the misspecification of the egg apparatus in the *iowa* mutant was also related to abnormal migration of nuclei before cellularization of the embryo sac when the oxidative burst is visualized in the central domain of the embryo sac. At this stage, mutant embryo sacs displayed two nuclei instead of one at the region in which the egg cell nucleus should be specified (Martin et al. 2013b, 2014). The atypical positioning of the nuclei resulted in an abnormal read-out of auxin signals in such nuclei along the auxin gradient, which was not disturbed by the deficiency on the MSD1 enzyme as revealed by the activity of the auxin reporter DR5:GFP (Martin et al. 2013b, Pagnussat et al. 2009). This finding suggests that although the ROS domain exerts central cell specification, it still requires the assistance of the auxin cytoplasm domain to accomplish sufficiently this role. Additionally, the ROS domain shows that it is capable of regulating organelle transport, at least the nuclei, within the embryo sac. Moreover such nuclei transport may not be mediated by auxin, as the auxin cytoplasm domain *per se* is not competent in organelle motion (Martin et al. 2013b, 2014, Pagnussat et al. 2009). The mechanism by which ROS steers nuclei movement is still unknown, but many reports have showed that ROS in cells can affect the organization of microtubule, a cytoskeletal element that drives organelle transport by the function of their cargo mobile proteins (Livanos et al. 2012a,b, 2014a,b). ROS are known to stimulate calcium ion uptake in cells, which in turn activates many regulatory proteins of microtubule formation, including cyclin-dependent kinases, mitogen-activated proteins and aurora kinases. Even though more research work is required (Martin et al. 2013b), it is likely that the ROS domain may regulate the mitotic progression within the syncytial embryo sac by means of altering cytoskeleton organization. This hypothesis may be demonstrated in the near future in studies of the cytoskeleton dynamics by using the *oiwa* and other mitochondrial mutants, for instance the *fiona*, *gamete cell defected* and *aac2* mutants. These mutants exhibit impairment of mitotic division and, consequently, arresting embryo sac development (Martin et al. 2013b, Wu et al. 2012, Kägi et al. 2010, Hauser et al. 2006).

The position of various cytoplasm domains during embryo sac development is illustrated in Figure. 1. In summary, the auxin and cytokinin cytoplasm domains play an essential role in establishing a polarized micropylar-chalaza axis and on determining micropylar and chalazal cell fate, respectively. The establishment of the poles progresses gradually throughout gametophyte development generating cell polarity. The chalazal pole is set up as early as megasporogenesis through the action of the CCD that induces the polarization of the cytoplasm in the MMC and during megaspore selection to allow the survival of the chalazal-most megaspore. During gametophyte development, the CCD regulates chalazal fate. Immediately after ovule primordium formation, the auxin pole is also established at the micropylar pole by the function of the auxin cytoplasm domain. However, it appears that the CCD assists the auxin cytoplasm domain in the establishment of auxin maxima region by regulating the position of PIN1. Later, the auxin pole promotes the formation of an auxin gradient to exert micropylar cell fate determination, according to the auxin read-outs by nuclei superpositioned along the gradient. Nonetheless, it seems that the auxin and cytokinin cytoplasm domains act antagonistically along the gradient to confer cell specification of the embryo sac, with the micropylar fate determined by the auxin pole whilst the chalazal cell fate is defined by the cytokinin pole. Additionally, the late-formed ROS domain confers central cell fate. Therefore, at the end of embryo sac development, a highly polarized embryo sac is built with two synergids and one egg cell localized in the micropylar pole whilst the chalazal pole hosts three antipodals. Moreover, individual cells are also polarized at the subcellular levels, notably the egg cell which is the precursor of the zygote after fertilization. The series of cell specification events fulfill the essential function of loading gene programs to an egg cell necessary to become a zygote and later regulates its development.

Asymmetric Cell Division in the Zygote is Primarily Regulated by Ca^{2+} and Auxin

In flowering plants, the zygote is the ultimate structure in reproduction since it develops into the embryo, the precursor of a new sporophyte. Therefore, it not surprising that the zygote follows pre-determined and newly programmed events to ensure successes in its development into a plant. The most notable feature is that the first cell division of a zygote is asymmetric, resulting in two cells with distinct cell fates. How does the zygote structure cytologically and genetically ensuring that the conditions for the first asymmetric cell division are met? As presented in previous sections,

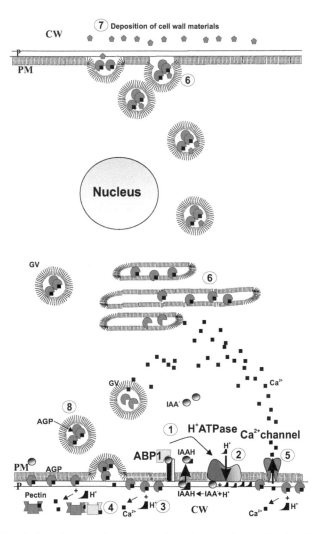

FIGURE 2   Regulation of zygote elongation by the Ca^{2+}-dependent mechanism. First, the auxin receptor ABP1 localizes at the plasma membrane (PM) of the micropylar pole. Upon auxin IAA binding, the receptor ABP1 generates auxin responses, including the phosphorylation of the $H^+ATPase$ (1). Later, the activated $H^+ATPase$ exports H^+ protons (black rectangle with meniscus) to the periplasm (P) (2). Once in the periplasm, H^+ protons promotes the release of Ca^{2+} (black squares) from the arabinogalactan proteins (AGPs; grey pies) polarized at the periplasm of micropylar end (3). The free Ca^{2+} promotes pectin polymerization and cell wall (CW) rigidity increases at the micropylar end of the zygote (4). This localized cell wall rigidity stops cell growth at the micropylar end and promotes elongation of the zygote at the chalazal end through turgor pressure. The increased tension at the micropylar region of the cell is also sensed by the mechanosensory Ca^{2+} channels resulting in its activation. Once functional, the Ca^{2+} channels promote the influx of this ion into the cytosol of the zygote (5). The Ca^{2+} is then sequestered by AGPs localized in the Golgi complex, notably Golgi vesicles (GV) (6); exocytotic deposition of membranes and cell wall materials (pentagons) occur primarily through the plasma membrane (PM) at the chalazal pole (7). This cycling of materials is reinitiated by the deposition of AGPs containing Ca^{2+} at the micropylar pole (8). Adapted from Lamport and Varnai (2013) and Chen et al. (2010).

a perfectly polarized and orderly cell fate-determined embryo sac is essential for achieving fertilization of the ovule and formation of the zygote (Pagnussat et al. 2009). However, the current body of knowledge has also indicated that the zygote undergoes changes in gene expression patterns as well as cytological features immediately after fertilization. Consequently, the maternal cues provided for the egg cell cannot account for a total regulation of an asymmetric division in the zygote (Musielak and Bayer 2014, Ueda and Laux 2012, Xin et al. 2012). Although the egg cell is structurally polarized after fertilization (Zhang and Laux 2011), the zygote enters into a morphogenetic event called 'egg cell activation' in which the original structure of the egg cell is altered through changes in number and shapes of mitochondria, ribosomes and Golgi vesicles. More strikingly, many reports have shown that upon fertilization, the zygote adopts a transient change in symmetry. The nucleus becomes more centrally located and small vacuoles, derived from the fragmentation of the inherited egg vacuole, spread throughout the cell (Ueda et al. 2011, Faure et al. 2002). Soon after, the zygote elongates, repolarization of the cytoplasm occurs with the relocation of the nucleus to the apical end of the cell and the vacuoles regroup into one large vacuole located at the basal end, near the micropyle. At that time, the zygote can experience a 3-fold increase in length. The change in shape of the zygote parallels changes in microtubule orientation. Microtubules shift from a random orientation in the young zygote into a transverse orientation prior to the asymmetric division. A regulatory mechanism which drives changes in the zygote leading to an asymmetric division is presented. We gathered and analyzed available evidence to formulate a functional model and proposed that Ca^{2+} is the major regulator of such a mechanism. The essential features of the model are summarized in Fig. 2.

The calcium ion dynamics within an embryo sac at the time of fertilization

Successes in fertilization, zygote formation and development depend on the correct specification of cells inside the embryo sac (Susaki et al. 2015). During fertilization, the correct specification of the synergids is critical as they guide the pollen tube to them and later facilitate the fusion of gametes. Initially, the transcription factor *MYB98*, a R2R3-type MYB protein, accumulates specifically in the synergids and modulates pollen tube guidance by an unknown molecular route (Kasahara et al. 2005). However, the egg cell can regulate pollen tube growth as it dictates the synergids to acquire the pollen attraction function. In this event, the egg cell induces the expression of AtLURE genes in the synergids to produce peptides which function as pollen tube attractants (Takeuchi & Hiyashiyama 2012). These pollen attractants are then secreted by the synergids to induce an increase in calcium (Ca^{2+}) in the pollen tube tips and sperm cells. The accumulation of the attractants serves as a signal to guide the pollen tube towards the synergids. Surprisingly no change in the cytoplasmic Ca^{2+} ($[Ca^{2+}]$cyt) occurs in the synergids during this event. However, upon pollen tube arrival at the synergids, the $[Ca^{2+}]$cyt increases at the micropylar pole and spreads to the chalazal pole (Iwano et al. 2012). This high $[Ca^{2+}]$cyt concentration may represent a threshold to activate the FERONIA/SIRENE (FER) receptor kinase which is localized exclusively in the filiform apparatus/synergids region and afterward interacts with ROPGEFs, guanine nucleotide exchange factors that activate RAC/ROP GTPases (plant Rho GTPases). The FER functions as a cell surface regulator in the FER-ROPGEF-RAC/ROP complex to stimulate nicotinamide adenine dinucleotide phosphate (NADPH) oxidase to produce reactive oxygen species (ROS). In this manner, Ca^{2+} promotes high ROS levels in the filiform apparatus area through the FER-ROPGEF-RAC/ROP complex to induce pollen tube rupture and sperm cell release (Duan et al. 2014). Subsequently, a transient maximum of $[Ca^{2+}]$ cyt in the synergids is reached upon pollen tube rupture which results in the programed cell death of the receptive synergid, most probably due to the increased levels of ROS (Bleckmann et al. 2014, Duan et al. 2014, Iwano et al. 2012). Thus, the fusion of the egg cell with a sperm cell is enabled by a high Ca^{2+} concentration environment, originated from the burst of the pollen tube and the death of the receptive synergid (Iwano et al. 2012). In fact, the egg cell experiences two transient increases in $[Ca^{2+}]$ cyt: the first sharp increase coincides with the rupture of the pollen tube and synergid cell death and this makes the egg cell competent to fusion. A second $[Ca^{2+}]$cyt peak occurs during fertilization, at the fusion site and then spreads over the entire cell as a wave-front. In the later increase, Ca^{2+} activates bundling proteins to mediate the formation of an extension of the filiform apparatus, a F-actin based cytoskeleton network, which directs the transport of the sperm cell nucleus to the egg cell nucleus (Chen et al. 2015, Kawashima et al. 2014a,b). Such transport of nuclei is assisted by myosins, calcium-activated motor proteins, which direct the nuclei through actin filaments during karyogamy, and thus achieves the fusion of gametes and the formation of the zygote (Chen et al. 2015, Kawashima et al. 2014a,b, Antoine et al. 2001). In fact, the F-actin based cytoskeleton connects both egg cell and central cell in order to achieve a synchronized fertilization and this example illustrates the importance of proper positioning of these cells in zygote and endosperm formation (Kawashima et al. 2014b). The varied Ca^{2+} concentrations within the embryo sac predispose the fertilized egg cell to changes that follow.

A calcium-dependent mechanism which regulates zygote polarity and its elongation

The newly formed zygote is governed by two sequential molecular systems, the maternally controlled system at an early developmental stage and later, a zygotic genome activation system. The latter system is ultimately responsible for the asymmetric division of the zygote (Xin et al. 2012). Both systems are highly dependent on the action of Ca^{2+}. The effects of Ca^{2+} can be seen immediately after egg cell and sperm fusion, as the increased level of $[Ca^{2+}]$cyt induces cell wall formation on the naked fertilized egg cell. This is most likely through exocytosis of cell wall materials after the activation of the maternally-derived annexin p35 by Ca^{2+} (Okamoto et al. 2004, Antoine et al. 2001). However, tracking $[Ca^{2+}]$cyt after fertilization in the zygote becomes more challenging as Ca^{2+} can be stored and retrieved dynamically in many organelles and calcium binding proteins, notably arabinogalactan proteins (AGP) which also coincidently regulate cell polarity, vesicle transport, cytoskeleton dynamics and cell expansion (Lamport and Varnai 2013, Qin and Zhao 2006, Shi et al. 2003).

In fact, recent evidence indicates that AGPs are the main regulators of zygote polarity and asymmetric cell division by controlling Ca^{2+} dynamics within the zygote. Such a function of AGPs is supported by the recent finding that AGPs work as a calcium capacitor (i.e. a source that can release and retrieve Ca^{2+} in a pH dependent fashion) in the periplasm and cell walls; releasing Ca^{2+} from its glycan motif to the cell cytoplasm (Lamport and Varnai 2013). However, it is likely that AGPs are embedded within a mechanism and are required to function in a concerted manner with other components, including elements of auxin signaling, the plasma membrane H^+-ATPase, and Ca^{2+} channels, which are reported to regulate zygote development (Chen et al. 2015, Chen et al. 2010, 2012, Yu and Zhao 2012, Qin and Zhao 2006, Antoine et al. 2000). Moreover, the action of these components must be timely and spatially determined inside the zygote. If this is the case, the zygote must undertakes many events immediately after fertilization to proper localize all these calcium-interacting factors in a specific region, for instance the micropylar end. In line with such an idea, many reports have shown that the AGPs and H^+-ATPase are strongly and evenly localized in the recently fertilized egg cell (i.e. globular zygote) but later they become polarized to the micropylar end by an unknown mechanism (Chen et al. 2015, 2012, 2010, Qin and Zhao 2006).

Similarly, the auxin receptor ABP1s also polarize in the cell membrane at the micropylar region although they are still active in other regions of the zygote (Chen et al. 2010). One can easily argue for the nonexistence of an auxin gradient in such specific region due to the broad activity of ABP1 in the zygote. However, disruption of IAA distribution with the IAA transport inhibitor, triiodobenzoic acid led to the delay of zygote asymmetric division and caused subsequently morphological aberrations of embryos (Chen et al. 2010, 2012). These findings confirm that an auxin gradient is formed at the micropylar end of the elongated zygote. This gradient depends mainly on the localization of the ABP1 receptor, which may initiate a signaling cascade that switches on auxin-dependent gene expression. The first ABP1-mediated event could be the activation of H^+-ATPase at this same location. Upon auxin binding, the ABP1 receptor promotes the activation of genes encoding regulatory proteins for H^+-ATPase phosphorylation. A potential candidate for such function would be the gene encoding a *SMALL AUXIN UP RNA* (SAUR) protein, which promotes the phosphorylation of C-terminal auto-inhibitory domain of the H^+-ATPase (Hohm et al. 2014). Once activated, the H^+-ATPase exports H^+ protons to the periplasm to decrease the pH and promote Ca^{2+} release from the Ca^{2+} compartments of AGPs, which are also abundantly present in the periplasm (Lamport and Varnai 2013, Qin and Zhao 2006, Shi et al. 2003).

As a result, a vast amount of Ca^{2+} released from AGPs in the periplasm, diffuses to the vicinity of the cell wall at the micropylar end in order to increase cross-linking of pectin chains. This cross-linking increases cell wall rigidity and favors cell elongation in the opposite direction to AGP localization (Lamport et al. 2014, Lamport and Varnai 2013, Pickard 2013, Ding and Pickard 1993). Subsequently, the localized cell wall rigidity and increased growth at the chalazal end generate oscillations in cellular tensions which are transmitted to the membrane at the micropylar pole to activate Ca^{2+} mechanosensory channels. Once functional, these Ca^{2+} channels drive the influx of this ion into the zygote cytoplasm to be subsequently sequestered by AGPs localized in the Golgi complex, notably Golgi vesicles (Lamport et al. 2014, Lamport and Varnai 2013). In the Golgi compartment, Ca^{2+} may activate a collection of enzymes which coordinate Golgi secretion, vesicle transport and exocytosis of membrane and cell wall materials towards the chalazal end to favor polarized growth of the zygote. This cycling of materials is reinitiated by the deposition of AGPs containing Ca^{2+} at the micropylar pole through Golgi vesicle transport as well (Lamport et al. 2014, Lamport and Varnai 2013, Qin and Zhao 2006). .

Therefore, given four of its components, AGPs, auxin, H$^+$-ATPase and mechano-sensory Ca^{2+} channels, are found to have a dynamic distribution in the zygote, we argue favorably for the existence of a Ca^{2+}-dependent mechanism responsible for the regulation of zygote development. The general model is shown in Figure 2. The components of this mechanism can play active roles in regulating zygote elongation and its first asymmetric division (Chen et al. 2010, 2012, 2015, Yu and Zhao 2012, Qin and Zhao 2006, Antoine et al. 2000). Moreover, recent studies focus mainly on the importance of AGPs in zygote elongation. These reports are convincing in that zygote elongation occurs only after polarization of AGPs with a high concentration at the micropylar end and little to no deposition at the chalazal end of the zygote (Chen et al. 2010, 2012, Yu and Zhao 2012, Qin and Zhao 2006). The high level of AGPs at the micropylar pole might be important to restrict cell expansion by increasing Ca^{2+}-mediated cell wall rigidity, while a depletion favors cell elongation at the opposite end of the zygote. However, it has been more striking to observe that zygote elongation is concurrent to the migration of the nucleus towards its chalazal pole (Qin and Zhao 2006). Whether AGPs affect the formation of the cytoskeleton to establish and stabilize the polarity of the zygote remain to be investigated. Additionally, more efforts are needed to demonstrate the functional integration of other components of the Ca^{2+}-dependent mechanism with the action AGPs in the polarized growth of the zygote. Furthermore, the action of other regulators of osmotic balance which induce increases in cell turgor for promoting zygote elongation also needs to be addressed. It is of paramount importance to demonstrate the existence of K$^+$ channels in the zygote which are known for facilitating H$^+$/ K$^+$ symport and promoting K$^+$ uptake and its accumulation in vacuoles to increase cell turgor leading to zygote elongation (Rigas et al. 2001).

Zygote asymmetric cell division is a calcium-dependent event modulated by AGPs

The Ca^{2+}-dependent mechanism underlying cell polarity and elongation in plants represents an essential strategy for achieving symmetry breaking leading to an asymmetric cell division. Such a mechanism seems to be conserved in many organisms and algal species have been considered as model systems for its deciphering (for details, see Bothwell et al. 2008, Homble and Leonetti 2007). In algal zygotes, Ca^{2+} establishes an ion pole immediately after fertilization and two Ca^{2+} transcellular current loops in parallel directions around the zygote soon appear. Both the Ca^{2+} pole and the transcellular ion currents promote an outgrowth of the zygote at this developmental pole to break the initial spherical symmetry and create a single axis allowing the generation of a polarized zygote (Bothwell et al. 2008, Homble and Leonetti 2007). Soon after, the Ca^{2+} pole fixes an F-actin patch which later marks the formation of a zone of microtubules. This zone serves to guide and localize the division plane and coordinate the formation of the cell plate, leading to an asymmetric division (Homble and Leonetti 2007, Bisgrove et al. 2003). In higher plants, similar series of events to those of the algae leading to the establishment of the division plane can be found. However, the ontogeny becomes more complex due to the fact that plants build distinct structures, in a timely fashion to direct the asymmetric division of the zygote (Yu and Zhao 2012).

Unlike algae, plants form an extra transient band of microtubules and microfilaments, the preprophase band (PPB), which encircles the nucleus near the cell membrane before mitosis. The PPB disappears immediately after the nuclear envelope breaks, but the cortical actin filaments persist outside of the PPB. Such degradation generates an actin-depleted zone at the previous PPB position which continues to exist throughout cytokinesis and serves as a positional clue to define the position of subsequent cell division related structures, i.e. mitotic spindle, phragmoplast and cell plate (Reddy 2001). Indeed, PPB is critical in orienting a bipolar mitotic spindle as microtubules in plants are organized in the absence of a centrosome, an organelle of algae and animals, constituted of centrioles and an amorphous mass of proteins which are responsible for microtubule organization through its nucleation and anchoring to the centrioles (Bisgrove et al. 2003). Thus, it seems that the PPB leaves positional signals around the nucleus to define the equatorial plane of the future mitotic spindle when dividing cells enter the G2 phase of the cell cycle after completing DNA replication (Bannigan et al. 2008). Later during anaphase and telophase, the phragmoplast is formed from the mitotic spindle conserving its equatorial position. The phragmoplast is an array of microtubules, microfilaments and endoplasmic reticulum elements that assemble in two opposite sets perpendicular to the division plane. This structure resembles a barrel with the plus ends of the microtubules and microfilaments towards the equatorial plane (Rasmussen et al. 2011a,b, Muller et al. 2009). The phragmoplast is implicated in the transport of vesicles and other cell wall materials to the cell plate through the function of microtubule-based motor proteins as demonstrated by pharmacological approaches (Reddy 2001). Therefore deciphering the mechanisms governing PPB formation and degradation is indispensable for understanding the molecular mechanisms underlying asymmetric cell division of the zygote. This task is challenging due to the complexity of processes regulating the formation of the cell wall-membrane-cytoskeleton system and the practical limitations in working with a microscopic zygote embedded in ovule tissues. However, reverse genetics has been the single approach that yielded positive outcomes by discovering two genes *TONNEAU1* (*TON1*) and *FASS/TONNEAU2* (*FASS/TON2*), whose disruption lead to a complete abrogation of PPB formation (Azimzadeh et al. 2008, Torres-Ruiz and Jürgens 1994). The *ton1* and *fass* mutants display identical phenotype and indicate that the first asymmetric division of the zygote is abnormal as the newly formed cell wall is oblique instead of being horizontal as in the wild type. Subsequently, cell divisions are also atypical since they exhibit random positions of cell division plane and abnormal cell elongation. Such a cell division pattern is accompanied by alterations in the interphase microtubule arrays which lack the normal parallel organization as the wild type cells. Such an abnormal array of microtubules leads to abnormal embryogenesis. The derived plants are dwarfed, revealing the biological significance of the first asymmetric division of the zygote in plant development (Spinner et al. 2013, Azimzadeh et al. 2008, Torres-Ruiz and Jürgens 1994).

Consequently, the definition of the function of the TON1 and FASS proteins in the formation and positioning of the PPB may represent a rational avenue to dissect the mechanism that regulates the asymmetric division of the zygote. A sequel of works showed that the *TON1* gene encodes a homolog protein to the animal centrosomal proteins, FOB and OFD1 and interacts in its N-terminal domain with the Arabidopsis CENTRIN, which in animal cells is the major component of the microtubule organizing center (MTOC) or centrosome (Spinner et al. 2013, 2010, Azimzadeh et al. 2008). The other gene *FASS* encodes the "B" subunit of protein phosphatase 2A which in turn interacts with the TON1 protein (Spinner et al. 2013, Camilleri et al. 2002, Torres-Ruiz and Jürgens 1994). More interestingly, it has been recently demonstrated that both TON1 and *FASS* are integrated into a phosphatase 2A complex (TTP complex), with FASS as regulatory subunit that is recruited to the cortical microtubules via the TONNEAU1-recruiting motif (TRM) family of proteins (Spinner et al. 2013). Such TTP complexes are of particular importance for understanding PPB formation since they are the primary structures in cytoskeleton formation as they function as nucleation sites of tubulin units, which occur after these bodies recruit γ-tubulin complexes (Drevensek et al. 2012, Kirik et al. 2012). Thus, giving their functional characteristics, the TTP complexes may be regarded as animal MTOCs. However, unlike animals in which MTOCs are centralized structures, the plant MTOCs are entities dispersed throughout the zygote, notably in the cell cortex where they can regulate PPB formation (Muller et al. 2009). It is precisely in the cell cortex that the TTP complexes assist by restricting the spreading of interphase microtubules and localizing them to the future cell division plane or the PPB zone. In this process, TTP complexes through a dephosphorylation-dependent mechanism selectively nucleate and stabilize microtubules at the zone of the PPB using non-PPB cortical microtubules derived from a process of selective depolymerization, probably mediated by phosphorylated kinesins (Walczak et al. 2013, Ambrose et al. 2008, 2011, Muller et al. 2009, Wright et al. 2009). Other enzymes, like CLASPs may enforce the attachment of microtubules to the cell cortex of PPB zone and induce PPB narrowing via modulation of microtubule dynamics in the PPB zone and/or by mediating microtubule-cortex interactions (Ambrose et al. 2008, 2011, Muller et al. 2009). However, the function of these enzymes in plant zygotes still needs to be demonstrated. Furthermore, there is a need to discern the mechanism by which PPB guide the formation of the mitotic spindle and phragmoplast. Some clues become available from the discovery of the TANGLED (TAN) protein which colocalizes with the PPB and persists at the cell division site as cells proceed through metaphase and cytokinesis (Walker et al. 2007). TAN protein is not required for formation of the PPB but rather it provides precise information on how the PPB determines positioning of the phragmoplast and cell plate, especially since evidence has shown that FASS is required for proper localization of TAN in the PPB (Rasmussen et al. 2011a,b, Walker et al. 2007). Whether a similar mechanism works during asymmetric cell division of the zygote needs to be addressed in the near future.

Despite good progress in our understanding of the molecular basis of PPB formation, the deciphering of molecular mechanisms that govern the positioning of PPB in the zygote is still obscure. Histological studies have shown that the position of the nucleus influences the placement of the PPB in some plant cells (Muller et al. 2009, Van Damme 2009) and this can also be true in positioning of the PPB in the zygote (Qin and Zhao 2006). After zygote elongation in tobacco, the nucleus migrates to the apical region of the cell where the future PPB is positioned. Coincidently AGPs are also abundant near the zygote nucleus, as indicated by a fluorescence labelled anti-AGP antibody. More importantly, treatments with an inhibitor of AGP activities, a synthetic phenylglycoside (bGlcY) that specifically binds AGPs, an increase in atypical symmetric divisions is observed, suggesting a strong role of AGPs in the positioning of PPB and cell plate (Qin and Zhao 2006). If this pattern of AGP distribution occurs in a similar manner in *Arabidopsis*, a potential explanation about the positioning of the PPB in zygote asymmetric cell division can be generated, based on an interaction of AGPs with TTP complexes. In expanding this idea, TTP complexes are constituted by Ca^{2+}-responsive proteins, for instance CENTRINs, therefore they can interact with Ca^{2+} compartments of AGPs to initiate microtubule nucleation and consequently, PPB formation within the cytoplasm. So far, the function of AGPs in asymmetric division and zygote development has been demonstrated (Geshi et al. 2013, Hu et al. 2006, Shi et al. 2003). Therefore, PPB positioning regulated by an interaction AGP-TTP complex is possible.

In summary, we have proposed a Ca^{2+}-dependent mechanism underlying zygote polarity and elongation as well as PPB positioning, which ultimately regulates the asymmetric cell division. This proposed mechanism will be more significant if future research can establish a cross-talk between Ca^{2+}-dependent mechanisms with classical routes, for instance the *WOX* pathway and the *YODA* MAPK phosphorylation cascade, which can also regulate zygote asymmetric cell division. The latter are briefly presented below.

WOX and *YODA* may cross-talk with the Ca^{2+}-dependent mechanisms to regulate zygote asymmetric cell division

The *WOX* and *YODA* pathways contain two groups of proteins which regulate elongation and asymmetric division of the zygote as well as apical and basal cell fate lineages (extensively reviewed by Musielak and Bayer 2014, Zhang and Laux 2011). The WOX (*WUSCHEL-RELATED HOMEOBOX*) pathway is constituted by proteins WOX2, WOX8 and WOX9 which are homeodomain transcription factors that interact in an ordered fashion for functioning in the zygote. In the first segment of the route, WOX8 together with the redundant WOX9 are activated by a zinc-finger transcription factor WRKY2 to regulate polar organelle localization and asymmetric division of the zygote. Genetic studies have demonstrated the function and interaction of these transcription factors during asymmetric zygote division. In such studies, zygotes of the *wrky2* mutants failed to establish polar positioning of organelles that resulted in an equal cell division and distorted embryo development. The downstream position of WOX8 in this gene network has been confirmed by rescuing the *wrky2* defects through the overexpression of WOX8 (Ueda et al. 2011). In the second segment of the WOX route, both WOX8 and WOX9 activate WOX2 to generate a proper balance between the basal regulators WOX8/9 with the apical regulator WOX2 that coordinates the asymmetric zygote division (Breuninger et al. 2008).

The entire collection of components of the YODA signaling cascade has not been completely elucidated since it acts in various plant structures and distinct molecular players are involved in a giving modality of the system (Musielak and Bayer 2014). In the zygote, YODA [YODA MITOGEN ACTIVATED PROTEIN KINASE KINASE KINASE 4 (MAPKKK4)] acts upstream of MITOGEN ACTIVATED PROTEIN KINASE 3 (MPK3) and MPK6 to regulate redundantly the elongation of the zygote. Consequently, mutant zygotes of any of these kinases fail to elongate, resulting in an almost symmetric division of the zygote (Ueda and Laux 2012, Jeong et al. 2011a,b, Bayer et al. 2009, Lukowitz et al. 2004).

At present, how the WOX pathway and the YODA signaling cascade interact with the Ca^{2+}-dependent mechanism, underlying the elongation of the zygote and formation of the PPB is not known. Mitogen activated protein kinases are known for activating H^+-ATPase (Zhang et al. 2013, Lew et al. 2006). One of the components of the YODA route may play a role in generating the H^+-ATPase-mediated Ca^{2+} pole during zygote elongation. Moreover, there is a possibility that the YODA pathway functions in PPB formation for the reason that MPK6 is localized in the PPB and phragmoplasts. Hence, it can be involved in the positioning of the cell division plane through the regulation of TANGLED1 (TAN1), PHRAGMOPLAST ORIENTING KINESIN 1 (POK1) and other cytoskeleton regulators (Smekalova et al. 2014).

Conclusions and Perspectives

In this review, we emphasized the implications of proper specification and polarity of the embryo sac components, notably the egg cell, synergids and central cell, in the formation, development and asymmetric division of the zygote. However, there are still many unanswered questions concerning the genetic control of the embryo sac in zygote development since the latter is housed within it. For example, auxin responses are observed immediately after fertilization (Chen et al. 2010), therefore it is likely that the majority of components of the auxin responses have an egg cell origin, instead of being *de novoly* synthesized. Although some auxin-dependent molecular players have been localized in the zygote, the deciphering of new ones as well as their function in zygote development should be addressed. More importantly, it will be very beneficial to determine a potential role of auxin-dependent molecular factors in the regulation of cytoskeleton dynamics during asymmetric cell division (Zhang et al. 2011). An additional related research avenue would be represented by studies on the effects of PIN1 and GNOM proteins in the positioning of the cell division plane (Rasmussen et al. 2011a,b). The ROS domain exhibits an import function of central cell specification; however, it may exert effects beyond embryo sac development, such as zygote development. In line of this assumption, it would be useful to identify novel NADP(H) oxidases which may assist the Ca^{2+}-dependent mechanisms in regulating zygote elongation and asymmetric cell division, as reported in a similar mechanism in pollen grains (Wudick and Feijo 2014). Undoubtedly, all investigations would promote the discovery of new genes that are intrinsic to the egg cell but may have a direct control in asymmetric cell division of the zygote.

We also document that essential molecular clues are recognized by a Ca^{2+}-dependent mechanism which modulates egg cell fertilization and regulates zygote elongation as well as asymmetric cell division. In this process, the Ca^{2+}-dependent mechanism is potentially assisted by the *WOX* and *YODA* pathways. The importance of AGPs is also emphasized. Specifically regarding the asymmetric division, the Ca^{2+}-dependent mechanism gathers most of the regulators involved in PPB formation. Further studies will lead us to understand the mechanism underlying the positioning of the cell plate during the asymmetric cell division of the zygote. For instance, it would be of paramount importance to determine if AGPs are positioned in the PPB and whether they interact with TON1 and the Ca^{2+}-sensitive CENTRIN proteins to direct PPB formation and positioning within the zygote during asymmetric division (Spinner et al. 2013, 2010, Azimzadeh et al. 2008). Furthermore, giving the reported role of Ca^{2+} in microfilament organization and its potential interaction with the microtubule associated proteins, it appears that such ion is a key signaling molecule involves in cytoskeleton dynamics

(Chen et al. 2015, Spinner et al. 2013, Kawashima et al. 2014a,b, Antoine et al. 2001). Therefore it would be extremely beneficial to develop novel non-destructive techniques to study and detect the distribution of trace amounts of this ion in the diminutive structure of the zygote (Antoine et al. 2000). These techniques would be very valuable to define the function of Ca^{2+} in cytoskeleton dynamics as well as the discovery of novel Ca^{2+}-dependent proteins associated with microtubule nucleation and microfilament polymerization (Spinner et al. 2013). A more sensitive Ca^{2+} tracking would also assists in a more precise characterization of mechanosensory Ca^{2+} channels to better define their function, not only in zygotic asymmetric division but also in zygote elongation, which is also an essential event for establishment of the plane of cell division (Lamport et al. 2014, Ueda and Laux 2012). Finally, it can be envisioned that the advent of the emerging techniques in cell biology, gene expression analysis and proteomics would make possible the deciphering of the molecular mechanisms underlying asymmetric cell division of the zygote in higher plants in the near future.

References

Ambrose, C., J.F. Allard, E.N. Cytrynbaum and G.O. Wasteneys. 2011. A CLASP-modulated cell edge barrier mechanism drives cell-wide cortical microtubule organization in Arabidopsis. Nature Commun. 2: 430.

Ambrose, J.C. and G.O. Wasteneys. 2008. CLASP modulates microtubule-cortex interaction during self-organization of acentrosomal microtubules, Mol. Biol Cell 19: 4730-4737.

Antoine, A.F., J.E. Faure, S. Cordeiro, C. Dumas, M. Rougier and J.A. Feijo. 2000. A calcium influx is triggered and propagates in the zygote as a wavefront during *in vitro* fertilization of flowering plants. Proc. Natl. Acad. Sci. USA. 97: 10643-10648.

Antoine, A.F., J.E. Faure, C. Dumas and J.A. Feijo. 2001. Differential contribution of cytoplasmic Ca^{2+} and Ca^{2+} influx to gamete fusion and egg activation in maize. Nat. Cell Biol. 3: 1120-1123, 2001.

Azimzadeh, J., P. Nacry, A. Christodoulidou, S. Drevensek, C. Camilleri, N. Amiour, F. Parcy, M. Pastuglia and D. Bouchez. 2008. *Arabidopsis* TONNEAU1 proteins are essential for preprophase band formation and interact with centrin. Plant Cell 20: 2146-2159.

Balasubramanian, S. and K. Schneitz. 2000. NOZZLE regulates proximal-distal pattern formation, cell proliferation and early sporogenesis during ovule development in *Arabidopsis thaliana*. Development 127: 4227-4238.

Balasubramanian, S. and K. Schneitz. 2002. NOZZLE links proximal-distal and adaxial-abaxial pattern formation during ovule development in *Arabidopsis thaliana*. Development 129: 4291-4300.

Bannigan, A., M. Lizotte-Waniewski, M. Riley and T.I. Baskin. 2008. Emerging molecular mechanisms that power and regulate the anastral mitotic spindle of flowering plants. Cell Motil. Cytoskeleton 65: 1-11.

Bayer, M., T. Nawy, C. Giglione, M. Galli, T. Meinnel and W. Lukowitz. 2009. Paternal control of embryonic patterning in *Arabidopsis thaliana*. Science 323: 1485-1488.

Bencivenga, S., L. Colombo and S. Masiero. 2011. Cross talk between the sporophyte and the megagametophyte during ovule development. Sex Plant Reprod. 24: 113-121.

Bencivenga, S., S. Simonini, E. Benkova and L. Colombo. 2012. The transcription factors BEL1 and SPL are required for cytokinin and auxin signaling during ovule development in *Arabidopsis*. Plant Cell 24: 2886-2897.

Bisgrove, S.R., D.C. Henderson and D.L. Kropf. 2003. Asymmetric division in fucoid zygotes is positioned by telophase nuclei. Plant Cell 15: 854-862.

Bleckmann, A., S. Alter and T. Dresselhaus. 2014. The beginning of a seed: regulatory mechanisms of double fertilization. Front Plant Sci. 5: 452.

Bothwell, J.H., J. Kisielewska, M.J. Genner, M.R. McAinsh and C. Brownlee. 2008. Ca^{2+} signals coordinate zygotic polarization and cell cycle progression in the brown alga *Fucus serratus.* Development 135: 2173-2181.

Brambilla, V., R. Battaglia, M. Colombo, S. Masiero, S. Bencivenga, M.M. Kater and L. Colombo. 2007. Genetic and molecular interactions between BELL1 and MADS box factors support ovule development in *Arabidopsis.* Plant Cell 19: 2544-2556.

Breuninger, H., E. Rikirsch, M. Hermann, M. Ueda and T. Laux. 2008. Differential expression of WOX genes mediates apical-basal axis formation in the *Arabidopsis* embryo. Dev. Cell 14: 867-876.

Camilleri, C., J. Azimzadeh, M. Pastuglia, C. Bellini, O. Grandjean and D. Bouchez. 2002. The *Arabidopsis* TONNEAU2 gene encodes a putative novel protein phosphatase 2A regulatory subunit essential for the control of the cortical cytoskeleton. Plant Cell 14: 833-845.

Ceccato, L., S. Masiero, R.D. Sinha, S. Bencivenga, I. Roig-Villanova, F.A. Ditengou, K. Palme, R. Simon and L. Colombo. 2013. Maternal control of PIN1 is required for female gametophyte development in *Arabidopsis.* PLoS. One 8: e66148.

Chen, D., Y. Deng and J. Zhao. 2012. Distribution and change patterns of free IAA, ABP 1 and PM H(+)-ATPase during ovary and ovule development of *Nicotiana tabacum* L. J. Plant Physiol. 169: 127-136.

Chen, D., Y. Ren, Y. Deng and J. Zhao. 2010. Auxin polar transport is essential for the development of zygote and embryo in *Nicotiana tabacum* L. and correlated with ABP1 and PM H^+-ATPase activities. J. Exp. Bot. 61: 1853-1867.

Chen, J., C. Gutjahr, A. Bleckmann and T. Dresselhaus. 2015. Calcium signaling during reproduction and biotrophic fungal interactions in plants. Mol. Plant 8: 595-611.

Cheng, C.Y., D.E. Mathews, G.E. Schaller and J.J. Kieber. 2013. Cytokinin-dependent specification of the functional megaspore in the *Arabidopsis* female gametophyte. Plant J. 73: 929-940.

Christensen, C.A., E.J. King, J.R. Jordan and G.N. Drews. 1997. Megagametogenesis in *Arabidopsis* wild type and the *Gf* mutant. Sex. Plant Reprod. 10: 49-64.

Cucinotta, M., L. Colombo and I. Roig-Villanova. 2014. Ovule development, a new model for lateral organ formation. Front Plant Sci. 5: 117.

Ding, J.P. and B.G. Pickard. 1993. Modulation of mechanosensitive calcium-selective cation channels by temperature. Plant J. 3: 713-720.

Drevensek, S., M. Goussot, Y. Duroc, A. Christodoulidou, S. Steyaert, E. Schaefer, E. Duvernois, O. Grandjean, M. Vantard, D. Bouchez and M. Pastuglia. 2012. The *Arabidopsis* TRM1-TON1 interaction reveals a recruitment network common to plant cortical microtubule arrays and eukaryotic centrosomes. Plant Cell 24: 178-191.

Drews, G.N. and A.M. Koltunow. 2011. The female gametophyte. Arabidopsis Book 9: e0155.

Duan Q., D. Kita, E.A. Johnson, M. Aggarwal, L. Gates, H.M. Wu and A.Y. Cheung. 2014. Reactive oxygen species mediate pollen tube rupture to release sperm for fertilization in *Arabidopsis.* Nat. Commun. 5: 3129.

Endress, P.K. 2011. Angiosperm ovules: diversity, development, evolution. Ann. Bot. 107: 1465-1489.

Faure, J.E., N. Rotman, P. Fortune and C. Dumas. 2002. Fertilization in *Arabidopsis thaliana* wild type: developmental stages and time course. Plant J. 30: 481-488.

Geshi, N., J.N. Johansen, A. Dilokpimol, A. Rolland, K. Belcram, S. Verger, T. Kotake, Y. Tsumuraya, S. Kaneko, T. Tryfona, P. Dupree, H.V. Scheller, H. Hofte and G. Mouille. 2013. A galactosyltransferase acting on arabinogalactan protein glycans is essential for embryo development in *Arabidopsis*. Plant J. 76: 128-137.

Glover, B.J., K. Torney, C.G. Wilkins and D.E. Hanke. 2008. CYTOKININ INDEPENDENT-1 regulates levels of different forms of cytokinin in *Arabidopsis* and mediates response to nutrient stress. J Plant Physiol. 165: 251-261.

Grieneisen, V.A., A.F. Maree and L. Ostergaard. 2013. Juicy stories on female reproductive tissue development: coordinating the hormone flows. J. Integr. Plant Biol. 55: 847-863.

Hackbusch, J., K. Richter, J. Muller, F. Salamini and J.F. Uhrig. 2005. A central role of *Arabidopsis thaliana* ovate family proteins in networking and subcellular localization of 3-aa loop extension homeodomain proteins. Proc. Natl. Acad. Sci. USA. 102: 4908-4912.

Hauser, B.A., K. Sun, D.G. Oppenheimer and T.L. Sage. 2006. Changes in mitochondrial membrane potential and accumulation of reactive oxygen species precede ultrastructural changes during ovule abortion. Planta 223: 492-499.

Hejatko, J., M. Pernisova, T. Eneva, K. Palme and B. Brzobohaty. 2003. The putative sensor histidine kinase CKI1 is involved in female gametophyte development in *Arabidopsis*. Mol. Genet. Genomics 269: 443-453.

Higuchi, M., M. S. Pischke, A. P. Mahonen, K. Miyawaki, Y. Hashimoto, M. Seki, M. Kobayashi, K. Shinozaki, T. Kato, S. Tabata, Y. Helariutta, M. R. Sussman and T. Kakimoto. 2004. *In planta* functions of the *Arabidopsis* cytokinin receptor family. Proc. Natl. Acad. Sci. USA 101: 8821-8826.

Hohm, T., E. Demarsy, C. Quan, P.L. Allenbach, T. Preuten, T. Vernoux, S. Bergmann and C. Fankhauser. 2014. Plasma membrane H⁺-ATPase regulation is required for auxin gradient formation preceding phototropic growth. Mol. Syst. Biol. 10: 751.

Homble, F. and M. Leonetti. 2007. Emergence of symmetry breaking in fucoid zygotes. Trends Plant Sci. 12: 253-259.

Hu Y., Y. Qin and J. Zhao. 2006. Localization of an arabinogalactan protein epitope and the effects of Yariv phenylglycoside during zygotic embryo development of *Arabidopsis thaliana*. Protoplasma 229: 21–31.

Iwano, M., Q.A. Ngo, T. Entani, H. Shiba, T. Nagai, A. Miyawaki, A. Isogai, U. Grossniklaus and S. Takayama. 2012. Cytoplasmic Ca²⁺ changes dynamically during the interaction of the pollen tube with synergid cells. Development 139: 4202-4209.

Jeong, S., M. Bayer and W. Lukowitz. 2011a. Taking the very first steps: from polarity to axial domains in the early *Arabidopsis* embryo. J Exp. Bot. 62: 1687-1697.

Jeong, S., T.M. Palmer and W. Lukowitz. 2011b. The RWP-RK factor GROUNDED promotes embryonic polarity by facilitating YODA MAP kinase signaling. Curr. Biol. 21: 1268-1276.

Kägi, C., N. Baumann, N. Nielsen, Y.D. Stierhof and R. Gross-Hardt. 2010. The gametic central cell of *Arabidopsis* determines the lifespan of adjacent accessory cells. Proc. Natl. Acad. Sci. USA. 107: 22350-22355.

Kajala, K., P. Ramakrishna, A. Fisher, D.C. Bergmann, I. De Smet, R. Sozzani, D. Weijers and S.M. Brady. 2014. Omics and modelling approaches for understanding regulation of asymmetric cell divisions in *Arabidopsis* and other angiosperm plants. Ann. Bot. 113: 1083-1105.

Kasahara, R.D., M.F. Portereiko, L. Sandaklie-Nikolova, D.S. Rabiger and G.N. Drews. 2005. MYB98 is required for pollen tube guidance and synergid cell differentiation in *Arabidopsis*. Plant Cell 17: 2981-2992.

Kawashima, T. and F. Berger. 2014a. Epigenetic reprogramming in plant sexual reproduction. Nat. Rev. Genet. 15: 613-624.

Kawashima, T., D. Maruyama, M. Shagirov, J. Li, Y. Hamamura, R. Yelagandula, Y. Toyama and F. Berger. 2014b. Dynamic F-actin movement is essential for fertilization in *Arabidopsis thaliana*. eLIFE 3: e04501.

Kirik, A., D.W. Ehrhardt and V. Kirik. 2012. TONNEAU2/FASS regulates the geometry of microtubule nucleation and cortical array organization in interphase *Arabidopsis* cells. Plant Cell 24: 1158-1170.

Kliebenstein, D.J., R.A. Monde and R.L. Last. 1998. Superoxide dismutase in *Arabidopsis*: an eclectic enzyme family with disparate regulation and protein localization. Plant Physiol. 118: 637-650.

Koszegi, D., A.J. Johnston, T. Rutten, A. Czihal, L. Altschmied, J. Kumlehn, S.E. Wust, O. Kirioukhova, J. Gheyselinck, U. Grossniklaus and H. Baumlein. 2011. Members of the RKD transcription factor family induce an egg cell-like gene expression program. Plant J. 67: 280-291.

Lamport, D.T. and P. Varnai. 2013. Periplasmic arabinogalactan glycoproteins act as a calcium capacitor that regulates plant growth and development. New Phytol. 197: 58-64.

Lamport, D.T., P. Varnai and C.E. Seal. 2014. Back to the future with the AGP-Ca^{2+} flux capacitor. Ann. Bot. 114: 1069-1085.

Leon, G., L. Holuigue and X. Jordana. 2007. Mitochondrial complex II is essential for gametophyte development in *Arabidopsis*. Plant Physiol. 143: 1534-1546.

Lew, R.R., N.N. Levina, L. Shabala, M.I. Anderca and S N. Shabala. 2006. Role of a mitogen-activated protein kinase cascade in ion flux-mediated turgor regulation in fungi. Eukaryot. Cell 5: 480-487.

Lieber, D., J. Lora, S. Schrempp, M. Lenhard and T. Laux. 2011. *Arabidopsis* WIH1 and WIH2 genes act in the transition from somatic to reproductive cell fate. Curr. Biol. 21: 1009-1017.

Lituiev, D. S., N.G. Krohn, B. Muller, D. Jackson, B. Hellriegel, T. Dresselhaus and U. Grossniklaus. 2013. Theoretical and experimental evidence indicates that there is no detectable auxin gradient in the angiosperm female gametophyte. Development 140: 4544-4553.

Livanos, P., P. Apostolakos and B. Galatis. 2012a. Plant cell division: ROS homeostasis is required. Plant Signal. Behav. 7: 771-778.

Livanos, P., B. Galatis and P. Apostolakos. 2014a. The interplay between ROS and tubulin cytoskeleton in plants. Plant Signal. Behav. 9: e28069.

Livanos, P., B. Galatis, C. Gaitanaki and P. Apostolakos. 2014b. Phosphorylation of a p38-like MAPK is involved in sensing cellular redox state and drives atypical tubulin polymer assembly in angiosperms. Plant Cell Environ. 37: 1130-1143.

Livanos, P., B. Galatis, H. Quader and P. Apostolakos. 2012b. Disturbance of reactive oxygen species homeostasis induces atypical tubulin polymer formation and affects mitosis in root-tip cells of *Triticum turgidum* and *Arabidopsis thaliana*. Cytoskeleton 69: 1-21.

Lukowitz, W., A. Roeder, D. Parmenter and C. Somerville. 2004. A MAPKK kinase gene regulates extra-embryonic cell fate in *Arabidopsis*. Cell 116: 109-119.

Martin, M.V., A.M. Distefano, E.J. Zabaleta and G.C. Pagnussat. 2013a. New insights into the functional roles of reactive oxygen species during embryo sac development and fertilization in *Arabidopsis thaliana*. Plant Signal. Behav. 8: 10, e25714.

Martin, M.V., A.M. Distefano, A. Bellido, J.P. Cordoba, D. Soto, G.C. Pagnussat and E. Zabaleta. 2014. Role of mitochondria during female gametophyte development and fertilization in *A. thaliana*. Mitochondrion 19 (Pt B): 350-356.

Martin, M.V., D.F. Fiol, V. Sundaresan, E.J. Zabaleta and G.C. Pagnussat. 2013b. *oiwa*, a female gametophytic mutant impaired in a mitochondrial manganese-superoxide dismutase, reveals crucial roles for reactive oxygen species during embryo sac development and fertilization in *Arabidopsis*. Plant Cell 25: 1573-1591.

Menke, F.L. and B. Scheres. 2009. Plant asymmetric cell division, vive la difference! Cell 137: 1189-1192.

Mercier, R., D. Vezon, E. Bullier, J. C. Motamayor, A. Sellier, F. Lefevre, G. Pelletier and C. Horlow. 2001. SWITCH1 (SWI1): a novel protein required for the establishment of sister chromatid cohesion and for bivalent formation at meiosis. Genes Dev. 15: 1859-1871.

Motamayor, J.C., D. Vezon, C. Bajon, A. Sauvanet, O. Grandjean, M. Marchand, N. Bechtold, G. Pelletier and C. Horlow. 2000. Switch (swi1), an *Arabidopsis thaliana* mutant affected in the female meiotic switch. Sex Plant Reprod, 12: 209-218.

Muller, S., A.J. Wright and L.G. Smith. 2009. Division plane control in plants: new players in the band. Trends Cell Biol. 19: 180-188.

Musielak, T.J. and M. Bayer, 2014. YODA signaling in the early *Arabidopsis* embryo. Biochem. Soc. Trans. 42: 408-412.

Okamoto, T., K. Higuchi, T. Shinkawa, T. Isobe, H. Lorz, T. Koshiba and E. Kranz. 2004. Identification of major proteins in maize egg cells. Plant Cell Physiol. 45: 1406-1412.

Olmedo-Monfil, V., N. Duran-Figueroa, M. Arteaga-Vazquez, E. Demesa-Arevalo, D. Autran, D. Grimanelli, R.K. Slotkin, R.A. Martienssen and J.-P. Vielle-Calzada. 2010. Control of female gamete formation by a small RNA pathway in *Arabidopsis*. Nature 464: 628-632.

Paciorek, T. and D.C. Bergmann. 2010. The secret to life is being different: asymmetric divisions in plant development. Curr. Opin. Plant Biol. 13: 661-669.

Pagnussat, G.C., H.J. Yu and V. Sundaresan. 2007. Cell-fate switch of synergid to egg cell in *Arabidopsis eostre* mutant embryo sacs arises from misexpression of the BEL1-like homeodomain gene BLH1. Plant Cell 19: 3578-3592.

Pagnussat, G.C., M. Alandete-Saez, J.L. Bowman and V. Sundaresan. 2009. Auxin-dependent patterning and gamete specification in the *Arabidopsis* female gametophyte. Science 324: 1684-1689.

Petricka, J.J., J.M. Van Norman and P.N. Benfey. 2009. Symmetry breaking in plants: molecular mechanisms regulating asymmetric cell divisions in *Arabidopsis*. Cold Spring Harbor Perspect. Biol. 1: a000497.

Pickard, B.G. 2013. Arabinogalactan proteins-becoming less mysterious. New Phytol. 197: 3-5.

Qin, Y. and J. Zhao. 2006. Localization of arabinogalactan proteins in egg cells, zygotes, and two-celled proembryos and effects of beta-D-glucosyl Yariv reagent on egg cell fertilization and zygote division in *Nicotiana tabacum* L. J Exp. Bot. 57: 2061-2074.

Rabiger, D.S. and G.N. Drews. 2013. MYB64 and MYB119 are required for cellularization and differentiation during female gametogenesis in *Arabidopsis thaliana*, PLoS. Genet. 9: e1003783.

Rasmussen, C.G., J.A. Humphries and L G. Smith. 2011a. Determination of symmetric and asymmetric division planes in plant cells. Annu. Rev. Plant Biol. 62: 387-409.

Rasmussen, C.G., B. Sun and L.G. Smith. 2011b. Tangled localization at the cortical division site of plant cells occurs by several mechanisms. J. Cell Sci. 124: 270-279.

Reddy, A.S. 2001. Molecular motors and their functions in plants. Int Rev. Cytol. 204: 97-178.

Rigas, S., G. Debrosses, K. Haralampidis, F. Vicente-Agullo, K.A. Feldmann, A. Grabov, L. Dolan and P. Hatzopoulos. 2001. TRH1 encodes a potassium transporter required for tip growth in *Arabidopsis* root hairs. Plant Cell 13: 139-151.

Schneitz, K., M. Hulskemp and R.E. Pruitt. 1995. Wild-type ovule development in *Arabidopsis thaliana*: a light microscope study of cleared whole-mount tissue. Plant J. 7: 731-749.

Schneitz, K., S.C. Baker, C.S. Gasser and A. Redweik. 1998. Pattern formation and growth during floral organogenesis: HUELLENLOS and AINTEGUMENTA are required for the formation of the proximal region of the ovule primordium in *Arabidopsis thaliana*. Development 125: 2555-2563.

Shi, H., Y. Kim, Y. Guo, B. Stevenson and J.K. Zhu. 2003. The *Arabidopsis* SOS5 locus encodes a putative cell surface adhesion protein and is required for normal cell expansion. Plant Cell 15: 19-32.

Siddiqi, I., G. Ganesh, U. Grossniklaus and V. Subbiah. 2000. The dyad gene is required for progression through female meiosis in *Arabidopsis*. Development 127: 197-207.

Smekalova, V., I. Luptovciak, G. Komis, O. Samajova, M. Ovecka, A. Doskocilova, T. Takac, P. Vadovic, O. Novak, T. Pechan, A. Ziemann, P. Kosutova and J. Samaj. 2014. Involvement of YODA and mitogen activated protein kinase 6 in *Arabidopsis* post-embryogenic root development through auxin up-regulation and cell division plane orientation. New Phytol. 203: 1175-1193.

Spinner, L., M. Pastuglia, K. Belcram, M. Pegoraro, M. Goussot, D. Bouchez and D.G. Schaefer. 2010. The function of TONNEAU1 in moss reveals ancient mechanisms of division plane specification and cell elongation in land plants. Development 137: 2733-2742.

Spinner, L., A. Gadeyne, K. Belcram, M. Goussot, M. Moison, Y. Duroc, D. Eeckhout, N. De Winne, E. Schaefer, E. Van De Slijke, G. Persiau, E. Witters, K. Gevaert, G. De Jaeger, D. Bouchez, D.Van Damme and M. Pastuglia. 2013. A protein phosphatase 2A complex spatially controls plant cell division. Nat. Commun. 4: Article 1863.

Sundaresan, V. and M. Alandete-Saez. 2010. Pattern formation in miniature: the female gametophyte of flowering plants. Development 137: 179-189.

Susaki, D., H. Takeuchi, H. Tsutsui, D. Kurihara and T. Higashiyama. 2015. Live imaging and laser disruption reveal the dynamics and cell-cell communication during *Torenia fournieri* female gametophyte development. Plant Cell Physiol. 56: 1031-1041.

Takeuchi, H. and T. Higashiyama. 2012. A species-specific cluster of defensin-like genes encodes diffusible pollen tube attractants in *Arabidopsis*. PLoS. Biol. 10: e1001449.

Torres-Ruiz, R.A. and G. Jürgens. 1994. Mutations in the FASS gene uncouple pattern formation and morphogenesis in *Arabidopsis* development. Development 120: 2967-2978.

Truernit, E., K.R. Siemering, S. Hodge, V. Grbic and J. Haseloff. 2006. A map of KNAT gene expression in the *Arabidopsis* root. Plant Mol. Biol. 60: 1-20.

Tucker, M.R., T. Okada, Y. Hu, A. Scholefield, J.M. Taylor and A.M. Koltunow. 2012. Somatic small RNA pathways promote the mitotic events of megagametogenesis during female reproductive development in *Arabidopsis*. Development 139: 1399-1404.

Ueda, M., Z. Zhang and T. Laux. 2011. Transcriptional activation of *Arabidopsis* axis patterning genes WOX8/9 links zygote polarity to embryo development. Dev. Cell 20: 264-270.

Ueda, M. and T. Laux. 2012. The origin of the plant body axis. Curr. Opin. Plant Biol. 15: 578-584.

Van Damme, D. 2009. Division plane determination during plant somatic cytokinesis. Curr. Opin. Plant Biol. 12: 745-751.

Walczak, C.E., S. Gayek and R. Ohi. 2013. Microtubule-depolymerizing kinesins. Annu. Rev. Cell Dev. Biol. 29: 417-441.

Walker, K.L., S. Muller, D. Moss, D.W. Ehrhardt and L.G. Smith. 2007. *Arabidopsis* TANGLED identifies the division plane throughout mitosis and cytokinesis. Curr. Biol. 17: 1827-1836.

Wei, B., J. Zhang, C. Pang, H. Yu, D. Guo, H. Jiang, M. Ding, Z. Chen, Q. Tao, H. Gu, L. J. Qu and G. Qin. 2015. The molecular mechanism of sporocyteless/nozzle in controlling *Arabidopsis* ovule development. Cell Res. 25: 121-134.

Willemsen, V. and B. Scheres. 2004. Mechanisms of pattern formation in plant embryogenesis. Annu. Rev. Genet. 38: 587-614.

Wright, A. J., K. Gallagher and L.G. Smith. 2009. discordial and alternative discordial function redundantly at the cortical division site to promote preprophase band formation and orient division planes in maize. Plant Cell 21: 234-247.

Wu, J.J., X.B. Peng and M.X. Sun. 2013. Mitochondria-responsive signaling between egg and central cell controls their coordinated maturation. Plant Signal. Behav. 8: e24076.

Wu, J.J., X.B. Peng, W.W. Li, R. He, H.P. Xin and M.X. Sun. 2012. Mitochondrial GCD1 dysfunction reveals reciprocal cell-to-cell signaling during the maturation of *Arabidopsis* female gametes. Dev. Cell 23: 1043-1058.

Wudick, M.M. and J.A. Feijo. 2014. At the intersection: merging Ca^{2+} and ROS signaling pathways in pollen. Mol. Plant 7: 1595-1597.

Xin, H.P., J. Zhao and M.X. Sun. 2012. The maternal-to-zygotic transition in higher plants. J. Integr. Plant Biol. 54: 610-615.

Yadegari, R. and G.N. Drews. 2004. Female gametophyte development. Plant Cell 16 Suppl: S133-S141.

Yamada, H., T. Suzuki, K. Terada, K. Takei, K. Ishikawa, K. Miwa, T. Yamashino and T. Mizuno. 2001. The *Arabidopsis* AHK4 histidine kinase is a cytokinin-binding receptor that transduces cytokinin signals across the membrane. Plant Cell Physiol. 42: 1017-1023.

Yang Z. and I. Lavagi. 2012. Spatial control of plasma membrane domains: ROP GTPase-based symmetry breaking. Curr. Opin. Plant Biol. 15: 601–607.

Yang, W.C., D.Q. Shi and Y.H. Chen. 2010. Female gametophyte development in flowering plants. Annu. Rev. Plant Biol. 61: 89-108.

Yu, M. and J. Zhao. 2012. The cytological changes of tobacco zygote and proembryo cells induced by beta-glucosyl Yariv reagent suggest the involvement of arabinogalactan proteins in cell division and cell plate formation. BMC Plant Biol. 12: 126.

Zhang, H., X. Niu, J. Liu, F. Xiao, S. Cao and Y. Liu. 2013. RNAi-directed downregulation of vacuolar H^+-ATPase subunit a results in enhanced stomatal aperture and density in rice. PLoS One 8: e69046.

Zhang, Z. and T. Laux. 2011. The asymmetric division of the *Arabidopsis* zygote: from cell polarity to an embryo axis. Sex Plant Reprod. 24: 161-169.

Zhang, Z., H. Friedman, S. Meir, E. Belausov and S. Philosoph-Hadas. 2011. Actomyosin mediates gravisensing and early transduction events in reoriented cut snapdragon spikes. J. Plant Physiol. 168: 1176-1183.

16

Plant Stem Cells

*Samuel Leiboff and Michael J. Scanlon**

Introduction

Plant stem cells must balance two essential functions: 1) maintenance of pluripotent cells and 2) regulation of stem cell differentiation during organogenesis (Steeves and Sussex 1972). By these two processes, plant stem cells generate complex and diverse adult plant morphologies. Plant stem cells are primarily housed within the root, shoot, and cambial meristems, highly regulated structures that are the topic of Chapters 11 and 13. Many of the positional cues required for plant stem cell differentiation are intimately linked to meristem morphology and will not be discussed in this chapter. This chapter will instead focus on general mechanisms of plant stem cell maintenance, a prerequisite to the formation of the plant body plan (Fig. 1). Stem cell maintenance can further be divided into two major processes: 1) assignment of stem cell identity and 2) stem cell proliferation. Genetic analyses and studies of natural variation in various model organisms (*Arabidopsis*, rice, maize, tomato, petunia, moss, and others) have begun to elucidate the fascinating molecular mechanisms underlying these processes.

As in other eukaryotes, stem cell identity in plants is marked by the expression of homeobox domain proteins (Mukherjee et al. 2009). These transcriptional regulators assign the various morphological identities that comprise the complex body plans of multicellular eukaryotes (Carroll 2000). Through their controlled expression and interactions with other regulatory proteins, homeobox genes nucleate gene modules, which manage stem cell reservoirs during morphogenesis. In plants, transcription factors from the *KNOX* and *WOX* gene families are required to organize and maintain stem cell identity (Mukherjee et al. 2009). These proteins possess 'atypical' DNA-binding homeodomains of between 60 and 66 amino acids; recent research shows that additional protein domains found in KNOX and WOX proteins allow complex protein-protein interactions that define their specific functions as plant stem cell regulators.

---
* Corresponding author : mjs298@cornell.edu

I. KNOTTED1 (KN1) and KNOX

The KNOTTED1-like homeobox proteins (KNOXs) comprise two phylogenetic classes (Chan et al. 1998). Class I KNOX proteins possess two KNOX domains, an ELK domain, as well as a 63 amino acid (three amino acid loop extension, TALE) homeodomain and are required for the establishment and maintenance of stem cell identity (Mukherjee et al. 2009). Homologs of the well-studied Class I *KNOX* genes have been identified in most plant lineages (Cronk 2001). In contrast, Class II *KNOXs* are less studied. Recent work in the moss *Physcomitrella patens* demonstrates that Class II *KNOX* genes regulate the transition from diploid to haploid body plans during the alternation of generations (Sakakibara et al. 2013); more recent work in angiosperms suggests that that Class I and Class II KNOXs have antagonistic functions (Furumizu et al. 2015).

The classical dominant mutation of maize, *KNOTTED1* (*KN1*), causes ectopic 'knots' of cells to grow on the blade of mutant leaves and was cloned by transposon tagging in 1989 (Hake et al. 1989). The ectopic outgrowths found on gain-of-function *kn1-o* plants are believed to result from the projection of indeterminate stem cell identity onto a field of determinate leaf cells, ultimately forming organized clusters of rapidly-dividing tissues in the leaf blade (Vollbrecht et al. 1991, Smith et al. 1992). Loss of *KN1* in maize or *SHOOT MERISTEMLESS1* (*STM1*), its homolog in *Arabidopsis thaliana*, causes a lethal depletion of stem cells in some genetic backgrounds and environmental conditions. (Long et al. 1996, Vollbrecht et al. 2000). In

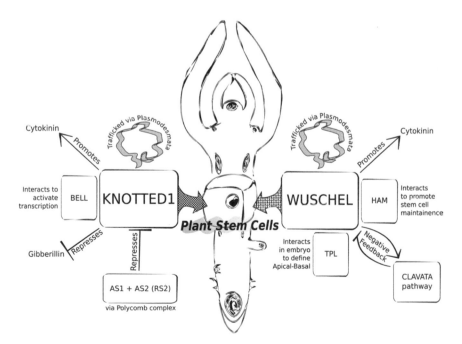

FIGURE 1 Plant stem cells are maintained by *KNOX* (ex. *KNOTTED1*) and *WOX* (ex. *WUSCHEL*) regulatory pathways. Reservoirs of plant stem cells can be found in shoot/branch apical meristems, the vascular cambium, and root apical meristems.

kn1 or *stm1* loss-of-function mutants, stem cell identity is organized during embryo-genesis, but is lost soon after germination. This failure in stem cell maintenance causes the shoot meristem to be depleted during the formation of lateral organs; ulti-mately the meristem is consumed, and organogenesis ceases (Vollbrecht et al. 2000).

KNOX expression domains and KN1 movement

In situ hybridization experiments show that *KNOX* transcript accumulation is restricted to the inner layers of the shoot meristem, and is conspicuously absent from the outer layers (Jackson et al. 1994, Chuck et al. 1996, Long et al. 1996). *KNOX* transcripts are downregulated in initiating leaf primordia, which correlates with the transition from indeterminate to determinate growth. Although *KNOTTED1* mRNA from is not found in the outermost, 'L1' meristem layer, immunohistolocalizations show that KN1 protein is found in all meristem layers (Jackson 2002). This discrepancy in the accumulation pattern of *KN1* RNA and protein has led to exciting research support-ing the hypothesis that KNOX proteins can traffic between cells via plasmodesmata, symplastic pores that connect the cytoplasm of adjacent plant cells (Jackson 2001). The first evidence for KNOX trafficking came from *in planta* assays of recombinant protein movement (Lucas et al. 1995, Kim et al. 2002, 2003,). Starting with the full-length maize KNOTTED1, several recombinant alleles with mutated protein domains were tested for movement after microinjection into plant cells (Lucas et al. 1995). The full-length *KN1* sequence and several mutated recombinant alleles were success-fully trafficked through the plant cell system. However, accumulation of the KN1-M6 protein, which contained a defective homeodomain, was restricted to injected cells. Transient expression assays have shown that full-length maize KN1 is able to transit between cells, with similar efficiency as observed in viral movement proteins. In con-trast smaller proteins with no identified movement mechanisms are unable to traffic between cells (Kim et al. 2002). A genetic screen for interacting proteins that facilitate KN1 movement identified a chaperonin that is required for full function of trafficked KN1 in *Arabidopsis* (Xu et al. 2011). These data suggest that during trafficking to the Arabidopsis epidermis the KN1 protein is unfolded, transported through a plasmodes-matal pore, and then refolded within the destination cell (Xu et al. 2011).

KNOXs as transcription factors

Comparative RNAseq and ChIPseq analyses in *KN1* mutant and wild type back-grounds revealed that more than 600 maize genes are bound and regulated by KN1 (Bolduc et al. 2012). Interestingly, RNAseq read counts in KN1 loss-of-function and wild-type backgrounds suggest that KN1 can function as either a transcriptional acti-vator, or as a repressor of target genes. Subsequent *in vitro* assays of KN1 and other KNOX proteins showed that they are weak transcriptional activators (Smith et al. 2002). To provide transcriptional activation, KN1 and KNOX proteins interact with other TALE homeobox proteins, including members of the *BELL* family (Smith et al. 2002, Ragni et al. 2008). *In vitro* and *in planta*, KNOXs and BELLs interact as het-erodimer complexes to activate target gene expression. This interaction is facilitated by two highly conserved MEINOX domains found on KNOX proteins, which inter-act with the MID domains (named SKY and BELL) of BELL proteins (Mukherjee et al. 2009).

The Arabidopsis gene, *KNATM*, encodes a truncated KNOX that lacks a functional homeodomain, and therefore cannot bind DNA (Magnani and Hake 2008, Mukherjee et al. 2009). Intriguingly, KNATM retains the ability to interact with BELL proteins, and the resulting heterodimer can act as a strong transcriptional activator *in vitro* (Magnani and Hake 2008). These studies suggest that interactions with other TALE homeobox proteins may be at least as important as autonomous KNOX/DNA binding during KNOX-mediated stem cell regulation (Hay and Tsiantis 2010).

KNOXs and plant hormones

KNOXs maintain stem cell identity in part by the regulation of plant hormone biosynthesis. In several species, KNOXs directly regulate gibberellin (GA) levels through the activation of gibberellin oxidases such as *GA2ox1*, which function in GA catabolism. Likewise, KNOXs repress the expression of the GA biosynthetic gene *GA20ox1* (Sakamoto et al. 2001, Hay et al. 2002, Shani ct al. 2006, Bolduc and Hake 2009). Thus, KNOX-mediated repression of GA accumulation inhibits determinate growth by preventing cell expansion and maturation. In addition, KNOXs promote cytokinin (CK) accumulation via activation of *ADENOSINE PHOSPHATE ISOPENTENYLTRANSFERASE* (*IPT*) CK-biosynthetic genes, and of the cytokinin activation gene *LONELY GUY* (*LOG*), which encodes a phosphoribohydrolase that converts inactive CK conjugates into biologically-active CK (Jasinski et al. 2005, Scofield et al. 2013). Although some of the stem cell organizing function of KNOXs is independent of cytokinin, high cytokinin levels repress endoreduplication and prevent cellular differentiation. Thus, KNOXs maintain an undifferentiated, pluripotent stem cell fate by direct upregulation of CK and downregulation of GA (Shani et al. 2006).

KNOXs are regulated by the POLYCOMB REPRESSIVE COMPLEX (PRC)

Whereas the expression of *KNOXs* is required for stem cell maintenance, the proper patterning of determinate lateral organs requires repression of stem cell identity (Micol et al. 2003). Multiple transcription factors act together to repress *KNOXs* during organogenesis (Hay and Tsiantis 2010). Whereas *KNOX* genes are expressed in stem cell niches and are absent from the differentiating cells of young leaf primordia, the MYB-domain transcription factor, *ASYMMETRIC LEAVES1* (*AS1*) and the lateral organ boundary domain (LBD) protein *AS2* both show KNOX-complimentary expression in differentiating cells of young leaf primorida (Theodoris et al. 2003). Loss of *AS1* in *Arabidopsis*, or its maize homologue *ROUGH SHEATH2* (*RS2*), leads to ectopic *KNOX* expression in developing leaf primordium, ultimately yielding dramatic mutant leaf phenotypes (Tsiantis et al. 1999, Phelps-Durr et al. 2005,). AS1 and AS2 heterodimerize and recruit the POLYCOMB REPRESSIVE COMPLEX to epigenetically repress *KNOX* expression in determinate lateral organ primordia (Xu and Shen 2008).

KNOXs at the center of organogenesis

In the shoot meristem, stem cell identity is marked by KNOX accumulation, whereas developing leaf primordia do not accumulate KNOX proteins. The site of new leaf initiation is marked by local accumulation of the plant hormone auxin (Reinhardt 2000, Deb et al. 2015), which results in downregulation of KNOX accumulation in the incipient leaf (Scanlon 2003, Hay et al. 2006). This localized auxin maxima is created by the convergence of polar auxin transporters from the *PIN-formed* (*PIN*) and *AUXIN-INSENSITIVE / LIKE AUXIN-INSENSITIVE* (*AUX/LAX*) families, and is followed by the expression of *LBD* genes such as *LATERAL ORGAN BOUNDARY* (*LOB*) and *AS2* (Hay and Tsiantis 2010, Johnston et al. 2014). AS2 and LBD proteins then repress KNOXs and stem cell identity in the developing leaf primordium, as described above. In this simple model for lateral organ specification, auxin-induced *LBD* gene expression plays a central role in repression of KNOX-mediated, stem cell identity (Hay and Tsiantis 2010, Johnston et al. 2014). This model has been success-fully extrapolated to multiple organ boundaries throughout plant development during the morphogenesis of leaves, branches, flowers, and ligules (Johnston et al. 2014).

Conserved *KNOX* function is required in several species for development of dis-sected (compound) leaves and lobed leaf margins (Bharathan et al. 2002, Chuck et al. 1996, Rupp et al. 1999, Scanlon 2003). In tomato, *KNOX* genes and auxin/*LBD* mod-ules mark the location of pinnae in dissected leaves, similar to what is observed dur-ing initiation of lateral organ primordia from the shoot meristem (Kimura et al. 2008). Similarly, overexpression of the *Arabidopsis KNOX* genes *BREVIPEDICELLUS* (*BP*) or *KNAT2* yields a highly lobed leaf lamina covered with ectopic meristems, com-plete with fully functional stem cells (Lincoln et al. 1994, Hake et al. 1995, Chuck et al. 1996). These data, and related studies, suggest that reactivation of stem cell regulatory networks in the marginal blastozones of otherwise determinate leaf pri-mordia can activate a secondary stage of morphogenesis in plants with dissected or highly lobed leaves (Barkoulas et al. 2008, Blein et al. 2008, Koenig et al. 2009).

II. WUSCHEL (WUS) and WOXs

The WUSCHEL-like homeobox proteins (WOXs) comprise a large family of pro-teins with a single homeodomain and an 8 amino-acid WUS-box domain (Mukherjee et al. 2009, van der Graaff et al. 2009). There are three commonly accepted sub-families of *WOXs*: the '*WUSCHEL*-like' subfamily that contains a two amino acid T-L motif at the beginning of the WUS-box, the ancient WOX subfamily with con-served members identified in algae and moss, and the intermediate WOX subfam-ily that establishes apical-basal auxin transport during embryogenesis (van der Graaff et al. 2009). The '*WUSCHEL*-like' subfamily is the most studied and best understood WOX clade. Within this subfamily of *WOXs* there are many members with tissue-specific expression domains, acting as organizers of stem cell identity throughout plant development (van der Graaff et al. 2009). Promoter fusion assays in *wox* mutant backgrounds show that the functions of several WOX proteins are cross-complementary, suggesting that WOX family evolution involved the subfunc-tionalization of conserved stem-cell organizing functions via adoption of distinct, tissue-specific promoters (Sarkar et al. 2007, Shimizu et al. 2009, Lin et al. 2013). Additionally, *WOXs* may act as persistent organizers of stem cell reservoirs such as *WUS* in the shoot apical meristem and *WOX5* in the root (Laux et al. 1996, Schoof et al. 2000, Zhang et al. 2013); other *WOXs* may act as ephemeral founder cell recruit-ment factors for organogenesis such as *NARROW SHEATH1* (*NS1*) and *NS2* in maize and *PRESSED FLOWER1* (*PRS1*)/*WOX3* and *WOX1* in *Arabidopsis* (Scanlon 2000, Matsumoto and Okada 2001, Shimizu et al. 2009, Vandenbussche et al. 2009, van der Graaff et al. 2009).

Antagonism with the *CLAVATA* pathway

WUS, the founding member of the *WOXs* is acclaimed for its antagonistic interaction with the *CLAVATA* genes in *Arabidopsis thaliana* (Laux et al. 1996, Müller et al. 2006, Durbak and Tax 2011). Whereas mutations in *WUS* lead to the loss of stem cells in the vegetative or inflorescence meristem, mutations in *CLAVATA* (*CLV*) genes lead to stem cell over-proliferation, which distorts the morphology of the shoot meristem. Although the phenotypes of loss-of-function *CLV* mutants are similar, *CLV1*, *CLV2*, and *CLV3* encode proteins of distinctly dissimilar molecular function (Fletcher and Meyerowitz 2000). *CLV3* encodes a small, secreted peptide-ligand that moves between cells through the extracellular apoplast (Clark et al. 1996, Rojo 2002, Wong et al. 2013). Studies of meristem size mutants in tomato reveal that proper recognition of this peptide is enhanced by post-translational arabinosylation (Xu et al. 2015). The small CLV3 peptide is in turn recognized by the leucine-rich-repeat (LRR) receptor kinase, CLV1 (Clark et al. 1996, Nimchuk et al. 2011). CLV1 can from dimers with several other LRR proteins including CLV2, a LRR that lacks a functional kinase domain (Durbak and Tax 2011). While CLV1 has been shown to bind CLV3 and respond by an unknown kinase cascade, the mechanism underlying the *CLV2* phenotype is still contested (Deyoung and Clark 2008, Durbak and Tax 2011, Nimchuk et al. 2011). However, it is generally accepted that WUS activates *CLV3* expression, while the CLAVATA signaling pathway in turn acts to repress *WUS*. This negative-feedback loop thereby preserves a steady-state level of stem cells in the shoot meristem (Yadav et al. 2010, 2011, 2013).

Regulatory loops consisting of *WOX* and *CLAVATA*-like (*CLE*) genes have been demonstrated in distinct processes during plant development as a widely-employed signaling module in plant morphogenesis (Wu et al. 2007). Notably, *WOX/CLE* stem cell regulation is found in root meristem maintenance (*WOX5/CLE40*) via the receptor kinase *CRINKLY4* (*CR4*) (Sarkar et al. 2007, Jun et al. 2010), and vascular differentiation employs *WOX4* in conjunction with *CLE41* and *CLE44* (Miyawaki et al. 2013). The conservation of *WOX/CLE* negative feedback loops in multiple plant species suggests that *WOX/CLE* interaction may be an ancestral, plant patterning toolkit (Miyawaki et al. 2013).

There is strong evidence for dynamic evolution of the WOX gene family among the species examined to date (Nardmann and Werr 2013). Although maize and *Arabidopsis* have 14 and 15 WOX family genes, respectively, not all WOX genes are represented in both lineages (Vandenbussche et al. 2009, van der Graaff et al. 2009). It may come as no surprise then, that several plant lineages show unique elaboration of the ancestral *WOX/CLE* regulatory pathway. For example, although mutation of the maize *CLV1* homolog, *THICK TASSEL DWARF1* (*TD1*) generates an inflorescence phenotype similar to the stem cell over proliferation observed in Arabidopsis, no interacting *WOX* has yet been identified (Bommert et al. 2005, Lunde and Hake 2009). Similar investigations with rice *CLV1* and *CLV3* homologs *FLORAL ORGAN NUMBER1* (*FON1*) and *FON2*, still seek an interacting *WOX* (Miyawaki et al. 2013). Recent research in maize has identified previously unrelated genetic mechanisms involving G-proteins, amongst others, that contribute to the stem cell maintenance and may indeed participate in *WOX/CLE* regulation (Bommert et al. 2013).

WOX interactions

ChIP-microarray data demonstrate that WUS directly binds more than 100 targets in the *Arabidopsis* genome (Busch et al. 2010). Several studies reported that WOXs generally act as transcriptional repressors (Leibfried et al. 2005, van der Graaff et al. 2009, Lin et al. 2013), however WUS is a known activator of *AGAMOUS* (*AG*) expression in floral meristems (Lenhard et al. 2001, Lohmann et al. 2001). Moreover, combined ChIP-microarray and RNAseq analyses suggest that WUS may either activate or repress its direct targets, which indicates that WUS regulatory functions are complex and multivariate (Busch et al. 2010).

WOXs interact with a variety of other proteins (van der Graaff et al. 2009). In petunia and Arabidopsis WOXs interact with the *HAIRY MERISTEM* (*HAM*) genes, GRAS-family transcription factors (Engstrom et al. 2011, Zhou et al. 2014). HAMs dimerize with WOX proteins and are required for stem cell maintenance. *WOXs* additionally interact with the transcriptional corepressor *TOPLESS* (*TPL*) genes during specification of the apical-basal axis in Arabidopsis embryos (Long et al. 2002, 2006).

WOX proteins influence cytokinin signaling via direct transcriptional repression of the stem cell repressor *RESPONSE REGULATOR7* (*RR7*), a negative regulator of cytokinin signaling (Leibfried et al. 2005). Models of WUS homeostasis in the shoot meristem predict that WUS, CLV, and cytokinin response are all required to maintain a steady state number of stem cell initials in the growing shoot (Gordon et al. 2009, Chickarmane et al. 2012).

WUSCHEL moves

Exciting new research reveals that WUS, much like KN1, is capable of intercellular trafficking, which enables non-cell-autonomous function (Yadav et al. 2011, Daum et al. 2014). WUS fusion proteins migrate several cell layers, from the interior of the *Arabidopsis* shoot meristem to the outer layers of the meristem, wherein *CLV* genes are activated. Careful dissection of WUSCHEL protein domains shows that the *WUS*-box domain promotes widespread, plasmodesmata-dependent protein movement—to a higher degree than observed with the native, full-length protein (Daum et al. 2014). Interestingly, conserved sequences in the C-terminus act to restrict WUS cell-to-cell trafficking to levels seen in wild-type *Arabidopsis* plants. Fusion proteins created from WOX5, another *WUS*-like WOX, can also traffic through the shoot meristem (Daum et al. 2014). In contrast, constructs containing WOX13, from the ancient WOX clade, did not show significant cell-to-cell-movement.

Genetic evidence indicates that *WUS*, *WOX3*, and *WOX5* all exhibit non-cell autonomous function to maintain stem/initial cell homeostasis (Scanlon 2000, Reddy and Meyerowitz 2005, Sarkar et al. 2007, van der Graaff et al. 2009). These data, when considered together with the cross-complementarity of WOX function, suggest that movement may be a key feature of *WUS*-like WOX proteins. However, analysis of PRESSED FLOWER1 (PRS1/WOX3) function revealed no evidence of cell-to-cell trafficking in floral meristems (Shimizu et al. 2009). Thus, whether the WUS-like clade of WOX proteins shares ancestral mechanisms enabling non-cell autonomous function is an unresolved question.

III. KNOXs and WOXs Underlie Persistent and Flexible Plant Forms

Plants are noted for their indeterminate growth pattern (Steeves and Sussex 1972). They comprise both the largest and oldest living things on our planet. Whereas animals undergo organogenesis only at key times during their lifecycle and most patterns are established in the embryo, plants continuously and reiteratively generate new organs across their lifetime (Walbot 1985). As sessile organisms, plants must additionally integrate information from dynamic aboveground and belowground environments to establish suitable organ form and an adaptive overall morphology (Sultan 2000). It comes as no surprise then, that plants maintain several sources of pluripotent stem cells (Steeves and Sussex 1972). Marked by the expression of KNOX and WOX proteins, plant stem cells are tightly regulated by a suite of partially redundant and intricately connected molecular pathways, integrating hormone signaling, epigenetic regulation, and even mechanical forces (Mukherjee et al. 2009, van der Graaff et al. 2009, Hay and Tsiantis 2010, Hamant 2013).

The similarities of these two gene families are striking (Fig. 1). Both families of proteins use homeobox domains to directly bind and regulate hundreds of genes (Busch et al. 2010, Bolduc et al. 2012). KNOX and WOX direct targets may either be activated or repressed. Gene regulatory modules involving KNOX/auxin/LBD and WOX/CLE interactions have been deployed in diverse plant tissues across distinct evolutionary clades (Hay et al. 2006, Johnston et al. 2014). KNOXs and WOXs participate in plasmodesmata-dependent cell-to-cell movement for proper function (Xu et al. 2011, Yadav et al. 2011, Daum et al. 2014). Both regulate key plant hormones, including CK to suppress cell differentiation (Shani et al. 2006). Localization assays find KNOXs and WOXs in overlapping domains in the shoot meristem (Lucas et al. 1995, Long et al. 1996). Yet, *KNOX* and *WOX* genes act independently of each other (Lenhard et al. 2002).

Such exquisite complexity in stem cell regulation points to the necessity of proper activation and deactivation of indeterminate growth programs. Without indeterminate growth, the plant cannot generate new organs (Long et al. 1996, Mayer et al. 1998, Vollbrecht et al. 2000). This phenotype is seen in the seedling-lethal loss-of-function mutants of *kn1* and *wus*. On the other hand, without deactivation of indeterminate growth, cell division will continue and differentiation will not occur (Hake et al. 1989, Timmermans et al. 1999, Brand 2000, Schoof et al. 2000, Phelps-Durr et al. 2005). Animal cancers, where cells divide indefinitely without differentiation, may be an example of such a syndrome (Shah and Sukumar 2010). Despite the many reservoirs of indeterminate stem cells that plants maintain, plants do not have a similar endogenous disease (Doonan and Sablowski 2010). Tumor-forming pathogens of plants must rely on reprogramming molecular machinery, by genetic transformation in some cases (Smith et al. 1912, Doonan and Sablowski 2010). By maintaining separate highly redundant and complex *KNOX* and *WOX* stem cell regulation pathways, plants protect themselves from disastrous adventitious cellular growth.

References

Barkoulas, M., A. Hay, E. Kougioumoutzi and M. Tsiantis. 2008. A developmental framework for dissected leaf formation in the Arabidopsis relative *Cardamine hirsuta*. Nat. Genet. 40: 1136-1141.

Bharathan, G., T.E. Goliber, C. Moore, S. Kessler, T. Pham and N.R. Sinha. 2002. Homologies in leaf form inferred from KNOXI gene expression during development. Science 296: 1858-1860.

Blein, T., A. Pulido, A. Vialette-Guiraud, K. Nikovics, H. Morin, A. Hay, I.E. Johansen, M. Tsiantis and P. Laufs. 2008. A conserved molecular framework for compound leaf development. Science 322: 1835-1839.

Bolduc, N. and S. Hake. 2009. The maize transcription factor KNOTTED1 directly regulates the gibberellin catabolism gene ga2ox1. Plant Cell 21: 1647-1658.

Bolduc, N., A. Yilmaz, M.K. Mejia-Guerra, K. Morohashi, D. O'Connor, E. Grotewold and S. Hake. 2012. Unraveling the KNOTTED1 regulatory network in maize meristems. Genes Dev. 26: 1685-1690.

Bommert, P., B. Il Je, A. Goldshmidt and D. Jackson. 2013. The maize Gα gene COMPACT PLANT2 functions in CLAVATA signalling to control shoot meristem size. Nature 502: 555-558.

Bommert, P., C. Lunde, J. Nardmann, E. Vollbrecht, M. Running, D. Jackson, S. Hake and W. Werr. 2005. thick tassel dwarf1 encodes a putative maize ortholog of the Arabidopsis CLAVATA1 leucine-rich repeat receptor-like kinase. Development 132: 1235-1245.

Brand, U. 2000. Dependence of stem cell fate in arabidopsis on a feedback loop regulated by CLV3 activity. Science 289: 617-619.

Busch, W., A. Miotk, F.D. Ariel, Z. Zhao, J. Forner, G. Daum, T. Suzaki, C. Schuster, S.J. Schultheiss, A. Leibfried, S. Haubeiss, N. Ha, R.L. Chan and J.U. Lohmann. 2010. Transcriptional control of a plant stem cell niche. Dev. Cell 18: 849-861.

Carroll, S.B. 2000. Endless Forms: The evolution of gene regulation and morphological diversity. Cell 101: 577-580.

Chan, R.L., G.M. Gago, C.M. Palena and D.H. Gonzalez. 1998. Homeoboxes in plant development. Biochim. Biophys. Acta 1442: 1-19.

Chickarmane, V.S., S.P. Gordon, P.T. Tarr, M.G. Heisler and E.M. Meyerowitz. 2012. Cytokinin signaling as a positional cue for patterning the apical-basal axis of the growing Arabidopsis shoot meristem. Proc. Natl. Acad. Sci. 109: 4002-4007.

Chuck, G., C. Lincoln and S. Hake. 1996. KNAT1 induces lobed leaves with ectopic meristems when overexpressed in Arabidopsis. Plant Cell 8: 1277-1289.

Clark, S.E., S.E. Jacobsen, J.Z. Levin and E.M. Meyerowitz. 1996. The CLAVATA and SHOOT MERISTEMLESS loci competitively regulate meristem activity in Arabidopsis. Development 122: 1567-1575.

Cronk, Q.C. 2001. Plant evolution and development in a post-genomic context. Nat. Rev. Genet. 2: 607-619.

Daum, G., A. Medzihradszky, T. Suzaki and J.U. Lohmann. 2014. A mechanistic framework for noncell autonomous stem cell induction in Arabidopsis. Proc. Natl. Acad. Sci. 111: 14619-14624.

Deb, Y., D. Marti, M. Frenz, C. Kuhlemeier and D. Reinhardt. 2015. Phyllotaxis involves auxin drainage through leaf primordia. Development 1-10.

Deyoung, B.J. and S.E. Clark. 2008. BAM receptors regulate stem cell specification and organ development through complex interactions with CLAVATA signaling. Genetics 180: 895-904.

Doonan, J.H. and R. Sablowski. 2010. Walls around tumours - why plants do not develop cancer. Nat. Rev. Cancer 10: 794-802.

Durbak, A.R. and F.E. Tax. 2011. CLAVATA signaling pathway receptors of Arabidopsis regulate cell proliferation in fruit organ formation as well as in meristems. Genetics 189: 177-1794.

Engstrom, E.M., C.M. Andersen, J. Gumulak-Smith, J. Hu, E. Orlova, R. Sozzani and J.L. Bowman. 2011. Arabidopsis homologs of the petunia hairy meristem gene are required for maintenance of shoot and root indeterminacy. Plant Physiol. 155: 735-7350.

Fletcher, J.C. and E.M. Meyerowitz. 2000. Cell signaling within the shoot meristem. Curr. Opin. Plant Biol. 3: 23-30.

Furumizu, C., J.P. Alvarez, K. Sakakibara and J.L. Bowman. 2015. Antagonistic roles for KNOX1 and KNOX2 genes in patterning the land plant body plan following an ancient gene duplication. PLOS Genet. 11: e1004980.

Gordon, S.P., V.S. Chickarmane, C. Ohno and E.M. Meyerowitz. 2009. Multiple feedback loops through cytokinin signaling control stem cell number within the Arabidopsis shoot meristem. Proc. Natl. Acad. Sci. USA. 106: 16529-16534.

Hake, S., B.R. Char, G. Chuck, T. Foster, J.A. Long and D.P. Jackson. 1995. Homeobox genes in the functioning of plant meristems. Philos. Trans. R. Soc. Lond. B. Biol. Sci. 350: 45-51.

Hake, S., E. Vollbrecht and M. Freeling. 1989. Cloning Knotted, the dominant morphological mutant in maize using Ds2 as a transposon tag. EMBO J. 8: 15-22.

Hamant, O. 2013. Widespread mechanosensing controls the structure behind the architecture in plants. Curr. Opin. Plant Biol. 16: 654-660.

Hay, A., M. Barkoulas and M. Tsiantis. 2006. ASYMMETRIC LEAVES1 and auxin activities converge to repress BREVIPEDICELLUS expression and promote leaf development in Arabidopsis. Development 133: 3955-3961.

Hay, A., H. Kaur, A. Phillips, P. Hedden, S. Hake and M. Tsiantis. 2002. The gibberellin pathway mediates KNOTTED1-type homeobox function in plants with different body plans. Curr. Biol. 12: 1557-1565.

Hay, A. and M. Tsiantis. 2010. KNOX genes: versatile regulators of plant development and diversity. Development 137: 3153-3165.

Jackson, D.P. 2001. The long and the short of it: signaling development through plasmodesmata. Plant Cell 13: 2569-2572.

Jackson, D.P. 2002. Double labeling of KNOTTED1 mRNA and protein reveals multiple potential sites of protein trafficking in the shoot apex. Plant Physiol. 129: 1423-1429.

Jackson, D.P., B. Veit and S. Hake. 1994. Expression of maize KNOTTED1 related homeobox genes in the shoot apical meristem predicts patterns of morphogenesis in the vegetative shoot. Development 413: 405-413.

Jasinski, S., P. Piazza, J. Craft, A. Hay, L. Woolley, I. Rieu, A. Phillips, P. Hedden and M. Tsiantis. 2005. KNOX action in Arabidopsis is mediated by coordinate regulation of cytokinin and gibberellin activities. Curr. Biol. 15: 1560-1565.

Johnston, R., M. Wang, Q. Sun, A.W. Sylvester, S. Hake and M.J. Scanlon. 2014. Transcriptomic Analyses Indicate That maize ligule development recapitulates gene expression patterns that occur during lateral organ initiation. Plant Cell Online 26: 4718-4732.

Jun, J., E. Fiume, A.H.K. Roeder, L. Meng, V.K. Sharma, K.S. Osmont, C. Baker, C.M. Ha, E.M. Meyerowitz, L.J. Feldman and J.C. Fletcher. 2010. Comprehensive analysis of CLE polypeptide signaling gene expression and overexpression activity in Arabidopsis. Plant Physiol. 154: 1721-1736.

Kim, J.Y., Z. Yuan, M. Cilia, Z. Khalfan-Jagani and D. Jackson. 2002. Intercellular trafficking of a KNOTTED1 green fluorescent protein fusion in the leaf and shoot meristem of Arabidopsis. Proc. Natl. Acad. Sci. USA. 99: 4103-4108.

Kim, J.Y., Z. Yuan and D. Jackson. 2003. Developmental regulation and significance of KNOX protein trafficking in Arabidopsis. Development 130: 4351-4362.

Kimura, S., D. Koenig, J. Kang, F.Y. Yoong and N.R. Sinha. 2008. Natural Variation in Leaf Morphology Results from Mutation of a Novel KNOX Gene. Curr. Biol. 18: 672-677.

Koenig, D., E. Bayer, J. Kang, C. Kuhlemeier and N.R. Sinha. 2009. Auxin patterns *Solanum lycopersicum* leaf morphogenesis. Development 136: 2997-3006.

Laux, T., K.F. Mayer, J. Berger and G. Jürgens. 1996. The WUSCHEL gene is required for shoot and floral meristem integrity in Arabidopsis. Development 122: 87-96.

Leibfried, A., J.P.C. To, W. Busch, S. Stehling, A. Kehle, M. Demar, J.J. Kieber and J.U. Lohmann. 2005. WUSCHEL controls meristem function by direct regulation of cytokinin-inducible response regulators. Nature 438: 1172-1175.

Lenhard, M., A. Bohnert, G. Jürgens and T. Laux. 2001. Termination of stem cell maintenance in Arabidopsis floral meristems by interactions between WUSCHEL and AGAMOUS. Cell 105: 805-814.

Lenhard, M., G. Jürgens and T. Laux. 2002. The WUSCHEL and SHOOTMERISTEMLESS genes fulfil complementary roles in Arabidopsis shoot meristem regulation. Development 129: 3195-3206.

Lin, H., L. Niu, N. a McHale, M. Ohme-Takagi, K.S. Mysore, and M. Tadege. 2013. Evolutionarily conserved repressive activity of WOX proteins mediates leaf blade outgrowth and floral organ development in plants. Proc. Natl. Acad. Sci. USA. 110: 366-371.

Lincoln, C., J.A. Long, J. Yamaguchi, K. Serikawa and S. Hake. 1994. A knotted1-like homeobox gene in Arabidopsis is expressed in the vegetative meristem and dramatically alters leaf morphology when overexpressed in transgenic plants. Plant Cell 6: 1859-1876.

Lohmann, J. U., R.L. Hong, M. Hobe, M.A. Busch, F. Parcy, R. Simon and D. Weigel. 2001. A molecular link between stem cell regulation and floral patterning in Arabidopsis. Cell 105: 793-803.

Long, J.A., E. Moan, J. Medford and M. Barton. 1996. A member of the KNOTTED class of homeodomain proteins encoded by the STM gene of Arabidopsis. Nature 379: 66-69.

Long, J.A., C. Ohno, Z.R. Smith and E.M. Meyerowitz. 2006. TOPLESS Regulates Apical Embryonic Fate in Arabidopsis. Science 312: 1520-1522.

Long, J.A., S. Woody, S. Poethig, E.M. Meyerowitz and M.K. Barton. 2002. Transformation of shoots into roots in Arabidopsis embryos mutant at the TOPLESS locus. Development 129: 2797-2806.

Lucas, W.J., S. Bouché-Pillon, D.P. Jackson, L. Nguyen, L. Baker, B. Ding and S. Hake. 1995. Selective trafficking of KNOTTED1 homeodomain protein and its mRNA through plasmodesmata. Science 270: 1980-1983.

Lunde, C. and S. Hake. 2009. The interaction of knotted1 and thick tassel dwarf1 in vegetative and reproductive meristems of maize. Genetics 181: 1693-1697.

Magnani, E. and S. Hake. 2008. KNOX lost the OX: the Arabidopsis KNATM gene defines a novel class of KNOX transcriptional regulators missing the homeodomain. Plant Cell 20: 875-887.

Matsumoto, N. and K. Okada. 2001. A homeobox gene, PRESSED FLOWER, regulates lateral axis-dependent development of Arabidopsis flowers. Genes Dev. 15: 3355-3364.

Mayer, K.F., H. Schoof, A. Haecker, M. Lenhard, G. Jürgens and T. Laux. 1998. Role of WUSCHEL in regulating stem cell fate in the Arabidopsis shoot meristem. Cell 95: 805-815.

Micol, L., S. Hake, J.J.L. Micol and S. Hake. 2003. The Development of Plant Leaves. Plant Physiol. 131: 389-394.

Miyawaki, K., R. Tabata and S. Sawa. 2013. Evolutionarily conserved CLE peptide signaling in plant development, symbiosis, and parasitism. Curr. Opin. Plant Biol. 16: 598-606.

Mukherjee, K., L. Brocchieri and T.R. Bürglin. 2009. A comprehensive classification and evolutionary analysis of plant homeobox genes. Mol. Biol. Evol. 26: 2775-2794.

Müller, R., L. Borghi, D. Kwiatkowska, P. Laufs and R. Simon. 2006. Dynamic and compensatory responses of Arabidopsis shoot and floral meristems to CLV3 signaling. Plant Cell 18: 1188-1198.

Nardmann, J. and W. Werr. 2013. Symplesiomorphies in the WUSCHEL clade suggest that the last common ancestor of seed plants contained at least four independent stem cell niches. New Phytol. 199: 1081-1092.

Nimchuk, Z.L., P.T. Tarr, C. Ohno, X. Qu and E.M. Meyerowitz. 2011. Plant stem cell signaling involves ligand-dependent trafficking of the CLAVATA1 receptor kinase. Curr. Biol. 21: 345-352.

Phelps-Durr, T.L., J. Thomas, P. Vahab and M.C.P. Timmermans. 2005. Maize rough sheath2 and its Arabidopsis orthologue ASYMMETRIC LEAVES1 interact with HIRA, a predicted histone chaperone, to maintain knox gene silencing and determinacy during organogenesis. Plant Cell 17: 2886-2898.

Ragni, L., E. Belles-Boix, M. Günl, and V. Pautot. 2008. Interaction of KNAT6 and KNAT2 with BREVIPEDICELLUS and PENNYWISE in Arabidopsis inflorescences. Plant Cell 20: 888-900.

Reddy, G.V. and E.M. Meyerowitz. 2005. Stem-cell homeostasis and growth dynamics can be uncoupled in the Arabidopsis shoot apex. Science 310: 663-667.

Reinhardt, D. 2000. Auxin regulates the initiation and radial position of plant lateral organs. Plant Cell Online 12: 507-518.

Rojo, E. 2002. CLV3 is localized to the extracellular space, where it activates the arabidopsis clavata stem cell signaling pathway. Plant Cell Online 14: 969-977.

Rupp, H.M., M. Frank, T. Werner, M. Strnad, and T. Schmülling. 1999. Increased steady state mRNA levels of the STM and KNAT1 homeobox genes in cytokinin overproducing *Arabidopsis thaliana* indicate a role for cytokinins in the shoot apical meristem. Plant J. 18: 557-563.

Sakakibara, K., S. Ando, H.K. Yip, Y. Tamada, Y. Hiwatashi, T. Murata, H. Deguchi, M. Hasebe and J.L. Bowman. 2013. KNOX2 genes regulate the haploid-to-diploid morphological transition in land plants. Science 339: 1067-1070.

Sakamoto, T., N. Kamiya, M. Ueguchi-Tanaka, S. Iwahori and M. Matsuoka. 2001. KNOX homeodomain protein directly suppresses the expression of a gibberellin biosynthetic gene in the tobacco shoot apical meristem. Genes Dev. 15: 581-590.

Sarkar, A.K., M. Luijten, S. Miyashima, M. Lenhard, T. Hashimoto, K. Nakajima, B. Scheres, R. Heidstra and T. Laux. 2007. Conserved factors regulate signalling in Arabidopsis thaliana shoot and root stem cell organizers. Nature 446: 811-814.

Scanlon, M.J. 2000. NARROW SHEATH1 functions from two meristematic foci during founder-cell recruitment in maize leaf development. Development 127: 4573-4585.

Scanlon, M.J. 2003. The polar auxin transport inhibitor N-1-naphthylphthalamic acid disrupts leaf initiation, KNOX protein regulation, and formation of leaf margins in maize. Plant Physiol. 133: 597-605.

Schoof, H., M. Lenhard, A. Haecker, K.F. Mayer, G. Jürgens and T. Laux. 2000. The stem cell population of Arabidopsis shoot meristems in maintained by a regulatory loop between the CLAVATA and WUSCHEL genes. Cell 100: 635-644.

Scofield, S., W. Dewitte, J. Nieuwland and J.A.H. Murray. 2013. The Arabidopsis homeobox gene SHOOT MERISTEMLESS has cellular and meristem-organisational roles with differential requirements for cytokinin and CYCD3 activity. Plant J. 75: 53-66.

Shah, N. and S. Sukumar. 2010. The Hox genes and their roles in oncogenesis. Nat. Rev. Cancer 10: 361-371.

Shani, E., O. Yanai, and N. Ori. 2006. The role of hormones in shoot apical meristem function. Curr. Opin. Plant Biol. 9: 484-489.

Shimizu, R., J. Ji, E. Kelsey, K. Ohtsu, P.S. Schnable and M.J. Scanlon. 2009. Tissue specificity and evolution of meristematic WOX3 function. Plant Physiol. 149: 841-850.

Smith, E.F., N.A. Brown and L. McCulloch. 1912. The structure and development of crown gall: A plant cancer. US Government Printing Office.

Smith, H.M. S., I. Boschke and S. Hake. 2002. Selective interaction of plant homeodomain proteins mediates high DNA-binding affinity. Proc. Natl. Acad. Sci. USA. 99: 9579-9584.

Smith, L.G., B. Greene, B. Veit and S. Hake. 1992. A dominant mutation in the maize homeobox gene, Knotted-1, causes its ectopic expression in leaf cells with altered fates. Development 116: 21-30.

Steeves, T.A. and I.M. Sussex. 1972. Patterns in Plant Development. Prentice-Hall, Englewood Cliffs, N.J.

Sultan, S.E. 2000. Phenotypic plasticity for plant development, function and life history. Trends Plant Sci. 5: 537-542.

Theodoris, G., N. Inada and M. Freeling. 2003. Conservation and molecular dissection of ROUGH SHEATH2 and ASYMMETRIC LEAVES1 function in leaf development. Proc. Natl. Acad. Sci. USA. 100: 6837-6842.

Timmermans, M.C.P., A. Hudson, P.W. Becraft and T. Nelson. 1999. ROUGH SHEATH2: A Myb Protein That Represses knox Homeobox Genes in Maize Lateral Organ Primordia. Science 284: 151-153.

Tsiantis, M., R. Schneeberger, J.F. Golz, M. Freeling and J.A. Langdale. 1999. The maize rough sheath2 gene and leaf development programs in monocot and dicot plants. Science 284: 154-156.

van der Graaff, E., T. Laux, and S.A. Rensing. 2009. The WUS homeobox-containing (WOX) protein family. Genome Biol. 10: 248.

Vandenbussche, M., A. Horstman, J. Zethof, R. Koes, A.S. Rijpkema, and T. Gerats. 2009. Differential recruitment of WOX transcription factors for lateral development and organ fusion in Petunia and Arabidopsis. Plant Cell 21: 2269-2283.

Vollbrecht, E., L. Reiser and S. Hake. 2000. Shoot meristem size is dependent on inbred background and presence of the maize homeobox gene, knotted1. Development 127: 3161-3172.

Vollbrecht, E., B. Veit, N.R. Sinha and S. Hake. 1991. The developmental gene Knotted-1 is a member of a maize homeobox gene family. Nature 350: 241-243.

Walbot, V. 1985. On the life strategies of plants and animals. Trends Genet. 1: 165-169.

Wong, C.E., M.B. Singh and P.L. Bhalla. 2013. Spatial expression of CLAVATA3 in the shoot apical meristem suggests it is not a stem cell marker in soybean. J. Exp. Bot: 64: 5641-5649.

Wu, X., J. Chory and D. Weigel. 2007. Combinations of WOX activities regulate tissue proliferation during Arabidopsis embryonic development. Dev. Biol. 309: 306-316.

Xu, C., K.L. Liberatore, C.A. MacAlister, Z. Huang, Y.-H. Chu, K. Jiang, C. Brooks, M. Ogawa-Ohnishi, G. Xiong, M. Pauly, J. Van Eck, Y. Matsubayashi, E. van der Knaap and Z. B. Lippman. 2015. A cascade of arabinosyltransferases controls shoot meristem size in tomato. Nat. Genet. 47: 784-792.

Xu, L. and W.-H. Shen. 2008. Polycomb silencing of KNOX genes confines shoot stem cell niches in Arabidopsis. Curr. Biol. 18: 1966-1971.

Xu, X.M., J. Wang, Z. Xuan, A. Goldshmidt, P.G. M. Borrill, N. Hariharan, J.Y. Kim and D. Jackson. 2011. Chaperonins facilitate KNOTTED1 cell-to-cell trafficking and stem cell function. Science 333: 1141-1144.

Yadav, R.K., M. Perales, J. Gruel, T. Girke, H. Jönsson and G.V. Reddy. 2011. WUSCHEL protein movement mediates stem cell homeostasis in the Arabidopsis shoot apex. Genes Dev. 25: 2025-2030.

Yadav, R.K., M. Perales, J. Gruel, C. Ohno, M.G. Heisler, T. Girke, H. Jönsson and G.V. Reddy. 2013. Plant stem cell maintenance involves direct transcriptional repression of differentiation program. Mol. Syst. Biol. 9: 654.

Yadav, R.K., M. Tavakkoli and G.V. Reddy. 2010. WUSCHEL mediates stem cell homeostasis by regulating stem cell number and patterns of cell division and differentiation of stem cell progenitors. Development 137: 3581-3589.

Zhang, W., R. Swarup, M. Bennett, G.E. Schaller and J.J. Kieber. 2013. Cytokinin induces cell division in the quiescent center of the Arabidopsis root apical meristem. Curr. Biol. 23: 1979-1989.

Zhou, Y., X. Liu, E.M. Engstrom, Z.L. Nimchuk, J.L. Pruneda-Paz, P.T. Tarr, A. Yan, S.A. Kay and E.M. Meyerowitz. 2014. Control of plant stem cell function by conserved interacting transcriptional regulators. Nature 517: 377-380.

17

Transdifferentiation: A Plant Perspective

Suong T.T. Nguyen and David W. McCurdy*

Introduction

Transdifferentiation is defined as the irreversible switch of one differentiated cell type into another (Okada 1991). This process occurs in both plants and animals, but transdifferentiation as a biological phenomenon has attracted much greater attention in animal systems compared to plants. In animals, this process, together with dedifferentiation and reprogramming, reflects the flexibility in cell differentiation and morphogenesis and has brought a new view of differentiated cells and their determined states. Traditional understanding of embryonic development favoured the concept of lineage-based differentiation, where a differentiated cell was viewed as the last of a progressive sequence of binary choices and once acquired, a stable differentiated state could not change its phenotype (Okada 1991, Grafi 2004, Tosh and Horb 2013). Therefore, transdifferentiation represented a reversal of developmental progression and hence received much criticism in the early literature. To some critics, examples of such cell type switching were attributed to tissue culture artefacts or cell fusion, or merely an exceptional phenomenon and thus of little significance for further investigation. However, since the first discovery of transdifferentiation, known as Wolffian lens generation (formation of a lens from the iris of the eye) was reported by G. Wolff in 1895 (see Okada 1991 and references therein), increasing numbers of reliable systems of transdifferentiation have been demonstrated, thus arguing for biological reality of this process (Eguchi and Okada 1973, Itoh and Eguchi 1986, Okada 1991). Since then, numerous studies have been devoted to transdifferentiation and other related subjects of flexibility in animal cell differentiation, not only because of their increasingly recognized significance in developmental biology, but more importantly their therapeutic potential in regenerative medicine (Tosh and Slack 2002, Burke and Tosh 2005, Jopling et al. 2011).

Although the occurrence of transdifferentiation either in nature, during regeneration or in tissue culture has been unequivocally established, it is still a matter of some controversy as to how individual cases can be confirmed or dismissed as a true transdifferentiation event. A set of prerequisites proposed by Eguchi and Kodama (1993)

School of Environmental and Life Sciences, The University of Newcastle, NSW 2308, Australia.
* Corresponding author : David.McCurdy@newcastle.edu.au

has been widely accepted among reproductive biologists to distinguish transdifferentiation from other types of modulation in cell differentiation. First, the differentiated states before and after the transdifferentiation must be clearly distinguished from one another based on morphological criteria accompanied by molecular and/or biochemical characterization. Second, the cell lineage relationship between the two cell types must be substantiated (Okada 1986, Eguchi and Kodama 1993). Transdifferentiation frequently involves the intervention of dedifferentiation and cell division, but these intermediate processes are not obligatory in all cases (Okada 1986, Beresford 1990, Eguchi and Kodama 1993).

To the best of our knowledge, there has been no comparable set(s) of criteria to evaluate a transdifferentiation process in plants. This is not surprising because plant cells are considered to have much greater developmental plasticity compared to animal cells, therefore observations of switches between two differentiated cells types, namely transdifferentiation, does not necessarily invoke criticism among plant biologists and thus the use of the term "transdifferentiation" tends to be easily accepted without much attention to the nature of the actual process. Furthermore, apart from the contribution toward general understanding of plant cell differentiation, transdifferentiation in plants seems not to have significant practical implications such as the therapeutic potential in regenerative medicine resulting from transdifferentiation in animals. Therefore, in some cases, transdifferentiation in plants is exchangeably described as redifferentiation or simply differentiation (Grafi et al. 2011b, Bhojwani and Dantu 2013), without significant impacts on developmental outcomes of the organism.

In this Chapter, we review representative examples of transdifferentiation in plants and evaluate whether they represent true examples of this phenomena in light of the criteria used in animal research, and further, whether these criteria are indeed relevant to plant biology. We also discuss current understanding of molecular pathways underlying these processes, and describe other examples occurring naturally or in regeneration which fulfil the criteria for transdifferentiation but have been more or less neglected as examples of this process in plants.

Historical Review: Xylem Cells as a Spotlight for Research on Plant Cell Transdifferentiation

The conversion of various differentiated non-vascular cells into vascular cells observed in nature or in experimental systems was well documented long before the phenomenon was referred to as transdifferentiation. In 1908, Simon first demonstrated the regeneration of xylem strands resulting from a switch of existing parenchyma cells into tracheary elements (TEs) around wounding sites of *Coleus* internodes (see Jacobs 1952 and references therein). Since then, wound-induced vascular regeneration, particularly xylem regeneration, has been systematically studied in *Coleus* (Jacobs 1952, Aloni and Jacobs 1977a,b) and other species including tobacco (Sussex et al. 1972), pea (Robbertse and McCully 1979, Rana and Gahan 1983) and pine (Kalev and Aloni 1999). The *in vitro* differentiation of TEs from, for example, cell suspension cultures of *Centaurea* (Torrey 1975), pith parenchyma cells of cultured lettuce (Dalessandro and Roberts 1971, Wilson et al. 1982), Jerusalem artichoke (Phillips and Dodds 1977) or carrot explants (Mizuno et al. 1971), and in particular mesophyll cells of *Zinnia elegans* (Fukuda and Komamine 1980a,

Church and Galston 1988), are even more dramatic examples of this process and have been extensively studied in the early literature. The remarkable *Zinnia* system, first reported in 1975 by Kohlenbach and Schmidt (see Fukuda and Komamine 1980a and references therein), has been the main focus for research on this phenomenon in plant physiology, with numerous publications appearing from 1980 to 1989 from laboratories studying different aspects of the *in vitro* conversion of isolated *Zinnia* mesophyll cells into TEs. Over this period, this phenomenon was variously referred to as "formation", "transformation", "redifferentiation", "*in vitro* differentiation" or simply "differentiation" of TEs.

Despite the long history of investigation of such switches from various differentiated cell types into TEs either in regeneration or *in vitro*, it was not until 1990 when the term "transdifferentiation" was first used in the plant literature by Sugiyama and Komamine to describe the phenomenon of isolated *Zinnia* mesophyll cells converting to TEs, but without, however, any definitive definition of the term or explanation of its usage (Sugiyama and Komamine 1990). The term "transdifferentiation" has since been accepted and used, relatively widely but not systematically, in subsequent studies of this phenomenon.

The Importance of Terminology

Transdifferentiation is generally categorized as either direct or indirect. Originally this subdivision was based on the absence or presence of cell division. Cells not dividing before altering their differentiated state were considered to undergo direct transdifferentiation, while indirect transdifferentiation required the intervention of cell division (Okada 1986, Beresford 1990). This subdivision is still widely applied in both animal (Tosh and Slack 2002, Wang et al. 2015) and plant research (Sugiyama and Komamine 1990, McManus et al. 1998, Reusche et al. 2012). However, a more recent definition in the animal literature for direct transdifferentiation describes conversion of a differentiated cell without undergoing dedifferentiation back to a pluripotent state or progenitor cell type (Jopling et al. 2011, Ma et al. 2013, Xu et al. 2014, Cai et al. 2007). These two definitions would coincide if dedifferentiation is coupled with cell division; however, in several cases cells can undergo dedifferentiation independent of cell division (Grafi 2004, Srivastava 2002, Sugiyama 2015), hence causing potential confusion when using these terms. For example, a transdifferentiation process occurring without cell division but with dedifferentiation (to a progenitor cell type) is defined as direct transdifferentiation according to Tosh and Slack (2002), but as indirect transdifferentiation according to the definition of Jopling et al. (2011). This confusion in terminology can be elevated when transdifferentiation is uncoupled from the dedifferentiation process entirely (Hanna et al. 2010, Sugimoto et al. 2011, Mora and Raya 2013), or to a lesser extent cell division (McManus et al. 1998, Reyes and Verfaillie 2004), where the term "transdifferentiation" is applied to describe only direct conversions of cell fate that do not involve dedifferentiation or cell division, respectively.

In our view, the definition of direct transdifferentiation by Jopling et al. (2011) and others, i.e., the conversion of differentiated cells without dedifferentiation, provides a more precise description of this process, incorporating current understanding of stem cells and reprogramming. Therefore, as stated previously, we apply this

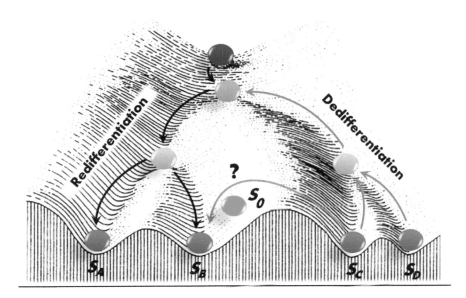

FIGURE 1 Transdifferentiation process in the Waddington's epigenetic landscape shows cell populations with different developmental potentials. Coloured marbles correspond to differentiation states: red (totipotent), orange (pluripotent), green (multipotent) and dark blue (singlepotent). The four dark blue marbles depict four differentiated cell states S_A–S_D. Normal differentiation is illustrated by black arrows. Indirect transdifferentiation involves a dedifferentiation step (blue arrows) requiring differentiated cells reverting back to a pluripotent state before undergoing a process of redifferentiation (black arrows). Direct transdifferentiation does not require cells to undergo dedifferentiation but potentially transit through an explicit intermediate stable state S_0 (light blue marble) as proposed by Xu et al. (2014).

definition to review examples of transdifferentiation in plants. We also have adapted the elegant approach of biophysicists (Wang et al. 2011, Xu et al. 2014) with regard to cell development and differentiation, where Waddington's epigenetic landscape (Waddington 1957) was explored to describe normal cell differentiation as well as the possibility of reverse differentiation of differentiated cells. Figure 1 is a combined modification of the original epigenetic landscape (Waddington 1957) and the schemes of cell type switching proposed by Xu and colleagues (see Fig. 1A in Xu et al. 2014). In this figure, coloured marbles represent cells at different states of differentiation occupying different valleys of the landscape; the deeper the valley the more highly differentiated state of the cell. The red marble corresponding to a stem cell or totipotent cell sits at the top of the landscape, follows the chreode-paths, gradually loses its totipotency and ultimately gives rise to a differentiated (singlepotent) cell (blue marbles) at the bottom of the landscape. Normal differentiation of cells is depicted by the black arrows, whereas blue and orange arrows illustrate dedifferentiation and direct transdifferentiation, respectively. The fate of a plant cell can be determined to be at, for example, the differentiated state S_B following normal differentiation. In tissue culture or regeneration in response to injury or pathogen attack, differentiated cell state S_B can be formed via direct and/or indirect transdifferentiation. Indirect transdifferentiation of S_B from S_C involves dedifferentiation of S_C to a progenitor or pluripotent state (orange marble) followed by normal differentiation to the new cell identity (S_B). The stepwise dedifferentiation requires the cell to move to a higher level on Waddington's landscape to reach less differentiated (multipotent/pluripotent) states represented by the green/orange marbles, and this can occur with or without cell division. In other cases S_C can transdifferentiate directly (orange arrow) to S_B skipping the dedifferentiation step, probably through an explicit intermediate stable state S_0 (light blue marble, Fig. 1) (Xu et al. 2014).

Examples of Transdifferentiation in Plant Biology

In Vitro Transdifferentiation of Tracheary Elements

The Zinnia elegans system

The best studied example of transdifferentiation is that of isolated mesophyll cells from *Zinnia elegans* leaves which transdifferentiate in culture to form TEs (Fukuda and Komamine 1980a, 1980b). This transdifferentiation process occurs in the absence of cell division, hence has been referred to as "direct transdifferentiation", first by the animal biologist Beresford (1990), and subsequently by Sugiyama and Komamine (1990) and others in the plant literature (McManus et al. 1998, Reusche et al. 2012). However, transdifferentiation in the *Zinnia* system requires a dedifferentiation step in which cells revert to a pluripotent procambial-like state (Fukuda 1996, 1997); therefore, according to the use of the term as described above, we propose that "indirect transdifferentiation" is a more appropriate term to describe this phenomenon.

The *in vitro* system for TE transdifferentiation in *Zinnia* was first established as an effective experimental system with high transdifferentiation frequency by Fukuda and Komamine (1980a) and has been used widely by various laboratories with minor modifications. Single mesophyll cells of *Zinnia* can be mechanically isolated by macerating pieces of surface-sterilized leaves. After extensive washing, the isolated cells are placed in a culture medium in the presence of auxin and cytokinin, and by three to four days culture, while some cells transdifferentiate to become xylem parenchyma, 30% of the population transdifferentiate to form TEs as defined by characteristic secondary wall thickenings (Fig. 2A) (Fukuda and Komamine 1980a). Early work on this system reported differential requirements for the hormones, with Church and Galston (1988) showing that auxin was required for the first 56 h of culture but cytokinin for only the first 24 h, whereas Fukuda and Komamine (1985) demonstrated that auxin and cytokinin were not required at all in the first 12 h of culture. Milioni et al. (2001) conducted a detailed time-course study of this response and made the remarkable discovery that the isolated mesophyll cells responded maximally to added auxin and cytokinin across a window from 46 to 50 h of culture, and within this window the hormones need only be present for 10 min to achieve a maximum response of 50% of the cells transdifferentiating to form TEs.

Further experimentation established that the transdifferentiation process occurred across three discrete stages. Figure 2B depicts these three stages on the Waddington's landscape in terms of dedifferentiation (Stage I), restriction of developmental potential (Stage II), and specific TE development (Stage III) (Fukuda 1996, 1997, 2004). Stage I can be regarded as a preparation stage to establish the

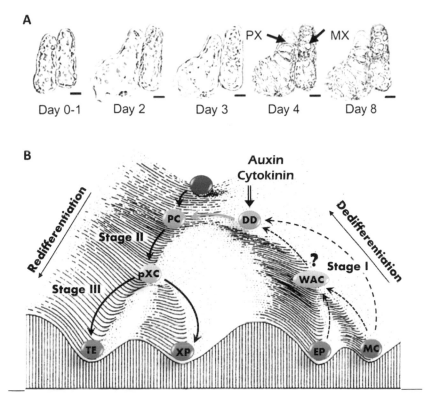

FIGURE 2 Transdifferentiation of tracheary elements from isolated mesophyll cells in the *Zinnia elegans* culture system. **(A)** Morphological changes of transdifferentiating mesophyll cells when cultured in semisolid medium. The same two mesophyll cells were monitored over time, with secondary wall deposition clearly seen by 4 days culture. Remarkably, one cell transdifferentiated into a protoxylem-like TE (PX), whereas the other into a metaxylem-like TE (MX). The culturing in semisolid medium established that the transdifferentiation process occurred without cell division. Scale bars = 10 μm. Figure reproduced from Beňová-Kákošová et al. (2006) with permission. **(B)** TE transdifferentiation from isolated mesophyll cells is illustrated in the Waddington's epigenetic landscape. Stage I corresponds to the dedifferentiation process, whereby isolated mesophyll cells adapt to a new environment in the culture medium, lose mesophyll identity and acquire pluripotent potential to progress further differentiation. During this reverse process, isolated mesophyll cells are proposed to transition through intermediate cell states, including wound-activated cell (WAC) and/or dedifferentiated cell (DD) states (Fukuda 2004). The transition from Stage I to Stage II is marked by the addition of auxin and cytokinin to the culture medium, which can be as little as a 10-min window at 48 h culture (Milioni et al. 2001, 2002), inducing DD cells to commit to a procambial-like cell (PC) state. Stages II and III can be referred to as a redifferentiation process, where PC can differentiate to form a xylem precursor-like cell (pXC) state, which then differentiates to form a tracheary element (TE), or less commonly xylem parenchyma cell (XP). In other culture systems, different cell types such as epidermal cells (EP) can transdifferentiate to become TEs.

competency of mesophyll cells becoming responsive to inductive signals to undergo redifferentiation (stages II and III). Mesophyll cells possess a highly differentiated phenotype with a large central vacuole and numerous lens-shaped chloroplasts occupying most parts of the cytoplasm and functioning in photosynthesis. These morphological and functional specifications remain stable during the lifespan of the cells *in planta*, hence in the Waddington's landscape (Fig. 2B) mesophyll cells metaphorically occupy a valley comparable to one of the four valleys representing four differentiated cell states (S_A to S_D) illustrated in Figure 1. In order to regain pluripotency mesophyll cells rely upon a sufficient push to escape their differentiated valley and progressively climb up the Waddington's landscape. Once directed to follow the correct reverse paths (dashed arrows, Fig. 2B) to reach a less-differentiated state, these mesophyll cells, now regarded as dedifferentiated cells (orange marble DD, Fig. 2B), are competent to respond to the inductive signals auxin and cytokinin in the culture medium to further progress differentiation. The transition from Stage I to Stage II is determined by a combination of these signals, whereby dedifferentiated cells become re-specified to a new fate, in this case xylem cell fate, via a pluripotent procambial-like state (orange marble PC, Fig. 2B). Stage II corresponds to the process of differentiation from procambial-like cells into xylem cell precursors (green marble pXC, Fig. 2B), and Stage III involves dramatic changes in cell morphology such as deposition of patterned secondary cell walls and the autolysis of cell content, ultimately resulting in xylem cell precursors becoming dead TEs characterized by hollow water-conducting tubes (blue marble TE, Fig. 2B). Collectively, stages II and III resemble the process of *in planta* xylem cell fate determination, in which procambial cells, differentiating from the meristem via procambial initials, metaphorically roll down the valley in the Waddington's landscape, gradually lose their pluripotency and finally reach the zone of single potency of xylem parenchyma (XP) cells or tracheary elements (TE) (Fig. 2B). In this example, meristematic stem cells are represented by the red marble sitting at the top of the Waddington's landscape, and procambial initials are not shown.

If one is satisfied with the Waddington's landscape as a depiction of TE transdifferentiation, then Stage I can be interpreted as being associated with a gravity-defeating push which directs mesophyll cells to move higher up the landscape, in a process known as dedifferentiation. Very little is known about the molecular events constituting this "push", despite the fact that cell dedifferentiation in general has long attracted interest from both animal and plant biologists. It is believed that this process is a prerequisite for the whole transdifferentiation process to occur (Fukuda 1997, McCann 1997) but seems not to be specific to the step of TE cell fate determination (Fukuda 1994). Indeed, auxin and cytokinin, which are absolute requirements for dedifferentiated cells restricting their potency of differentiation to a TE cell fate in the *Zinnia* system, are not required in the first 12 h (Fukuda and Komamine 1985) or more precisely the first 48 h (Milioni et al. 2001) of culture. Milioni and colleagues (2001) have proposed that the first 48 h correspond to a developmental course in which mesophyll cells adapt to a new environment and acquire competency to respond to inductive signals. They found that a narrow window of 10 min of exposure of auxin and cytokinin at 48 h is both necessary and sufficient to induce TE differentiation by 96 h. Furthermore, by adding these two growth hormones at 48 h instead of 0 h of culture, they improved the synchrony of the *Zinnia* system significantly (Milioni et al. 2001). These findings suggest different levels of competency in different mesophyll cells undergoing the dedifferentiation process. This situation can be interpreted based on the Waddington's landscape: it is critical for dedifferentiating mesophyll cells to reach a certain height on the landscape, here illustrated by the position of a dedifferentiated cell (DD - orange marble, Fig. 2B), to be competent to respond to auxin and cytokinin. Any cell states below this threshold height are not sufficiently competent to respond to these signals. Different isolated mesophyll cells in the culture may climb the landscape at different rates, or may start this process later than neighbouring cells, which collectively may account for the relatively asynchronous TE formation if inductive signals are present at the start of the culture. By 48 h, most mesophyll cells in the culture become fully competent. In an induction-free medium, all cells that can attain competency have done so by 48 h, and thus subsequent exposure to inductive signals induces this population to collectively commit to a new fate TE fate. Delaying the addition of auxin and cytokinin for more than 48 h reduced the final number of TEs, although a small number of cells remain competent for TE formation at even 90 h of culture (Milioni et al. 2001). This observation suggests that cells may be able to hold the dedifferentiated state for only a defined period of time, after which the "gravity" force pulls these cells down the Waddington's landscape, probably to the valleys of apoptosis if no hormone(s) is added.

Interpretation of transdifferentiation based on this uphill-climbing model may raise a question about the nature of the gravity-defeating push, which bought about the dedifferentiated state of mesophyll cells in Stage I. Wounding, auxin and cytokinin are the three factors involved in initiating the transdifferentiation of TEs from mesophyll cells in the *Zinnia* system (Fukuda 1997, McCann 1997). The work done by Milioni et al. (2001), as discussed above, plus other evidence (Fukuda and Komamine 1985), seems to exclude the participation of auxin and cytokinin in early events associated with cell adapting to new environments and acquiring competency, hence pointing to wounding as potentially the only factor controlling dedifferentiation in this system. However, this hypothesis has not been investigated further at the molecular level, partially due to the fact that the redifferentiation process, corresponding to stages II and III, has attracted more interest compared to dedifferentiation occurring during Stage I. Indeed Milioni and colleagues took advantage of their highly synchronous *Zinnia* system to perform large-scale expression analysis based on a cDNA-amplified fragment length polymorphism approach, however, just focusing on the time points at 48 h onward and overlooking the dedifferentiation process presumably occurring before the addition of auxin and cytokinin (Milioni et al. 2002). The only molecular work published on Stage I was done by Demura et al. (2002), where a comprehensive analysis using a large cDNA array with more than 8000 *Zinnia* cDNA clones was performed to examine transcriptional profiles throughout the transdifferentiation process. This study revealed two main groups of events during Stage I: wound-induced events and those of down-regulation of photosynthetic activity. Wound-induced events were believed to occur during the 1 h preparation of isolated mesophyll cells and to be induced by wound stress. These events are potentially associated with genes encoding six protein kinases and three transcription factors (Demura et al. 2002). Down-regulation of photosynthetic activity is evidenced by the down-regulation of 29 out of 33 cDNAs potentially encoding plastid ribosomal proteins and chlorophyll a/b-binding proteins (Demura et al. 2002), which presumably accounts for mesophyll cells losing the ability to perform photosynthesis (Fukuda 1997). In this culture system, however, unlike the system used by Milioni and colleagues (2001, 2002), auxin and cytokinin were administered to isolated mesophyll cells at the beginning of culture, hence the influence of these hormones was not necessarily separated from events occurring during Stage I. Therefore, it still remains unclear if the down-regulation of photosynthetic genes, among other early events associated with cells adapting to a new environment and acquiring competency, is attributable solely to wounding or the combination of wounding, auxin and cytokinin. Further analysis of gene expression focusing on Stage I (from 0 to 48 h of culture without the addition of auxin and cytokinin; Milioni et al. 2001, 2002) may shed light on the dedifferentiation process, of which understanding at the molecular level is still in its infancy.

Although, it has been unequivocally established that cell division or replication of DNA in the S-phase of the cell cycle is not a prerequisite for the dedifferentiation step of TE transdifferentiation in the *Zinnia* system (Fukuda and Komamine 1980b, 1981a), some minor amount of DNA synthesis is essential for the sequence of transdifferentiation (Fukuda and Komamine 1981b, Sugiyama and Komamine 1990). This minor DNA synthesis was found to be DNA-repair events including repair-type synthesis of DNA, the synthesis of poly(ADP-ribose) and the rejoining of DNA strand breaks (Sugiyama et al. 1995, Shoji et al. 1997). This conclusion was further supported by the fact that culturing *Zinnia* mesophyll cells with PhT, an inhibitor of repair-type DNA synthesis, inhibited TE transdifferentiation via preventing the expression of TED (tracheary element differentiation) genes in Stage II (Demura and Fukuda 1993).

While various events occurring during the dedifferentiation process (Stage I) seem not to be specific to TE transdifferentiation, the transition from Stage I to Stage II can be regarded as the first specific step of TE cell fate determination. This transition presumably corresponds to the 10-minute window of auxin and cytokinin exposure in the system developed by Milioni et al. (2001, 2002) as discussed above (Fig. 2B). Withdrawal of auxin and cytokinin from the culture medium at 48 h after 10 min exposure does not affect the later processes of transdifferentiation (Milioni et al. 2001, 2002). This is strong evidence of a true transdifferentiation as it is consistent with observations of transdifferentiation in animals; for example, in *Caenorhabditis elegans*, brief expression of a single transcription factor can induce stable transdifferentiation of intestinal cells from fully differentiated pharynx cells (Riddle et al. 2013). This is an important characteristic of transdifferentiation, namely the phenotype of newly differentiated cells must remain stable (Okada 1991). This observation is also another testament to the excellence of the highly synchronous *Zinnia* system developed by the McCann group. It is also reasonable to envisage that, if another hormone or sets of hormones other than auxin and cytokinin are added at this specific window, mesophyll cells may be diverted to a different developmental pathway such as xylem parenchyma cells or even phloem sieve elements.

Stages II and III mimic *in planta* xylogenesis (Fukuda 2004) and thus have attracted greater interest from plant biologists. Thus, molecular events occurring in these two stages have been more thoroughly investigated compared to Stage I. Stage II corresponds to the *in vivo* differentiation from procambial initials to xylem cell precursors and is featured by the accumulation of *TED2*, *TED3* and *TED4* transcripts - novel markers for early stages of vascular differentiation (Demura and Fukuda 1993, 1994, Fukuda 1994). *In situ* hybridization revealed that these three transcripts accumulate in cells involved in vascular differentiation in intact plants. *TED2* transcripts, which encode a probable quinone oxidoreductase, were restricted to immature phloem and immature xylem cells and procambial cells of roots, whereas *TED3*, which encodes a putative cell wall protein (Demura and Fukuda 1994), and *TED4* which encodes a lipid transfer protein, were expressed mainly in TE precursor cells and immature xylem cells, respectively. Both *in situ* and *in vitro* analyses confirmed that TED transcripts were expressed sequentially, namely *TED2*, *TED4* and then *TED3* (Demura and Fukuda 1994). These markers were also identified in the cDNA-AFLP (Milioni et al. 2002) and microarray (Demura et al. 2002) analyses of gene expression programs during TE transdifferentiation in *Zinnia*. These two comprehensive studies also revealed that several genes involved in auxin signalling were expressed during Stage 2 (Demura et al. 2002), or within 30 minutes of the addition of auxin and cytokinin (Milioni et al. 2002). A NAC transcription factor was also upregulated within 30 minutes (Milioni et al. 2002).

Much later after the expression of *TED* transcripts, dramatic morphological changes including secondary wall formation and autolysis occur in transdifferentiating cells, indicating the commencement of Stage III. As noted by Demura et al. (2002), early Stage III gene expression is characterised by a large number of genes associated with secondary wall formation such as *CesA* and *EGase* genes as well as primary wall degradation including genes encoding *O*-glucosyl hydrolases. Later in Stage III, genes involved in lignin biosynthesis such as laccases- and peroxidases-genes were differentially expressed. This lignification process is believed to proceed after cell death, when autolysis is completed (Demura et al. 2002, Turner et al. 2007). Programmed cell death-specific genes, such as genes encoding cysteine proteases, serine proteases, aspartic peptidase and lipolytic acyl hydrolase, were differentially expressed during Stage III (Demura et al. 2002).

The Arabidopsis thaliana system

To take advantage of the wealth of publicly accessible molecular resources of the model plant *Arabidopsis thaliana*, various *Arabidopsis* xylogenic culture systems have been established. When *Arabidopsis* suspension cells were cultured with auxin, cytokinin, brassinosteroid and high concentration of boric acid, around 50% of cells transdifferentiated to form TEs after 7 days culture (Kubo et al. 2005). Utilizing subsequent GeneChip microarrays these workers identified a family of seven NAC transcription factors, *VND1-7*, which were upregulated during TE transdifferentiation. Among these genes, *VND6* and *VND7* were identified as master switches for the formation of metaxylem and protoxylem vessels, respectively (Kubo et al. 2005). Overexpression of either *VND6* or *VND7* causes ectopic TE development (Kubo et al. 2005, Yamaguchi et al. 2010). To examine microtubule dynamics during TE formation, Oda et al. (2005) developed a culture system using transformable *Arabidopsis* cell suspension cells. After 96 h culture approximately 30% of suspension cells of the AC-GT13 line (expressing a green fluorescent protein-tubulin fusion protein) were induced to form TEs by removal of auxin from and addition of brassinosteroid to the culture media (Oda et al. 2005). TEs derived from these cells were morphologically indistinguishable from those derived from non-transformed Col-0 cells, and also had distinct patterns of secondary wall deposition similar to those of mesophyll cell-derived TEs in the *Zinnia* system (Oda et al. 2005). Instead of hormones, master-switch transcription factors can be used as effective triggers to induce TE transdifferentiation (Oda et al. 2010), presumably involving direct transdifferentiation because no up-regulation of procambial-related genes was detected (Kondo et al. 2015).

Very recently a novel system for TE transdifferentiation using *Arabidopsis* leaf disks has been described (Kondo et al. 2015). In this system, bikinin was used to promote xylem cell differentiation via its inhibitive effects on glycogen synthase kinase

3 proteins, the central regulators for differentiation of procambial cells into xylem cells (Kondo et al. 2014). Leaf disks of *Arabidopsis* rosette leaves were cultured with auxin and cytokinin in the presence of bikinin. Three days after culture a large number of mesophyll cells transdifferentiated to form TEs in a bikinin dose-dependent manner (Kondo et al. 2015). This cell-type switching occurred via a dedifferentiation step to form procambial cells before converting to xylem cells, hence, according to our criteria described above, is an indirect transdifferentiation process. Bikinin was found to accelerate the formation of procambial cells, however, the underlying mechanism is unknown (Kondo et al. 2015). The advantages of this leaf-disk culture, as stated by the authors, are its synchronous differentiation of both procambial cells and TEs suitable for exploring the whole process of xylem transdifferentiation, its utilization of various *Arabidopsis* mutants and *Arabidopsis* transgenic lines harbouring vascular cell markers enabling spatiotemporal monitoring of cell fate switching, and its applicability to different species besides *Arabidopsis*. Such advantages promise a greater understanding of development of vascular tissues, particularly procambial cell formation and xylem cell differentiation.

Transdifferentiation of transfer cells

Transfer cells (TCs) represent a particularly striking and rare case of transdifferentiation associated with programmed development. TCs are anatomically specialised to achieve enhanced rates of nutrient transport across apoplasmic/symplasmic bottlenecks throughout the nutrient acquisition, distribution and utilization pathways in plants. TCs develop extensive wall ingrowths (McCurdy et al. 2008, McCurdy 2015) which provide a scaffold to increase plasma membrane surface area and thus enable enhanced densities of nutrient transporters to facilitate their capacity for enhanced rates of solute exchange (Offler et al. 2003). TCs also develop localised densities of mitochondria and organelles of the endomembrane system to accommodate increased energy demands to drive nutrient transport and vesicle secretion for wall ingrowth building, respectively (Gunning and Pate 1969, Pate and Gunning 1972). Numerous differentiated cell types, such as companion cells, xylem and phloem parenchyma cells, root cortical cells and epidermal cells can develop TC characteristics (Pate and Gunning 1972). This process is considered to be an example of transdifferentiation, whereby a given cell type, for example a phloem parenchyma cell, transdifferentiates to become a phloem parenchyma TC (Offler et al. 2003). It is important to clarify, however, that not all TCs originate via a transdifferentiation process. TCs in the basal endosperm of cereal seeds, such as barley, are a component of the triploid endosperm, and develop as part of a cell fate specification pathway during endosperm cellularization (Thiel 2014), a process which is distinct from transdifferentiation. Also, giant cells and syncytia resulting from nematode infection of roots have been described as TC-like due to the presence of wall ingrowth-like structures (Jones and Northcote 1972a,b). In this case, however, these cells form from undifferentiated procambial cells and pericycle, and involve the fusion of many cells into a single, larger giant cell or syncytia, a process which seems to be quite different from transdifferentiation of a single cell from one type to another of different function.

Two prominent examples of the transdifferentiation of TCs in plant biology are considered here. Abaxial epidermal cells of *Vicia faba* cotyledons transdifferentiate to become epidermal TCs and thus facilitate nutrient transport across the maternal/

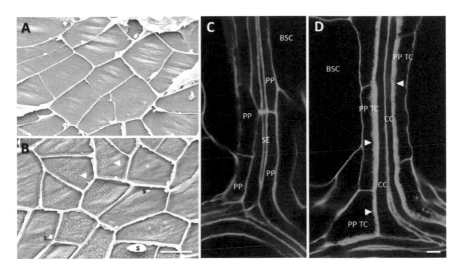

FIGURE 3 Transdifferentiation of transfer cells. **(A, B)** Scanning electron microscopy views of the cytoplasmic face of adaxial epidermal cells of cultured *Vicia faba* cotyledons. The cytoplasmic face of the outer periclinal wall of epidermal cells of cotyledons was revealed by a dry-cleaving procedure (Talbot et al. 2001). **(A)** Adaxial epidermal cells at 0 h culture. In normal development of the cotyledons, adaxial epidermal cells do not transdifferentiate into transfer cells, as evidenced by their smooth primary cell walls. Remnant anticlinal walls remaining after the dry-cleaving procedure reveal the outline of each epidermal cell. **(B)** Adaxial epidermal cells after 15 h culture undergoing transdifferentiation to form transfer cells, as evidenced by numerous individual papillate wall ingrowths (arrowheads) deposited on the cytoplasmic face of the outer periclinal primary wall. Note that the extent of wall ingrowth deposition varies in each epidermal cell, with one cell (asterisk) yet to initiate this process. S - starch grains. Scale bar = 20 μm. Images provided by Dr Mark Talbot. **(C, D)** Phloem parenchyma cells in *Arabidopsis* leaf minor veins. **(C)** In immature juvenile leaves phloem parenchyma (PP) cells in vascular bundles have not developed wall ingrowths. **(D)** In mature juvenile leaves phloem parenchyma cells transdifferentiate into TCs, as evidenced by the polarised deposition of an extensive layer of reticulate wall ingrowths (arrowheads) along the face of PP TCs adjacent to cells of the SE/CC complex. PP - phloem parenchyma; BSC – bundle sheath cell; CC – companion cell; TC – transfer cell; SE – sieve element. Scale bar = 5 μm.

filial interface of developing seeds (Harrington et al. 1997). Importantly from an experimental perspective, adaxial epidermal cells, which normally do not form TCs *in planta*, can be induced to transdifferentiate into adaxial epidermal TCs by placing isolated cotyledons adaxial-surface down on culture medium (Farley et al. 2000, Offler et al. 2003). This process induces a semi-synchronous development of a reticulate wall ingrowth network (Fig. 3) and ultimately TC functionality (Farley et al. 2000). This experimental induction of adaxial epidermal TCs has been used extensively to study the morphology of reticulate wall ingrowth deposition (Talbot et al. 2001, 2002, 2007), signalling pathways that induce wall ingrowth deposition (Zhou et al. 2010, Andriunas et al. 2011, 2012) and transcriptional changes associated with this process (Dibley et al. 2009, Zhang et al. 2015). Upon transfer of cotyledons to culture, cell division is transiently induced in about 10% of the adaxial epidermal cells (Dibley et al. 2009, Zhang et al. 2015), but nearly 90% of all adaxial epidermal cells go on to form a TC morphology by 24 h of culture (Wardini et al. 1997). This observation indicates that cell division is not a mandatory step in this example of transdifferentiation, and questions whether dedifferentiation to a mitotic state is a prerequisite event (Dibley et al. 2009). Epidermal-specific changes in expression of numerous transcription factors occur across TC transdifferentiation (Dibley et al. 2009, Zhang et al. 2015), including genes associated with chromatin remodelling (Dibley et al. 2009), but definitive molecular evidence for dedifferentiation is absent. Consequently, the transdifferentiation of epidermal TCs in *V. faba* cotyledons may represent a case of direct transdifferentiation not involving cell division.

Phloem parenchyma cells also transdifferentiate into phloem parenchyma TCs in veins of *Arabidopsis* leaves (Haritatos et al. 2000, Amiard et al. 2007) and cotyledons (Nguyen and McCurdy 2015). Wall ingrowths in phloem parenchyma cells are deposited across the region of cell wall abutting adjacent cells of the sieve element/companion cell (SE/CC) complex (Fig. 3). These phloem parenchyma TCs are proposed to be involved in a two-step phloem loading strategy whereby sucrose delivered by an apoplasmic pathway from mesophyll cells is unloaded into the apoplasm by AtSWEET11/12 uniporters located on the plasma membrane of the phloem parenchyma TCs. This apoplasmic sucrose is subsequently uploaded into the SE/CC complex via CC-localised AtSUC2 (Chen et al. 2012). The highly localized deposition of wall ingrowths to the face of phloem parenchyma cells directly neighbouring cells of the SE/CC complex is proposed to spatially restrict the unloading of apoplasmic sucrose to sites immediately adjacent to the SE/CC complex, thus minimizing access to this sucrose by pathogens (Chen et al. 2012). In addition to this role in facilitating phloem loading, Amiard et al. (2007) have proposed that the wall ingrowths may provide a physical barrier to pathogen entry into SEs. This conclusion is based in part from the observation that wall ingrowth deposition in phloem parenchyma TCs is enhanced in response to stresses involving jasmonic acid signalling (Amiard et al. 2007). However, while it appears that wall ingrowth deposition in phloem parenchyma TCs can be enhanced in response to stress hormones, the signals that cause induction of the transdifferentiation process itself are not known. No evidence is available suggesting that a preceding step of cell division of phloem parenchyma occurs prior to the formation of wall ingrowths, therefore, similar to the epidermal cells of *V. faba* cotyledons, this transdifferentiation process may result from direct transdifferentiation without cell division.

Transdifferentiation in regeneration

Transdifferentiation occurring as part of regeneration of plant tissues represents a common example of transdifferentiation in plants. The disturbance of the microenvironment of cells is a critical trigger for transdifferentiation to occur (Okada 1991). Sources of this disturbance may come from mechanical injury or attack by pathogens, potentially providing positional signals to convert existing differentiated cells to form new cell types to compensate for loss of functional cells. Investigation at the molecular level of transdifferentiation associated with regeneration, however, is difficult due to the highly asynchronous nature of regeneration. Unlike isolated cell culture systems (*Zinnia, Arabidopsis*) or culture of *V. faba* cotyledons or *Arabidopsis* leaf disks, in which reasonably synchronous transdifferentiation of a single cell type occurs, transdifferentiation associated with regeneration may involve various cell types, often occurring in a sequential manner and thus compromising the interpretation of molecular events associated with the transdifferentiation process specific to each cell type. Therefore, current understanding of molecular pathways regulating transdifferentiation occurring during regeneration is limited.

Since vascular strand continuity is essential for plant development, plants employ several strategies including transdifferentiation to re-establish this continuity when it is interrupted by various disturbances. Consequently, transdifferentiation of vascular cells from various cell types is the most frequently studied example of transdifferentiation in regeneration. Transdifferentiation of vascular cells, particularly of primary vascular tissues, has a long history of investigation in a variety of species (although not always being referred to as a transdifferentiation process in the early literature; see Historical Review section). When the primary vascular tissue in a stem or petiole is severed, for example by a wound, new xylem and phloem cells are generated by transdifferentiation of existing parenchyma cells, bridging new vascular tissue with old to overcome the discontinuity. Presumably the molecular mechanisms of wound-induced transdifferentiation of vascular cells, *in vitro* transdifferentiation of TEs and *in planta* differentiation of TEs during normal vascular development, are similar if not identical. Indeed the processes of transdifferentiation of TEs from pith parenchyma cells induced *in situ* by vascular bundle interruption in *Zinnia* internodes (Nishitani et al. 2002) and from isolated mesophyll cells in suspension culture (Demura and Fukuda 1993), both involve the expression of *TED3*, a molecular marker of TE precursor cells in *Zinnia* (Demura and Fukuda 1994). When vascular strands of the first internodes of Zinnia seedlings are injured with a blade, transcripts of *TED3* accumulate in pith parenchyma cells surrounding the wound site 48 h after wounding (Nishitani et al. 2002). Similarly, *TED3* transcripts accumulated by 36 h culture of isolated mesophyll cells (Demura and Fukuda 1993). Nishitani et al. (2002) also reported that transcripts of *ZeHB3* (*Zinnia elegans* homeobox gene 3), a molecular marker of early stages of phloem differentiation in intact plants (Nishitani et al. 2001), increased by 36 h post-wounding in immature phloem layers in and between severed vascular bundles (Nishitani et al. 2002). It is interesting to note that, in severed *Zinnia* internodes, wound-induced transdifferentiation of pith parenchyma cells into TE precursors, akin to the *in vitro* TE transdifferentiation from isolated mesophyll cells, does not involve the intervention of cell division, whereas wound-induced transdifferentiation of pith parenchyma cells into immature phloem cells requires cell division (Nishitani et al. 2002). Whether a dedifferentiation step is a prerequisite of these transdifferentiation processes is currently unknown.

Transdifferentiation of vascular cells has also been observed in secondary vascular tissue during bark regeneration. When bark of *Eucommia ulmoides* Oliv. and many other species is removed at a suitable developmental stage of the tree, new bark is rebuilt within 1-2 months (Pang et al. 2008, Zhang et al. 2011, Chen et al. 2014). During this regeneration of secondary vascular tissue in *E. ulmoides*, immature xylem cells in several cell surface layers formed callus at six days after girdling (DAG), whereas immature xylem cells in layers beneath these surface layers transdifferentiated to form phloem cells including SEs and CCs at 12 DAG, and cells in the deeper layers of immature xylem cells dedifferentiated into cambium cells at 2-3 weeks after girdling (Pang et al. 2008). Similar processes were also observed during bark regeneration upon girdling in *Populus tomentose* (Zhang et al. 2011). Although the transdifferentiation of phloem cells from xylem cells in both *E. ulmoides* and *P. tomentose* is preceded by cell division, it is unclear if re-entry into the cell cycle is a prerequisite of this process. It has been proposed that, in *P. tomentose*, epigenetic regulation and cell cycle re-entry may play important roles in the switching of xylem cell fate into phloem or cambium cell fate, evidenced by the significant changes in gene expression of DNA methyltransferases, histone methyltransferases, chromatin remodelling-related proteins, polycomb group (PcG) proteins, type A and type B cyclins (CYCAs, CYCBs), cyclin-dependent kinases (CDKs) and CDK-like (CKL) genes (Zhang et al. 2011). The authors also reported down-regulation of xylem-specific genes as well as an activation of phloem and cambium development during bark regeneration (Zhang et al. 2011). It is not easy, however, to separate differential gene expression associated with the transdifferentiation process of phloem cells from that of the dedifferentiation of cambial cells, whereby progenitor cells of both examples are differentiating xylem cells left on the trunk after bark girdling. Mimicking these processes under more controlled laboratory conditions may provide a means to investigate the transdifferentiation process with higher resolution. An *in vitro* system of secondary vascular tissue regeneration in poplar has recently been established (unpublished data in Chen et al. 2014), promising a leap in our understanding of secondary vascular development, particularly in regeneration, in the near future.

In addition to physical assaults, pathogen attacks also induce transdifferentiation of vascular cells to substitute infected and non-functional vascular cells. A striking example of this scenario is in leaves of *Arabidopsis* seedlings following inoculation with *Verticilium longisporum* whereby chloroplast-containing bundle sheath cells transdifferentiate to form TEs characterized by secondary cell wall modifications (Fig. 4) (Reusche et al. 2012). Loss of chloroplasts in bundle sheath cells surrounding the vascular bundles in leaves results in a phenomenon known as vein clearing (Fradin and Thomma 2006), among other above-ground symptoms of infected plants including stunted growth, loss of fresh weight and chloroses (Reusche et al. 2012). *Verticilium*-infections frequently lead to wilting (Fradin and Thomma 2006); however, that is not the case for the invasion of *V. longisporum* in its *Brassicaceae* hosts including *Brassica napus* (Floerl et al. 2008) and *Arabidopsis* (Reusche et al. 2012). In these species, the infection of *V. longisporum* brought about enhanced drought tolerance of infected plants, which was attributed to *de novo* xylem formation resulting from the transdifferentiation of bundles sheath cells and xylem parenchyma cells and the reinitiation of cambial activity (Reusche et al. 2012). Histological analysis of bundle sheath cell-derived TEs revealed that some of these elements featured annular thickenings typical of protoxylem vessels while others exhibited reticulate thickenings typical of metaxylem vessels. Consistent with the finding that *VND6* and *VND7*, two NAC-domain transcription factors, are necessary and sufficient to induce transdifferentiation of various types of cells into metaxylem- and protoxylem-like vessel elements, respectively (Kubo et al. 2005), Reusche et al. (2012) showed that expression levels of *VND6* and *VND7* and their positive feedback loop regulators *ASL19* and *ASL20,* were significantly induced in infected plants at 14 day after inoculation. Moreover, in recently transdifferentiated bundle sheath cells, enhanced transcriptional activity of *VND7* in response to *V. longisporum* infection was also observed (Reusche et al. 2012). It is interesting to note that in this system the transdifferentiation occurred exclusively in bundle sheath cells located on the adaxial side (abutting xylem) but not on the abaxial side (abutting phloem) of vascular bundles of leaves, and that this transdifferentiation was not caused by *in situ* fungal colonization. These observations collectively suggest that the cue triggering this transdifferentiation may be derived from *V. longisporum*-colonized proximal vascular bundles (Reusche et al.

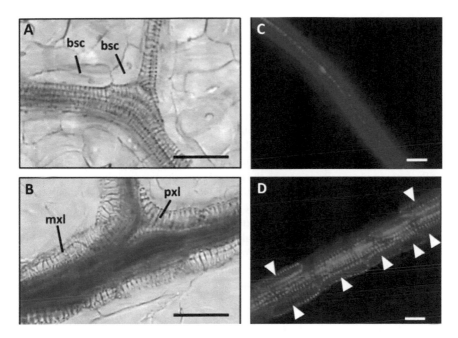

FIGURE 4 Transdifferentiation of tracheary elements from bundle sheath cells in *Arabidopsis* leaf veins upon *Verticillium* infection. (**A, C**) Vascular bundles in leaves of non-infected plants. (**B, D**) Vascular bundles in leaves of *Verticillium*-infected plants. (**A, B**) Bright-field images of leaf vein. Leaves were stained with trypan blue. (**A**) Leaf vein of non-infected plant showing clear bundle sheath cells (bsc) at the periphery of vascular bundles. (**B**) Leaf vein of infected plant at 21 days after infection showing bundle sheath cells transdifferentiated into protoxylem-like (pxl) and metaxylem-like (mxl) cells. (**C, D**) Epifluorescence images of leaf vein of non-infected plant (**C**) and leaf vein of infected plant at 21 days after infection (**D**). Arrowheads in (**D**) show *de novo*-formed TEs. bsc - bundle sheath cell; mxl - metaxylem-like; pxl - protoxylem-like. Scale bars = 50 µm for A and B, and 25 µm for C and D. Images reproduced from Reusche et al. (2012) with permission.

2012), transported through the transpiration stream and released into the vicinity of adaxial/xylem-associated bundle sheath cells.

The transdifferentiation of TEs from bundle sheath cells occurred without preceding cell division, and thus Reusche et al. (2012) considered this to be an example of direct transdifferentiation. However, as discussed previously, we define direct transdifferentiation as a process occurring without dedifferentiation; a process that was not investigated in the Reusche et al. (2012) study. In the *Zinnia* system the transdifferentiation of TEs from isolated mesophyll cells occurred via a dedifferentiation step where mesophyll cells reverted back to a pluripotent procambial-like stage before redifferentiating into TEs (Demura and Fukuda 1993, Fukuda 1997). The formation of procambial cells was also demonstrated as an intervening step during the transdifferentiation of mesophyll cells into TEs in the *Arabidopsis* leaf-disk culture system (Kondo et al. 2015). Similar phenomenon may occur during the conversion from bundle sheath cells into TEs, in which pathogen-induced senescence may allow cells to undergo dedifferentiation (Grafi et al. 2011a; Reusche et al. 2012). Alternatively, Reusche et al. (2012) also discussed the possibility that certain perivascular totipotent stem cells (Sugimoto et al. 2010, 2011) may have a positional and developmental relationship to the bundle sheath cells which transdifferentiated into TEs (Reusche et al. 2012).

Conclusion

Transdifferentiation is the remarkable ability of both plant and animal cells to undergo an irreversible switch from one cell type into another. Due to the long recognised capacity for developmental plasticity in plants, transdifferentiation has not received as much attention as that afforded in animal systems. Indeed, the first use of the term in plant science was by the animal reproductive biologist W.A. Beresford (1990), describing the transdifferentiation of TEs in the *Zinnia* mesophyll cell culture system, a phenomenon first described by Kohlenbach and Schmidt in 1975. Four decades later this remains the most extensively studied example of transdifferentiation at the morphological, biochemical and molecular levels in plant biology. Plant biologists have used the term transdifferentiation somewhat glibly to describe often different phenomena. We propose that transdifferentiation is best described as either indirect, involving dedifferentiation back to a progenitor cell state, or indirect, not involving dedifferentiation, or at least not to a progenitor cell state. Both direct and indirect transdifferentiation can occur with or without cell division. Examples of transdifferentiation in the plant literature mostly involve the formation of cells of the vascular system, in particular TEs. Presumably this reflects the importance of water transport in vascular plants, and the capacity of this system for regeneration after wounding. The *in vitro* transdifferentiation of TEs in both the *Zinnia* system and more recently in *Arabidopsis*, has been extremely valuable in identifying molecular switches responsible for TE differentiation *in planta*. More emphasis is now needed to understand at the molecular level the reverse step of transdifferentiation, namely dedifferentiation and potential re-entry into the cell cycle. Given the new ideas now emerging regarding dedifferentiation associated with *in vitro* reprogramming (Rose 2016, Chapter 18), it will be interesting to compare at the molecular level partial dedifferentiation associated with reversion to a pluripotent state (transdifferentiation) (Srivastava 2002), compared to complete dedifferentiation into callus tissue as can occur during a process such as somatic embryogenesis.

References

Aloni, R. and W. Jacobs. 1977a. Polarity of tracheary regeneration in young internodes of coleus (*Labiatae*). Amer. J. Bot. 64: 395-403.

Aloni, R. and W. Jacobs. 1977b. The time course of sieve tube and vessel regeneration and their relation to phloem anastomoses in mature internodes of coleus. Amer. J. Bot. 64: 615-621.

Amiard, V., B. Demmig-Adams, K.E. Mueh, R. Turgeon, A.F. Combs and W.W. Adams. 2007. Role of light and jasmonic acid signaling in regulating foliar phloem cell wall ingrowth development. New Phytol. 173: 722-731.

Andriunas, F.A., C.E. Offler, D.W. McCurdy and J.W. Patrick. 2012. Reactive oxygen species form part of a signalling pathway that initiates *trans*-differentiation of epidermal transfer cells in *Vicia faba* cotyledons. J. Exp. Bot. 63: 3617-3630.

Andriunas, F.A., H.-M. Zhang, H. Weber, D.W. McCurdy, C.E. Offler and J.W. Patrick. 2011. Glucose and ethylene signaling pathways converge to regulate *trans*-differentiation of epidermal transfer cells in *Vicia narbonensis* cotyledons. Plant J. 68: 987-998.

Beňová-Kákošová, A., C. Digonnet, F. Goubet, P. Ranocha, A. Jauneau, E. Pesquet, O. Barbier, Z. Zhang, P. Capek, P. Dupree, D. Lišková and D. Goffner. 2006. Galactoglucomannans increase cell population density and alter the protoxylem/metaxylem tracheary element ratio in xylogenic cultures of zinnia. Plant Physiol. 142: 696-709.

Beresford, W.A. 1990. Direct transdifferentiation: can cells change their phenotype without dividing? Cell Differ. Dev. 29: 81-93.

Bhojwani, S.S. and P.K. Dantu. 2013. Plant tissue culture: an introductory text. Springer. India.

Burke, Z.D. and D. Tosh. 2005. Therapeutic potential of transdifferentiated cells. Clinical Sci. 108: 309-321.

Cai, S., X. Fu and Z. Sheng. 2007. Dedifferentiation: a new approach in stem cell research. BioScience 57: 655-662.

Chen, L.Q., X.Q. Qu, B.H. Hou, D. Sosso, S. Osorio, A.R. Fernie and W.B. Frommer. 2012. Sucrose efflux mediated by SWEET proteins as a key step for phloem transport. Science 335: 207-210.

Chen, J.J., J. Zhang and X.Q. He. 2014. Tissue regeneration after bark girdling. Physiol. Plant. 151:147-155.

Church, D.L. and A.W. Galston. 1988. Kinetics of determination in the differentiation of isolated mesophyll cells of *Zinnia elegans* to tracheary elements. Plant Physiol. 88: 92-96.

Dalessandro, G. and L.W. Roberts. 1971. Induction of xylogenesis in pith parenchyma explants of Lactuca. Amer. J. Bot. 58: 378-385.

Demura, T. and H. Fukuda. 1993. Molecular cloning and characterization of cDNAs associated with tracheary element differentiation in cultured *Zinnia* cells. Plant Physiol. 103: 815-821.

Demura, T. and H. Fukuda. 1994. Novel vascular cell-specific genes whose expression is regulated temporally and spatially during vascular system development. Plant Cell 6: 967-981.

Demura T., G. Tashiro, G. Horiguchi, N. Kishimoto, M. Kubo, N. Matsuoka, A. Minami, M. Nagata-Hiwatashi, K. Nakamura, Y. Okamura, N. Sassa, S. Suzuki, J. Yazaki, S. Kikuchi and H. Fukuda. 2002. Visualization by comprehensive microarray analysis of gene expression programs during transdifferentiation of mesophyll cells into xylem cells. Proc. Natl. Acad. Sci. USA 99: 15794-15799.

Dibley, S.J., Y. Zhou, F.A. Andriunas, M.J. Talbot, C.E. Offler, J.W. Patrick and D.W. McCurdy. 2009. Early gene expression programs accompanying *trans*-differentiation of epidermal cells of *Vicia faba* cotyledons into transfer cells. New Phytol. 182: 863-877.

Eguchi, G. and R. Kodama. 1993. Transdifferentiation. Curr. Opin. Cell Biol. 5: 1023-1028.

Eguchi, G. and T.S. Okada. 1973. Differentiation of lens tissue from the progeny of chick retinal pigment cells cultured *in vitro*: a demonstration of a switch of cell types in clonal cell culture. Proc. Natl. Acad. Sci. USA. 70: 1495-1499.

Fradin, E.F. and B.P. Thomma. 2006. Physiology and molecular aspects of *Verticillium* wilt diseases caused by *V. dahliae* and *V. alboatrum*. Mol. Plant Pathol. 7: 71-86.

Farley, S.J., J.W. Patrick and C.E. Offler. 2000. Functional transfer cells differentiate in cultured cotyledons of *Vicia faba* L. seeds. Protoplasma 214: 102-117.

Floerl, S., C. Druebert, A. Majcherczyk, P. Karlovsky, U. Kües and A. Polle. 2008. Defence reactions in the apoplastic proteome of oilseed rape (*Brassica napus* var. *napus*) attenuate *Verticillium longisporum* growth but not disease symptoms. BMC Plant Biol. 8: 129.

Fukuda, H. 1994. Redifferentiation of single mesophyll cells into tracheary elements. Int. J. Plant Sci. 155: 262-271.

Fukuda, H. 1996. Xylogenesis: Initiation, progression and cell death. Ann. Rev. Plant Physiol. Plant Mol. Biol. 47: 299-325.

Fukuda, H. 1997. Tracheary element differentiation. Plant Cell 9: 1147-1156.

Fukuda, H. 2004. Signals that control plant vascular cell differentiation. Nature Rev. Mol. Cell Biol. 5: 379-391.

Fukuda, H. and A. Komamine. 1980a. Establishment of an experimental system for the tracheary element differentiation from single cells isolated from the mesophyll of *Zinnia elegans*. Plant Physiol. 65: 57-60.

Fukuda, H. and A. Komamine. 1980b. Direct evidence for cytodifferentiation to tracheary elements without intervening mitosis in a culture of single cells isolated from the mesophyll of *Zinnia elegans*. Plant Physiol. 65: 61-64.

Fukuda, H. and A. Komamine. 1981a. Relationship between tracheary element differentiation and the cell cycle in single cells isolated from the mesophyll of *Zinnia elegans*. Physiol. Plant. 52: 423-430.

Fukuda, H. and A. Komamine. 1981b. Relationship between tracheary element differentiation and DNA synthesis in single cells isolated from the mesophyll of *Zinnia elegans* - analysis by inhibitors of DNA synthesis. Plant Cell Physiol. 22: 41-49.

Fukuda, H. and A. Komamine. 1985. Cytodifferentiation. pp. 149-212. *In*: I.K. Vasil (ed.). Cell Culture and Somatic Cell Genetics of Plants: Cell Growth, Nutrient, Cytodifferentiation and Cryopreservation. Volume 2. Academic Press, London.

Grafi, G. 2004. How cells dedifferentiate - a lesson from plants. Dev. Biol. 268: 1-6.

Grafi, G., V. Chalifa-Caspi, T. Nagar, I. Plaschkes, S. Barak and V. Ransbotyn. 2011a. Plant response to stress meets dedifferentiation. Planta 233: 433-438.

Grafi, G., A. Florentin, V. Ransbotyn and Y. Morgenstern. 2011b. The stem cell state in plant development and in response to stress. Front. Plant Sci. 2: 53. doi: 10.3389/fpls.2011.00053.

Gunning, B.E.S. and J.S. Pate. 1969. Transfer cells: plant cells with wall ingrowths, specialized in relation to short distance transport of solutes - their occurrence, structure, and development. Protoplasma 68: 107-133.

Hanna, J.H., S. Krishanu and J. Rudolf. 2010. Pluripotency and cellular reprogramming - facts, hypotheses, unresolved issues. Cell 143: 508-525.

Haritatos, E., R. Medville and R. Turgeon. 2000. Minor vein structure and sugar transport in *Arabidopsis thaliana*. Planta 211: 105-111.

Harrington, G.N., Y. Nussbaumer, X.-D. Wang, M. Tegeder, V.R. Franceschi, W.B. Frommer, C.E. Offler and J.W. Patrick. 1997. Spatial and temporal expression of sucrose transport-related genes in developing cotyledons of *Vicia faba* L. Protoplasma 200: 35-50.

Itoh, Y. and G. Eguchi. 1986. *In vitro* analysis of cellular metaplasia from pigmented epithelial cells to lens phenotypes. Dev. Biol. 1986. 115: 353-362.

Jacobs, W.P. 1952. The role of auxin in differentiation of xylem around a wound. Amer. J. Bot. 39: 301-309.

Jones, M.G.K. and D.H. Northcote. 1972a. Nematode induced syncytium - a multinucleate transfer cell. J. Cell Sci. 10: 789-809.

Jones, M.G.K. and D.H. Northcote. 1972b. Multinucleate transfer cells induced in *Coleus* roots by the root-knot nematode, *Meloidogyne arenaria*. Protoplasma 75: 381-395.

Jopling, C., S. Boue and J.C. Izpisua Belmonte. 2011. Dedifferentiation, transdifferentiation and reprogramming - three routes to regeneration. Nat. Rev. Mol. Cell. Biol. 2011. 12: 79-89.

Kalev, N. and R. Aloni. 1999. Role of ethylene and auxin in regenerative differentiation and orientation of tracheids in *Pinus pinea* seedlings. New Phytol. 142: 307-313.

Kondo, Y., T. Ito, H. Nakagami, Y. Hirakawa, M. Saito, T. Tamaki, K. Shirasu and H. Fukuda. 2014. Plant GSK3 proteins regulate xylem cell differentiation downstream of TDIF-TDR signalling. Nat. Commun. 5: 3504.

Kondo, Y., T. Fujita, M. Sugiyama and H. Fukuda. 2015. A novel system for xylem cell differentiation in *Arabidopsis thaliana*. Mol. Plant. 8: 612-621.

Kubo, M., M. Udagawa, N. Nishikubo, G. Horiguchi, M. Yamaguchi, J. Ito, T. Mimura, H. Fukuda and T. Demura. 2005. Transcription switches for protoxylem and metaxylem vessel formation. Genes Dev. 19: 1855-1860.

Ma, T., M. Xie, T. Laurent and S. Ding. 2013. Progress in the reprogramming of somatic cells. Circ. Res. 2013. 112: 562-574.

McCann, M.C. 1997. Tracheary element formation - building up to a dead end. Trends Plant Sci. 2: 333-338.

McCurdy, D.W. 2015. Transfer cells - novel cell types with unique wall ingrowth architecture designed for optimized nutrient transport. pp. 287-317. *In*: H. Fukuda (ed.). Plant Cell Patterning and Cell Shape. Wiley, Hoboken, New Jersey.

McCurdy, D.W., J.W. Patrick and C.E. Offler. 2008. Wall ingrowth formation in transfer cells: novel examples of localized wall deposition in plant cells. Curr. Opin. Plant Biol. 11: 653-661.

McManus, M.T., D.S. Thompson, C. Merriman, L. Lyne and D.J. Osborne. 1998. Transdifferentiation of mature cortical cells to functional abscission cells in *Phaseolus vulgaris* (L.). Plant Physiol. 116: 891-899.

Milioni, D., P.E. Sado, N.J. Stacey, C. Domingo, K. Roberts and M.C. McCann. 2001. Differential expression of cell-wall-related genes during the formation of tracheary elements in the *Zinnia* mesophyll cell system. Plant Mol. Biol. 47: 221-238.

Milioni, D., P.E. Sado, N.J. Stacey, K. Roberts and M.C. McCann. 2002. Early gene expression associated with the commitment and differentiation of a plant tracheary element is revealed by cDNA-amplified fragment length polymorphism analysis. Plant Cell 14: 2813-2824.

Mizuno, K., A. Komamine and M. Shimokoriyama. 1971. Vessel element formation in cultured carrot-root phloem slices. Plant Cell Physiol. 12: 823-830.

Mora, S. and A. Raya. 2013. Dedifferentiation, transdifferentiation, and reprogramming. pp. 152-163. *In*: C. Simón, A. Pellicer and R.R. Pera (eds.). Stem Cells in Reproductive Medicine. Cambridge University Press, New York.

Nguyen, S.T.T. and D.W. McCurdy. 2015. High-resolution confocal imaging of wall ingrowth deposition in plant transfer cells: Semi-quantitative analysis of phloem parenchyma transfer cell development in leaf minor veins of *Arabidopsis*. BMC Plant Biol. 15: 109. doi: 10.1186/s12870-015-0483-8.

Nishitani, C., T. Demura and H. Fukuda. 2001. Primary phloem-specific expression of a *Zinnia elegans* homeobox gene. Plant Cell Physiol. 42: 1210-1218.

Nishitani, C., T. Demura and H. Fukuda. 2002. Analysis of early processes in wound-induced vascular regeneration using *TED3* and *ZeHB3* as molecular markers. Plant Cell Physiol. 43: 79-90.

Oda, Y., T. Mimura and S. Hasezawa. 2005. Regulation of secondary cell wall development by cortical microtubules during tracheary element differentiation in *Arabidopsis* cell suspensions. Plant Physiol. 137: 1027-1036.

Oda, Y., Y. Iida, Y. Kondo and H. Fukuda. 2010. Wood cell-wall structure requires local 2D-microtubule disassembly by a novel plasma membrane-anchored protein. Curr. Biol. 20: 1197-1202.

Offler, C.E., D.W. McCurdy, J.W. Patrick and M.J. Talbot. 2003. Transfer cells: cells specialized for a special purpose. Ann. Rev. Plant Biol. 54: 431-454.

Okada, T.S. 1986. Transdifferentiation in animal cells - fact or artifact. Dev. Growth Differ. 28: 213-221.

Okada, T.S. 1991. Transdifferentiation: Flexibility in Cell Differentiation. Clarendon Press, Oxford.

Pang, Y., J. Zhang, J. Cao, S.Y. Yin, X.Q. He and K.M. Cui. 2008. Phloem transdifferentiation from immature xylem cells during bark regeneration after girdling in *Eucommia ulmoides* Oliv. J. Exp. Bot. 59: 1341-1351.

Pate, J.S. and B.E.S. Gunning. 1972. Transfer cells. Ann. Rev. Plant Physiol. 23: 173-196.

Phillips, R. and J.H. Dodds. 1977. Rapid differentiation of tracheary elements in cultured explants of Jerusalem artichoke. Planta 135: 207-212.

Rana, M.A. and R.B. Gahan. 1983. A quantitative cytochemical study of determination for xylem-element formation in response to wounding in roots of *Pisum sativum* L. Planta 157: 307-316.

Reyes, M. and C.M. Verfaillie. 2004. Adult stem cell plasticity. pp. 90-100. *In*: K. Atkinson, R. Champlin, J. Ritz, W. Fibbe, P. Ljungman and M.K. Brenner (eds.). Clinical Bone Marrow and Blood Stem Cell Transplantation. Cambridge University Press, New York.

Reusche, M., K. Thole, D. Janz, J. Truskina, S. Rindfleisch, C. Drübert, A. Polle, V. Lipka and T. Teichmann. 2012. *Verticillium* infection triggers *VASCULAR-RELATED NAC DOMAIN7*-dependent *de novo* xylem formation and enhances drought tolerance in *Arabidopsis*. Plant Cell 24: 3823-3837.

Riddle, M.R., A. Weintraub, K.C. Nguyen, D.G. Hall and J.H. Rothman. 2013. Transdifferentiation and remodeling of post-embryonic *C. elegans* cells by a single transcription factor. Development. 40: 4844-4849.

Robbertse, P.J. and M.E. McCully. 1979. Regeneration of vascular tissue in wounded pea roots. Planta. 145: 167-173.

Rose, R.J. 2016. Genetic reprogramming of plant cells *in vitro* via dedifferentiation or pre-existing stem cells. Chapter 18. *In*: R.J. Rose (ed.) Molecular Cell Biology of the Growth and Differentiation of Plant Cells. CRC Press, Boca Raton.

Shoji, Y., M. Sugiyama and A. Komamine. 1997. Involvement of poly(ADP-ribose) synthesis in transdifferentiation of isolated mesophyll cells of *Zinnia elegans* into tracheary elements. Plant Cell Physiol. 38: 36-43.

Srivastava, L.M. 2002. Plant Growth and Development: Hormones and Environment. Academic Press, Burlington.

Sugiyama, M. 2015. Historical review of research on plant cell dedifferentiation. J. Plant Res. 128: 349-359.

Sugiyama, M. and A. Komamine. 1990. Transdifferentiation of quiescent parenchymatous cells into tracheary elements. Cell Differ. Dev. 31: 77-87.

Sugiyama, M., E.C. Yeung, Y. Shoji and A. Komamine. 1995. Possible involvement of DNA-repair events in the transdifferentiation of mesophyll cells of *Zinnia elegans* into tracheary elements. J. Plant Res. 108: 351-361.

Sugimoto, K., S.P. Gordon and E.M. Meyerowitz. 2011. Regeneration in plants and animals-dedifferentiation, transdifferentiation, or just differentiation. Trends Cell Biol. 21: 212-218.

Sugimoto, K., Y. Jiao and E.M. Meyerowitz. 2010. Arabidopsis regeneration from multiple tissues occurs via a root development pathway. Dev. Cell 18: 463-471.

Sussex, I.M., M.E. Clutter and M.H.M. Goldsmith. 1972. Wound recovery by pith cell redifferentiation: structural changes. Am. J. Bot. 59: 797-804.

Talbot, M.J., V.R. Franceschi, D.W. McCurdy and C.E. Offler. 2001. Wall ingrowth architecture in epidermal transfer cells of *Vicia faba* cotyledons. Protoplasma 215: 191-203.

Talbot, M.J., C.E. Offler and D.W. McCurdy. 2002. Transfer cell wall architecture: a contribution towards understanding localized wall deposition. Protoplasma 219: 197-209.

Talbot, M.J., G.O. Wasteneys, C.E. Offler and D.W. McCurdy. 2007. Cellulose synthesis is required for deposition of reticulate wall ingrowths in transfer cells. Plant Cell Physiol. 48: 147-158.

Thiel, J. 2014. Development of endosperm transfer cells in barley. Front. Plant Sci. 5: 108. doi: 10.3389/fpls.2014.00108.

Tosh, D. and J.M.W. Slack. 2002. How cells change their phenotype. Nat. Rev. Mol. Cell. Biol. 3: 187-194.

Tosh, D. and M.E. Horb. 2013. How cells change their phenotype. pp. 95-100. *In*: A. Atala and R. Lanza (eds.). Handbook of Stem Cells. Volume 1. Academic Press, Burlington.

Torrey, J.G. 1975. Tracheary element formation from single isolated cells in culture. Physiol. Plant. 35: 158-165.

Turner, S., P. Gallois and D. Brown. 2007. Tracheary element differentiation. Ann. Rev. Plant. Biol. 58: 407-433.

Xu, L., K. Zhang and J. Wang. 2014. Exploring the mechanisms of differentiation, dedifferentiation, reprogramming and transdifferentiation. PLoS ONE 9: e105216.

Yamaguchi, M., N. Goue, H. Igarashi, M. Ohtani, Y. Nakano, J.C. Mortimer, N. Nishikubo, M. Kubo, Y. Katayama, K. Kakegawa, P. Dupree and T. Demura. 2010. *VASCULAR-RELATED NAC-DOMAIN6* and *VASCULAR-RELATED NAC-DOMAIN7* effectively induce transdifferentiation into xylem vessel elements under control of an induction system. Plant Physiol. 153: 906-914.

Waddington, C.H. 1957. The Strategy of the Genes. Allen & Unwin, London.

Wang, J., K. Zhanga, L. Xua and E. Wang. 2011. Quantifying the Waddington landscape and biological paths for development and differentiation. Proc. Natl. Acad. Sci. USA. 108: 8257-8262.

Wang, T., R. Chai, G.S. Kim, N. Pham, L. Jansson, D.H. Nguyen, B. Kuo, L. May, J. Zuo, L.L. Cunningham and A.G. Cheng. 2015. Lgr5+ cells regenerate hair cells via proliferation and direct transdifferentiation in damaged neonatal mouse utricle. Nat. Comm. 6: 6613.

Wardini, T., X.-D. Wang, C.E. Offler and J.W. Patrick. 1997. Induction of wall ingrowths of transfer cells occurs rapidly and depends upon gene expression in cotyledons of developing *Vicia faba* seeds. Protoplasma 231: 15-23.

Wilson, J.W., L.W. Roberts, P.M. Gresshoff and S.J. Dircks. 1982. Tracheary element differentiation induced in isolated cylinders of lettuce pith: a bipolar gradient technique. Ann. Bot. 50: 605-614.

Zhang, J., G. Gao, J.J. Chen, G. Taylor, K.M. Cui and X.Q. He. 2011. Molecular features of secondary vascular tissue regeneration after bark girdling in *Populus*. New Phytol. 192: 869-884.

Zhang, H.-M., S. Wheeler, X. Xia, R. Radchuk, H. Weber, C.E. Offler and J.W. Patrick. 2015. Differential transcriptional networks associated with key phases of ingrowth wall construction in *trans*-differentiating epidermal transfer cells of *Vicia faba* cotyledons. BMC Plant Biol. 15: 103. doi: 10.1186/s12870-015-0486-5.

Zhou, Y., F. Andriunas, C.E. Offler, D.W. McCurdy and J.W. Patrick. 2010. An epidermal-specific ethylene signal cascade regulates *trans*-differentiation of transfer cells in *Vicia faba* cotyledons. New Phytol. 185: 931-943.

18

Genetic Reprogramming of Plant Cells In Vitro via Dedifferentiation or Pre-existing Stem Cells

Ray J Rose*

Introduction

When plant cells or tissues are cultured *in vitro* with suitable culture media they can be directed into different developmental pathways. This capacity has been demonstrated in numerous species with many different cell and tissue types and different culture media. Plants are more readily directed into regeneration than animal cells but this capacity to genetically reprogram *in vitro* under suitable conditions is characteristic of eukaryotic cells. In plants, the capacity to reprogram cell fate using plant hormones, the auxins and cytokinins, was demonstrated many years ago (Skoog and Miller 1957). In this Chapter three types of development reprogramming are considered: somatic embryogenesis (SE), root organogenesis and shoot organogenesis. Somatic embryogenesis involves the formation of both shoot and root meristems and the capacity to produce a normal flowering plant (Steward et al. 1958). With root organogenesis a root meristem develops and differentiates into a root with a capacity to produce lateral roots; and with shoot organogenesis there is the development of a shoot apical meristem with the capacity to differentiate into the tissues of the stem, its buds and reproductive tissues (Skoog and Miller 1957). Callus is a form of differentiation and this is addressed in the SE and organogenesis context.

A major focus in this Chapter is on the genetic reprogramming to form stem cells *in vitro* from which the embryos or organs are formed or reprogramming from pre-existing stem cells (Verdeil et al. 2007, Wang et al. 2011); with particular emphasis on SE.

School of Environmental and Life Sciences, The University of Newcastle, University Drive, Callaghan, NSW 2308. Australia.
* Corresponding author : Ray.Rose@newcastle.edu.au

Cellular Origins of Somatic Embryos and *De Novo* Organs

Somatic embryogenesis from differentiated cells

In a number of species somatic embryos are produced from differentiated cells (Rose et al. 2010, Wang et al. 2011, Fehér 2015, Rocha et al. 2015) that have their existing genetic program erased and a new genetic pathway programmed. This can occur in both tissue explants and in isolated protoplasts. While in both these cases the developmental events are similar they are not identical. In *Medicago truncatula* (Medicago) tissue explants it is the edge of the explant that is in direct contact with the medium and it is cells near the cut surface that produce embryos (Fig. 1). In these cases there are strong stress effects (Imin et al. 2005, Wang et al. 2011) and auxin gradients across the tissue (Su et al. 2009). In protoplasts the medium surrounds the single cell and due to the isolation procedure the wound stress is even greater (Pasternak et al. 2002, Imin et al. 2004, Tiew et al. 2015). This is reflected in the difficulty of regenerating plants from protoplasts from diverse species (Papadakis and Roubelakis-Angelakis 1999). Isolated haploid microspores can be induced by stress to form somatic embryos (Soriano et al. 2014). In this case, there is a change in the developmental pathway producing a somatic embryo rather than a pollen grain. There is a change in the fate of the vegetative nucleus and is in a different category to a differentiated cell, such as a mesophyll cell, that is at the end point of differentiation. The microspore is involved in establishing a new generation and will be discussed further in relation to S.E. induction mechanisms.

The cellular changes that are required to produce an embryo from a differentiated cell can be described in the following way:

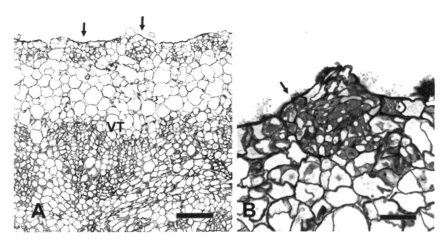

FIGURE 1   Somatic embryos (arrows) initiated near the surface of a Medicago leaf explant (A and B) cultured with 10 µM NAA (naphthalene acetic acid) and 4 µM BAP (6-benzylamino purine). Serial sections showed no connection to vascular tissue. Higher magnification of globular somatic embryo in B. VT = vascular tissue. Scale bars A = 100 µm, B = 25 µm. Reproduced with permission from Wang et al. (2011) Annals of Botany.

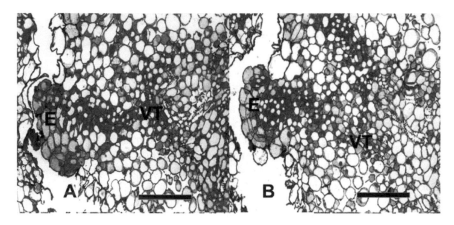

FIGURE 2   Serial sections (A and B) showing the lineage of cells from the vascular tissue to the somatic embryo forming on the surface of callus from a Medicago leaf explant cultured with 10 μM NAA and 4 μM BAP. E = embryo, VT = vascular tissue. Scale bars = 100 μm. Reproduced with permission from Wang et al. (2011), Annals of Botany.

Differentiated cell → Dedifferentiation → Cell Division(s) → Totipotent Stem Cell → Embryo

What is less predictable in the above schematic is the number of cell divisions that occur prior to a totipotent stem cell being produced. However callusing and subsequently groups of cells known as proembryogenic masses (PEMs) are commonly present in somatic embryo regeneration systems (indirect SE) compared to direct SE with few cell divisions. Fehér et al. (2003) in their review, have concluded that cell division is necessary to initiate the new genetic program and for subsequent embryo formation in both direct and indirect SE. Genetic reprogramming in this SE pathway must first facilitate an erasure of the differentiation state to produce a meristematic cell and then institute a new program to produce a totipotent embryonic cell.

Somatic embryogenesis from pre-existing stem cells associated with the vasculature

Another common pathway in the formation of somatic embryos has the progenitor cells coming from existing stem cells that exist in the vasculature - the procambial, cambial or pericycle cells (Fig. 2, Guzzo et al. 1994, Rose et al. 2010, Wang et al. 2011, Fehér 2015). The model then shifts from that in the dedifferentiation schematic to the following:

Vascular - associated stem cell* → Cell Division(s) → Totipotent Stem Cell → Embryo

* Procambium, pericycle or similar

What is clear here is that a major chromatin reorganisation of the same order to that involved in dedifferentiation is likely not required. Again the number of cell divisions involved before a new program is produced is variable. Similarly the common successful use of immature embryos as explants may often be related to the enrichment of pre-existing stem cells.

Root organogenesis

Adventitious roots in leaf explants of Medicago (Rose et al. 2006) and Arabidopsis are derived from the stem cells of the procambium (Liu et al. 2014). This is similar to the situation for somatic embryos derived from vascular-associated stem cells but in this case the stem cell rather than forming totipotent stem cells form the root stem cells (or founder cells) that will form the root meristem. This is very clear in Medicgao where numerous root meristems emanate from procambial-derived cells (Fig. 3, Rose et al. 2006). This *in vitro* developmental pathway is similar to the way lateral roots form *in vivo* where the pericycle cells act as the precursor stem cells (Casimiro et al. 2003). Again these are vasculature-associated cells as is the procambium and cambium and it is likely that depending on the explant any one of these cell types would have the capacity to act as the precursor stem cells.

As has been pointed out callus can arise from these vasculature associated cells and in this sense is similar to cells involved in the root regeneration pathway (Sugimoto et al. 2010, 2011, Liu et al. 2014). This argues for callus being persistent stem cells derived from a stem cell population. However, as discussed under SE callus can be derived from differentiated cells – notably mesophyll cells in leaf explants of Medicago and mesophyll protoplasts. Whether it is possible for the dedifferentition pathway to lead to root formation in explants under some circumstances is unclear. Again, protoplasts from plants such as *Nicotiana tabacum* can generate roots after

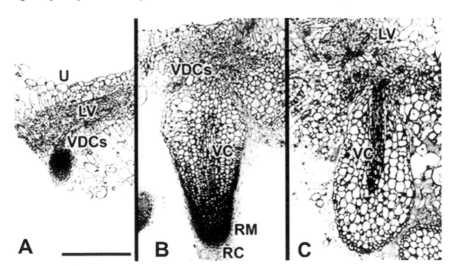

FIGURE 3 Transvers sections through roots forming from Medicago leaf explants cultured with 10 μM NAA. These sections are from the *sickle* (*skl*) mutant which produce more roots than wild-type but are morphologically the same as wild-type. (A) Root meristems formed from VDCs. (B) Central vascular cylinder formed from the root meristem. (C) Root vascular cylinder connected to vascular tissue of the original leaf explant. LV, leaf vein; RC, root cap; RM, root meristem; U, upper surface of leaf explant; VC, vascular cylinder; VDCs, vein-derived cells. Scale bar = 200 μm. Reproduced with permission from Rose et al. (2006), Journal of Experimental Botany.

callusing using the hormone ratios described by Skoog and Miller (1957); and cellular origins of the root meristems is worthy of more detailed study.

In both Medicago (Nolan et al. 2003) and Arabidopsis (Liu et al. 2014) there is evidence that auxin is the prime driver of *de novo* root organogenesis; as in the original studies by Skoog and Miller (1957) and cytokinin will inhibit root organogenesis (Nolan and Rose 1998).

Shoot organogenesis

Shoot organogenesis goes back to the Skoog and Miller (1957) studies but it is only relatively recently that the cellular origins of shoots in cultured tissues have started to be questioned in relation to stem cell concepts. There is now substantive data using systems based on organogenesis of cultured Arabidopsis root explants (He et al. 2012) indicating that vascular associated pericycle cells and lateral root primordia can be induced by cytokinin to form shoots, without dedifferentiation and a callus phase involving reprogramming to organogenesis. Atta et al. (2009) used Arabidopsis root and hypocotyl explants for detailed morphological studies with marker lines to show that regenerated shoots developed directly and indirectly from the xylem pole pericycle. The callus was in fact lateral root primordia originating from the pericycle on callus inducing medium (CIM) and shoot apical meristems (SAMs) developed from these cells on shoot inducing medium (SIM) with cytokinin or even if cultured directly on cytokinin without prior growth on CIM. Similar results were obtained by Sugimoto et al. (2010) in their study primarily addressing callus formation in Arabidopsis. They found that with increased cytokinin (required for shoot organogenesis) in callus inducing medium there was still callus enriched in root-meristem-like cells. When the callus is transferred to shoot induction medium there is rapid initiation of shoot genes. It appears that in the shoot induction medium small patches of the root-meristem-like cells initiate new shoot meristems (Gordon et al. 2007). In terms of the patches observation this is very similar to somatic embryogenesis and the PEMs of SE and the proliferating procambial cells that form roots (Wang et al. 2011). Chatfield et al. (2013) using a synchronous system of large scale shoot organogenesis showed quite dramatically that shoot meristems could be formed from lateral root primordia/meristems or callus expressing markers for this identity.

It is unclear if in tissue explants specific dedifferentiated cells can be the initial progenitors in some cases. We do however know that callus from mesophyll protoplasts can form shoots (Thomas and Rose 1983).

Callus

Given the above considerations of SE and organogenesis it is apparent that there is callus that derives from pre-existing stem cells without dedifferentiation. However there can be callus formation, in explants and protoplasts, as a result of dedifferentiation, and these cells are capable of forming totipotent stem cells.

The mechanism for these cell fate changes involved in SE or *de novo* organogenesis of callus is still far from clear. However, there is progress in formulating some conceptual understanding of the mechanism of SE and *de novo* organogenesis, particularly based on stem cell concepts, and these are now discussed.

Mechanism of Somatic Embryogenesis and *De Novo* Organogenesis

The dedifferentiation pathway to somatic embryogenesis

(a) Dedifferentiation and chromatin remodelling

The study by Zhao and co-workers (2001) represented an initial approach to analyse the chromatin changes during dedifferentiation and the induction of cell division using Nicotiana mesophyll protoplasts. They argue for two sequential changes. First, isolated protoplasts undergo chromatin decondensation compared to the starting leaf tissue. This conclusion of increased DNA accessibility was based on the increased fluorescence of propidium iodide stained nuclei and increased sensitivity of chromatin to micrococcal nuclease. Second, was the response to the transfer of the protoplasts to auxin and cytokinin in the culture medium. There was a dependence on the ubiquitin proteolytic system as the cells progressed into S phase and there needed to be protein turnover. The argument was that this represents competence for a cell fate change followed by a commitment to cell division. Both points are conceptually reasonable; a change in chromatin organisation followed by the increased activity of the proteasome system to degrade proteins of the previous gene expression state. It is widely accepted that stress has an important role in dedifferentiation and the change in chromatin state (recent reviews by Fehér 2015, Grafi and Barak 2015) ultimately leading to changes in cell fate. What is also clear is that the changes in chromatin state are associated with DNA methylation changes and histone modifications such as methylation and acetylation (Fehér 2015, Grafi and Barak 2015). What is less clear is the underlying mechanism causing these chromatin changes. It is known that one of the very rapid changes is the generation of reactive oxygen species (ROS, Fehér et al. 2008, Rose et al. 2013, Tiew et al. 2015) but how these link to the chromatin changes are unclear.

Changes in DNA methylation as a result of *in vitro* culture of plant cells are well documented (Loschiavo et al. 1989, Miguel and Marum 2011) and these changes occur throughout the genome (Kurdyukov et al. 2014a). One approach to the significance of DNA methylation has been to utilise the methylation inhibitor 5'- azacytidine. In carrot (Yamamoto et al. 2005) and Medicago (Santos and Fevereiro 2002) 5'-azacytidine is disruptive to somatic embryogenesis. Recent studies by Solis et al. (2015) in barley and *Brassica napus* showed that low concentrations of 5'-azacytidine for short periods induced hypomethylation and chromatin decondensation and increased microspore SE initiation. However methylation was necessary for embryo development as longer incubations in 5'- azacytidine inhibited SE. This is consistent with particular patterns of methylation for differentiation (Solis et al. 2015).

It does seem that for a given SE system that the induction of SE requires more open chromatin and the correct DNA methylation state. This is illustrated in Medicago where the highly embryogenic line Jemalong 2HA has many methylation changes, including some that affect the expression of signaling genes (Kurdyukov et al. 2014a).

Histone modification is the other epigenetic factor important in regulating the chromatin state in somatic embryogenesis and has been considered in recent reviews (Ikeuchi et al. 2013, Mahdavi-Darvar et al. 2015, Fehér 2015). An example of histone modification comes from Nicotiana protoplasts. Histone H3K9/14 acetylation and chromatin decondensation occurred during the dedifferentiation of tobacco leaf-derived protoplasts, priming for transcriptional activation of the cell cycle (Williams et al. 2003). The histone deacetylase inhibitor trichostatin (TSA) enhances somatic embryo induction from male gametophytes of Brassica and Arabidopsis because of hyperacetylation of histones (Li et al. 2014). In the case of the male gametophyte it is important that the microspore not continue on its normal developmental pathway (Soriano et al. 2014) and the prevention of deacetylation becomes important to maintain cell divisions leading to SE. Like DNA methylation it again shows the direct significance of the chromatin state, in this case with histone acetylation.

Histone methylation is another regulatory mechanism influencing the chromatin state. H3K27me3 is a repressive chromatin mark for reprogramming in both plants and animals (Makarevitch et al. 2013, Nashun et al. 2015). The H3K27 methyltransferases are PRC2 (POLYCOMB REPRESSIVE COMPLEX 2) members of the Polycomb Group (PcG) - type regulator complex. In the Arabidopsis leaf to callus transition there is global reprogramming of H3K27me3 such that H3K27me3 levels are negatively correlated with gene expression (He et al. 2012). These conclusions were based on data obtained from PRC2 mutants. Results obtained indicated that H3K27me3 levels decreased in the auxin regulation network (He et al. 2012). Genes that are required to be repressed such as those specifying leaf characteristics had high H3K27me3 methylation probably due to PcG-mediated modification.

The changes in DNA and histone marks are reflective of the major changes in gene expression that result in the reprogramming to the totipotent state. But how can this be related to changes to specific genes linked to the embryogenic pathway? An interesting gene in this regard is the *PICKLE* (*PKL*) gene which acts as a negative regulator of SE. This was discovered with the *pkl* mutant in Arabidopsis where roots from *pkl* plants underwent SE without the application of plant hormones (Ogas et al. 1997). Gibberellic acid was able to inhibit this SE of the *pkl* mutant. In carrots GA is also antagonistic to SE (Tokuji and Kuriyama 2003). This is not the case in *M. truncatula,* but what is important here is that the *PKL* gene is also down regulated (Nolan et al. 2014). In all these cases however ABA which is frequently antagonistic to GA, promotes SE. What appears to be the case is that it is the interaction between GA and ABA regulation that is able to derepress the embryogenic pathway, a necessary condition for SE. The *PKL* gene is a chromatin remodelling ATPase and studies by Zhang et al. (2008, 2012) supports a specific role of *PKL* in promoting the trimethylation of histone H3 lysine 27. Further in Arabidopsis there are links to *AGL15* and *LEC* genes, important SE genes in regulating GA metabolism and *PKL* (Braybrook and Harada 2008, Zhang et al. 2008). *PKL* when down-regulated is a candidate gene for making cells competent for SE, unblocking the embryogenic pathway.

(b) Connecting chromatin remodelling to the activation of specific genes for embryogenesis

The chromatin remodelling required for SE because of the need to repress and derepress genes is not so surprising, though the details of how such changes occur during dedifferentiation remain critical considerations. The chromatin loosening and remodelling involved in dedifferentiation would allow access for the transcriptional regulation by hormones, in the culture medium, required for the activation or repression of specific genes. The hormones utilised to produce somatic embryos are auxin with cytokinin sometimes also being required. Auxins are able to influence transcription by removing AUX/IAA proteins freeing the ARF transcription factors (Shan et al. 2012) while cytokinins are able to influence transcription via signalling the phosphorylation of the type-B and type-A ARR transcription factors (Shan et al. 2012). Initiation of the cell cycle involves the regulation of many genes. Entry into G1 requires activation of the E2F transcription factor complex by the phosphorylation of the retinoblastoma-related (RBR) protein, which relieves the inhibition of the E2F transcription factors. The CDKA (cyclin dependent kinase A)/ CYCD (cyclin D) complex phosphorylates the RBR protein (De Veylder et al. 2007, Schaller et al. 2014). E2FB has been shown to be regulated by auxin (Magyar et al. 2005) while cytokinins have also been shown to influence the G1/S transition by regulating the expression of CYCD3 genes (Schaller et al. 2014). Associated with the dedifferentiation of protoplasts there is decondensing of the chromatin structure of the E2F target genes by histone H3 modification (Williams et al. 2003). This illustrates a possible dynamic of chromatin remodelling and hormones activating key gene sets to initiate cell division.

Another gene that is critical for SE is *WUSCHEL* (*WUS*), the gene required for stem cell maintenance in the shoot apical meristem (Mayer et al. 1998). Early expression of WUS is characteristic of a number of species and there is a strong case that this is a critical gene required for the production of embryonic stem cells (Zuo et al. 2002, Chen et al. 2009, Su et al. 2009, Elhiti et al. 2010). So while dedifferentiation and cell division occurs this needs to lead to some new cells becoming embryonic stem cells. Auxin is able to stimulate *WUS* expression in Arabidopsis (Su et al. 2009) while in *M. truncatula* it is cytokinin (Chen et al. 2009). However in shoot organogenesis *in vitro WUS* is induced in a cytokinin enriched medium (Gordon et al. 2007). An important focus for future work is to understand how different hormone(s) are able to regulate *WUS* in different *in vitro* developmental contexts.

With *WUS* it appears likely that as for cell cycle genes the activation of this specific gene is associated with chromatin remodelling to facilitate its transcription. These data are from *de novo* shoot regeneration in Arabidopsis rather than SE. *WUS* was regulated by both DNA methylation and histone modification (Li et al. 2011). *WUS* transcription was associated with reduced DNA methylation and there were histone marks associated with euchromatin transcription. There were increased levels of histone H3K4me3 and H3K9ac, and reduced levels of H3K9me2 at the *WUS* genomic sequences (Li et al. 2011).

(c) The proembryogenic mass (PEM)

The PEMs are a group or patch of densely cytoplasmic cells in the callus from which embryos are derived. It would be expected that some or all of these cells would be embryonic stem cells. In Medicago (Chen et al. 2009) there is initially strong *WUS* expression at the callused edges of the explant, but as callusing progresses and *CLAVATA3* (*CLV3*) expression increases, WUS expression becomes restricted to patches of expression. In the Arabidopsis study (Su et al. 2009) there were also small centres of *WUS* expression in the callus. Zuo et al. (2002) studies on overexpression in Arabidopsis suggest that WUS can act to maintain the identity of embryonic stem cells. Similarly there is evidence that WUS can act to establish cells with shoot stem identity in Arabidopsis (Gallois et al. 2004). Therefore, it is plausible that WUS expression patches indicate the site of the embryonic stem cells. How is it then that only some embryonic stem cells actually produce embryos? One possibility is that it is only some cells where the hormonal conditions are optimal, as gradients are formed from the cells near the callus edge to deeper within the callus. The *PINFORMED (PIN)* auxin *t*ransport genes are critical in this regard. In Arabidopsis auxin gradients are associated with *WUS* induction and *PIN1* expression. PIN1 is subsequently associated with polar auxin transport as the embryo develops (Su et al. 2009). PIN1 like WUS is essential for SE in Arabidopsis.

Why is it that WUS with its central role in stem cell maintenance in the apical meristem can influence embryonic stem cell identity and the bipolar embryo with shoot and root meristems? An interesting point about WUS is that it can act interchangeably with *WUSCHEL-RELATED HOMEOBOX5 (WOX5)*, the stem cell maintenance gene in the root. A *WOX5–WUS*cDNA transgene restores stem cells in the root meristem of a *wox5* mutant in Arabidopsis (Sarkar et al. 2007). There is also evidence that WUS is involved in ovule development and early embryonic cell divisions (Groß-Hardt et al. 2002, Kurdyukov et al. 2014b).

While it is known that isolated single cells can produce embryos, as in the tracking studies of carrot (Schmidt et al. 1997), in callus there is the proposition that embryos can have a multicellular origin with groups of cells organising to form an embryo (Williams and Maheswaran 1986). This is hard to visualise in PEMs unless during protoderm formation some cells, not from an embryonic stem cell, are captured to form a chimaera type situation.

The pathway to somatic embryogenesis from pre-existing stem cells

In a number of cases it is apparent that somatic embryos derive from procambial (also known as provascular cells) or perhaps pericycle cells of the vascular tissue. These vascular associated cells are stem-cell-like (Jiang et al. 2015). Guzzo et al. (1994) using cyto -histological studies on carrot hypocotyl explants showed that PEMs derive from the procambium and these cells form embryos when the auxin is removed. From this study it appears that transfer of PEM cells to culture media produced somatic embryos from asymmetric divisions. While the *SOMATIC EMBRYO RECEPTOR-LIKE KINASE1 (SERK1)* expression is not necessarily a marker for SE, cells leading to SE will express *SERK1* and this is the case for the procambium in both Arabidopsis (Kwaaitaal and de Vries 2007) and in *M. truncatula* leaf explants (Wang et al. 2011), though in Medicago most embryos are derived from dedifferentiating cells. In recent Arabidopsis studies immature zygotic embryos (Su et al 2009, Gliwicka et al. 2013, Zheng et al. 2013) or SAM explants (Zheng et al. 2013) have been used. In both tissue types there is enrichment for stem cells, though no tracking studies have been done. In a recent SE study on peach, somatic embryos originated from procambium cells (De Almeida et al. 2012).

The first report on the significance of *WUS* in SE came from the study of Zuo et al. (2002) where overexpression of *WUS* caused by the activation of an inducible promoter in Arabidopsis seedlings allowed the induction of somatic embryos in a range of cultured tissues. There was no callus phase, although some cell division cannot be ruled out. It is possible that a range of tissues were used and that a common cell type was induced to follow an embryogenic pathway. Vascular associated stem cells would fit this criteria. *WUS* completely substituted for exogenous hormones.

In the previous section on the SE dedifferentiation pathway it is apparent that chromatin remodelling is an initial step in forming an embryonic stem cell. Where somatic embryos may derive from stem or stem-like cells, it would seem that these cells are already in a primed chromatin state where there is access to gene sets which can be influenced by the hormonal environment. Again, however, the key determining transcription factor genes such as *WUS* would need to be in a transcriptionally active chromatin region. The existing data would point to the *PIN* genes as being important in steps subsequent to *WUS* activation (Su et al. 2009).

Genes of special importance in SE

Genes that are particularly critical in SE such as *PKL, WUS, PIN,* and to a lesser extent *CLV3* and *SERK1*, have been discussed in the above sections in relation to stem cell concepts. Other significant genes have been investigated and these need to be placed in context. However, the emphasis in this chapter is on the stem cells that serve as the progenitors of the embryo or the *de novo* organs. There have been a number of transcriptomic (e.g. Thibaud-Nissen et al. 2003, Mantiri et al. 2008, Sharma et al. 2008, Gliwicka et al. 2013, Salvo et al. 2014) and proteomic studies (e.g. Imin et al. 2004, 2005) that have shown many genes are associated with the induction of SE. This is to be expected given the many changes that occur in the transition from one developmental state to another and the need to integrate metabolic, cellular and developmental changes. The discussion of specific genes is based on the model in Fig. 4.

The LEAFY COTYLEDON2 (LEC2) transcription factor is important given its overexpression induces SE in seedlings and the models that give it a key role (Stone et al. 2001, Braybrook and Harada 2008). In zygotic embryo development it is not an early embryo gene and is a master regulator of seed storage. However part of its role in zygotic embryogenesis is to regulate abscisic acid (ABA) and gibberellic acid (GA) levels in the maturation phase of embryogenesis (Braybrook and Harada 2008). This is where it could have a role in derepressing the embryonic pathway similar to *PKL* but targeting suitable stem cells. This would be consistent with over expression in seedlings targeting such cells. LEC1 and AGAMOUS-LIKE 1 (AGL1) are other transcription factors that influences GA levels and can stimulate SE (Braybrook and Harada 2008). The argument here is that they allow the embryogenic pathway to be activated. LEC2 has also been shown to influence auxin biosynthesis (Stone et al. 2008) which is important in organising embryo patterning.

Another extensively studied gene is *SERK1*, and while this gene is important in SE it is likely to be in its role as being associated with transition to different developmental stages throughout the plant (Nolan et al. 2009). In Medicago it is auxin-induced. SERK receptor biology is complex with various receptor interactions and ligands (Aan Den Toorn et al. 2015), so there may still be some surprises with SERK biology. Another gene that has been investigated is *SOMATIC EMBRYO RELATED FACTOR1 (SERF1)* which is a transcription factor and a member of the AP2/ERF superfamily. This is an early embryogenesis gene that has been associated to SE in Medicago (Mantiri et al. 2008), soybean and Arabidopsis (Zheng et al. 2013), is ethylene dependent, and has been linked to *AGL15* and *WUSCHEL* (Mantiri et al. 2008, Zheng et al. 2013). In *Medicago* it also requires auxin and cytokinin. While there is evidence that *SERF1* expression is an SE requirement it appears to be linked to

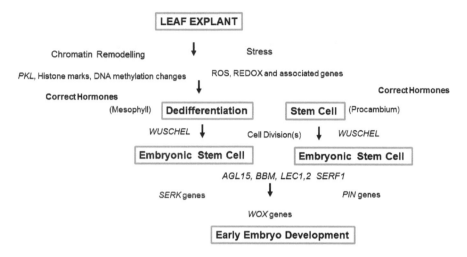

FIGURE 4 A model for SE formation via dedifferentiation or from pre-existing stem cells. Excision of explants or isolation of protoplasts leads to stress, ROS production, expression of stress-related genes and redox changes; with differentiated cells undergoing dedifferentiation. Chromatin decondenses with associated DNA methylation, histone methylation and acetylation changes allowing exogenous hormones (auxin or auxin plus cytokinin) and endogenous hormones (abscisic acid, ethylene, gibberellic acid) to access and activate or repress appropriate genes. This leads to *WUS* - induced embryonic stem cells. In the cases of pre-existing stem cells the chromatin state allows hormone-induced SE, similar to what ultimately occurs in the dedifferentiation pathway. *AGL1, BBM, LEC* and *SERF* transcription factors drive embryo development from the embryonic stem cells. *SERK* expression is connected to developmental transitions as mature embryos form. *PIN* and related genes determine the auxin gradients and *WOX* embryo patterning genes are important directors of embryo patterning (Haecker et al. 2004, Su et al. 2015). See Rose et al. (2010, 2013), Fehér (2015) and this Chapter for background to this model.

channelling the stress response (Rose et al. 2013). *BABY BOOM* (*BBM*) was initially discovered to induce SE in microspores and can also be induced by stress treatments (Boutilier et al. 2002, Li et al. 2014) and it is also a member of the AP2/ERF superfamily of transcription factors. It is not closely related to *SERF* but could also have a role in integrating stress and development.

It is not surprising that there are changes in expression of multiple genes as the stress, hormones and cell types interact to form the progenitor embryonic cells. A number of genes highlighted in the Karami et al. (2009) review such as *MtSTRESS KINASE1* (*MtSK1*), Glutathione-S-transferase genes (*GSTs*), Germin-like protein genes (*GLPs*) and Thioredoxin H (*TrxH*) genes are involved in stress related responses including redox regulation. The different hormones have their signaling modules and targets, and the different transcription factors are modulated and can also have a number of targets. However these different genes can be conceptually fitted to the framework in Fig. 4.

Small RNAs

Small RNAs represent another layer of regulation and there are still many unknowns. High throughput studies have shown that many small RNAs are expressed in SE

(Wu et al. 2015). A number of miRNAs such as miR156, have been linked to SE induction (Wu et al. 2011, Mahdavi-Darvari et al. 2015). In Medicago a miRNA has been shown to influence ethylene signalling and is differentially expressed in the highly embryogenic 2HA line (Kurdyukov et al. 2014a), but it is not clear how it is linked to SE. What was also evident in this latter study was that transposase genes were activated in the highly regenerable line (compared to the parent line) which likely reflects chromatin remodelling and its importance in SE. Of interest in the context of the mechanisms discussed here is the miRNA regulation of genes such as *LEC2* (Willmann et al. 2011).

Organogenesis - Root induction *in vitro*

Root induction was obtained *in vitro* in the classic Skoog and Miller (1957) experiments by using high auxin: cytokinin ratios. The study of root organogenesis *in vitro* has probably been less of a priority as the formation of new root meristems in lateral root formation can readily be studied *in vivo*. If, however we consider leaf tissue, often used as explants in SE, then the new auxin-induced root meristem clearly comes from the pre-existing vascular associated stem cells in model systems such as *M. truncatula* (Fig. 3, Rose et al. 2006) and Arabidopsis (Liu et al. 2014). This is as opposed to SE where whether pre-existing stem cells are involved in the induction of SE in cultured tissue, depends on the species and tissue type as well as stress and the inductive hormones.

In Medicago cells proliferate from the procambium and root meristems arise from these cells (Rose et al. 2006). The procambial cells are stem cells that produce the vasculature (Fukuda 2004) but need to be programmed in the direction of the root meristem. These cells presumably have a chromatin state that is more amenable to reprogramming. The development of the root meristem is associated with the root stem cell identity transcription factor WOX5 (Chen et al. 2009). Auxin has directed the pre-existing stem cells to form root stem (or founder) cells. Again regeneration involves the induction of a specific stem cell. Whether the culture stress contributes to this has not been specifically studied but it is clear that there is ROS production along the veins very early after leaf excision (Wang et al. 2011) and redox influences the root meristem formation (Imin et al. 2007). If cytokinin alone is added then this differentiation into root cells does not occur in Medicago; although rarely, it can form an embryonic stem cell (Nolan et al. 2003). Most of the embryonic stem cells in Medicago occur by the dedifferentiation pathway associated with massive callusing at the explant surface.

Work has also been carried out in Arabidopsis which has reached quite interesting conclusions in relation to callus formation and the root developmental pathway (Sugimoto et al. 2010). Their data show that callus formation from a range of different organs is due to the differentiation of pericycle-like stem cells. The WOX 5 reporter is expressed in all the calli formed irrespective of the cultured organ as is *PLETHORA1*, another root meristem gene. These data have similarities, but with some major differences, to Medicago. In the Sugimoto et al. (2010) study similar results were obtained with different auxin (2,4-D) and kinetin ratios. In Medicago there is some callus formed associated with the veins in NAA alone and NAA plus BAP, but once BAP is added to NAA there is massive callusing associated with dedifferentiating cells on the edge of the tissue.

In another study in Arabidopsis there was further support for callus having root meristem characteristics (Liu et al. 2014). In this case leaf explants were grown on medium without hormones and roots formed from the procambium as in Medicago. In this case there was a single adventitious root in which endogenous auxin was involved. However a new finding was that WOX11 AND WOX12 are involved in the transition to root organogenesis. These members of the WOX family act redundantly to mediate the first step of the stem cell state transition to root founder cells and *WOX5* expression (Liu et al. 2014).

There is a consensus that root organogenesis *in vitro* occurs from pre-existing stem cells of the procambium or pericycle –like cells and there is differentiation into root stem cells or founder cells involving WOX genes.

Organogenesis – shoot induction *in vitro*

Shoot formation *in vitro* was demonstrated by Skoog and Miller (1957) in callus by using high cytokinin:auxin ratios. Shoot formation *in vitro* is also a pathway for plant transformation. How does shoot meristem formation *in vitro* compare with what is known in relation to SE and root organogenesis? Gordon et al. (2007) have studied shoot induction in Arabidopsis where as expected shoot meristems are initiated in areas of low auxin and high cytokinin. Their data are consistent with a model where early *WUS* expression within callus is required for shoot meristem progenitor cell identity and later *WUS* expression is required for further shoot development. Both *WUS* and *PIN1* are necessary for efficient generation of shoot meristems. *CUP-SHAPED COTYLEDON 2 (CUC2)* is also an early developmental regulator. These data emphasise *WUS* as a stem cell identity transcription gene involved in inducing progenitor cells, and requiring *PIN* genes which are involved in patterning the emerging meristem (Aida et al. 2002).

There is further evidence that emphasises the significance of WUS in relation to shoot stem cell identity which comes from a study by Gallois et al. (2004). Induction of WUS expression in roots established cells which were able to produce shoots. These shoots came from near the root tip. Precisely which cells was not specified. Of further interest in this study was that root cells could be driven into SE if auxin (NAA) was added to the WUS expressing roots which is of particular interest as in Arabidopsis it is auxin and WUS that are very central to SE induction, but cytokinin and WUS in shoot induction (Gordon et al. 2007). In the absence of auxin in the Gallois et al. (2004) study there would quite likely be endogenous cytokinin as root tips produce cytokinins (Aloni et al. 2005). More recent studies by Chatfield et al. (2013) have shown that cytokinin-induced WUS is required for the conversion of lateral root primordia into shoot meristems.

A further study on WUS in relation to organogenesis throws light on the WUS relationship to chromatin state. Shoots were regenerated on media containing both auxin and cytokinin after transfer of the callus. *WUS* showed reduced DNA methylation and increased H3K4me3 and H3K9ac levels while there were reduced H3K9me2 levels; all consistent with gene activation. (Li et al. 2011). These data indicate that *WUS* derepression is required in order to facilitate shoot induction. Which cells were specifically involved was not clear. However it is clear from other Arabidopsis studies that the shoot meristem cells emanate originally from vascular associated stem cells (Atta et al. 2009, Chatfield et al. 2013).

Callus

Callus has been considered in the context of SE and organogenesis, particularly in relation to root organogenesis. There are two types of callus. Where embryos do not derive from pre-existing stem cells then the new dividing cells are the first step in totipotency and possibly pluripotency. In the case of cells derived from pre-existing stem cells then it is those cells or their progenitors from which embryos or organs are derived.

Conclusions : *In Vitro* Reprogramming and Chromatin States

From the array of studies of *in vitro* SE and organogenesis, while the details are clearly different between different species and different explant types there are some conceptual similarities that relate to understanding plant cells and their developmental fates. This can be thought of in the context of a number of chromatin states.

State 1 – Differentiated state: a specialised cell such as a mesophyll cell.

State 2 – Dedifferentiated dividing cell: not a stem cell and continues to divide as callus.

State 3 – A stem cell: derived from State 2 or pre-existing in the explant.

State 4 – A stem cell derived from State 2 or 3 and continues to divide as callus.

State 5 – Stem cell commits from State 3 or 4 to form an embryonic, root or shoot stem cell (or could be called a specific type of founder cell).

In the context of these cellular changes there is a need not only to understand the relationship between DNA methylation and histone modifications in relation to individual genes but how the whole genome is patterned to have a particular chromatin state accessible by regulators.

In terms of specific genes it is *WUS* and the related *WOX* genes that are particularly significant in determining *State 5* commitment to embryogenesis or shoot or root organogenesis. The hormone regulation remains an important part of this. In the case of SE auxin commonly induces *WUS* whereas it is cytokinin in the case of shoot organogenesis. This indicates co-expression of other genes are important and this could also be dependent on the cell location and development stage. In the case of SE it is now apparent that endogenous hormones relating to both stress and derepressing embryogenesis genes are also important. What is also of interest is that auxin can induce roots or embryos depending on species and explant type which also could provide clues as to interacting players with the WUS and WOX genes.

In terms of plant development it is clear that vascular associated stem cells are remarkable in being able to be the progenitors of callus, roots, shoots and embryos as well as veins. The recent research on the relationship between the HAIRY MERISTEM (HAM) family of transcription factors and WUS/WOX transcription factors indicates a common framework in stem cell function (Zhou et al. 2014). These HAM and WUS/WOX interactions appear to be important in driving downstream events. It also provides a basis for how *in vitro* these interactions might provide a basis for the different directions a stem cell population may take.

Once the correct committed stem cell forms *in vitro* and the developmental pathway initiates then the genetic regulation *in vitro* and *in vivo* should be quite similar.

Acknowledging that the environment and nutrient sources are different, related *in vivo* and *in vitro* experimentation should be able to provide new insights into plant development.

Acknowledgments

The author's research has been supported by the Australian Research Council project number CEO348212 and The University of Newcastle.

References

Aan Den Toorn, M., C. Albrecht and S. De Vries. 2015. On the origin of SERKS: bioinformatics analysis of the somatic embryogenesis receptor kinases. Mol. Plant 8: 762-782.

Aida, M., T. Vernoux, M. Furutani, J. Traas and M. Tasaka, M. 2002. Roles of PIN-FORMED1 and MONOPTEROS in pattern formation of the apical region of the Arabidopsis embryo. Development 129: 3965-3974.

Aloni, R., M. Langhans, E. Aloni, E. Dreieicher and C.I. Ullrich. 2005. Root-synthesised cytokinin in Arabidopsis is distributed in the shoot by the transpiration stream. J. Exp. Bot. 56: 1535-1544.

Atta, R., L. Laurens, E. Boucheron-Dubuisson, A. Guivarc'h, E. Carnero, V. Giraudat-Pautot, P. Rech, and D. Chriqui. 2009. Pluripotency of Arabidopsis xylem pericycle underlies shoot regeneration from root and hypocotyl explants grown *in vitro*. Plant J. 57: 626-644.

Boutilier, K., R. Offringa, V.K. Sharma, H. Kieft, T. Ouellet, L. Zhang, J. Hattori, C-M. Liu, A.A.M. van Lammeren, B.L.A. Miki, J.B.M. Custers and M.M. van Lookeren Campagne. 2002. Ectopic expression of BABY BOOM triggers a conversion from vegetative to embryonic growth. Plant Cell 14:1737-1749.

Braybrook, S.A. and J.J. Harada. 2008. LECs go crazy in embryo development. Trends in Plant Sci. 13: 624-630.

Casimiro, I., T. Beeckman, N. Graham, R. Bhalerao, H. Zhang, P. Casero, G. Sandberg and M. Bennett. 2003. Dissecting Arabidopsis lateral root development. Trends Plant Sci. 8: 165-171.

Chatfield, S.P., R. Capron, A. Severino, P.-A. Penttila, S. Alfred, H. Nahal and N.J. Provart. 2013. Incipient stem cell niche conversion in tissue culture: using a systems approach to probe early events in WUSCHEL-dependent conversion of lateral root primordia into shoot meristems. Plant J. 73: 798-813.

Chen, S.K., S. Kurdyukov, A, Kereszt, , X-D.Wang, P.M. Gresshoff and R.J. Rose. 2009. The association of homeobox gene expression with stem cell formation and morphogenesis in cultured *Medicago truncatula*. Planta 230: 827-840.

De Almeida, M., C.V. De Almeida, E.M. Graner, G.E. Brondani and M.F. De Abreu-Tarazi. 2012. Pre-procambial cells are niches for pluripotent and totipotent stem-like cells for organogenesis and somatic embryogenesis in the peach palm: a histological study. Plant Cell Rep. 31: 1495-1515.

De Veylder, L., T. Beeckman and D. Inzé. 2007. The ins and outs of the plant cell cycle. Nat. Rev. Mol Cell Biol. 8: 655-665.

Elhiti, M., M. Tahir, R.H. Gulden, K. Khamiss and C. Stasolla. 2010. Modulation of embryo-forming capacity in culture through the expression of *Brassica* genes involved in the regulation of the shoot apical meristem. J. Exp. Bot. 61: 4069-4085.

Fehér, A. 2015. Somatic embryogenesis – stress induced remodeling of plant cell fate. Biochim. et Biophys. Acta - Gene Regulatory Mechanisms 1849: 385-402.

Fehér, A., K. Ölvös, T.P. Pasternak and A.P. Szandtner. 2008. The involvement of reactive oxygen species (ROS) in the cell cycle activation (G_0-to-G_1 transition) of plant cells. Plant Signal. & Behav. 3: 823-826.

Fehér, A., T.P. Pasternak and D. Dudits. 2003. Transition of somatic plant cells to an embryogenic state. Plant Cell Tiss. Organ Cult. 74: 201-228.

Fukuda, H. 2004. Signals that control plant vascular cell differentiation. Nature Rev. Molec. Cell Biol. 5: 379-391.

Gallois, J-L., F.R. Nora, Y. Mizukami and R. Sablowski. 2004. WUSCHEL induces shoot stem cell activity and developmental plasticity in the root meristem. Genes & Dev. 18: 375-380.

Gliwicka, M., K. Nowak, S. Balazadeh, B. Mueller-Roeber and M.D. Gaj. 2013. Extensive modulation of the transcription factor transcriptome during somatic embryogenesis in *Arabidopsis thaliana*. PLoS ONE 8: e69261.

Gordon, S.P., M.G. Heisler, G.V. Reddy, C. Ohno, P. Das and E.M. Meyerowitz. 2007. Pattern formation during de novo assembly of the Arabidopsis shoot meristem. Development 134: 3539-3548.

Grafi, G. and S. Barak. 2015. Stress induces cell differentiation in plants. Biochim. Biophys. Acta - Gene Regulatory Mechanisms 184: 378-384.

Groß-Hardt, R., M. Lenhard and T. Laux. 2002. WUSCHEL signaling functions in interregional communication during Arabidopsis ovule development. Genes & Dev. 16: 1129-1138

Guzzo, F., B. Baldan, P. Mariani, F. Lo Schiavo and M.Terzi 1994. Studies on the origin of totipotent cells in explants of *Daucus carota* L. J. Exp. Bot. 45:1427-1432.

Haecker, A., R. Groß-Hardt, B. Geiges, A. Sarkar, H. Breuninger, M. Herrmann and T. Laux. 2004. Expression dynamics of WOX genes mark cell fate decisions during early embryonic patterning in *Arabidopsis thaliana*. Development 131: 657-668.

He, C., X. Chen, H. Huang and L. Xu. 2012. Reprogramming of H3K27me3 is critical for acquisition of pluripotency from cultured Arabidopsis tissues. PLOS Genetics 8: e1002911.

Ikeuchi, M., K. Sugimoto and I. Iwase. 2013. Plant callus: mechanisms of induction and repression. Plant Cell 25: 3159-3173.

Imin, N., F. De Jong, U. Mathesius, G. van Noorden, N.A. Saeed, X-D.Wang, R.J. Rose and B.G. Rolfe. 2004. Proteome reference maps of *Medicago truncatula* embryogenic cell cultures generated from single protoplasts. Proteomics 4: 1883-1896.

Imin, N., M. Nizamidin, D. Daniher, K.E. Nolan, R.J. Rose and B.G. Rolfe. 2005. Proteomic analysis of somatic embryogenesis in *Medicago truncatula*. Explant cultures grown under 6-benzylaminopurine and 1-naphthaleneacetic acid treatments. Plant Physiol. 137: 1250-1260.

Imin, N., M. Nizamidin, T.Wu and B.G. Rolfe. 2007. Factors involved in root formation in *Medicago truncatula*. J. Exp. Bot. 58: 439-451.

Jiang F., Z. Feng, H. Liu and Z. Jian. 2015. Involvement of plant stem cells or stem cell-like cells in dedifferentiation. Front. Plant Sci. 6: 1028.

Karami, O., B. Aghavaisi and A.M. Pour. 2009. Molecular aspects of somatic-to-embryogenic transition in plants. J. Chem. Biol. 2: 177-190.

Kwaaitaal, M.A.C.J. and S.C. De Vries. 2007. The *SERK1* gene is expressed in procambium and immature vascular cells. J. Exp. Bot. 58: 2887-2896.

Kurdyukov. S., U. Mathesius, K.E. Nolan, M.B. Sheahan, N. Goffard, B.J. Carroll and R.J. Rose. 2014a. The 2HA line of *Medicago truncatula* has characteristics of an epigenetic mutant that is weakly ethylene insensitive. BMC Plant Biol. 14: 174.

Kurdyukov, S., Y. Song, M.B. Sheahan and R.J. Rose. 2014b. Transcriptional regulation of early embryo development in the model legume *Medicago truncatula*. Plant Cell Rep. 33: 349-362.

Li, H., M. Soriano, J. Cordewener, J.M. Muiño, T. Riksen, H. Fukuoka, G.C. Angenent and K. Boutilier. 2014. The histone deacetylase inhibitor trichostatin A promotes totipotency in the male gametophyte. Plant Cell 26: 195-209.

Li, W., H. Liu, J.C. Zhi, Y.H. Su, H.N. Han, Y. Zhang and X.S. Zhang. 2011. DNA methylation and histone modifications regulate *de novo* shoot regeneration in Arabidopsis by modulating *WUSCHEL* expression and auxin signaling. PLoS Genetics 7: e1002243.

Liu, J., L. Sheng, Y. Xu, J. Li, Z.Yang, H. Huang and L. Xu. 2014. *WOX11* and *WOX12* are involved in the first-step cell fate transition during *de novo* root organogenesis in Arabidopsis. Plant Cell 26: 1089-1093.

LoSchiavo, F., L. Pitto, G. Giuliano, G. Torti, V. Nuti-Ronchi, D. Marazziti, R. Vergara, S. Orselli and M. Terzi. 1989. DNA methylation of embryogenic carrot cell cultures and its variations as caused by mutation, differentiation, hormones and hypomethylating drugs. Theor. Applied Genet. 77: 325-331.

Magyar, Z., L. De Veylder, A. Atanassova, L. Bako, D. Inzé and L.Bögre. 2005. The role of the Arabidopsis E2FB transcription factor in regulating auxin-dependent cell division. Plant Cell 17: 2527-2541.

Mahdavi-Darvari, F., N.M. Noor and I. Ismanizan. 2015. Epigenetic regulation and gene markers as signals of early somatic embryogenesis. Plant Cell Tiss. Organ Cult. 20: 407-422.

Makarevitch, I., S.R. Eichten, R. Briskine, A.J. Waters, O.N. Danilevskaya, R.B. Meeley, C.L. Myers, M.W. Vaughn and N.M. Springer. 2013. Genomic distribution of maize facultative heterochromatin marked by trimethylation of H3K27. Plant Cell 25: 780-793.

Mantiri, F.R., S. Kurdyukov, D.P. Lohar, N. Sharopova, N.A. Saeed, X.-D. Wang , K.A. Vandenbosch and R.J. Rose. 2008. The transcription factor MtSERF1 of the ERF subfamily identified by transcriptional profiling is required for somatic embryogenesis induced by auxin plus cytokinin in *Medicago truncatula*. Plant Physiol. 146: 1622-1636.

Mayer, K.F.X., H. Schoof, A. Haecker, M. Lenhard, G. Jürgens and T. Laux. 1998. Role of *WUSCHEL* in regulating stem cell fate in the Arabidopsis shoot meristem. Cell 95: 805-815.

Miguel, C. and L. Marum. 2011. An epigenetic view of plant cells cultured *in vitro*: somaclonal variation and beyond. J. Exp. Bot 62: 3713-3725.

Nashun, B., P.W.S. Hill and P. Hajkova. 2015. Reprogramming of cell fate: epigenetic memory and the erasures of memories past. Embo J. 34: 1296-1308.

Nolan, K.E., S. Kurdyukov and R.J. Rose. 2009. Expression of the *SOMATIC EMBRYOGENESIS RECEPTOR-LIKE KINASE* (*SERK1*) gene is associated with developmental changes in the life cycle of the model legume *Medicago truncatula*. J. Exp. Bot. 60: 1759-1771.

Nolan, K.E., R.R. Irwanto and R.J. Rose. 2003. Auxin up-regulates *MtSERK1* expression in both *Medicago truncatula* root-forming and embryogenic cultures. Plant Physiol. 133: 218-230.

Nolan, K.E. and R.J. Rose. 1998. Plant regeneration from cultured *Medicago truncatula* with particular reference to abscisic acid and light treatments. Aust. J. Bot. 46: 151-160.

Nolan, K.E., Y. Song, S. Liao, N.A. Saeed, X. Zhang and R.J. Rose. 2014. An unusual abscisic acid and gibberellic acid synergism increases somatic embryogenesis, facilitates its genetic analysis and improves transformation in *Medicago truncatula*. PLoS ONE 9: e99908.

Ogas, J., J-C. Cheng, Z.R. Sung and C. Somerville. 1997. Cellular differentiation regulated by gibberellin in the *Arabidopsis thaliana pickle* mutant. Science 277: 91-94.

Papadakis, A.K. and K.A. Roubelakis-Angelakis. 1999. The generation of active oxygen species differs in tobacco and grapevine mesophyll protoplasts. Plant Physiol. 121: 197-206.

Pasternak, T.P., E. Prinsen, F. Ayaydin, P. Miskolczi, G. Potters, H. Asard, H.A. Van Onckelen, D. Dudits and A. Féher. 2002. The Role of auxin, pH, and stress in the activation of embryogenic cell division in leaf protoplast-derived cells of alfalfa. Plant Physiol. 129: 1807-1819.

Rocha, D.I., D.L.P. Pinto, L.M. Vieira, F.A.O. Tanaka, M.C. Dornelas and W.C. Otoni. 2016. Cellular and molecular changes associated with competence acquisition during passion fruit somatic embryogenesis and analysis of SERK gene expression. Protoplasma 253: 595-609.

Rose, R.J., X-D. Wang, K.E. Nolan and B.G. Rolfe. 2006. Root meristems in *Medicago truncatula* tissue culture arise from vascular-derived procambial-like cells in a process regulated by ethylene. J. Exp. Bot. 57: 2227-2235.

Rose R.J., F.R. Mantiri, S. Kurdyukov, S-K. Chen, X-D. Wang, K.E. Nolan and M.B. Sheahan. 2010. Developmental biology of somatic embryogenesis. pp. 3-26. *In*: E.C. Pua and M.R. Davey (eds.). Plant Developmental Biology: Biotechnology Perspectives. Volume 2. Springer-Verlag, Berlin.

Rose, R.J., M.B. Sheahan and T.W-Y. Tiew. 2013. Connecting stress to development in the induction of somatic embryogenesis. pp. 146-165. *In*: J. Aslam, P.S. Srivastava and M.P. Sharma (eds.). Somatic Embryogenesis and Gene expression. Narosa Publishing House, New Delhi.

Salvo, S.A.G.D., C.N. Hirsch, C.R. Buell, S.M. Kaeppler and H.F. Kaeppler. 2014. Whole transcriptome profiling of maize during early somatic embryogenesis reveal altered expression of stress factors and embryogenesis-related genes. PloS ONE 9, e111407.

Santos, D. and P. Fevereiro. 2002. Loss of DNA methylation affects somatic embryogenesis in *Medicago truncatula*. Plant Cell Tiss. Organ Cult. 70: 155-161.

Sarkar, A.K., M. Luijten, S. Miyashima, M. Lenhard, T. Hashimoto, K. Nakajima, B. Scheres, R. Heidstra and T. Laux. 2007. Conserved factors regulate signalling in *Arabidopsis thaliana* shoot and root stem cell organizers. Nature 446: 811-814.

Schaller, G.E., I.H. Street and J.J. Kieber. 2014. Cytokinin and the cell cycle. Curr. Opin. Plant Biol. 21: 7-15.

Schmidt, E.D.L., F. Guzzo, M.A.J. Toonen and S.C. DeVries. 1997. A leucine-rich repeat containing receptor-like kinase marks somatic plant cells competent to form embryos. Development 124: 2049-2062.

Shan, X., J. Yan, and D. Xie. 2012. Comparison of phytohormone signalling mechanisms. Curr. Opinion in Plant Biol. 15: 84-91.

Sharma, S.K., S. Millam, P.E. Hedley, J. McNicol and G.J. Bryan. 2008. Molecular regulation of somatic embryogenesis in potato: an auxin led perspective. Plant Mol. Biol. 68: 185-201.

Skoog, F. and C.O. Miller. 1957. Chemical regulation of growth and organ formation in plant tissues cultured *in vitro*. Symp. Soc. Exp. Biol. 54: 118-130.

Solis, M-T., A-A. El-Tantawy, V. Cano, M.C. Risueño and P.S. Testillano. 2015. 5-azacytidine promotes microspore embryogenesis initiation by decreasing global methylation, but prevents subsequent embryo development in rapeseed and barley. Front. Plant Sci. 6: 472.

Soriano, M., H. Li, C. Jacquard, G.C. Angenent, J. Krochko, R. Offringa and K. Boutilier. 2014. Plasticity in cell division patterns and auxin transport dependency during *in vitro* embryogenesis in *Brassica napus*. Plant Cell 26: 2568-2581.

Steward, F.C., M.O. Mapes and K. Mears. 1958. Growth and organised development of cultured cells. II. Organization in cultures grown from freely suspended cells. Am J. Bot. 45: 705-708.

Stone, S. L., L.W. Kwong, K. M.Yee, J. Pelletier, L. Lepiniec, R.L. Fischer, R.B. Goldberg and J.J. Harada. 2001. LEAFY COTYLEDON2 encodes a B3 domain transcription factor that induces embryo development. Proc. Natl. Acad. Sci. USA 98: 11806-11811.

Stone, S.L., S.A. Braybrook, S.L. Paula, L.W. Kwong, J. Meuser, J. Pelletier, T-F Hsieh, R.L. Fischer, R.B. Goldberg and J.J. Harada. 2008. Arabidopsis LEAFY COTYLEDON2 induces maturation traits and auxin activity: Implications for somatic embryogenesis. Proc. Natl. Acad. Sci. USA 98: 3151-3156.

Su, Y.H., Y.B. Liu, B. Bai and X.S. Zhang. 2015, Establishment of embryonic shoot-root axis is involved in auxin and cytokinin response during Arabidopsis somatic embryogenesis. Front. Plant Sci. 5: 792.

Su, Y. H., X.Y. Zhao, Y.B. Liu, C.L. Zhang, S.D. O'Neill and X.S. Zhang. 2009. Auxin-induced WUS expression is essential for embryonic stem cell renewal during somatic embryogenesis in Arabidopsis. Plant J. 59: 448-460.

Sugimoto, K., S.P. Gordon and E.M. Meyerowitz. 2011. Regeneration in plants and animals: dedifferentiation, transdifferentiation or just differentiation. Trends in Cell Biol. 21: 212-218.

Sugimoto, K., Y. Jiao and E.M. Meyerowitz. 2010. Arabidopsis regeneration from multiple tissues occurs via a root development pathway. Developmental Cell 18: 463-471.

Thibaud-Nissen, F., R.T. Shealy, A. Khanna and L.O. Vodkin. 2003. Clustering of microarray data reveals transcript patterns associated with somatic embryogenesis in soybean. Plant Physiol. 132: 118-136.

Thomas, M.R. and R.J. Rose. 1983. Plastid number and plastid structural changes associated with tobacco mesophyll protoplast culture and plant regeneration. Planta 158: 329-338.

Tiew, T.W., M.B. Sheahan and R.J. Rose. 2015. Peroxisomes contribute to reactive oxygen species homeostasis and cell division induction in Arabidopsis protoplasts. Front. Plant Sci. 6: 658.

Tokuji, Y. and K. Kuriyama. 2003. Involvement of gibberellin and cytokinin in the formation of embryogenic cell clumps in carrot (*Daucus carota*). J. Plant Physiol. 160: 133-141.

Verdeil J-L, L. Alemanno, N. Niemenak and T.J. Tranbarger. 2007. Pluripotent versus totipotent plant stem cells: dependence versus autonomy? Trends Plant Sci. 12: 245-252.

Wang, X-D., K.E. Nolan, R.R. Irwanto, M.B. Sheahan and R.J. Rose. 2011. Ontogeny of embryogenic callus in *Medicago truncatula*: the fate of the pluripotent and totipotent stem cells. Ann. Bot. 107: 599-609.

Williams, E.G. and G. Maheswaran. 1986. Somatic embryogenesis: factors influencing coordinated behaviour of cells as an embryogenic group. Ann. Bot. 57: 443-462.

Williams, L., J. Zhao, N. Morozova, Y. Li, Y. Avivi and G. Grafi. 2003. Chromatin reorganization accompanying cellular dedifferentiation is associated with modifications of histone H3, redistribution of HP1, and activation of E2F-target genes. Developmental Dynamics 228: 113-120.

Willmann, M.R., A.J. Mehalick, R.L. Packer and P.D. Jenik. 2011. MicroRNAs regulate the timing of embryo maturation in Arabidopsis. Plant Physiol. 155: 1871-1884.

Wu, X-M., S-J. Kou, Y-L. Liu, Y-N. Fang, Q. Xu and W-W Guo. 2015. Genomewide analysis of small RNAs in nonembryogenic and embryogenic tissues of citrus: microRNA- and siRNA-mediated transcript cleavage involved in somatic embryogenesis. Plant Biotech. J. 13: 383-394.

Wu, X.-M., M.-Y. Liu, X.-X. Ge, Q. Xu and W.-W. Guo. 2011. Stage and tissue-specific modulation of ten conserved miRNAS and their targets during somatic embryogenesis of Valencia sweet orange. Planta 233: 495-505.

Yamamoto N, H. Kobayashi, T. Togashi Y. Mori, K. Kikuchi, K. Kuriyama and Y. Tokuji. 2005. Formation of embryogenic cell clumps from carrot epidermal cells is suppressed by 5-azacytidine, a DNA methylation inhibitor. J. Plant Physiol. 162: 47-54.

Zhao, J., N. Morozova, L. Williams, L. Libs, Y. Avivi and G. Grafi, G. 2001. Two phases of chromatin decondensation during dedifferentiation of plant cells. J. Biol. Chem. 276: 22772-22778.

Zhang, H., B. Bishop, W. Ringenberg, W.M. Muir and J. Ogas. 2012. The CHD3 remodeler PICKLE associates with genes enriched for trimethylation of histone H3 lysine 27. Plant Physiol 159: 418-432.

Zhang, H., S.D. Rider Jr., T.J. Henderson, M. Fountain, K. Chuang, V. Kandachar, A. Simons, H.J. Edenberg, J. Romero-Severson, W.M. Muir and J. Ogas. 2008. The CHD3 remodeler *PICKLE* promotes trimethylation of histone H3 lysine 27. J. Biol. Chem. 283: 22637-22648.

Zheng, Q.,Y. Zheng and S.E. Perry. 2013. AGAMOUS-Like15 promotes somatic embryogenesis in Arabidopsis and soybean in part by the control of ethylene biosynthesis and response. Plant Physiol 161: 2113-2127.

Zhou, Y., X. Liu, E.M. Engstro, Z. L. Nimchuck, J.L. Pruneda-Paz, P.T. Tarr, A. Yan, S.A. Kay and E.M. Meyerowitz. 2014. Control of plant stem cell function by conserved interacting transcriptional regulators. Nature 517: 377-380.

Zuo, J., Q-W. Niu, G. Frugis and N.H. Chua, 2002. The *WUSCHEL* gene promotes vegetative-to embryonic transition in Arabidopsis. Plant J. 30: 349-359.

19

Death and Rebirth: Programmed Cell Death during Plant Sexual Reproduction

David J.L. Hunt and Paul F. McCabe*

Introduction

Programmed cell death (PCD) is the phenomenon by which a plant may initiate the death of its cells in a highly-regulated and uniform manner. PCD is known to play a fundamental role in most developmental processes such as xylogenesis or senescence. Ironically, one of the developmental processes which rely most heavily upon PCD is the production of new life via sexual reproduction. This review will provide an overview of the key drivers involved in regulating PCD in plant cells and, subsequently, outline the evidence in support of a role for PCD in sexual reproduction.

Programmed Cell Death in Plants

In animal systems, cells which have been induced to die via a controlled pathway undergo a process known as apoptosis. First described by Kerr et al. (1972), apoptosis is characterised by the condensation of the nucleus, and the shrinkage and subsequent fragmentation of the cell into small 'apoptotic bodies' which are then recycled by phagocytes (Savill and Fadok 2000). Apoptosis is known to be a highly controlled process by which the cell actively chooses to initiate a death mechanism and must not be confused with necrosis, which is a term used to describe the unorganised form of death observed in cells which have been subject to excessive stress conditions resulting in rapid loss of osmosregulative capacity and the immediate rupture of the cell (Lennon et al. 1991).

At a molecular level, apoptosis has been well characterised and is primarily driven by a family of cysteine proteases known as caspases (Adrain and Martin 2001). Caspase activation can occur via the release of cytochrome c from the mitochondria and the subsequent formation of the apoptosome, a high molecular weight caspase-activating protein complex. Accordingly, a considerable number of the key regulators of apoptosis have been shown to interfere with mitochondrial function (Adrain and Martin 2001).

School of Biology and Environmental Science, University College Dublin, Belfield, Dublin 4.
* Corresponding author : paul.mccabe@ucd.ie

Plant PCD is a distinct form of cell death distinguished from apoptosis in several ways including the absence of cell degradation into apoptotic bodies and their recycling via phagocytes. The most striking characteristic of plant cells undergoing PCD is the retraction of the cytoplasm away from the cell wall (McCabe et al. 1997). Cells undergoing necrosis fail to display this morphology due to the disorganised destruction of the cell and rapid rupture of the cell membrane. While no caspase homologues have been identified in plants, caspase-like proteins have been observed and are known to show cleavage activities similar to their animal equivalents (Woltering 2004). Additionally, DNA cleavage into a 'ladder' of 180-200 bp multimers is a common feature of both apoptosis and plant PCD (Reape and McCabe 2008). Reape et al. (2008) present a simple model for distinguishing between live cells and those that have undergone either PCD or necrosis. Fluorescein diacetate (FDA) is cleaved by live plant cells and as such will fluoresce under light at 490 nM. In dead cells, no FDA fluorescence will be observed. In such instances, PCD can be detected via the characteristic retraction of the cytoplasm away from the plasma membrane. Conversely, necrotic cells will display neither cytoplasmic retraction nor FDA fluorescence.

Intracellular Drivers of PCD

Mitochondria

Although mitochondrial involvement in PCD was originally observed in mammalian systems, evidence now exists to suggest a role for their involvement in PCD in plants. Cucumber cells in which PCD has been induced by heat-shock at 55°C (McCabe et al. 1997) show translocation of cytochrome c from the mitochondria to the cytoplasm which begins during the actual heat treatment (Balk et al. 1999). This translocation of cytochrome c is not a result of mitochondrial disintegration as both fumarase (a 50 kDa mitochondrial membrane protein) and porin (an integral outer membrane protein) are still both present in the mitochondria following heat-shock-induced PCD (Balk et al. 1999, Balk et al. 2003). Additionally, mitochondrial membrane integrity showed only a small decrease following heat shock as detected via an assay examining the latency of cytochrome c oxidase activity (Balk et al. 1999).

The release of cytochrome c from the mitochondria during PCD has also been demonstrated in developmentally-regulated programmed cell death. Balk and Leaver (2001) describe the release of cytochrome c during the programmed cell death of sunflower tapetal cells during anther development. In this system, male sterility in the PET1-CMS line was the result of tapetal cell death occurring prematurely in the pachytene stage rather than the tetrad stage. This cell death was shown to coincide with the release of cytochrome c within the tapetal cells. Cytochrome c release has also been detected during the death of incompatible pollen tubes during self-incompatibility (SI) responses in *Papaver* (Thomas and Franklin-Tong 2004). This cytochrome c release was shown to be preceded by an increase in internal Ca^{2+} concentration, suggesting a possible intracellular signalling mechanism linking Ca^{2+} and cytochrome c release.

While in animal systems cytochrome c release is known to be a direct step in the activation of caspases, at present no evidence exists to suggest that it has the same role in the induction of PCD in plant cells. In an Arabidopsis cell-free system, which can be used to study intracellular activities in subcellular fractions, the addition of broken mitochondria induces chromatin condensation and fragmentation of nuclear DNA, an effect not seen when purified cytochrome *c* was added to the system (Balk et al. 2003). Despite this, the mitochondria is still thought to play an important role in plant PCD, possibly via modulation of cellular levels of reactive oxygen species (ROS) (Vacca et al. 2006) or the release other death inducing proteins (Balk et al. 2003).

Metacaspases

In plants, no caspase homologues have been identified but a number of proteins with caspase-like activity have been characterised These so-called metacaspases can be grouped into two groups, with type I metacaspases displaying a proline- or gluta-mine-rich N-terminal extension and type II having none (Vercammen et al. 2004) and appear to be lysine/arginine-specific cysteine proteases, thus differentiating their protease activity from traditional animal caspases which are aspartic acid-specific (Vercammen et al. 2004). At least one example of a metacaspase which is essential for PCD to occur during the death of the suspensor cell during somatic embryogen-esis has been characterised (Suarez et al. 2004). The activity of this metacaspase will be discussed in greater detail in a subsequent section.

Programmed Cell Death During Sexual Reproduction

As previously stated, PCD is a vital process for the correct development of numerous plant organs. This may involve the removal of certain tissues such as leaf or petal senescence or the production of structural or vascular tissues from dead cells such as bark or xylem vessels. During sexual reproduction PCD plays a particularly impor-tant role in order to allow for correct gamete development, sexual fertilisation and subsequent seed production. The following sections will give a description of the role of PCD in various tissues that are essential in sexual reproduction.

Sex determination via male and female organ abortion

Species of plants which develop unisexual flowers account for approximately 10% of the total number of flowering plants (Wu and Cheung 2000). Unisexuality dictates that the development of either the male or female reproductive organs is prevented either via the arrest of their initial development or the termination of their develop-ment subsequent to its induction. Unisexual flowers appear to have evolved indepen-dently in a number of species with some species adopting a method of organ abortion which is brought about via PCD. In maize, an agronomically important unisexual species, the abortion of the pistil primordia results in the production of a male flower, while in a female flower the later arrest of the staminate occurs (Cheng et al. 1983). In maize, this abortion of early sexual organs involves the vacuolation of their cells and the degradation of organelles contained therein. Pistil abortion has also shown to be accompanied by nuclear degradation and a loss of overall structural integrity (Calderon-Urrea and Dellaporta 1999).

This abortion of primordial male or female tissue appears to be governed by complex molecular pathways. In maize, nuclear degradation is mediated by the

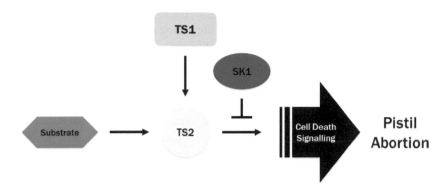

FIGURE 1   *tasselseed* genes play an important role in the abortion of developing pistils in unisexual maize flowers. In this model, proposed by Calderon-Urrea and Dellaporta (1999), TASSELSEED 2 (TS2) acts on an unknown substrate – suggested to be a steroid – to initiate a cell death programme in developing pistils. TS2 expression is positively regulated by TASSELSEED 1 (TS1) but its ability to initiate a cell death response can be negatively regulated by SILKLESS 1 (SK1). How TS2 initiates this cell death programme is still unknown.

tasselseed2 (*ts2*) gene (Calderon-Urrea and Dellaporta 1999), an alcohol-dehydrogenase which has been shown to accumulate in the gynoecium shortly prior to its abortion (DeLong et al. 1993). The expression of *ts2* is positively-regulated by *tasselseed1* (*ts1*) but the effect of this *tasselseed*-mediated cell death is in turn negatively regulated by the *silkless1* (*sk1*) gene (Calderon-Urrea and Dellaporta 1999) (Fig. 1). These observations led to the formation of a model whereby *sk1* may act as a moderator during sex determination by regulating *tasselseed*-mediated death, either directly or by influencing a downstream signalling cascade. *Ts6* and *ts4* mutants display a phenotype in which female organs fail to abort and male organs fail to develop, leading to the feminisation of the male flower (Chuck et al. 2007). Additionally, both mutations promote spikelet meristem branching, suggesting that the genetic control of sex determination and meristematic development are governed by the same gene networks. *ts4* appears to negatively regulate *indeterminate spikelet1* (*ids1*), an *APATELA2* (*AP2*) family gene which functions in spikelet meristem determinacy, a result which suggests that sex determination in maize may be determined via negative regulation of the floral homeotic pathway (Chuck et al. 2007).

The Arabidopsis *AGAMOUS* MADS box gene is known to act antagonistically against *AP2* in order to bring about the development of both male and female organs (Mizukami and Ma 1992). In dioecious species, disparities in MADS box gene expression have been shown to correlate to the formation of either male or female flowers (Hardenack et al. 1994, Ainsworth et al. 1995, Park et al. 2003). In cucumber, the genetic basis of sex determination can be linked to three loci, the *F/f*, *M/m* and *A/a* loci, which interact to produce three phenotypes: gyneocious, monoecious and andromonecious. Ethylene appears to mediate the interaction of these three loci and the resulting sex of the flowers produced by plants from the three phenotypes, possibly via negative regulation of the *M* locus (Yamasaki et al. 2001). Additionally, *CS-ACS1*, a 1-aminocyclopropane-1-carboxylate (ACC)-synthase gene, has been shown to be closely linked to the *F* locus (Trebitsh et al. 1997). ACC is known to play an important role in the biosynthesis of ethylene and reinforces the notion that ethylene plays an important regulatory role during sex determination in cucumber. CS-ACS1 expression is induced by auxin, suggesting the involvement of a complex interplay of hormonal stimuli in sex determination (Trebitsh et al. 1997). Indeed, several other hormones are known to play a direct role during sex determination. Abortion of the pistil can be inhibited through the ectopic expression of the cytokinin-synthesising isopentenyl transferase (IPT) enzyme under the control of SAG, a senescence-inducible promoter isolated from *Arabipdopsis* (Young et al. 2004). The aforementioned *ts1* gene from maize is now thought to encode a plastid-targeted lipoxygenase which functions in the synthesis of jasmonic acid (Acosta et al. 2009). This synthesised jasmonic acid is thought to positively regulate the formation of male flowers as the application of jasmonic acid (JA) to *ts1* and *ts2* mutants rescued stamen development. Mutation of *anther ear1*, a gene encoding an enzyme involved in an early step during gibberellic acid (GA) synthesis, results in the formation of perfect flowers on normally pistillate maize ears, an effect which can be reversed by GA supplementation (Bensen et al. 1995). The endogenous balance between GA and ethylene appears to play a critical role in sex determination in the flowering fern *Anemia phyllitidis,* which suggests a conserved relationship may exist in both higher plants as well as some pteridophytes (Kaźmierczak and Kaźmierczak 2007).

Other factors can also have a role in production of unisexual flowers in several species. For example, the effect of the external environment on sex determination has been addressed but remains very poorly understood. Differences in sex ratios can be observed in cucumber plants grown under differing soil and light conditions. For example, under reduced daylight intensity and duration, staminate flowers are preferentially produced by the plant (Tiedjens 1928). Additionally, evidence exists to suggest that epigenetic changes can influence the sex determination of a flower via transposon-induced changes in gene transcription (Martin et al. 2009).

Tissue death during anther maturation

During anther maturation, programmed cell death is always initiated in two specific tissue types: the tapetum and the stomium. The tapetum functions primarily as a layer of nutritive cells which secrete various molecules into the anther locule in order to sustain the developing microspores. This is initially achieved via active secretion but later is a result of the degeneration and collapse of the tapetal cells. Cell death within the stomium results in axes of weakness along the outer wall of the anther. The drying of the anther as a whole results in dehiscence via the controlled shattering of the anther along these axes and subsequent release of pollen grains.

Papini et al. (1999) report that tapetal cell death in angiosperms involves the organised condensation and disassembly of the cell as well as degradation of the nuclear material. In addition, the cell wall is dissolved in order to allow the release of nutritive molecules into the anther locule. The authors also note that the mitochondria persist longest within the cell, suggesting the cell death process is actively-driven. In sunflower, the PET1-CMS mutant displays premature programmed cell death in the tapetum and is consequently male-sterile. This death has been shown to be associated with the premature partial release of cytochrome *c* from the mitochondria during the pachytene, rather than the tetrad stage (Balk and Leaver 2001). Similar observations have been made in rice wherein male sterility occurs in the event of premature tapetum degeneration (Ku et al. 2003). Conversely, preventing cell death in the tetrad stage also results in male sterility, possibly due to the starvation of the microspore cells (Kawanabe et al. 2006). As such, the correct timing of the abolition of the tapetal cells is critical to normal pollen development and an intimate relationship exists between the two.

In rice, the *tapetum degeneration retardation (tdr)* mutant fails to exhibit PCD in the tapetum and is therefore male sterile (Li et al. 2006). The *TDR* transcript can be detected primarily in the tapetal cells during anther development. Its protein product is a putative transcription factor and is thought to upregulate Os *CP1* (a Cys protease) and Os *c6* (a putative protease inhibitor) suggesting that *TDR* is responsible for the regulation of a downstream cell death signalling pathway (Li et al. 2006). *TDR* also appears to alter the synthesis of lipidic compounds within the tapetal cells, particularly in relation to fatty acid synthesis and decarbonylation (Zhang et al. 2008). Numerous genes involved in lipid metabolism, processing and transfer to the microspore surface were altered in the *tdr* mutant, both positively and negatively, while a selection of PCD-related genes with functions including calcium signalling, heat shock responses and cysteine cleavage also had their expression altered (Zhang et al. 2008).

The *PERSISTANT TAPETAL CELL1* (*ptc1*) mutant also displays male sterility due to the inability of the tapetal cells to undergo PCD during the tetrad stage (Li et al. 2011a). In this instance, the tapetal cells continue to proliferate but then undergo a rapid death. As with *TDR*, suppression of *PTC1* led to changes in the expression of numerous genes involved in cell death, lipid processing and protein catalysis. Li et al. (2011b) also describe a signalling cascade responsible for tapetal PCD wherein APOPTOSIS INHIBITOR5 (API5), an orthologue of a human PCD-related gene, interacts with API5-INTERACTING PROTEIN1 (AIP1) and AIP2, two DEAD-box ATP-dependent RNA helices. These proteins in turn bind to the promoter region of CP1, a rice Cys protease, to induce cell death (Fig. 2). In humans, API5 is upregulated in tumour cells and its depletion results in tumour death, leading to the suggestion that it has a role in the maintenance of cell viability (Faye and Poyet 2010). How API5 attained a pro-PCD role in plants is unknown. Reactive oxygen species (ROS) are also known to play a critical role in rice tapetal cell PCD, the homeostasis of which is believed to be governed by the influence of MADS3 over MT-1-4b (Hu et al. 2011).

In Arabidopsis, the *male sterility1* (*ms1*) mutation results in blocked tapetum PCD and resulting male sterility (Vizcay-Barrena and Wilson 2006). This cell death is known to be driven by the Cys protease CEP1 which is produced in proenzyme form before being transported to the vacuole for processing and eventual release via the rupture of the tonoplast (Zhang et al. 2014). The MYB80 transcription factor also appears to regulate tapetum PCD via the induction of UNDEAD expression, an A1 aspartic protease with a mitochondrial targeting signal (Phan et al. 2011). This mechanism may provide a link between a mitochondrial death effector and tapetal death. Epigenetic changes also appear to play an important role in tapetal PCD, as an increase in DNA methylation and the activity of DNA methyl transferase 1 (MET1) increase significantly during the process of cell death (Solís et al. 2014).

In tomato stomial cells, cell death is preceded by the accumulation of ricinosomes, a subset of precursor protease vesicles (Senatore et al. 2009). These ricinosomes accumulate a KDEL-tailed Cys protease called SlCysEP which, upon the death of the cell, is released via the rupture of the membrane. In ethylene insensitive tobacco, gametophyte production is unaffected but dehiscence is delayed due to the failure of stomial cells to degenerate in a timely manner (Rieu et al. 2003). Stomium cell death appears to be regulated as part of a larger network of overall anther dehiscence,

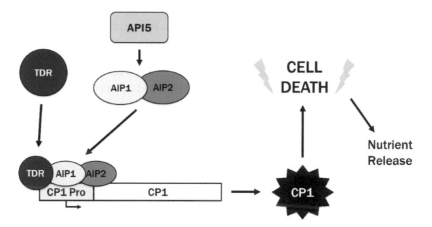

FIGURE 2 A simplified schematic of the API5- AIP1/2-CP1 pathway described by Li et al. (2011b) which brings about PCD and eventual nutrient release from tapetal cells in rice. TAPETUM DEGENERATION RETARDATION (TDR) is expressed and binds to the promoter region of the *CP1* Cys protease gene. APOPTOSIS INHIBITOR5 (API5) activates API5-INTERACTING PROTEIN1 & 2 (AIP1, AIP2) which form a dimer and also bind to the *CP1* promoter region to form a transcription initiation complex. This results in the activation of CP1, which is involved in inducing cell death, leading to the release of intracellular nutrients into the anther locule. Loss of function of any of the components involved in this pathway results in male sterility.

which in *Arabidopsis* is controlled at least to some extent by the interaction of the TGA9/10 transcription factors with ROXY1/2 glutaredoxins (Murmu et al. 2010). The aspartic protease *PROMOTION OF CELL SURVIVAL1* (*PCS1*) is known to play an important role in the maintenance of cell viability in both the male and female gametophytes as well as the embryo, while ectopic expression of PCS1 leads to the failure of the septum and stomial cells to degenerate which results in male sterility (Ge et al. 2005). *WUSCHEL* (*WUS*) is a gene previously characterised in meristematic cells. Its expression has also been detected in immature stomial cells prior to their differentiation and degeneration while *wus* mutant anthers fail to open as a result of stomial cells failing to degenerate (Deyhle et al. 2007).

Non-functional megaspore abortion

During the formation of the female gametophyte, the diploid megasporocyte cell undergoes meiosis to produce four haploid megaspores. Three of these megaspores then undergo programmed cell death while the fourth enters the megagametogenesis pathway in order to produce the gametophyte (Wu and Cheung 2000). In non-functional megaspores, cell death is accompanied by the typical characteristics of cellular shrinkage and nuclear and organelle degradation (Russell 1979). The degraded non-functional megaspores are eventually crushed by the expansion of the embryo sac. Qiu et al. (2008) observed changes in calcium distribution within cells during megasporogenesis. While, a relatively low level of calcium precipitate was observed within the megasporocyte, following meiosis this level increased within all four megaspores. However, the depletion of intracellular calcium precipitates, particularly in the nucleus, then preceded the degeneration of each of the non-functional megaspores. The functional megaspore maintains its high level of calcium precipitates throughout the degeneration of the micropylar megaspores and through-out its development into the embryo sac.

The molecular mechanisms involved in the selection of which megaspore cell will be retained in order to undergo megagametogenesis are very poorly understood. In Arabidopsis, evidence exists to suggest that the classical arabinogalactan-protein AGP18 positively regulates this selection (Demesa-Arévalo and Vielle-Calzada 2013). Overexpression of *AGP18* results in the continued survival of more than one megaspore cell, each having the capacity to form a functional megaspore. However, it appears that neither AGP18, nor post-translational modifications thereof, can induce the specification of the functional megaspore prior to meiosis. Similarities exist between these findings and numerous observations regarding the role of AGPs in stimulating somatic embryogenesis *in vitro* (for review see Seifert and Roberts 2007). In alfalfa, a Mps-one-binder (Mob)-like protein appears to be associated with PCD in reproductive tissues and expression of a Mob-like transcript was detected in degenerating megaspore cells, but also other degenerating tissues such as the tape-tum, leading to the suggestion that this Mob-like protein may at least be associated with, if not directly responsible for, PCD. (Citterio et al. 2005). Papini et al. (2011) present a model in which callose deposition at the cell wall determines which mega-spores will degenerate. Callose deposition has previously been suggested as a marker for somatic embryogenesis from epidermal and cortical root cells in chicory (Dubois et al. 1990).

Pollen tube death during self-incompatibility

The self-incompatibility (SI) response is a mechanism employed by some plants in which pollen grains are prevented from fertilising an egg from the same plant by inducing cell death within either the pollen grain upon the stylar surface or the pol-len tube shortly after its penetration. In many species the SI response is known to be governed primarily by a complex interaction between the male and female poly-morphic *S* loci (Nasrallah et al. 1985, McClure et al. 1989, McClure et al. 1990, Stein et al. 1991, Huang et al. 1994, Lee et al. 1994, Murfett et al. 1994, Schopfer et al. 1999, Takasaki et al. 2000, Takayama et al. 2001, Ushijima et al. 2003, Sijacic et al. 2004, Thomas and Franklin-Tong 2004, Wheeler et al. 2009). Some of the style S-locus are RNases (McClure et al. 1989) and bring about SI via the degradation of pollen RNA (McClure et al. 1990, Huang et al. 1994, Murfett et al. 1994). These S-RNases and associated mRNAs are significantly down-regulated in the event of a compatible pollen interaction (Liu et al. 2009). S proteins can also trigger DNA fragmentation, possibly in a $[Ca^{2+}]$-dependent manner (Jordan et al. 2000).

The *Papaver* SI response appears to be a biphasic process, consisting initially of the inhibition of pollen tube growth, followed by the activation of a DEVDase/caspase-like molecule which initiates the death response (Thomas and Franklin-Tong 2004). Incompatible pollen challenged with S-proteins express a protein kinase, termed p26.1, which appears to play a role in the SI response and is phosphorylated under the control of Ca^{2+} and calmodulin (Rudd et al. 1996). Analysis of p26.1 showed that it comprises two proteins, p26.1a and Pr-p26.1b, both of which are soluble inorganic pyrophospha-tases (sPPases) (de Graaf et al. 2006). These sPPases are thought to be involved in numerous biosynthetic pathways in both prokaryotes and eukaryotes and are neces-sary for pollen tube growth. In plants, p26.1 is Mg^{2+} dependent and can be inhibited by Ca^{2+}. Additionally, its activity is suppressed via phosphorylation. Together, these results suggest that inhibition of pollen tube growth may be the result of suppression of p26.1 activity via Ca^{2+}-mediated phosphorylation (de Graaf et al. 2006).

Cell death involves swelling of the mitochondria, Golgi and endoplasmic reticulum (ER), nuclear condensation and loss of mitochondrial cristae. In addition, electron density within the matrix of the mitochondria is reduced and significant callose deposition can be observed at the cell wall (Geitmann et al. 2004). DNA degradation and cytochrome *c* release can be detected, and an increase in intracellular $[Ca^{2+}]$ appears to occur upstream of this cytochrome *c* release (Thomas and Franklin-Tong 2004). Additionally, increases in $[Ca^{2+}]$ stimulate enhanced cleavage of PARP, a substrate of caspase-3 in animal cells, indicating a role for Ca^{2+} in stimulating the SI response. The activity of three caspase-like enzymes, a DEVDase, a VEIDase and a LEVDase can account for this cleavage activity (Bosch and Franklin-Tong 2007). DEVDase and VEIDase show rapid accumulation following the induction of SI, while LEVDase is accumulated more slowly. Cytosolic pH is reduced in order to correspond with the optimal working pH of these caspase-like enzymes. The signalling cascades involved in activating these caspase-like proteins are not yet understood, although in *Papaver* DEVDase activity is thought to be regulated by the MAP kinase p56 (Li et al. 2007).

Rapid increases in intracellular $[Ca^{2+}]$ can be detected within seconds of the SI-response being activated (Franklin-Tong et al. 1997, 2002). The inhibition of pollen tube growth is known to be controlled via a slow-moving Ca^{2+} wave under the control of inositol (1,4,5)-triphosphate (Franklin-Tong et al. 1996). Increases in internal $[Ca^{2+}]$ at the pollen tube tip are known to lead to actin depolymerisation (Snowman et al. 2002) and contributes to the overall death of the pollen tube (Thomas et al. 2006). Additionally, alterations to actin filaments were sufficient to activate DEVDase (Thomas et al. 2006). Increases in $[Ca^{2+}]_{cyt}$ lead to increases in ROS and nitric oxide (NO) within the pollen tubes, which in turn also act to induce DEVDase activity and actin reorganisation, suggesting a possible signalling relationship between Ca^{2+}, ROS/NO and PCD (Wilkins et al. 2011). In addition, microtubule depolymerisation is also induced during the SI response in manner that is at least partly dependent on actin depolymerisation (Poulter et al. 2008). However, while actin depolymerisation leads to microtubule depolymerisation, the opposite effect is not seen, nor does microtubule destabilisation lead to PCD. Nevertheless, microtubule depolymerisation is still a necessary step in the death of the pollen tube.

Cell death in the carpel associated with fertilisation

In order for a pollen tube to fertilise the egg cell, it must travel down through the style and penetrate the embryo sac from the bottom. For this to happen, cell death must occur in numerous tissues within the carpel. Pollination induces significant cell death in the carpel transmitting tissue in a manner that is pollination specific and involves ethylene-mediated poly(A) tail shortening of transmitting tissue-specific mRNAs (Wang et al. 1996). Glycoproteins are believed to be secreted from the extracellular matrix of the transmitting tissue to the growing pollen tube (Lind et al. 1996). Death of the transmitting tissue is thought to be induced in order to protect against pathogen invasion, which could hinder the fertilisation process. Nucellar cells also undergo PCD after coming into contact with the pollen tube, displaying DNA cleavage, cytoplasm collapse and organelle degeneration (Hiratsuka et al. 2002). The death of these cells is thought to be necessary for the nutrition of the growing pollen tube, and vesicles containing various cellular materials are known to be secreted by dying nucellar cells to be taken up by the pollen tube via endocytosis (Hiratsuka and Terasaka 2011).

Upon the pollen tube reaching the ovule, the sperm gains access to the embryo sac via the death of one of two synergid cells. In *Arabidopsis*, cell death does not occur until the pollen tube has reached the synergid cell and grown around it (Sandaklie-Nikolova et al. 2007). At this point, the synergid cell is induced to die and sperm release occurs. Cell death is accompanied by the hallmark characteristic of nuclear degradation, while in the persisting synergid, no changes to the nucleus are observed (An and You 2004). A significant increase in DNA methylation can also be detected in the degenerating synergid in comparison to the persisting synergid (Tagliasacchi et al. 2007). Ethylene signalling within the synergid cell is necessary to initiate death and prevent additional pollen tubes from penetrating the embryo sac. This effect can be induced by mimicking a post-fertilisation ethylene burst. Both ETHLYENE INSENSITIVE2 (EIN2) and EIN3 are responsible for mediating the ethylene signalling cascade responsible for synergid PCD (Völz et al. 2013). Which synergid cell is committed to die is determined by a complex interaction between the cells and the growing pollen tip, mediated by intracellular calcium signatures under the control of the *FERONIA* signalling pathway (Ngo et al. 2014). Mutation of the mitochondrial gene *GFA2* results in a failure of synergid cells to undergo cell death. *GFA2* is believed to act as a chaperone in the mitochondrial matrix and loss of function mutants may be unable to undergo PCD as a result of improper processing of mitochondrial proteins involved in PCD induction (Christensen et al. 2002).

While the death of one of the synergid cells has a clear function in allowing the pollen tube to enter the embryo sac, the reason behind the death of the antipodal cells is less obvious. It appears that antipodal cell death does not occur in all species, and little is known about the molecular mechanisms that drive this process. An and You (2004) present evidence that in wheat, nuclear degradation occurs in antipodal cells but their cytoplasm and organelles appear to function normally. However, no evidence is presented which supports a strong argument that the cells actually undergo PCD. In *Arabidopsis*, conflicting reports exist regarding the fate of the antipodal cells. Murgia et al. (1993) describe post-fertilisation antipodal cells with an electron-dense cytoplasm and mitochondria with few cristae. These antipodal cells were then seen to degrade and were undiscernible aside from the cell wall and some cellular remnants. Kägi et al. (2010) present evidence that mitochondrial proteins produced within the central cell govern the timing of antipodal cell death. Conversely, Song et al. (2014) employed a female gametophyte-specific fluorescent reporter system to show that the antipodal cells persist beyond fertilisation and are still maintained even after all of the cells contained within the original, unfertilised embryo sac are no longer present. Clearly, results such as these highlight our lack of understanding regarding the function of the antipodal cells and the role, if any, that their death may have in fertilisation.

Death of the suspensor during embryo development

The suspensor is a terminally differentiated embryonic region consisting of one or numerous cells which serves as a bridge between the developing embryo and surrounding tissues within the developing seed. The suspensor arises from the basal cell of the two-cell pro-embryo and has been shown to be involved in nutrient transport to the developing embryo as well as hormone biosynthesis (Kawashima and Goldberg 2010). Later in embryo development, the suspensor undergoes programmed cell death and is not a component of the resulting seedling. Suspensor death is associated with the hallmark feature of DNA degradation (Wredle et al. 2001, Giuliani et al. 2002, Lombardi et al. 2007b, Zhao et al. 2013). The timing of this cell death can vary greatly even among members of the same family (Endo 2012), while the pattern in which death occurs is not uniform among all species. In tobacco, as in the majority of species, suspensor death begins in the basal cell and progresses throughout the suspensor in the direction of the developing embryo (Zhao et al. 2013). Conversely, in *Phaseolus coccineus*, cell death begins in the neck region, closest to the embryo, and proceeds towards the basal end of the suspensor (Lombardi et al. 2007a).

Ricinosomes are known to accumulate in the Quinoa suspensor prior to its degradation (López-Fernández and Maldonado 2013), similar to the manner in which they accumulate in stomial cells that are destined to die. Ricinosome accumulation correlates with the rise of mature cysteine-endopeptidase (Cys-EP), which is highly homologous to other cysteine endopeptidases known to be expressed in senescing tissues undergoing programmed cell death (Schmid et al. 1998). In *Phaseolus coccineus,* cytochrome *c* release is detected in the dying suspensor, along with the activation of caspase-1-like and caspase-3-like proteins which are induced at an acidic pH (Lombardi et al. 2007a). In Arabidopsis, mutants with a knockout of the *kiss of death* (*KOD*) display a reduced rate of PCD in their suspensors as well as in heat-shocked root hairs (Blanvillain et al. 2011). *KOD* expression initiates ion leakage and caspase-3-like activation and results in mitochondrial depolarisation, but is localised to the cytosol. PCD induced by overexpression of *KOD* could be blocked by the caspase inhibitor p35, as well as *BAX-inhibitor 1* (*AtBi-1*), which is involved in calcium efflux from the ER and interacts with the calcium binding-protein CALMODULIN. These results suggest that KOD may function upstream of the activation of caspase-like molecules during cell death.

Artificially induced down-regulation of the tobacco protease inhibitor, cystatin NtCYS, results in precocious basal cell death and the abortion of seeds. NtCYS down-regulation leads to the expression of caspase-1-like, caspase-3-like and caspase-6-like proteins. Conversely, NtCP14 is a pro-death protease which induces death in both the basal and apical lineages. Zhao et al. (2013) propose a model whereby NtCYS acts as an inhibitor of NtCP14. While NtCYS expression remains high, NtCP14 is deactivated and the cell remains viable. Upon the initiation of PCD in the suspensor, NtCYS expression is reduced and NtCP14 is activated.

Somatic embryogenesis (SE) is the term used to describe the development of a viable, non-zygotic embryo from a somatic plant cell. SE can be induced *in vitro* and is a useful technique for studying the development of embryos in a freely observable manner. In Norway spruce, two waves of PCD are observed during SE (Filonova et al. 2000). The first is associated with the disposal of unwanted cells in the pro-embryogenic mass upon the initiation of SE. The second results in the death of the terminally differentiated suspensor cell. This death involves the degradation of the microtubule network, while the F-actin network persists but displays significant changes in orientation, suggesting that it has a key role in the death of the cell (Smertenko et al. 2003). Expression of a VEIDase with caspase-6-like protease activity can be detected in these cultures and the proteolytic activity of the enzyme is known to be highly sensitive to changes in pH, ionic strength, $[Zn^{2+}]$ and temperature (Bozhkov et al. 2004). Expression of this VEIDase is highest during the death of the suspensor and silencing of the protease results in a failure of PEMs to progress through the correct embryogenic pathway.

Norway spruce suspensor cell death is also dependent on the proteolytic activity of mcII-Pa, a metacaspase that translocates from the cytoplasm to the nucleus during PCD (Bozhkov et al. 2005). McII-Pa is thought to have a direct role in the initiation of nuclear degradation which commits the suspensor to die and allows normal embryo development to occur. Silencing mcII-Pa results in a culture which continues to proliferate but fails to form embryos (Suarez et al. 2004). Additionally, VEIDase activity and the occurrence of fragmented nuclei is reduced in this mutant line, indicating a reduction in PCD-related processes in the culture. Like VEIDase, the proteolytic activity of mcII-Pa is highly sensitive to $[Zn^{2+}]$ (Bozhkov et al. 2005).

The observation that $[Zn^{2+}]$ influences VEIDase activity is of interest as $[Zn^{2+}]$ has been shown to be an effective marker with which to determine cell fate in embryogenic cultures. Free intracellular zinc appears to be necessary for cell viability and the depletion of intracellular $[Zn^{2+}]$ induces cell death within the embryo. Conversely, zinc supplementation maintains the viability of the suspensor cell and therefore leads to abnormal embryo development (Helmersson et al. 2008). The role of zinc in suppressing cell death is thought to involve the suppression of metacaspase proteases, possibly by binding to the inactive form of the protein. Nitrous oxide has also been found to preferentially accumulate in the embryo when compared to the suspensor in response to putrescine supplementation (Silveira et al. 2006). However, it is not known whether this amounts to a protective effect in the same manner as with zinc accumulation

Cell death in the endosperm and aleurone layer

The endosperm is a tissue which is formed within developing seeds in order to provide nutrition to the embryo. During seed development, the endosperm initiates a cell death programme which ultimately leads to the death of the entire tissue. As with PCD in all other tissue types, endosperm cell death includes the fragmentation of nuclear DNA (Young et al. 1997, Young and Gallie 1999, Wredle et al. 2001, He and Kermode 2003, DeBono and Greenwood 2006). Endosperm death is accompanied by the up-regulation of lipid transfer proteins which are thought to facilitate the recycling of endosperm lipids and may also protect the developing seedling from damage from proteases involved in the degradation of endosperm tissue (Eklund and Edqvist 2003). In maize, ABA and ethylene signalling appear to play a vital role in the initiation of PCD and DNA degradation in endosperm cells (Young and Gallie 2000, Young et al. 1997). Mutants which are ABA-insensitive or deficient display increased ethylene production during the development of the kernel and accordingly exhibit precocious DNA fragmentation and cell death in the endosperm (Young and Gallie 2000). Little is known about the molecular mechanisms involved in this process. Down-regulation of the *RBR1* gene results in increased cytochrome *c* release, DNA fragmentation and cell death in the endosperm (Sabelli et al. 2013). However, it is not known whether RBR1 has a direct role in the initiation of cell death and whether its activity is mediated via an ABA/ethylene signalling pathway.

Cell death in the endosperm appears to follow different patterns among various species, even those within the same family such as wheat and maize. While endosperm death in maize begins in the oldest, central cells and radiates outwards towards the youngest tissues, in wheat, endosperm death does not follow any observable pattern

(Young and Gallie 1999). Additionally, discrepancies in spikes in ethylene production can be seen between the two species. In tomato, GA appears to play a regulatory role in initiating endosperm cell death (DeBono and Greenwood 2006). Dying endosperm cells accumulate ricinosomes which contain the KDEL-tailed C1A cysteine proteinase, SlCysEP, and burst upon the death of the cell (Trobacher et al. 2013). GA supplementation resulted in the induction and processing of SlCysEP, while supplementation with both GA and ABA led to delayed SlCysEP accumulation and prevents its processing to a mature form. Cells supplemented with both GA and ABA showed higher viability compared with those incubated in the presence of GA only. Together, these results suggest a model whereby GA induces endosperm PCD in a manner that is antagonistically mediated by ABA.

Ricinosomes have also been observed during cell death in the endosperm of both *Ricinus communis* and *Chenopodium quinoa* (Schmid et al. 1999, Than et al. 2004, López-Fernández and Maldonado 2013). Characterisation of the CysEP produced within *Ricinus communis* ricinosomes confirmed it is an endopeptidase with preference for His, Arg, Lys, Leu, Phe, Tyr, Met, Thr and Trp at the P_1 site and Leu, Val, Ile, Met, Phe, Tyr and Trp at the P_2 site (Than et al. 2004). CysEP appears to be able to accept Pro at both the P_1 and P_1' positions, thus allowing it to cleave at both the N- and C-terminal sides of the amino acid, an activity which is unusual for a protease. These results suggest that CysEP has the capability to cleave a large number of proteins upon its activation, particularly those present within the cell wall. In the megagametophyte tissue, the primary nutritive tissue within the developing white spruce seed, the expression of a CysEP-related protein can also be detected following germination (He and Kermode 2003). The presence of ricinosome-like bodies could also be detected in dying megagametophyte tissue. Additionally, caspase-like proteases were detected in dying megagametophyte tissue and caspase-3 inhibitors could delay the degradation of nuclear DNA compared with untreated tissue.

The aleurone layer, which surrounds the endosperm and is morphologically and developmentally different from it, typically contains proteins and fats but little to no starch. In barley, DNA fragmentation begins near the embryo and advances towards the distal end of the seed (Wang et al. 1998). This DNA fragmentation occurs only after the expression of α-amylase mRNA, which is thought to be induced by GA produced by the embryo. The synthesis of α-amylase appears to be hormonally regulated as α-amylase fails to accumulate in the culture medium of aleurone protoplasts when ABA is added to the medium, an effect which is reversible by adding GA (Bethke et al. 1999). Concurrently, the supplementation of culture medium with GA led to the death of the aleurone protoplasts contained therein, an effect which could be delayed to a significant extent by addition of ABA. This was despite the accumulation of α-amylase in the medium, suggesting that GA-stimualted α-amylase production alone is not sufficient to induce PCD and that ABA does not prevent α-amylase synthesis. The incubation of isolated aleurone layers in GA and their subsequent exposure to H_2O_2 resulted in a high percentage of dead cells, while almost all those incubated in ABA and then exposed to H_2O_2 remained alive (Bethke and Jones 2001, Fath et al. 2001). The light-induced production of H_2O_2 can also induce cell death in aleurone cells treated with GA - but not those treated with ABA - in a manner that can be slowed by antioxidants and reductants (Bethke and Jones 2001). This is due to the down-regulation of the H_2O_2-metabolising enzyme CAT by GA and its up-regulation by ABA (Fath et al. 2001). In addition, ABA supplementation results in the maintenance of APX and SOD, two enzymes important for ROS scavenging and processing, while GA depletes their expression.

Clearly, the interaction of GA, ABA and ROS is a major regulator of PCD in the aleurone layer. Beligni et al. (2002) report that endogenously produced nitric oxide (NO) delays PCD in the barley aleurone layer, in a manner that does not inhibit the GA-induced synthesis of α-amylase. In this system, NO appears to act as an antioxidant and the GA-induced decline in activity of ROS metabolising enzymes CAT and SOD could be slowed via the endogenous production of NO. The transcription factor GAMyb is required for the activation of GA-induced PCD, which in turn is inhibited by the downstream expression of the ABA-inducible protein HVA22 (Guo and Ho 2008). HVA22 appears to be localised to the ER and Golgi and may negatively regulate PCD by preventing the redistribution of nutrients and degradation of protein vesicles necessary for cell death to occur. Haem oxygenase-1 (HO-1) appears to delay PCD in the aleurone layer and the HO inducers haematin and carbon monoxide (CO) can reduce PCD rates via HO-1 activation, an effect mimicked by antioxidants (Wu et al. 2011). This reduction in PCD rates is thought to be the result of the antioxidant activity of HO and its ability to metabolise H_2O_2. GA-treated aleurone tissue display a reduction in hydrogen sulfide (H_2S) production and the activity of the H_2S-synthesising enzyme L-cysteine desulfhydrase, while ABA-treatment stimulates the opposite effect (Xie et al. 2014). The reversal of this reduction in endogenous H_2S via H_2S donors also results in the prevention of GA-induced PCD. This prevention of GA-induced PCD was brought about via the H_2S-induced expression of HO-1, suggesting that H_2S plays an integral signalling role in the maintenance of aleurone viability. However, despite all the evidence for the role of GA-induced ROS in mediating PCD in the aleurone, Aoki et al. (2014) used a mutant study to determine that PCD in the aleurone is not directly induced by ROS but through DELLA protein, a negative regulator of GA signalling.

As previously stated, DNA fragmentation is a hallmark feature of PCD in all species. In wheat, a GA-induced nuclease is localised to the nucleus of aleurone cells undergoing PCD and is not expressed when GA synthesis was inhibited or in GA-insensitive mutants (Domínguez et al. 2004). This nuclease appears to be activated by Ca^{2+} and Mg^{2+}, but inhibited by Zn^{2+}. The importance of Ca^{2+} in mediating aleurone layer death has long been postulated. Kuo et al. (1996) show that okadaic acid prevents a GA-induced increase in cytosolic $[Ca^{2+}]$ but not those induced by hypoxia, suggesting a different signalling pathway mediates this increase. The authors suggest that while the signalling pathways governing the various PCD-inducing mechanisms are distinct from each other that they, along with their ABA-mediated counter signals, must diverge at the critical step of changes in cytosolic $[Ca^{2+}]$. As such, $[Ca^{2+}]$ changes may act as a central signalling hub in determining cellular fate similar to its role in other cell types. However, the mechanisms governing this process are poorly understood and require further investigation.

Conclusions and Future Perspectives

Plant sexual reproduction is a vital process for the support of our global ecosystem and, consequently, human life. While it may appear paradoxical that the death of specific groups of cells plays such an important role in plant reproduction, research has provided us with a wealth of knowledge regarding the numerous and critical functions performed by PCD in this process. As food security becomes a more pressing issue for society, it is important that our understanding of the mechanisms governing plant sexual reproduction and associated PCD are furthered in order to meet the demands of the future.

References

Acosta, I.F., H. Laparra, S.P. Romero, E. Schmelz, M. Hamberg, J.P. Mottinger, M. Moreno and S.L. Dellaporta. 2009. *tasselseed1* is a lipoxygenase affecting jasmonic acid signaling in sex determination of maize. Science 323: 262-265.

Adrain, C. and S.J. Martin. 2001. The mitochondrial apoptosome: A killer unleashed by the cytochrome seas. Trends Biochem. Sci. 26: 390-397.

Ainsworth, C., S. Crossley, V. Buchanan-Wollaston, M. Thangavelu and J. Parker. 1995. Male and female flowers of the dioecious plant sorrel show different patterns of MADS box gene expression. Plant Cell 7: 1583-1598.

An, L.-H. and R.-L. You. 2004. Studies on nuclear degeneration during programmed cell death of synergid and antipodal cells in *Triticum aestivum*. Sex. Plant Reprod. 17: 195-201.

Aoki, N., Y. Ishibashi, K. Kai, R. Tomokiyo, T. Yuasa and M. Iwaya-Inoue. 2014. Programmed cell death in barley aleurone cells is not directly stimulated by reactive oxygen species produced in response to gibberellin. J. Plant Physiol. 171: 615-618.

Balk, J., S.K. Chew, C.J. Leaver and P.F. McCabe. 2003. The intermembrane space of plant mitochondria contains a DNase activity that may be involved in programmed cell death. Plant J. 34: 573-583.

Balk, J. and C.J. Leaver. 2001. The PET1-CMS mitochondrial mutation in sunflower is associated with premature programmed cell death and cytochrome *c* release. Plant Cell 13: 1803-1818.

Balk, J., C.J. Leaver and P.F. McCabe. 1999. Translocation of cytochrome *c* from the mitochondria to the cytosol occurs during heat-induced programmed cell death in cucumber plants. FEBS Lett. 463: 151-154.

Beligni, M.V., A. Fath, P.C. Bethke, L. Lamattina and R.L. Jones. 2002. Nitric oxide acts as an antioxidant and delays programmed cell death in barley aleurone layers. Plant Physiol. 129: 1642-1650.

Bensen, R.J., G.S. Johal, V.C. Crane, J.T. Tossberg, P.S. Schnable, R.B. Meeley and S.P. Briggs. 1995. Cloning and characterization of the maize *An1* gene. Plant Cell 7: 75-84.

Bethke, P. and R. Jones. 2001. Cell death of barley aleurone protoplasts is mediated by reactive oxygen species. Plant J. 25: 19-29.

Bethke, P., J. Lonsdale, A. Fath and R. Jones. 1999. Hormonally regulated programmed cell death in barley aleurone cells. Plant Cell 11: 1033-1045.

Blanvillain, R., B. Young, Y. Cai, V. Hecht, F. Varoquaux, V. Delorme, J.-M. Lancelin, M. Delseny and P. Gallois. 2011. The Arabidopsis peptide kiss of death is an inducer of programmed cell death. EMBO J. 30: 1173-1183.

Bosch, M. and V.E. Franklin-Tong. 2007. Temporal and spatial activation of caspase-like enzymes induced by self-incompatibility in *Papaver* pollen. Proc. Natl. Acad. Sci. USA. 104: 18327-18332.

Bozhkov, P.V., L.H. Filonova, M.F. Suarez, A. Helmersson, A.P. Smertenko, B. Zhivotovsky and S. von Arnold. 2004. VEIDase is a principal caspase-like activity involved in plant programmed cell death and essential for embryonic pattern formation. Cell Death Differ. 11: 175-182.

Bozhkov, P.V., M.F. Suarez, L.H. Filonova, G. Daniel, A.A. Zamyatnin, S. Rodriguez-Nieto, B. Zhivotovsky and A. Smertenko. 2005. Cysteine protease mcII-Pa executes programmed cell death during plant embryogenesis. Proc. Natl. Acad. Sci. USA. 102: 14463-14468.

Calderon-Urrea, A. and S.L. Dellaporta. 1999. Cell death and cell protection genes determine the fate of pistils in maize. Development 126: 435-441.

Cheng, P.C., R.I. Greyson and D.B. Walden. 1983. Organ initiation and the development of unisexual flowers in the tassel and ear of *Zea mays*. Am. J. Bot. 70: 450-462.

Christensen, C.A., S.W. Gorsich, R.H. Brown, L.G. Jones, J. Brown, J.M. Shaw and G.N. Drews. 2002. Mitochondrial GFA2 is required for synergid cell death in Arabidopsis. Plant Cell 14: 2215-2232.

Chuck, G., R.Meeley, E. Irish, H. Sakai and S. Hake. 2007. The maize *tasselseed4* microRNA controls sex determination and meristem cell fate by targeting *Tasselseed6/indeterminate spikelet1*. Nat. Genet. 39: 1517-1521.

Citterio, S., E. Albertini, S. Varotto, E. Feltrin, M. Soattin, G. Marconi, S. Sgorbati, M. Lucchin and G. Barcaccia. 2005. Alfalfa Mob1-like genes are expressed in reproductive organs during meiosis and gametogenesis. Plant Mol. Biol. 58: 789-807.

de Graaf, B.H.J., J.J. Rudd, M.J. Wheeler, R.M. Perry, E.M. Bell, K. Osman, F.C.H. Franklin and V.E. Franklin-Tong. 2006. Self-incompatibility in *Papaver* targets soluble inorganic pyrophosphatases in pollen. Nature 444: 490-493.

DeBono, A.G. and J.S. Greenwood. 2006. Characterization of programmed cell death in the endosperm cells of tomato seed: two distinct death programs. Can. J. Bot. 84: 791-804.

DeLong, A., A. Calderon-Urrea and S.L. Dellaporta. 1993. Sex determination gene *TASSELSEED2* of maize encodes a short-chain alcohol dehydrogenase required for stage-specific floral organ abortion. Cell 74: 757-768.

Demesa-Arévalo, E. and J.-P. Vielle-Calzada. 2013. The classical arabinogalactan protein AGP18 mediates megaspore selection in *Arabidopsis*. Plant Cell 25: 1274-1287.

Deyhle, F., A.K. Sarkar, E.J. Tucker and T. Laux. 2007. *WUSCHEL* regulates cell differentiation during anther development. Dev. Biol. 302: 154-159.

Domínguez, F., J. Moreno and F.J. Cejudo. 2004. A gibberellin-induced nuclease is localized in the nucleus of wheat aleurone cells undergoing programmed cell death. J. Biol. Chem. 279: 11530-11536.

Dubois, T., M. Guedira, J. Dubois and J. Vasseur. 1990. Direct somatic embryogenesis in roots of *Cichorium*: Is callose an early marker? Ann. Bot. 65: 539-545.

Eklund, D.M. and J. Edqvist. 2003. Localization of nonspecific lipid transfer proteins correlate with programmed cell death responses during endosperm degradation in *Euphorbia lagascae* seedlings. Plant Physiol. 132: 1249-1259.

Endo, Y. 2012. Characterization and systematic implications of the diversity in timing of programmed cell death of the suspensors in Leguminosae. Am. J. Bot. 99: 1399-1407.

Fath, A., P.C. Bethke and R.L. Jones. 2001. Enzymes that scavenge reactive oxygen species are down-regulated prior to gibberellic acid-induced programmed cell death in barley aleurone. Plant Physiol. 126: 156-166.

Faye, A. and J.-L. Poyet. 2010. Targeting AAC-11 in cancer therapy. Expert Opin. Ther. Targets. 14: 57-65.

Filonova, L.H., P.V. Bozhkov, V.B. Brukhin, G. Daniel, B. Zhivotovsky and S. von Arnold. 2000. Two waves of programmed cell death occur during formation and development of somatic embryos in the gymnosperm, Norway spruce. J. Cell Sci. 113: 4399-4411.

Franklin-Tong, V., B. Drobak, A. Allan, P. Watkins and A. Trewavas. 1996. Growth of pollen tubes of *Papaver rhoeas* is regulated by a slow-moving calcium wave propagated by inositol 1, 4, 5-trisphosphate. Plant Cell 8: 1305-1321.

Franklin-Tong, V.E., G. Hackett and P.K. Hepler. 1997. Ratio-imaging of Ca^{2+}_i in the self-incompatibility response in pollen tubes of *Papaver rhoeas*. Plant J. 12: 1375-1386.

Franklin-Tong, V.E., T.L. Holdaway-Clarke, K.R. Straatman, J.G. Kunkel and P.K. Hepler. 2002. Involvement of extracellular calcium influx in the self-incompatibility response of *Papaver rhoeas*. Plant J. 29: 333-345.

Ge, X., C. Dietrich, M. Matsuno, G. Li, H. Berg and Y. Xia. 2005. An *Arabidopsis* aspartic protease functions as an anti-cell-death component in reproduction and embryogenesis. EMBO Rep. 6: 282-288.

Geitmann, A., V.E. Franklin-Tong and A.C. Emons. 2004. The self-incompatibility response in *Papaver rhoeas* pollen causes early and striking alterations to organelles. Cell Death Differ. 11: 812-822.

Giuliani, C., G. Consonni, G. Gavazzi, M. Colombo and S. Dolfini. 2002. Programmed cell death during embryogenesis in maize. Ann. Bot. 90: 287-292.

Guo, W.-J. and T.-H.D Ho. 2008. An abscisic acid-induced protein, HVA22, inhibits gibberellin-mediated programmed cell death in cereal aleurone cells. Plant Physiol. 147: 1710-1722.

Hardenack, S., D. Ye, H. Saedler and S. Grant. 1994. Comparison of MADS box gene expression in developing male and female flowers of the dioecious plant white campion. Plant Cell 6: 1775-1787.

He, X. and A. Kermode. 2003. Nuclease activities and DNA fragmentation during programmed cell death of megagametophyte cells of white spruce (*Picea glauca*) seeds. Plant Mol. Biol. 51: 509-521.

Helmersson, A., S. von Arnold and P.V. Bozhkov. 2008. The level of free intracellular zinc mediates programmed cell death/cell survival decisions in plant embryos. Plant Physiol. 147: 1158-1167.

Hiratsuka, R. and O. Terasaka. 2011. Pollen tube reuses intracellular components of nucellar cells undergoing programmed cell death in *Pinus densiflora*. Protoplasma 248: 339-351.

Hiratsuka, R., Y. Yamada and O. Terasaka. 2002. Programmed cell death of *Pinus* nucellus in response to pollen tube penetration. J. Plant Res. 115: 141-148.

Hu, L., W. Liang, C. Yin, X. Cui, J. Zong, X. Wang, J. Hu and D. Zhang. 2011. Rice MADS3 regulates ROS homeostasis during late anther development. Plant Cell 23: 515-533.

Huang, S., H.S. Lee, B. Karunanandaa and T.H. Kao. 1994. Ribonuclease activity of *Petunia inflata* S proteins is essential for rejection of self-pollen. Plant Cell 6: 1021-1028.

Jordan, N.D., F.C.H. Franklin and V.E. Franklin-Tong. 2000. Evidence for DNA fragmentation triggered in the self-incompatibility response in pollen of *Papaver rhoeas*. Plant J. 23: 471-479.

Kägi, C., N. Baumann, N. Nielsen, Y.-D. Stierhof and R. Gross-Hardt. 2010. The gametic central cell of *Arabidopsis* determines the lifespan of adjacent accessory cells. Proc. Natl. Acad. Sci. USA. 107: 22350-22355.

Kawanabe, T., T. Ariizumi, M. Kawai-Yamada, H. Uchimiya and K. Toriyama. 2006. Abolition of the tapetum suicide program ruins microsporogenesis. Plant Cell Physiol. 47: 784-787.

Kawashima, T. and R.B. Goldberg. 2010. The suspensor: not just suspending the embryo. Trends Plant Sci. 15: 23-30.

Kaźmierczak, A. and J.M. Kaźmierczak. 2007. The level of endogenous 1-aminocyclopropane-1-carboxylic acid in gametophytes of *Anemia phyllitidis* is increased during GA3-induced antheridia formation. Acta Physiol. Plant. 29: 211-216.

Kerr, J., A. Wyllie and A. Currie. 1972. Apoptosis: A basic biological phenomenon with wide-ranging implications in human disease. Br. J. Cancer 26: 239-257.

Ku, S., H. Yoon, H.S. Suh and Y.-Y. Chung. 2003. Male-sterility of thermosensitive genic male-sterile rice is associated with premature programmed cell death of the tapetum. Planta 217: 559-565.

Kuo, A., S. Cappelluti, M. Cervantes-Cervantes, M. Rodriguez and D.S. Bush. 1996. Okadaic acid, a protein phosphatase inhibitor, blocks calcium changes, gene expression, and cell death induced by gibberellin in wheat aleurone cells. Plant Cell 8: 259-269.

Lee, H.-S., S. Huang and T. Kao. 1994. S proteins control rejection of incompatible pollen in *Petunia inflata*. Nature, 367: 560-563.

Lennon, S., S. Martin and T. Cotter. 1991. Dose-dependent induction of apoptosis in human tumour cell lines by widely diverging stimuli. Cell Prolif. 24: 203-214.

Li, H., Z. Yuan, G. Vizcay-Barrena, C. Yang, W. Liang, J. Zong, Z.A. Wilson and D. Zhang. 2011a. *PERSISTENT TAPETAL CELL1* encodes a PHD-finger protein that is required for tapetal cell death and pollen development in rice. Plant Physiol. 156: 615-630.

Li, N., D.-S. Zhang, H.-S. Liu, C.-S. Yin, X.-X. Li, W.-Q. Liang, Z. Yuan, B. Xu, H.-W. Chu, J. Wang, T.-Q. Wen, H. Huang, D. Luo, H. Ma and D.-B. Zhang. 2006. The rice tapetum degeneration retardation gene is required for tapetum degradation and anther development. Plant Cell, 18: 2999-3014.

Li, S., J. Samaj and V.E. Franklin-Tong. 2007. A mitogen-activated protein kinase signals to programmed cell death induced by self-incompatibility in *Papaver* pollen. Plant Physiol. 145: 236-245.

Li, X., X. Gao, Y. Wei, L. Deng, Y. Ouyang, G. Chen, X. Li, Q. Zhang and C. Wu. 2011b. Rice APOPTOSIS INHIBITOR5 coupled with two DEAD-box adenosine 5'-triphosphate-dependent RNA helicases regulates tapetum degeneration. Plant Cell 23: 1416-1434.

Lind, J.L., I. Bönig, A.E. Clarke and M.A. Anderson. 1996. A style-specific 120-kDa glycoprotein enters pollen tubes of *Nicotiana alata in vivo*. Sex. Plant Reprod. 9: 75-86.

Liu, B., D. Morse and M. Cappadocia. 2009. Compatible pollinations in *Solanum chacoense* decrease both S-RNase and S-RNase mRNA. PLoS One, 4: e5774

Lombardi, L., N. Ceccarelli, P. Picciarelli and R. Lorenzi. 2007a. Caspase-like proteases involvement in programmed cell death of *Phaseolus coccineus* suspensor. Plant Sci. 172: 573-578.

Lombardi, L., N. Ceccarelli, P. Picciarelli and R. Lorenzi. 2007b. DNA degradation during programmed cell death in *Phaseolus coccineus* suspensor. Plant Physiol. Biochem. 45: 221-227.

López-Fernández, M.P. and S. Maldonado. 2013. Ricinosomes provide an early indicator of suspensor and endosperm cells destined to die during late seed development in quinoa (*Chenopodium quinoa*). Ann. Bot. 112: 1253-1262.

Martin, A., C. Troadec, A. Boualem, M. Rajab, R. Fernandez, H. Morin, M. Pitrat, C. Dogimont and A. Bendahmane. 2009. A transposon-induced epigenetic change leads to sex determination in melon. Nature 461: 1135-1138.

McCabe, P.F., A. Levine, P.-J. Meijer, N. Tapon and R. Pennell. 1997. A programmed cell death pathway activated in carrot cells cultured at low cell density. Plant J. 12: 267-280.

McClure, B.A., V. Haring, P.R. Ebert, M.A. Anderson, R.J. Simpson, F. Sakiyama and A.E. Clarke. 1989. Style self-incompatibility gene products of *Nicotiana alata* are ribonucleases. Nature 342: 955-957.

McClure, B.A., J.E. Gray, M.A. Anderson and A.E. Clarke. 1990. Self-incompatibility in *Nicotiana alata* involves degradation of pollen rRNA. Nature 347: 757-760.

Mizukami, Y. and H. Ma. 1992. Ectopic expression of the floral homeotic gene *AGAMOUS* in transgenic *Arabidopsis* plants alters floral organ identity. Cell 71: 119-131.

Murfett, J., T.L. Atherton, B. Mou, C.S. Gassert and B.A. McClure. 1994. S-RNase expressed in transgenic *Nicotiana* causes S-allele-specific pollen rejection. Nature 367: 563-566.

Murgia, M., B.-Q. Huang, S.C. Tucker and M.E. Musgrave. 1993. Embryo sac lacking antipodal cells in *Arabidopsis thaliana* (Brassicaceae). Am. J. Bot. 80: 824-838.

Murmu, J., M.J. Bush, C. DeLong, S. Li, M. Xu, M. Khan, C. Malcolmson, P.R. Fobert, S. Zachgo, S.R. Hepworth. 2010. *Arabidopsis* basic leucine-zipper transcription factors TGA9 and TGA10 interact with floral glutaredoxins ROXY1 and ROXY2 and are redundantly required for anther development. Plant Physiol. 154: 1492-1504.

Nasrallah, J.B., T.-H. Kao, M.L. Goldberg and M.E. Nasrallah. 1985. A cDNA clone encoding an S-locus-specific glycoprotein from *Brassica oleracea*. Nature 318: 263-267.

Ngo, Q.A., H. Vogler, D.S. Lituiev, A. Nestorova and U. Grossniklaus. 2014. A calcium dialog mediated by the *FERONIA* signal transduction pathway controls plant sperm delivery. Dev. Cell 29: 491-500.

Papini, A., S. Mosti and L. Brighigna. 1999. Programmed-cell-death events during tapetum development of angiosperms. Protoplasma 207: 213-221.

Papini, A., S. Mosti, E. Milocani, G. Tani, P. Di Falco and L. Brighigna. 2011. Megasporogenesis and programmed cell death in *Tillandsia* (Bromeliaceae). Protoplasma 248: 651-662.

Park, J.-H., Y. Ishikawa, R. Yoshida, A. Kanno and T. Kameya. 2003. Expression of *AODEF*, a B-functional MADS-box gene, in stamens and inner tepals of the dioecious species *Asparagus officinalis* L. Plant Mol. Biol. 51: 867-875.

Phan, H.A., S. Iacuone, S.F. Li and R.W. Parish. 2011. The MYB80 transcription factor is required for pollen development and the regulation of tapetal programmed cell death in *Arabidopsis thaliana*. Plant Cell 23: 2209-2224.

Poulter, N.S., S. Vatovec and V.E. Franklin-Tong. 2008. Microtubules are a target for self-incompatibility signaling in *Papaver* pollen. Plant Physiol. 146: 1358-1367.

Qiu, Y.L., R.S. Liu, C.T. Xie, S.D. Russell and H.Q. Tian. 2008. Calcium changes during megasporogenesis and megaspore degeneration in lettuce (*Lactuca sativa* L.). Sex. Plant Reprod. 21: 197-204.

Reape, T.J. and P.F. McCabe. 2008. Apoptotic-like programmed cell death in plants. New Phytol. 180: 13-26.

Reape, T.J., E. Molony and P.F. McCabe. 2008. Programmed cell death in plants: distinguishing between different modes. J. Exp. Bot. 59: 435-444.

Rieu, I., C. Mariani and K. Weterings. 2003. Expression analysis of five tobacco EIN3 family members in relation to tissue-specific ethylene responses. J. Exp. Bot. 54: 2239-2244.

Rudd, J., F. Franklin, J. Lord and V.E. Franklin-Tong. 1996. Increased phosphorylation of a 26-kD pollen protein is induced by the self-incompatibility response in *Papaver rhoeas*. Plant Cell 8: 713-724.

Russell, S. 1979. Fine structure of megagametophyte development in *Zea mays*. Can. J. Bot. 57: 1093-1110.

Sabelli, P.A., Y. Liu, R.A. Dante, L.E. Lizarraga, H.N. Nguyen, S.W. Brown, J.P. Klingler, J. Yu, E. LaBrant, T.M. Layton, M. Feldman and B.A. Larkins. 2013. Control of cell proliferation, endoreduplication, cell size, and cell death by the retinoblastoma-related pathway in maize endosperm. Proc. Natl. Acad. Sci. USA. 110: E1827-1836.

Sandaklie-Nikolova, L., R. Palanivelu, E.J. King, G.P. Copenhaver and G.N. Drews. 2007. Synergid cell death in Arabidopsis is triggered following direct interaction with the pollen tube. Plant Physiol. 144: 1753-1762.

Savill, J. and V. Fadok. 2000. Corpse clearance defines the meaning of cell death. Nature 407: 784-788.

Schmid, M., D. Simpson and C. Gietl. 1999. Programmed cell death in castor bean endosperm is associated with the accumulation and release of a cysteine endopeptidase from ricinosomes. Proc. Natl. Acad. Sci. USA. 96: 14159-14164.

Schmid, M., D. Simpson, F. Kalousek and C. Gietl. 1998. A cysteine endopeptidase with a C-terminal KDEL motif isolated from castor bean endosperm is a marker enzyme for the ricinosome, a putative lytic compartment. Planta 206: 466-475.

Schopfer, C.R., M.E. Nasrallah and J.B. Nasrallah. 1999. The male determinant of self-incompatibility in *Brassica*. Science 286: 1697-1700.

Seifert, G. and K. Roberts. 2007. The biology of arabinogalactan proteins. Annu. Rev. Plant Biol. 58: 137-161.

Senatore, A., C.P. Trobacher and J.S. Greenwood. 2009. Ricinosomes predict programmed cell death leading to anther dehiscence in tomato. Plant Physiol. 149: 775-790.

Sijacic, P., X. Wang, A.L. Skirpan, Y. Wang, P.E. Dowd, A.G. McCubbin, S. Huang and T.-H. Kao. 2004. Identification of the pollen determinant of S-RNase-mediated self-incompatibility. Nature 429: 302-305.

Silveira, V., C. Santa-Catarina, N.N. Tun, G.F.E. Scherer, W. Handro, M.P. Guerra and E.I.S Floh. 2006. Polyamine effects on the endogenous polyamine contents, nitric oxide release, growth and differentiation of embryogenic suspension cultures of *Araucaria angustifolia* (Bert.) O. Ktze. Plant Sci. 171: 91-98.

Smertenko, A.P., P.V. Bozhkov, L.H. Filonova, S. von Arnold and P.J. Hussey. 2003. Re-organisation of the cytoskeleton during developmental programmed cell death in *Picea abies* embryos. Plant J. 33: 813-824.

Snowman, B.N., D.R. Kovar, G. Shevchenko, V.E. Franklin-Tong and C.J. Staiger. 2002. Signal-mediated depolymerization of actin in pollen during the self-incompatibility response. Plant Cell 14: 2613-2626.

Solís, M.T., N. Chakrabarti, E. Corredor, J. Cortés-Eslava, M. Rodríguez-Serrano, M. Biggiogera, M.C. Risueño and P.S. Testillano. 2014. Epigenetic changes accompany developmental programmed cell death in tapctum cells. Plant Cell Physiol. 55: 16-29.

Song, X., L. Yuan and V. Sundaresan. 2014. Antipodal cells persist through fertilization in the female gametophyte of *Arabidopsis*. Plant Reprod. 27: 197-203.

Stein, J.C., B. Howlett, D.C. Boyes, M.E. Nasrallah and J.B. Nasrallah. 1991. Molecular cloning of a putative receptor protein kinase gene encoded at the self-incompatibility locus of *Brassica oleracea*. Proc. Natl. Acad. Sci. USA. 88: 8816-8820.

Suarez, M.F., L.H. Filonova, A. Smertenko, E.I. Savenkov, D.H. Clapham, S. von Arnold, B. Zhivotovsky and P.V. Bozhkov. 2004. Metacaspase-dependent programmed cell death is essential for plant embryogenesis. Curr. Biol. 14: R339-340.

Tagliasacchi, A.M., A.C. Andreucci, E. Giraldi, C. Felici, F. Ruberti and L.M.C. Forino. 2007. Structure, DNA content and DNA methylation of synergids during ovule development in *Malus domestica* Borkh. Caryologia 60: 290-298.

Takasaki, T., K. Hatakeyama, G. Suzuki, M. Watanabe, A. Isogai and K. Hinata. 2000. The S receptor kinase determines self-incompatibility in *Brassica* stigma. Nature 403: 913-916.

Takayama, S., H. Shimosato, H. Shiba, M. Funato, F.S. Che, M. Watanabe, M. Iwano and A. Isogai. 2001. Direct ligand-receptor complex interaction controls *Brassica* self-incompatibility. Nature 413: 534-538.

Than, M.E., M. Helm, D.J. Simpson, F. Lottspeich, R. Huber and C. Gietl. 2004. The 2.0Å crystal structure and substrate specificity of the KDEL-tailed cysteine endopeptidase functioning in programmed cell death of *Ricinus communis* endosperm. J. Mol. Biol. 336: 1103-1116.

Thomas, S.G. and V.E. Franklin-Tong. 2004. Self-incompatibility triggers programmed cell death in *Papaver* pollen. Nature 429: 305-309.

Thomas, S.G., S. Huang, S. Li, C.J. Staiger and V.E. Franklin-Tong. 2006. Actin depolymerization is sufficient to induce programmed cell death in self-incompatible pollen. J. Cell Biol. 174: 221-229.

Tiedjens, V.A. 1928. Sex ratios in cucmber flowers as affected by different conditions of soil and light. J. Agric. Res. 36: 721-746.

Trebitsh, T., J.E. Staub and S.D. O'Neill. 1997. Identification of a 1-aminocyclopropane-1-carboxylic acid synthase gene linked to the *Female* (*F*) locus that enhances female sex expression in cucumber. Plant Physiol. 113: 987-995.

Trobacher, C.P., A. Senatore, C. Holley and J.S. Greenwood. 2013. Induction of a ricinosomal-protease and programmed cell death in tomato endosperm by gibberellic acid. Planta 237: 665-679.

Ushijima, K., H. Sassa, A.M. Dandekar, T.M. Gradziel, R. Tao, and H. Hirano. 2003. Structural and transcriptional analysis of the self-incompatibility locus of almond : Identification of a pollen-expressed F-Box gene with haplotype-specific polymorphism. Plant Cell 15: 771-781.

Vacca, R., D. Valenti, A. Bobba, R. Merafina, S. Passarella and E. Marra. 2006. Cytochrome *c* is released in a reactive oxygen species-dependent manner and is degraded via caspase-like proteases in tobacco bright-yellow 2 cells en route to heat shock-induced cell death. Plant Physiol. 141: 208-219.

Vercammen, D., B. van de Cotte, G. De Jaeger, D. Eeckhout, P. Casteels, K. Vandepoele, I. Vandenberghe, J. Van Beeumen, D. Inzé and F. Van Breusegem. 2004. Type II metacaspases Atmc4 and Atmc9 of *Arabidopsis thaliana* cleave substrates after arginine and lysine. J. Biol. Chem. 279: 45329-45336.

Vizcay-Barrena, G. and Z.A. Wilson. 2006. Altered tapetal PCD and pollen wall development in the *Arabidopsis ms1* mutant. J. Exp. Bot. 57: 2709-2717.

Völz, R., J. Heydlauff, D. Ripper, L. von Lyncker and R. Groß-Hardt. 2013. Ethylene signaling is required for synergid degeneration and the establishment of a pollen tube block. Dev. Cell 25: 310-316.

Wang, H., H.M. Wu and A.Y. Cheung. 1996. Pollination induces mRNA poly(A) tail-shortening and cell deterioration in flower transmitting tissue. Plant J. 9: 715-727.

Wang, M., B.J. Oppedijk, M.P.M Caspers, G.E.M. Lamers, M.J. Boot, D.N.G. Geerlings, B. Bakhuizen, A.H. Meijer and B. Van Duijn. 1998. Spatial and temporal regulation of DNA fragmentation in the aleurone of germinating barley. J. Exp. Bot. 49: 1293-1301.

Wheeler, M.J., B.H.J. de Graaf, N. Hadjiosif, R.M. Perry, N.S. Poulter, K. Osman, S. Vatovec, A. Harper, F.C.H. Franklin and V.E. Franklin-Tong. 2009. Identification of the pollen self-incompatibility determinant in *Papaver rhoeas*. Nature 459: 992-995.

Wilkins, K.A., J. Bancroft, M. Bosch, J. Ings, N. Smirnoff and V.E. Franklin-Tong. 2011. Reactive oxygen species and nitric oxide mediate actin reorganization and programmed cell death in the self-incompatibility response of *Papaver*. Plant Physiol. 156: 404-416.

Woltering, E.J. 2004. Death proteases come alive. Trends Plant Sci. 9: 469-472.

Wredle, U., B. Walles and I. Hakman. 2001. DNA fragmentation and nuclear degradation during programmed cell death in the suspensor and endosperm of *Vicia faba*. Int. J. Plant Sci. 162: 1053-1063.

Wu, H.M. and A.Y. Cheung. 2000. Programmed cell death in plant reproduction. Plant Mol. Biol. 44: 267-281.

Wu, M., J. Huang, S. Xu, T. Ling, Y. Xie and W. Shen. 2011. Haem oxygenase delays programmed cell death in wheat aleurone layers by modulation of hydrogen peroxide metabolism. J. Exp. Bot. 62: 235-248.

Xie, Y., C. Zhang, D. Lai, Y. Sun, M.K. Samma, J. Zhang and W. Shen. 2014. Hydrogen sulfide delays GA-triggered programmed cell death in wheat aleurone layers by the modulation of glutathione homeostasis and heme oxygenase-1 expression. J. Plant Physiol. 171: 53-62.

Yamasaki, S., N. Fujii, S. Matsuura, H. Mizusawa and H. Takahashi. 2001. The *M* locus and ethylene-controlled sex determination in andromonoecious cucumber plants. Plant Cell Physiol. 42: 608-619.

Young, T. and D. Gallie. 1999. Analysis of programmed cell death in wheat endosperm reveals differences in endosperm development between cereals. Plant Mol. Biol. 39: 915-926.

Young, T. and D. Gallie. 2000. Programmed cell death during endosperm development. pp. 39-57 In: Lam, E., Fukuda, H., Greenberg, J. (Eds.), Programmed cell death in higher plants. Springer, the Netherlands.

Young, T., J. Giesler-Lee and D. Gallie. 2004. Senescence-induced expression of cytokinin reverses pistil abortion during maize flower development. Plant J. 38: 910-922.

Young, T.E., D.R. Gallie and D.A. DeMason. 1997. Ethylene-mediated programmed cell death during maize endosperm development of wild-type and *shrunken2* genotypes. Plant Physiol. 115: 737-751.

Zhang, D., D. Liu, X. Lv, Y. Wang, Z. Xun, Z. Liu, F. Li and H. Lu. 2014. The cysteine protease CEP1, a key executor involved in tapetal programmed cell death, regulates pollen development in *Arabidopsis*. Plant Cell 26: 2939-2961.

Zhang, D.-S., W.-Q. Liang, Z. Yuan, N. Li, J. Shi, J. Wang, Y.-M. Liu, W.-J. Yu and D.-B. Zhang. 2008. Tapetum degeneration retardation is critical for aliphatic metabolism and gene regulation during rice pollen development. Mol. Plant 1: 599-610.

Zhao, P., X.M. Zhou, L.Y. Zhang, W. Wang, L.G. Ma, L.B. Yang, X.B. Peng, P.V. Bozhkov and M.X. Sun. 2013. A bipartite molecular module controls cell death activation in the basal cell lineage of plant embryos. PLoS Biol. 11: e1001655.

20

Storage Cells - Oil and Protein Bodies

Karine Gallardo,[1] *Pascale Jolivet,*[2] *Vanessa Vernoud,*[1]
Michel Canonge,[2] *Colette Larré*[3] *and Thierry Chardot*[2]*

Introduction

In addition to their central role in the higher plant life cycle, seeds such as those of legumes, oil crops and cereals constitute major sources of carbohydrates, proteins and oil for food and feed. Legume and oil crops store high amounts of proteins and oil, respectively, in specialized intracellular structures, called oil and protein bodies. In this chapter, we summarize the current knowledge on the biology of these storage intracellular structures, which are of great interest for engineering legume and oil crop seed composition in the context of a growing demand for products from plant origin and for added end-user

Oil Bodies

Definition, localization, actual status

Animals, microbes and plants can store lipids in their tissues and organs as subcellular inclusions. In plants, the presence of these inclusions was reported more than one century ago (Price 1912, Dangeard 1922). They are called under different names depending on authors: oleosomes (Dangeard 1922), spherosomes (Yatsu and Jacks 1972), or oil bodies (Price 1912, Huang 1996). For the sake of clarity, we will refer to these inclusions as oil bodies (OBs) in this chapter. Plant cytosolic OBs are found in various organs and tissues, among them seeds (Yatsu and Jacks 1972), leaves (Price 1912), meristems (Dangeard 1922), roots (Barimahasare and Bal 1994), and pollen (Piffanelli and Murphy 1998). In mature seeds from oleaginous or proteaginous (Siloto et al. 2006, Wang et al. 2012b, Boulard et al. 2015) OBs surround

[1] INRA, UMR1347 Agroécologie, 17 rue Sully, 21065 Dijon, France.
[2] Institut Jean-Pierre Bourgin, UMR1318 INRA-AgroParisTech INRA Centre de Versailles-Grignon Route de St-Cyr (RD10), 78026 Versailles Cedex, France.
[3] INRA UR1268 BIA (Biopolymères, Interactions Assemblages) Rue de la Géraudière BP 71627 44316 Nantes CEDEX 3, France.
* Corresponding author : thierry.chardot@versailles.inra.fr

protein bodies. OB structure is unique among organelles and highly conserved. The diameter of OBs is between 0.65 and 2 μm (Tzen et al. 1993). They consist of a core of neutral lipids (triglycerides, sterol esters) surrounded by phospholipids (PLs) in which various proteins are found (Yatsu and Jacks 1972, Huang 1996). The amount of PLs covering OBs is consistent with the existence of a monolayer (Beisson et al. 1996). The number of identified proteins at the surface of OBs increased significantly (Jolivet et al. 2013) since the first pioneering work of Huang (1996). This may reflect different roles (storage of energy, building blocks) for these sub-cellular inclusions, depending on their localization and the physiological state of the organ where they are found. After being considered during a long time as inert balls of fat, OBs are now considered by most of authors as organelles with their own dynamics (Beckman 2006) and numerous roles. The data on OB dynamics derive mostly from studies performed on yeast and mammalian models (Murphy 2012) but some recent works prove that seed OBs are also dynamic organelles in plants (Jolivet et al. 2011, Miquel et al. 2014, Deruyffelaere et al. 2015).

Composition (proteins, lipids)

Proteins

The study of the protein complement of OBs has been favoured during the last fifteen years, thanks to the development of proteomics. Number and nature of proteins found at the surface of OBs strongly depend on the model studied. In oil seed OBs, the number of proteins varies from 3 to more than 30 (Jolivet et al. 2013). *Saccharomyces cerevisiae* OBs contain at least 93 (Grillitsch et al. 2011) to 151 different proteins (Bouchez et al. 2015), and the OB proteome of the lipid-producing yeast *Rhodosporidium toruloides* consisted of 226 proteins, many of which are involved in lipid metabolism and OB formation and evolution (Zhu et al. 2015). Contrary to yeasts (Athenstaedt et al. 2006, Zhu et al. 2015), seed OBs from various species are covered mostly with oleosins and do not contain any enzyme belonging to the neutral lipid biosynthetic pathway (Kennedy pathway) (Jolivet et al. 2013). This might be due to the fact that OBs are mostly purified from mature seeds, which are in a quiescent state. It will be necessary to investigate OB proteomes from developing or germinating seeds to explore the dynamics of the organelle, and to identify new classes of proteins involved in OB dynamics.

Oleosins are abundant seed proteins, accounting for more than 10% of total seed proteins (Jolivet et al. 2004). Oleosins have a unique triblock (two amphiphilic regions and a hydrophobic core) structure. They possess the longest hydrophobic segment known to date, and are inserted in the PL monolayer, but the absence of high resolution structure data is a roadblock for understanding their role. Their N-terminal acetylation, phosphorylation and ubiquitination have been observed from various species and at different maturity states (Lin et al. 2005, Jolivet et al. 2009, Hsiao and Tzen 2011, Jolivet et al. 2011, Vermachova et al. 2011, Jolivet et al. 2013, Deruyffelaere et al. 2015). The extent of post translational modifications (PTMs) and their impact on OB dynamics is largely unknown. Literature data concerning their secondary structure are contradictory (either mainly α or β folded). Due to low solubility of oleosins in aqueous media, their high resolution structure is unknown. We have designed an approach aimed at maintaining oleosin soluble using amphiphilic

Structure of major OB proteins: oleosins
First coherent set of data at different scales

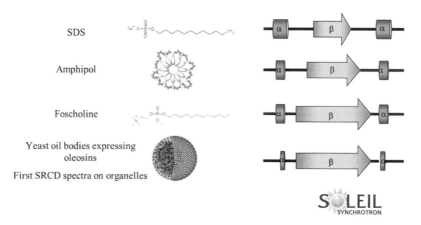

SDS

Amphipol

Foscholine

Yeast oil bodies expressing
oleosins

First SRCD spectra on organelles

S☼LEIL
SYNCHROTRON

FIGURE 1   Models of OB structural proteins maintained soluble in various environments. Adapted from Gohon et al. (2011) and Vindigni et al. (2013).

polymers, conventional detergents, or eukaryotic expression system targeting oleosins to OBs. The use of various biophysical methods (Synchrotron Radiation Circular Dichroism, Xray scattering, available at SOLEIL Synchrotron) permitted to characterize oleosin fold in both environments: surfactants or native-like yeast OBs. These approaches lead to a secondary structure model for oleosins in which the central hydrophobic region adopts an original β fold among eukaryotic proteins (Gohon et al. 2011, Vindigni et al. 2013) (Fig. 1).

Few proteins associated with OBs exhibit enzymatic activity. Among them are caleosins, which possess peroxygenase activity (Hanano et al. 2006, Partridge and Murphy 2009), possibly involved in plant defense, and steroleosins, which have a hydroxysteroid dehydrogenase reductase activity (Lin et al. 2002, D'Andrea et al. 2007). Oleosins are considered as structural proteins by most authors. They have been reported to have weak, and to date non-confirmed, monoacylglycerol acyltransferase and phospholipase activities (Parthibane et al. 2012). Sugar-Dependent 1 lipase is also associated with OBs (Eastmond 2006). The delivery mechanism of this enzyme by peroxisome extension has been reported very recently (Thazar-Poulot et al. 2015). The number of proteins reported as associated with OBs increases with time. It has been suggested that OBAP1, a member of the recently discovered family of OB associated proteins (OBAP) is involved in the stability of OBs (Lopez-Ribera et al. 2014).

Lipids

Triglycerides (TAGs) are the main compounds found in seed OBs, but wax esters are also found in some species, such as Jojoba shrub (*Simmondsia chinensis*) (Benzioni 1978). OB fatty acid (FA) composition varies between organs. *Arabidopsis thaliana* seed OBs contained characteristic 20:1 eicosenoic FA in their TAGs, whereas the OBs from leaves instead contained more 16:3 and 18:3 FAs that are most characteristic of leaf acyl lipids (Schmid and Patterson 1988, Horn et al. 2011). Phosphatidylcholine is the most abundant class of PLs found in OBs of seeds from brassicacea, such as rapeseed and mustard (Tzen et al. 1993, Boulard et al. 2015) and *A. thaliana* (this work), as well as in crop seed OBs (cotton, flax, maize) (Tzen et al. 1993). Diacyl glycerol, free FAs, esterified and free sterols (β-sitosterol, campesterol and brassi-casterols), and micronutriments such as tocopherols are also present in OBs (Huang 1992, Dyas and Goad 1994, Fisk et al. 2006, White et al. 2006, Boulard et al. 2015). We have performed a detailed study of the minor compounds found in the unsaponifiable fraction of OBs from mature *A. thaliana* seeds using the method described by Beveridge et al. (2002). Using GC-MS, we have identified more than 10 different compounds (δ-tocopherol, cholesterol, brassicasterol, campesterol, stigmasterol, cleosterol, β-sitosterol, Δ5 avenasterol, Δ 5'-24 (25) stigmasterol, cycloartenol) in the latter fraction. Apparently, β-sitosterol is the most abundant product found in *A. thaliana* and rapeseed OB extracts (Boulard et al. 2015). Cycloartenol is the first cyclic compound in the plant steroid biosynthesis. δ-tocopherol is an anti-oxidant already found in OBs from sunflower (Fisk et al. 2006) and rapeseed (Boulard et al. 2015). Taken together, these results reveal that seed OBs act as a sink for various lipids and that their composition is far more complex than generally reported. They could serve as a source for valuable molecules, such as vitamins, sterols.

Role

Despite having similar morphology, OBs seem to have different functions depending on their locations. Leaves contain generally few amounts of lipids in their OBs, which are likely to have a different function from the long term storage function of TAGs in seeds (Horn et al. 2011). Leaf OB lipids could serve, along with carbohydrates as a diurnal photosynthetic store (Lin and Oliver 2008). It is possible to increase the number of OBs and the amount of oil stored by heterologous expression of DGAT (acyl CoA : diacylglycerol acyltransferase) (Zhou et al. 2013), of oleosins (Winichayakul et al. 2013) or by disruption of CGI-58 (James et al. 2010). Such increase in oil amount is of obvious interest for biotechnological purposes.

It has also been suggested that root OBs can be utilized for membrane and rhizobial proliferation during establishment of symbiosis in the infected cells (Barimahasare and Bal 1994, Pateraki et al. 2014). The recent finding that the *Coleus forshohlii* roots contain OBs enriched with terpenes confer to the organelle a role for lipophilic bioactive metabolites storage (Pateraki et al. 2014).

Dynamics

Exalbuminous seeds like *Brassica napus* (rapeseed) or *A. thaliana* store their oil mainly in the embryo. Albuminous (endospermic) seeds of *Ricinus communis* (castor bean) accumulate their lipid reserve in the large endosperm tissue that surrounds the small embryo (Voelker and Kinney 2001). Oil accumulation is part of the seed maturation process, a highly controlled developmental program that occurs in embryo tissues once morphogenesis is achieved (Baud and Lepiniec 2010). In higher plants, the biosynthetic pathways for FAs and TAGs involve different sub-cellular

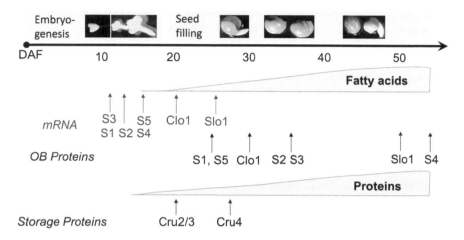

FIGURE 2 Kinetics of accumulation of fatty acids and proteins in rapeseed. The time points corresponding to the start of accumulation of oil bodies structural proteins, of their mRNAs and of cruciferins are shown. Adapted from Jolivet et al. (2011) and Hajduch et al. (2006). DAF: Day After Flowering, Cru: Cruciferin, Clo: Caleosin, Slo: Steroleosin, S: Oleosins. Numbers refer to protein isoforms.

compartments; *de novo* FA synthesis occurs in the plastids and TAGs assemble in the endoplasmic reticulum (ER); the overall process being under the control of transcription factors. Accumulation of lipids and proteins follow a sigmoid pattern during seed development in *B. napus* (Murphy et al. 1989, Eastmond and Rawsthorne 2000, Jolivet et al. 2011) and in *A. thaliana* (Focks and Benning 1998, Baud et al. 2002). FA composition evolves from being similar to that of vegetative tissues at initial stages of seed development to a final stage where TAGs are the predominant lipid components (Baud et al. 2002, Jolivet et al. 2011). Genes encoding FA synthesis enzymes display a bell-shaped pattern of expression at the onset of the maturation phase, whereas the expression level of several FA modifying enzymes and oleosins increases later (Baud and Lepiniec 2009).

Oleosins belong to a multigenic family (Kim et al. 2002) in *B. napus* and *A thaliana*. The existence of an ordered expression scheme for these proteins during seed maturation has been revealed at both mRNA and protein levels (Jolivet et al. 2011, Miquel et al. 2014) (Fig. 2). Oleosin maximal accumulation was concomitant with oil deposition (Jolivet et al. 2011). More recent work (Miquel et al. 2014) suggests that the oleosin isoforms found at the surface of seed OBs could have specific functions in the dynamics of lipid accumulation. Mutants of the core retromer (a multiprotein complex involved in protein trafficking) display defects in OB biogenesis and storage oil breakdown during postgerminative growth in *A. thaliana*, revealing communication of peroxisomes with OBs (Thazar-Poulot et al. 2015).

A gradual disappearance of oleosins concomitant with TAG mobilization has been reported in several species, but the role of oleosin degradation during OB mobilization has remained speculative. It has been suggested that disruption of the integrity of the protein coat may be required to facilitate or promote access of lipolytic enzymes to the TAGs. Ubiquitination of oleosins has been observed in sesame postgerminative seedlings, but the type of modification has not been determined (Hsiao and Tzen

2011). We have very recently demonstrated that the sequential proteolysis of the five *A. thaliana* seed specific oleosins begins just prior to lipid degradation (Deruyffelaere et al. 2015). We have observed for the first time phosphorylation and ubiquitination of *A thaliana* oleosins: these PTMs were concomitant with oleosin degradation. Ubiquitination topology was surprisingly complex. OLE1 and OLE2, the major seed OB oleosins were modified by three distinct and predominantly exclusive motifs: monoubiquitin, K48-linked diubiquitin (K48Ub2) and K63-linked diubiquitin (Deruyffelaere et al. 2015). Ubiquitinated oleosins may be channeled towards specific degradation pathways according to ubiquitination type. One of these pathways was identified as the ubiquitin–proteasome pathway. K48Ub2-modified oleosins are selectively extracted from the OB coat and degraded by the proteasome. Proteasome inhibition also reduced lipid hydrolysis, providing *in vivo* evidence that oleosin degradation is required for lipid mobilization.

New insights

The seeds constitute the main vector of plant propagation, and are also a major source of nutrients for food and feed. The overall metabolic pathways of reserve accumulation are well known, however the way they take place in the cell are still largely unknown. The very recent results from our group and from others (Jolivet et al. 2011, Gidda et al. 2013, Jolivet et al. 2013, Miquel et al. 2014, Lopez-Ribera et al. 2014, Deruyffelaere et al. 2015, Thazar-Poulot et al. 2015) definitively stress the fact that OBs are organelles which possess a growing set of proteins, and have their own dynamics through the seed life cycle and collaborations with other organelles. In the context of a growing demand of products of vegetal origin, especially for food use, these new concepts will help the understanding of how plants store their reserves in specialized organelles, and how they mobilize their lipids during germination. This will permit a rationale engineering of reserve storage in seeds. Most of the studies on seed OBs have been done using oleaginous species. The studies on proteaginous species tend to prove that OBs from these species share common features with their oleaginous counterparts (Wang et al. 2012b).

Protein Bodies

Protein bodies (PBs) are sub-cellular compartments originating from the endoplasmic reticulum (ER) containing high amounts of proteins named seed storage proteins (SSPs). They are formed during the filling phase of seed development, concomitantly with the production of OBs (Wang et al. 2012a,b), through a series of well-orchestrated events implying extensive cross-talk between the seed coat, endosperm and embryo (Weber et al. 2005). PBs accumulate predominantly within the embryo in dicots and within both the endosperm and embryo in monocots. The structure and properties of the storage proteins found in PBs of cereal grains and legume seeds have been extensively studied (for reviews see Shewry et al. 1995, Shewry and Halford 2002). In this section, after presenting the classification and localization of storage proteins, we will report on the mechanisms governing their synthesis, regulation, trafficking and deposition in PBs and then in protein storage vacuoles (PSVs). We will also highlight recent findings obtained using the model plant *A. thaliana* and in legumes, whose seeds accumulate high protein levels ranging from 20 to 40% of the seed dry weight.

Classification, structure and localization of seed storage proteins

SSPs have been classified into four categories according to their solubility characteristics: albumins (water-soluble), globulins (salt-soluble), prolamins (alcohol/water soluble), and glutelins (insoluble in neutral solvents). Globulins and albumins, the major SSPs in dicots, have also been described on the basis of their sedimentation coefficient: 2S for the albumin-type, 11/12S (legumins/cruciferins) and 7S (vicilins) for the globulins (Shewry et al. 1995, Shewry and Casey 1999). These proteins accumulate in different proportions and are all encoded by multigene families, as shown in Table 1.

In their mature packed form, SSPs are oligomeric. The 11/12S globulins (300–400 kDa) are made of six subunits of 50-60 kDa composed of acidic and basic polypeptides linked by a disulphide bridge (Shewry et al. 1995, Jaworski and Aitken 2014). The 7S globulins (150–200 kDa) are trimeric and consist of α', α and β subunits (40–70 kDa). They are not disulphide-bonded and are susceptible to glycosylation. The 2S albumins (12–15 kDa) consist of two subunits (4–9 kDa) linked by four disulphide bonds and accumulate as heterodimers (Lawrence et al. 1994, Byczynska and Barciszewski 1999). These proteins accumulate in intracellular compartments, referred to as PSVs, delimited by a membrane marked by the presence of α-tonoplast intrinsic proteins (Johnson et al. 1989, Jiang et al. 2001). Easily recognizable in electronic microscopy because of their electron-opacity, PSVs have been observed in different species with various diameters: 6–9 μm in *A. thaliana* (Shimada et al. 2003), 10–20 μm in *B. napus* (Gillespie et al. 2005) and 15 μm in pumpkin (*Cucurbita* sp, var Small sugar) seeds (Jiang et al. 2000). They are characterized by a nearly pH-neutral lumen and the storage proteins represent most of the material accumulated in this organelle (Vitale and Chrispeels 1992, Wang and Perry 2013). They also contain low amounts of acidic proteases (Hara-Nishimura et al. 1993), glycoproteins (Hara-Nishimura et al. 1998, Elmer et al. 2003) and, in Brassicaceae, phytin, myrosinase-binding proteins and numerous proteins designated as PSV-embedded proteins (Gillespie et al. 2005).

Biogenesis of protein storage vacuoles

SSP are synthesized as single entities on the rough ER then enter the luminal space after removal of their signal peptide. Subsequently, intra-chain disulphide bonds are

TABLE 1 Number of genes encoding seed storage proteins in four species.

	2S	7S	11S/12S
Brassicaceae			
Arabidopsis thaliana	5	–	4
Brassica napus	16	–	9
Fabaceae or Leguminosae			
Medicago truncatula	2	3	4
Pisum sativum	2	5	6

formed, at least one for globulins and up to four in the case of 2S albumins. The pro-globulins undergo assembly in trimers before reaching the PSVs. For 2S albumins, additional peptide segments are removed in the lumen before they leave the ER. The precursor forms of SSPs reach the PSVs *via* three possible routes. Major PSV-localized proteins, such as 11S (Vitale and Raikhel 1999, Jolliffe et al. 2005) and 7S globulins, followed the conventional Golgi-mediated route to PSV although only the 7S undergo glycosylation. In pea (*Pisum sativum*), they reached PSVs *via* dense vesicles of 150–200 nm diameter, also called PBs, which may fused directly to PSV or develop into PSVs (Hinz et al. 1999). An alternative route including multivesicular bodies (MVB) as an intermediate step for the PBs before they reach the storage vacuole has been proposed (Robinson et al. 1998). These MVBs were assumed to constitute prevacuolar compartments for PSV (Jiang et al. 2001). A last route bypassing the Golgi complex has been shown in developing cereal grains for prolamins and in pumpkin seeds for 2S and 11S proteins. These proteins aggregate in the ER, bud out as large vesicles (200-400 nm) referred as precursor-accumulating vesicles (PACs), which can also receive glycoproteins such as ricin before being delivered to PSVs by autophagy (Hara-Nishimura et al. 1998).

Many vesicle carriers for storage protein transport have been identified, leading to the elaboration of a consensual PSV trafficking model (Jolliffe et al. 2005, Vitale and Hinz 2005). However, the issues of why one route or another is favoured or whether the pathway varies during seed development remain to be resolved (Abirached-Darmency et al. 2012). In addition, vacuolar sorting signals are required to target proteins to vacuoles. Some have been identified for members of the 2S and 7S families but are lacking in 11S globulin sequences (Vitale and Hinz 2005, Neuhaus and Rogers 1998). For these proteins, a sorting mechanism based on their capacity to aggregate into granules and trigger dense vesicle budding has been proposed (Arvan et al. 2002, Jolliffe et al. 2005). Furthermore, PSV biogenesis is governed by complex mechanisms which orchestrate and regulate the life of endomembrane compartments (Bassham et al. 2008). The carrier vesicles are surrounded by membranes which include components shown to act as receptor or marker in the transport of protein cargo and their delivery to target destination, as reviewed and discussed by Fujimoto and Ueda (2012), and Uemura and Ueda (2014).

Whatever the pathway, SSP reaching the PSV undergo a maturation process, leading to conformational changes and folding. Vacuolar processing enzymes are responsible for the proteolytic maturation of 11S globulins and 2S albumins (Hara-Nishimura et al. 1993). Gruis et al. (2002) demonstrated the existence of alternative proteolytic pathways implicated in SSP maturation in *A. thaliana*. The 11S are cleaved at a well conserved Asn–Gly peptide bond converting the pro-protein into two disulphide-linked alpha and beta polypeptides. In the case of 2S albumin, the processing is more complex with the removal of three peptide stretches at different positions of the sequence. Despite Asn residues being found several times at the P1 position of the cleaved bonds, conserved cleavage sequences were not found in the napin/2S family (Ericson et al. 1986). While the maturation of 11S globulins is essential to obtain their final hexameric structure, the role of maturation for 2S albumins is much less obvious: it substantially increases the isoelectric point from 7 to 11 and may affect its compactness. Unfolded proteins are often produced during this maturation process. These need to be degraded to maintain the integrity and functionality of ER components. In the rice endosperm, the quality of PB is maintained by polyubiquitination of unfolded SSPs through the Hrd1 ubiquitin ligase system (Ohta and Takaiwa 2015).

A good understanding of PSV formation is particularly important in the field of plant molecular farming, as underlined in recent papers (De Meyer and Depicker 2014, Hegedus et al. 2015). It is worth mentioning that PB formation depends on the protein storage capacity of legume seeds, which is determined by both the number of cotyledon cells, acquired during embryogenesis, and the duration of the filling phase that largely depends on nitrogen availability (Munier-Jolain et al. 1998, Munier-Jolain and Ney 1998, Salon et al. 2001). It also relies on a tight regulation of seed metabolism to switch the developmental program towards protein deposition and to ensure large amounts of proteins will be synthesized during seed filling to further support hetero-trophic growth of the seedling. The molecular mechanisms governing the metabolic control and regulation of SSP synthesis are presented in the following sections.

Shifting the metabolism towards protein deposition

The formation of PBs implies a switch from an embryogenesis-oriented program, characterized by intense cell divisions within the embryo, to a filling mode directed towards embryo cell expansion and reserve accumulation. Over the last decade, the availability of genomics and post-genomics sequences for the legume model spe-cies *Medicago truncatula* (Barrel Medic) and *Lotus japonicus* (Lotus) has boosted the use of omics approaches for studying key stages of seed development, including the switch towards protein deposition (Thompson et al. 2009 and references therein, Verdier et al. 2013). With the development of high-throughput sequencing, data are now acquired in related crop species, such as soybean (*Glycine max*) and chickpea (*Cicer arietinum*) (Collakova et al. 2013, Jones and Vodkin 2013, Pradhan et al. 2014). A study targeted to the nuclear proteome of developing *M. truncatula* seeds revealed that the most abundant nuclear proteins present just before seed filling belong to the functional class of ribosome biogenesis (Repetto et al. 2008). In particular, there was a large pool of ribosomal proteins, which are synthesized within the nucleolus and may further serve for translation of storage proteins, explaining how legume seeds can synthesize large amounts of proteins while desiccating and entering into a quiescent state. A question that arises is whether this pool of ribosomal proteins controls the homeostasis of protein amount per seed under challenging environ-mental conditions. In legumes and cereals, epigenetic components are also largely represented at this transition stage, some of them participating in transcriptional repression *via* the modification of gene accessibility, such as histone deacetylases (Li et al. 2008, Demetriou et al. 2009, Repetto et al. 2012, Collakova et al. 2013). Histone deacetylases are involved in the repression of embryo-specific transcription factors, some of them corresponding to master regulators of storage protein synthesis such as ABSCISIC ACID INSENSITIVE 3 (ABI3), FUSCA 3 (FUS3) and LEAFY COTYLEDON 1 (LEC1) (Vicient et al. 2000, Kagaya et al. 2005, Tanaka et al. 2008). The sharp decrease of histone deacetylase transcripts observed at the beginning of seed filling (Repetto et al. 2008) may therefore be crucial for the entry into the stor-age phase. In seeds, the epigenetic regulation of the genome, which modulates chro-matin structure to limit the expression of genes to a particular tissue at a specific developmental stage, could be part of the currently unclear mechanisms shifting the metabolism towards protein deposition.

Metabolic control of seed protein deposition: a coordinated interaction between seed tissues

Among the metabolites particularly abundant at the beginning of protein deposition are asparagine and sucrose (Collakova et al. 2013, Li et al. 2015), the latter acting not only as a carbon source but also as a signal to control the entry of the legume embryo into the storage mode (Weber et al. 2005). In soybean seeds, the level of free asparagine was found to be positively correlated with protein content, whereas no significant relationships were observed between sucrose concentration and protein concentration at maturity (Pandurangan et al. 2012). Hence, the relatively high level of free asparagine at this stage may ensure the provision of nitrogen backbone for PB formation. The tissues surrounding the legume embryo play important roles in the transient store of these nutrients coming from the phloem and in their metabolism before translocation to the embryo through active transport systems (Miranda et al. 2001, Rolletschek et al. 2005, Sanders et al. 2009). The seed coat supports storage compound synthesis in the filial tissues by transmitting phloem-derived nutrients, mainly sugars, glutamine and asparagine (Rochat and Boutin 1991). Moreover, studies in *M. truncatula* and soybean have shown that the seed coat is characterized by high level and activity of asparaginase (Gallardo et al. 2007, Pandurangan et al. 2012), which metabolizes asparagine to provide the embryo with other amino acids, such as alanine and glutamine (Atkins et al. 1975). A specific asparaginase gene, *ASPGBla*, has been identified that may be responsible for the enhanced flux of nitrogen to the embryo of soybean cultivars displaying high seed protein levels (Pandurangan et al. 2012). The endosperm, whose role in the control of seed size at early stages of seed development has been well established (Garcia et al. 2005, Kondou et al. 2008, D'Erfurth et al. 2012, Noguero et al. 2015), plays a key role in controlling the flux of nutrients to nourish the embryo during the storage phase. It is able to accumulate high levels of soluble metabolites, such as sucrose and amino acids, in its vacuole during the pre-storage phase. Among the stored amino acids are alanine and glutamine whose subsequent decline in the endosperm vacuole coincides with an increase in protein storage activities within the embryo (Melkus et al. 2009). It is postulated that by accumulating high levels of soluble metabolites, the endosperm may act as a buffer to ensure enough metabolites reach the embryo even under unfavourable environmental conditions.

The transcriptional regulators of seed protein deposition

SSP-encoding genes are specifically induced in immature seeds and tightly regulated during seed filling. The *A. thaliana* 12S globulins and 2S albumins, are encoded respectively by four (*CRA1*, *CRB*, *CRC* and *CRU2*) and five (*At2S1* to *At2S5*) genes (Krebbers et al. 1988, Pang et al. 1988, van der Klei et al. 1993) which are coordinately transcribed during the early and mid-maturation phases (Fig. 2). In legumes, comparative studies of SSP mRNA and protein quantities have shown that they are highly correlated, suggesting that SSP expression is primarily controlled by a transcriptional regulation (Gatehouse et al. 1982, Walling et al. 1986, Gallardo et al. 2007). Legumins (11S) and vicilins (7S) accumulate in a sequential manner during the filling phase as synthesis of vicilins precedes that of legumins (Meinke et al. 1981, Gatehouse et al. 1982, Gallardo et al. 2007, Wang et al. 2012b). Although this phasing has been widely observed in legumes, such as pea, soybean and *M. truncatula*, the underlying regulation mechanisms, which may involve different transcription factors (TFs) described below, remain to be studied in this family.

Molecular basis of seed-specific SSP expression has been mainly focused on the identification of regulatory *cis*-acting elements in their promoters and of the corresponding TF both in *A. thaliana* and legumes (Ellerstrom et al. 1996, Chandrasekharan et al. 2003 and references therein). The best characterized motifs include the RY motif (CATGCA) and the G-box related ACGT element representing putative binding sites for TFs from the B3 and bZIP families, respectively (Baumlein et al. 1992, Ezcurra et al. 1999, Fujiwara et al. 2014, also reviewed in Vicente-Carbajosa and Carbonero 2005). Legumes SSP gene promoters are enriched with a third novel E2Fb-like motif, but the functional relevance of this cis-acting element has not been demonstrated yet (Fauteux and Stromvik 2009)

Components of the regulatory network governing SSP synthesis have been characterized, mainly in *A. thaliana*. LEC2 (LEAFY COTYLEDON2), FUS3 and ABI3 encode TFs from the B3 domain superfamily of DNA binding proteins. Together with LEC1, which encodes a protein homologous to the HAP3 subunit of the CAAT box binding proteins, they represent master regulators of the maturation phase. Genetic studies have shown that mutations in any of these genes lead to similar pleiotropic effects on seed maturation, including severe defects in SSP accumulation and in the acquisition of desiccation tolerance (reviewed in Santos-Mendoza et al. 2008). These genes represent a complex and intricate regulatory network which, combined with inputs from hormone, sugar and light signaling pathways, refines the program for seed maturation (reviewed in Weber et al. 2005 and Baud et al. 2008). Binding of ABI3, LEC2 and FUS3 to the RY element present in SSP promoters, mediated by the B3 DNA-binding domain, has been shown *in vitro* or *in vivo* in yeast (Reidt et al. 2000, Monke et al. 2004, Kroj et al. 2003). LEC2 triggers the expression of SSP-encoding genes such as *AT2S1-S4* and CRA1 (Braybrook et al. 2006), while more recent ChIP-Chip experiments identified several SSP genes as direct target of ABI3 and FUS3 (Monke et al. 2012, Wang and Perry 2013). Cooperation of the B3-domain TF with other TF from the MYB family is needed for correct temporal and spatial SSP expression in seed. bZIP10, bZIP25, and bZIP53 have been shown to bind the G-box, often associated with RY elements within SSP promoters, and to regulate SSP accumulation (Lara et al. 2003, Alonso et al. 2009). Heterodimerization of bZIP53 with bZIP10 or bZIP25 promotes strong activation of SSP genes and the presence of ABI3 increases the heterodimerization. The formation of a ternary complex, to achieve correct SSP gene expression has been proposed (Alonso et al. 2009). The contribution of chromatin remodeling and histone modification to SSP seed specific expression was illustrated by the work on the transcriptional activation by ABI3 of the *β-phaseolin* gene which encodes the most abundant SSP in *Phaseolus vulgaris* (Ng et al. 2006). In *M. truncatula*, gene expression profiling during seed development identified several TFs related to the seed filling phase (Verdier et al. 2008, Kurdyukov et al. 2014). Some TFs co-expressed with SSP mRNA represent orthologs of *A. thaliana* genes involved in SSP accumulation, while others may be more specifically related to the delayed expression of the legumin 11S class (Verdier et al. 2008).

Although in dicots the principal storage tissue is the embryo, the profiles of protein deposition (and oil accumulation) in the maturing embryo and endosperm was recently examined (Barthole et al. 2014). Different protein profiles within both compartments were highlighted, with the endosperm synthesizing less 12S globulins and thus being strongly enriched in 2S albumins. This profile was consistent with the expression level of the corresponding SSP-encoding genes within each seed compartment. A MYB TF, MYB118, was further identified as a repressor of the expression of maturation-related genes within the endosperm. The *myb118* mutant displayed a partial relocation of storage compounds from the embryo to the endosperm and a lower accumulation of reserves within the embryo, suggesting that repression of endosperm maturation is important to promote embryo filling (Barthole et al. 2014).

Acknowledgments

We wish to warmly thank Gilles Clément for expert analysis of the unsaponifiable fraction of oil bodies and Sabine d'Andréa for fruitful discussions.

References

Abirached-Darmency, M., F. Dessaint, E. Benlicha and C. Schneider. 2012. Biogenesis of protein bodies during vicilin accumulation in *Medicago truncatula* immature seeds. BMC Res. Notes 5: 409.

Alonso, R., L. Onate-Sanchez, F. Weltmeier, A. Ehlert, I. Diaz, K. Dietrich, J. Vicente-Carbajosa and W. Droge-Laser. 2009. A pivotal role of the basic leucine zipper transcription factor bZIP53 in the regulation of Arabidopsis seed maturation gene expression based on heterodimerization and protein complex formation. Plant Cell 21: 1747-1761.

Arvan, P., B.Y. Zhang, L. Feng, M. Liu and R. Kuliawat. 2002. Lumenal protein multimerization in the distal secretory pathway/secretory granules. Curr. Opin. Cell Biol. 14: 448-453.

Athenstaedt, K., P. Jolivet, C. Boulard, M. Zivy, L. Negroni, J.M. Nicaud and T. Chardot. 2006. Lipid particle composition of the yeast *Yarrowia lipolytica* depends on the carbon source. Proteomics 6: 1450-1459.

Atkins, C.A., J.S. Pate and P.J. Sharkey. 1975. Asparagine metabolism-key to the nitrogen nutrition of developing legume seeds. Plant Physiol. 56: 807-812.

Barimahasare, J. and A.K. Bal. 1994. Symbiotic nitrogen-fixing root-nodules of *lathyrus maritimus* (L) bigel (beach pea) from newfoundlane shore lines with special reference to oleosomes (Lipid bodies). Plant Cell Environ. 17: 115-119.

Barthole, G., A. To, C. Marchive, V. Brunaud, L. Soubigou-Taconnat, N. Berger, B. Dubreucq, L. Lepiniec and S. Baud. 2014. MYB118 represses endosperm maturation in seeds of Arabidopsis. Plant Cell 26: 3519-3537.

Bassham, D.C., F. Brandizzi, M.S. Otegui and A.A. Sanderfoot. 2008. The secretory system of Arabidopsis. Arabidopsis Book 6: e0116.

Baud, S., J.P. Boutin, M. Miquel, L. Lepiniec and C. Rochat. 2002. An integrated overview of seed development in *Arabidopsis thaliana* ecotype WS. Plant Physiol. Biochem. 40: 151-160.

Baud, S., B. Dubreucq, M. Miquel, C. Rochat and L. Lepiniec. 2008. Storage reserve accumulation in Arabidopsis: metabolic and developmental control of seed filling. Arabidopsis Book 6: e0113.

Baud, S. and L. Lepiniec. 2009. Regulation of de novo fatty acid synthesis in maturing oilseeds of Arabidopsis. Plant Physiol. Biochem. 47: 448-455.

Baud, S. and L. Lepiniec. 2010. Physiological and developmental regulation of seed oil production. Prog. Lipid Res. 49: 235-249.

Baumlein, H., I. Nagy, R. Villarroel, D. Inze and U. Wobus. 1992. Cis-analysis of a seed protein gene promoter: the conservative RY repeat CATGCATG within the legumin box is essential for tissue-specific expression of a legumin gene. Plant J. 2: 233-239.

Beckman, M. 2006. Cell biology. Great balls of fat. Science 311: 1232-1234.

Beisson, F., N. Ferte and G. Noat. 1996. Oil-bodies from sunflower (*Helianthus annuus* L.) seeds. Biochem. J. 317: 955-956.

Benzioni, A. 1978. Fruits development and wax biosynthesis in Jojoba. New Phytol. 81: 105-109.

Beveridge, T.H., T.S. Li and J.C. Drover. 2002. Phytosterol content in American ginseng seed oil. J Agric. Food Chem. 50: 744-750.

Bouchez, I., M. Pouteaux, M. Canonge, M. Genet, T. Chardot, A. Guillot and M. Froissard. 2015. Regulation of lipid droplet dynamics in *Saccharomyces cerevisiae* depends on the Rab7-like Ypt7p, HOPS complex and V1-ATPase. Biol. Open 4: 764-775.

Boulard, C., M. Bardet, T. Chardot, B. Dubreucq, M. Gromova, A. Guillermo, M. Miquel, N. Nesi, S. Yen-Nicolay and P. Jolivet. 2015. The structural organization of seed oil bodies could explain the contrasted oil extractability observed in two rapeseed genotypes. Planta 242: 53-68.

Braybrook, S.A., S.L. Stone, S. Park, A.Q. Bui, B.H. Le, R.L. Fischer, R.B. Goldberg and J.J. Harada. 2006. Genes directly regulated by LEAFY COTYLEDON2 provide insight into the control of embryo maturation and somatic embryogenesis. Proc. Natl. Acad. Sci. USA. 103: 3468-3473.

Byczynska, A. and J. Barciszewski. 1999. The biosynthesis, structure and properties of napin - the storage protein from rape seeds. J. Plant Physiol. 154: 417-425.

Chandrasekharan, M.B., K.J. Bishop and T.C. Hall. 2003. Module-specific regulation of the beta-phaseolin promoter during embryogenesis. Plant J. 33: 853-866.

Collakova, E., D. Aghamirzaie, Y. Fang, C. Klumas, F. Tabataba, A. Kakumanu, E. Myers, L.S. Heath and R. Grene. 2013. Metabolic and transcriptional reprogramming in developing Soybean (*Glycine max*) embryos. Metabolites 3: 347-372.

D'Andrea, S., M. Canonge, A. Beopoulos, P. Jolivet, M.A. Hartmann, M. Miquel, L. Lepiniec and T. Chardot. 2007. At5g50600 encodes a member of the short-chain dehydrogenase reductase superfamily with 11beta- and 17beta-hydroxysteroid dehydrogenase activities associated with *Arabidopsis thaliana* seed oil bodies. Biochimie 89: 222-229.

D'Erfurth, I., C. Le Signor, G. Aubert, M. Sanchez, V. Vernoud, B. Darchy, J. Lherminier, V. Bourion, N. Bouteiller, A. Bendahmane, J. Buitink, J.M. Prosperi, R. Thompson, J. Burstin and K. Gallardo. 2012. A role for an endosperm-localized subtilase in the control of seed size in legumes. New Phytol. 196: 738-751.

Dangeard, P. 1922. Sur la structure de la cellule chez les Iris. C.R. Acad. Sci. 175: 7-12.

De Meyer, T. and A. Depicker. 2014. Trafficking of endoplasmic reticulum-retained recombinant proteins is unpredictable in *Arabidopsis thaliana*. Front. Plant Sci. 5: 473.

Demetriou, K., A. Kapazoglou, A. Tondelli, E. Francia, M.A. Stanca, K. Bladenopoulos and A.S. Tsaftaris. 2009. Epigenetic chromatin modifiers in barley: I. Cloning, mapping and expression analysis of the plant specific HD2 family of histone deacetylases from barley, during seed development and after hormonal treatment. Physiol. Plant 136: 358-368.

Deruyffelaere, C., I. Bouchez, H. Morin, A. Guillot, M. Miquel, M. Froissard, T. Chardot and S. D'Andrea. 2015. Ubiquitin-mediated proteasomal degradation of oleosins is involved in oil body mobilization during post-germinative seedling growth in Arabidopsis. Plant Cell Physiol. 56: 1374-1387.

Dyas, L. and L.J. Goad. 1994. The occurrence of free and esterified sterols in the oil bodies isolated from Maize seed scutella and a Celery cell-suspension culture. Plant Physiol. Biochem. 32: 799-805.

Eastmond, P.J. 2006. SUGAR-DEPENDENT1 encodes a patatin domain triacylglycerol lipase that initiates storage oil breakdown in germinating Arabidopsis seeds. Plant Cell 18: 665-675.

Eastmond, P.J. and S. Rawsthorne. 2000. Coordinate changes in carbon partitioning and plastidial metabolism during the development of oilseed rape embryos. Plant Physiol. 122: 767-774.

Ellerstrom, M., K. Stalberg, I. Ezcurra and L. Rask. 1996. Functional dissection of a napin gene promoter: identification of promoter elements required for embryo and endosperm-specific transcription. Plant Mol. Biol. 32: 1019-1027.

Elmer, A., W. Chao and H. Grimes. 2003. Protein sorting and expression of a unique soybean cotyledon protein, GmSBP, destined for the protein storage vacuole. Plant Mol. Biol. 52:1089-1106.

Ericson, M.L., J. Rodin, M. Lenman, K. Glimelius, L.G. Josefsson and L. Rask. 1986. Structure of the rapeseed 1.7 S storage protein, napin, and its precursor. J. Biol. Chem. 261: 14576-14581.

Ezcurra, I., M. Ellerstrom, P. Wycliffe, K. Stalberg and L. Rask. 1999. Interaction between composite elements in the napA promoter: both the B-box ABA-responsive complex and the RY/G complex are necessary for seed-specific expression. Plant Mol. Biol. 40: 699-709.

Fauteux, F., and M.V. Stromvik. 2009. Seed storage protein gene promoters contain conserved DNA motifs in Brassicaceae, Fabaceae and Poaceae. BMC Plant Biol. 9: 126.

Fisk, I.D., D.A. White, A. Carvalho and D.A. Gray. 2006. Tocopherol - An intrinsic component of sunflower seed oil bodies. J. Am. Oil Chem. Soc. 83: 341-344.

Focks, N. and C. Benning. 1998. wrinkled1: A novel, low-seed-oil mutant of Arabidopsis with a deficiency in the seed-specific regulation of carbohydrate metabolism. Plant Physiol. 118: 91-101.

Fujimoto, M. and T. Ueda. 2012. Conserved and plant-unique mechanisms regulating plant post-Golgi traffic. Front. Plant Sci. 3: 197.

Fujiwara, M., T. Uemura, K. Ebine, Y. Nishimori, T. Ueda, A. Nakano, M.H. Sato and Y. Fukao. 2014. Interactomics of Qa-SNARE in *Arabidopsis thaliana*. Plant Cell Physiol. 55: 781-789.

Gallardo, K., C. Firnhaber, H. Zuber, D. Hericher, M. Belghazi, C. Henry, H. Kuster and R. Thompson. 2007. A combined proteome and transcriptome analysis of developing *Medicago truncatula* seeds. Mol. Cell. Proteomics 6: 2165-2179.

Garcia, D., J.N. Fitz Gerald and F. Berger. 2005. Maternal control of integument cell elongation and zygotic control of endosperm growth are coordinated to determine seed size in Arabidopsis. Plant Cell 17: 52-60.

Gatehouse, J.A., I.M. Evans, D. Bown, R.R. Croy and D. Boulter. 1982. Control of storage-protein synthesis during seed development in pea (*Pisum sativum* L.). Biochem. J. 208: 119-127.

Gidda, S.K., S. Watt, J. Collins-Silva, A. Kilaru, V. Arondel, O. Yurchenko, P.J. Horn, C.N. James D. Shintani, J.B. Ohlrogge, K.D. Chapman, R.T. Mullen and J.M. Dyer. 2013. Lipid droplet-associated proteins (LDAPs) are involved in the compartmentalization of lipophilic compounds in plant cells. Plant Signal Behav. 8: e27141.

Gillespie, J., S.W. Rogers, M. Deery, P. Dupree and J.C. Rogers. 2005. A unique family of proteins associated with internalized membranes in protein storage vacuoles of the Brassicaceae. Plant J. 41: 429-441.

Gohon, Y., J.D. Vindigni, A. Pallier, F. Wien, H. Celia, A. Giuliani, C. Tribet, T. Chardot and P. Briozzo. 2011. High water solubility and fold in amphipols of proteins with large hydrophobic regions: oleosins and caleosin from seed lipid bodies. Biochim. Biophys. Acta 1808: 706-716.

Grillitsch, K., M. Connerth, H. Kofeler, T.N. Arrey, B. Rietschel, B. Wagner, M. Karas and G. Daum. 2011. Lipid particles/droplets of the yeast *Saccharomyces cerevisiae* revisited: lipidome meets proteome. Biochim. Biophys. Acta 1811: 1165-1176.

Gruis, D.F., D.A. Selinger, J.M. Curran and R. Jung. 2002. Redundant proteolytic mechanisms process seed storage proteins in the absence of seed-type members of the vacuolar processing enzyme family of cysteine proteases. Plant Cell 14: 2863-2882.

Hajduch, M., J.E. Casteel, K.E. Hurrelmeyer, Z. Song, G.K. Agrawal and J.J. Thelen. 2006. Proteomic analysis of seed filling in *Brassica napus*. Developmental characterization of metabolic isozymes using high-resolution two-dimensional gel electrophoresis. Plant Physiol. 141: 32-46.

Hanano, A., M. Burcklen, M. Flenet, A. Ivancich, M. Louwagie, J. Garin and E. Blee. 2006. Plant seed peroxygenase is an original heme-oxygenase with an EF-hand calcium binding motif. J. Biol. Chem. 281: 33140-33151.

Hara-Nishimura, I., T. Shimada, K. Hatano, Y. Takeuchi and M. Nishimura. 1998. Transport of storage proteins to protein storage vacuoles is mediated by large precursor-accumulating vesicles. Plant Cell 10: 825-836.

Hara-Nishimura, I., Y. Takeuchi and M. Nishimura. 1993. Molecular characterization of a vacuolar processing enzyme related to a putative cysteine proteinase of *Schistosoma mansoni*. Plant Cell 5: 1651-1659.

Hegedus, D.D., C. Coutu, M. Harrington, B. Hope, K. Gerbrandt and I. Nikolov. 2015. Multiple internal sorting determinants can contribute to the trafficking of cruciferin to protein storage vacuoles. Plant Mol. Biol. 88: 3-20.

Hinz, G., S. Hillmer, M. Baumer and I.I. Hohl. 1999. Vacuolar storage proteins and the putative vacuolar sorting receptor BP-80 exit the golgi apparatus of developing pea cotyledons in different transport vesicles. Plant Cell 11: 1509-1524.

Horn, P.J., N.R. Ledbetter, C.N. James, W.D. Hoffman, C.R. Case, G.F. Verbeck and K.D. Chapman. 2011. Visualization of lipid droplet composition by direct organelle mass spectrometry. J. Biol. Chem. 286: 3298-3306.

Hsiao, E.S. and J.T. Tzen. 2011. Ubiquitination of oleosin-H and caleosin in sesame oil bodies after seed germination. Plant Physiol. Biochem. 49: 77-81.

Huang, A.H. 1996. Oleosins and oil bodies in seeds and other organs. Plant Physiol. 110: 1055-1061.

Huang, A.H.C. 1992. Oil bodies and oleosins in seeds. Ann. Rev. Plant Physiol. Plant Mol. Biol. 43:177-200.

James, C.N., P.J. Horn, C.R. Case, S.K. Gidda, D. Zhang, R.T. Mullen, J.M. Dyer, R.G.W. Anderson and K.D. Chapman. 2010. Disruption of the Arabidopsis CGI-58 homologue produces Chanarin-Dorfman-like lipid droplet accumulation in plants. Proc. Natl. Acad. Sci. USA. 107: 17833-17838.

Jaworski, A.F. and S.M. Aitken. 2014. Expression and characterization of the *Arabidopsis thaliana* 11S globulin family. Biochim. Biophys. Acta 1844: 730-735.

Jiang, L., T.E. Phillips, C.A. Hamm, Y.M. Drozdowicz, P.A. Rea, M. Maeshima, S.W. Rogers and J.C. Rogers. 2001. The protein storage vacuole: a unique compound organelle. J. Cell Biol. 155: 991-1002.

Jiang, L., T.E. Phillips, S.W. Rogers and J.C. Rogers. 2000. Biogenesis of the protein storage vacuole crystalloid. J. Cell Biol. 150: 755-770.

Johnson, K.D., E.M. Herman and M.J. Chrispeels. 1989. An abundant, highly conserved tonoplast protein in seeds. Plant Physiol. 91: 1006-1013.

Jolivet, P., F. Acevedo, C. Boulard, S. d'Andrea, J.D. Faure, A. Kohli, N. Nesi, B. Valot and T. Chardot. 2013. Crop seed oil bodies: from challenges in protein identification to an emerging picture of the oil body proteome. Proteomics 13: 1836-1849.

Jolivet, P., C. Boulard, A. Bellamy, C. Larre, M. Barre, H. Rogniaux, S. d'Andrea, T. Chardot and N. Nesi. 2009. Protein composition of oil bodies from mature *Brassica napus* seeds. Proteomics 9: 3268-3284.

Jolivet, P., C. Boulard, A. Bellamy, B. Valot, S. d'Andrea, M. Zivy, N. Nesi and T. Chardot. 2011. Oil body proteins sequentially accumulate throughout seed development in *Brassica napus*. J. Plant Physiol. 168: 2015-2020.

Jolivet, P., E. Roux, S. D'Andrea, M. Davanture, L. Negroni, M. Zivy and T. Chardot. 2004. Protein composition of oil bodies in *Arabidopsis thaliana* ecotype WS. Plant Physiol. Biochem. 42: 501-509.

Jolliffe, N.A., C.P. Craddock and L. Frigerio. 2005. Pathways for protein transport to seed storage vacuoles. Biochem. Soc. Trans. 33: 1016-1018.

Jones, S.I. and L.O. Vodkin. 2013. Using RNA-Seq to profile soybean seed development from fertilization to maturity. PLoS One 8: e59270.

Kagaya, Y., R. Toyoshima, R. Okuda, H. Usui, A. Yamamoto and T. Hattori. 2005. LEAFY COTYLEDON1 controls seed storage protein genes through its regulation of FUSCA3 and ABSCISIC ACID INSENSITIVE3. Plant Cell Physiol. 46: 399-406.

Kim, H.U., K. Hsieh, C. Ratnayake and A.H. Huang. 2002. A novel group of oleosins is present inside the pollen of Arabidopsis. J. Biol. Chem. 277: 22677-22684.

Kondou, Y., M. Nakazawa, M. Kawashima, T. Ichikawa, T. Yoshizumi, K. Suzuki, A. Ishikawa, T. Koshi, R. Matsui, S. Muto and M. Matsui. 2008. RETARDED GROWTH OF EMBRYO1, a new basic helix-loop-helix protein, expresses in endosperm to control embryo growth. Plant Physiol.147: 1924-1235.

Krebbers, E., L. Herdies, A. De Clercq, J. Seurinck, J. Leemans, J. Van Damme, M. Segura, G. Gheysen, M. Van Montagu and J. Vandekerckhove. 1988. Determination of the processing sites of an Arabidopsis 2S albumin and characterization of the complete gene family. Plant Physiol. 87: 859-866.

Kroj, T., G. Savino, C. Valon, J. Giraudat and F. Parcy. 2003. Regulation of storage protein gene expression in Arabidopsis. Development 130: 6065-6073.

Kurdyukov, S., Y. Song, M.B. Sheahan and R.J. Rose. 2014. Transcriptional regulation of early embryo development in the model legume *Medicago truncatula*. Plant Cell Rep. 33: 349-62.

Lara, P., L. Onate-Sanchez, Z. Abraham, C. Ferrandiz, I. Diaz, P. Carbonero and J. Vicente-Carbajosa. 2003. Synergistic activation of seed storage protein gene expression in Arabidopsis by ABI3 and two bZIPs related to OPAQUE2. J. Biol. Chem. 278: 21003-21011.

Lawrence, M.C., T. Izard, M. Beuchat, R.J. Blagrove and P.M. Colman. 1994. Structure of phaseolin at 2.2 A resolution. Implications for a common vicilin/legumin structure and the genetic engineering of seed storage proteins. J. Mol. Biol. 238: 748-776.

Li, G., B.R. Nallamilli, F. Tan and Z. Peng. 2008. Removal of high-abundance proteins for nuclear subproteome studies in rice (*Oryza sativa*) endosperm. Electrophoresis 29: 604-617.

Li, L., M. Hur, J.Y. Lee, W. Zhou, Z. Song, N. Ransom, C.Y. Demirkale, D. Nettleton, M. Westgate, Z. Arendsee, V. Iyer, J. Shanks, B. Nikolau and E.S. Wurtele. 2015. A systems biology approach toward understanding seed composition in soybean. BMC Genomics 16: S9.

Lin, L.J., P.C. Liao, H.H. Yang and J.T. Tzen. 2005. Determination and analyses of the N-termini of oil-body proteins, steroleosin, caleosin and oleosin. Plant Physiol. Biochem. 43: 770-776.

Lin, L.J., S.S. K. Tai, C.C. Peng and J.T.C. Tzen. 2002. Steroleosin, a sterol-binding dehydrogenase in seed oil bodies. Plant Physiol. 128: 1200-1211; erratum in Vol. 129: 1930-1930.

Lin, W. and D.J. Oliver. 2008. Role of triacylglycerols in leaves. Plant Sci. 175: 233-237.

Lopez-Ribera, I., J.L. La Paz, C. Repiso, N. Garcia, M. Miquel, M.L. Hernandez, J.M. Martinez-Rivas and C.M. Vicient. 2014. The evolutionary conserved oil body associated protein OBAP1 participates in the regulation of oil body size. Plant Physiol. 164: 1237-1249.

Meinke, D.W., J. Chen and R.N. Beachy. 1981. Expression of storage-protein genes during soybean seed development. Planta 153: 130-139.

Melkus, G., H. Rolletschek, R. Radchuk, J. Fuchs, T. Rutten, U. Wobus, T. Altmann, P. Jakob and L. Borisjuk. 2009. The metabolic role of the legume endosperm: a noninvasive imaging study. Plant Physiol. 151: 1139-1154.

Miquel, M., G. Trigui, S. d'Andrea, Z. Kelemen, S. Baud, A. Berger, C. Deruyffelaere, A. Trubuil, L. Lepiniec and B. Dubreucq. 2014. Specialization of oleosins in oil body dynamics during seed development in Arabidopsis seeds. Plant Physiol. 164: 1866-1878.

Miranda, M., L. Borisjuk, A. Tewes, U. Heim, N. Sauer, U. Wobus and H. Weber. 2001. Amino acid permeases in developing seeds of *Vicia faba* L.: expression precedes storage protein synthesis and is regulated by amino acid supply. Plant J. 28: 61-71.

Monke, G., L. Altschmied, A. Tewes, W. Reidt, H.P. Mock, H. Baumlein and U. Conrad. 2004. Seed-specific transcription factors ABI3 and FUS3: molecular interaction with DNA. Planta 219: 158-166.

Monke, G., M. Seifert, J. Keilwagen, M. Mohr, I. Grosse, U. Hahnel, A. Junker, B. Weisshaar, U. Conrad, H. Baumlein and L. Altschmied. 2012. Toward the identification and regulation of the *Arabidopsis thaliana* ABI3 regulon. Nucleic Acids Res. 40: 8240-8254.

Munier-Jolain, N.G., N.M. Munier-Jolain, R. Roche, B. Ney and C. Duthion. 1998. Seed growth rate in grain legumes - I. Effect of photoassimilate availability on seed growth rate. J. Exp. Bot. 49: 1963-1969.

Munier-Jolain, N.G. and B. Ney. 1998. Seed growth rate in grain legumes - II. Seed growth rate depends on cotyledon cell number. J. Exp. Bot. 49: 1971-1976.

Murphy, D.J. 2012. The dynamic roles of intracellular lipid droplets: from archaea to mammals. Protoplasma 249: 541-585.

Murphy, D.J., I. Cummins and A.S. Kang. 1989. Synthesis of the major oil-body membrane protein in developing rapeseed (*Brassica napus*) embryos. Integration with storage-lipid and storage-protein synthesis and implications for the mechanism of oil-body formation. Biochem J. 258 : 285-293.

Neuhaus, J.M. and J.C. Rogers. 1998. Sorting of proteins to vacuoles in plant cells. Plant Mol. Biol. 38: 127-144.

Ng, D.W., M.B. Chandrasekharan and T.C. Hall. 2006. Ordered histone modifications are associated with transcriptional poising and activation of the phaseolin promoter. Plant Cell 18: 119-132.

Noguero, M., C. Le Signor, V. Vernoud, K. Bandyopadhyay, M. Sanchez, C. Fu, I. Torres-Jerez, J. Wen, K.S. Mysore, K. Gallardo, M. Udvardi, R. Thompson and J. Verdier. 2015. DASH transcription factor impacts *Medicago truncatula* seed size by its action on embryo morphogenesis and auxin homeostasis. Plant J. 81: 453-466.

Ohta, M. and F. Takaiwa. 2015. OsHrd3 is necessary for maintaining the quality of endoplasmic reticulum-derived protein bodies in rice endosperm. J Exp Bot. 66: 4585-93.

Pandurangan, S., A. Pajak, S.J. Molnar, E.R. Cober, S. Dhaubhadel, C. Hernandez-Sebastia, W.M. Kaiser, R.L. Nelson, S.C. Huber and F. Marsolais. 2012. Relationship between asparagine metabolism and protein concentration in soybean seed. J Exp. Bot. 63: 3173-3184.

Pang, P.P., R.E. Pruitt and E.M. Meyerowitz. 1988. Molecular-cloning, genomic organization, expression and evolution of 12S-seed storage proteins genes of *Arabidopsis thaliana*. Plant Mol. Biol. 11: 805-820.

Parthibane, V., S. Rajakumari, V. Venkateshwari, R. Iyappan and R. Rajasekharan. 2012. Oleosin is bifunctional enzyme that has both monoacylglycerol acyltransferase and phospholipase activities. J. Biol. Chem. 287: 1946-1954.

Partridge, M. and D.J. Murphy. 2009. Roles of a membrane-bound caleosin and putative peroxygenase in biotic and abiotic stress responses in Arabidopsis. Plant Physiol. Biochem. 47: 796-806.

Pateraki, I., J. Andersen-Ranberg, B. Hamberger, A.M. Heskes, H.J. Martens, P. Zerbe, S.S. Bach, B.L. Moller and J. Bohlmann. 2014. Manoyl oxide (13R), the biosynthetic precursor of forskolin, is synthesized in specialized root cork cells in *Coleus forskohlii*. Plant Physiol. 164: 1222-1236.

Piffanelli, P. and D.J. Murphy. 1998. Novel organelles and targeting mechanisms in the anther tapetum. Trends Plant Sci. 3: 250-253.

Pradhan, S., N. Bandhiwal, N. Shah, C. Kant, R. Gaur and S. Bhatia. 2014. Global transcriptome analysis of developing chickpea (*Cicer arietinum* L.) seeds. Front Plant Sci. 5: 698.

Price, S.R. 1912. Note on the oil bodies in the mesophyl of the cherry laurel leaf. New Phytol. 11: 371-372.

Reidt, W., T. Wohlfarth, M. Ellerstrom, A. Czihal, A. Tewes, I. Ezcurra, L. Rask and H. Baumlein. 2000. Gene regulation during late embryogenesis: the RY motif of maturation-specific gene promoters is a direct target of the FUS3 gene product. Plant J. 21: 401-408.

Repetto, O., H. Rogniaux, C. Firnhaber, H. Zuber, H. Kuster, C. Larre, R. Thompson and K. Gallardo. 2008. Exploring the nuclear proteome of *Medicago truncatula* at the switch towards seed filling. Plant J. 56: 398-410.

Repetto, O., H. Rogniaux, C. Larre, R. Thompson and K. Gallardo. 2012. The seed nuclear proteome. Front. Plant Sci. 3: 289.

Robinson, D.G., G Hinz and S.E. Holstein. 1998. The molecular characterization of transport vesicles. Plant Mol. Biol. 38: 49-76.

Rochat, C. and J.P. Boutin. 1991. Metabolism of phloem-borne amino-acids in maternal tissues of fruits of nodulated or nitrate-fed pea-plants (*Pisum sativum* L.). J. Exp. Bot. 42: 207-214.

Rolletschek, H., F. Hosein, M. Miranda, U. Heim, K.P. Gotz, A. Schlereth, L. Borisjuk, I. Saalbach, U. Wobus and H. Weber. 2005. Ectopic expression of an amino acid transporter (VfAAP1) in seeds of *Vicia narbonensis* and pea increases storage proteins. Plant Physiol. 137: 1236-1249.

Salon, C., N.G. Munier-Jolain, G. Duc, A.S. Voisin, D. Grandgirard, A. Larmure, R.J.N. Emery and B. Ney. 2001. Grain legume seed filling in relation to nitrogen acquisition: A review and prospects with particular reference to pea. Agronomie 21: 539-552.

Sanders, A., R. Collier, A. Trethewy, G. Gould, R. Sieker and M. Tegeder. 2009. AAP1 regulates import of amino acids into developing Arabidopsis embryos. Plant J. 59: 540-552.

Santos-Mendoza, M., B. Dubreucq, S. Baud, F. Parcy, M. Caboche and L. Lepiniec. 2008. Deciphering gene regulatory networks that control seed development and maturation in Arabidopsis. Plant J. 54: 608-620.

Schmid, K.M. and G.W. Patterson. 1988. Distribution of cyclopropenoid fatty acids in malvaceous plant parts. Phytochemistry 27: 2831-2834.

Shewry, P.R. and R Casey. 1999. Seed proteins: Kluwer Academics, The Netherland.

Shewry, P.R. and N.G. Halford. 2002. Cereal seed storage proteins: structures, properties and role in grain utilization. J. Exp. Bot. 53: 947-958.

Shewry, P.R., J.A. Napier and A.S. Tatham. 1995. Seed storage proteins: structures and biosynthesis. Plant Cell 7: 945-956.

Shimada, T., K. Yamada, M. Kataoka, S. Nakaune, Y. Koumoto, M. Kuroyanagi, S. Tabata, T. Kato, K. Shinozaki, M. Seki, M. Kobayashi, M. Kondo, M. Nishimura and I. Hara-Nishimura. 2003. Vacuolar processing enzymes are essential for proper processing of seed storage proteins in *Arabidopsis thaliana*. J. Biol. Chem. 278: 32292-32299.

Siloto, R.M., K. Findlay, A. Lopez-Villalobos, E.C. Yeung, C.L. Nykiforuk and M.M. Moloney. 2006. The accumulation of oleosins determines the size of seed oilbodies in Arabidopsis. Plant Cell. 18: 1961-74.

Tanaka, M., A. Kikuchi and H. Kamada. 2008. The Arabidopsis histone deacetylases HDA6 and HDA19 contribute to the repression of embryonic properties after germination. Plant Physiol. 146: 149-161.

Thazar-Poulot, N., M. Miquel, I. Fobis-Loisy and T. Gaude. 2015. Peroxisome extensions deliver the Arabidopsis SDP1 lipase to oil bodies. Proc. Natl. Acad. Sci. U S A 112: 4158-4163.

Thompson, R., J. Burstin and K. Gallardo. 2009. Post-genomics studies of developmental processes in legume seeds. Plant Physiol. 151: 1023-1029.

Tzen, J., Y. Cao, P. Laurent, C. Ratnayake and A. Huang. 1993. Lipids, proteins, and structure of seed oil bodies from diverse species. Plant Physiol. 101: 267-276.

Uemura, T. and T. Ueda. 2014. Plant vacuolar trafficking driven by RAB and SNARE proteins. Curr Opin. Plant Biol. 22: 116-121.

van der Klei, H., J. Van Damme, P. Casteels and E. Krebbers. 1993. A fifth 2S albumin isoform is present in *Arabidopsis thaliana*. Plant Physiol. 101: 1415-1416.

Verdier, J., K. Kakar, K. Gallardo, C. Le Signor, G. Aubert, A. Schlereth, C.D. Town, M.K. Udvardi and R.D. Thompson. 2008. Gene expression profiling of *M. truncatula* transcription factors identifies putative regulators of grain legume seed filling. Plant Mol. Biol. 67: 567-580.

Verdier, J., I. Torres-Jerez, M. Wang, A. Andriankaja, S.N. Allen, J. He, Y. Tang, J.D. Murray and M.K. Udvardi. 2013. Establishment of the *Lotus japonicus* Gene Expression Atlas (LjGEA) and its use to explore legume seed maturation. Plant J. 74: 351-362.

Vermachova, M., Z. Purkrtova, J. Santrucek, P. Jolivet, T. Chardot and M. Kodicek. 2011. New protein isoforms identified within *Arabidopsis thaliana* seed oil bodies combining chymotrypsin/trypsin digestion and peptide fragmentation analysis. Proteomics 11: 3430-3434.

Vicente-Carbajosa, J. and P. Carbonero. 2005. Seed maturation: developing an intrusive phase to accomplish a quiescent state. Int. J. Dev. Biol. 49: 645-651.

Vicient, C.M., N. Bies-Etheve and M. Delseny. 2000. Changes in gene expression in the leafy cotyledon1 (lec1) and fusca3 (fus3) mutants of *Arabidopsis thaliana* L. J. Exp. Bot. 51: 995-1003.

Vindigni, J.D., F. Wien, A. Giuliani, Z. Erpapazoglou, R. Tache, F. Jagic, T. Chardot, Y. Gohon and M. Froissard. 2013. Fold of an oleosin targeted to cellular oil bodies. Biochim. Biophys. Acta 1828: 1881-1888.

Vitale, A. and M.J. Chrispeels. 1992. Sorting of proteins to the vacuoles of plant cells. BioEssays 14 : 151-160.

Vitale, A. and G. Hinz. 2005. Sorting of proteins to storage vacuoles: how many mechanisms? Trends Plant Sci. 10: 316-323.

Vitale, A. and N.V. Raikhel. 1999. What do proteins need to reach different vacuoles? Trends Plant Sci. 4: 149-155.

Voelker, T. and A.J. Kinney. 2001. Variations in the biosynthesis of seed-storage lipids. Annu. Rev. Plant Physiol. Plant Mol. Biol. 52: 335-361.

Walling, L., G.N. Drews and R.B. Goldberg. 1986. Transcriptional and post-transcriptional regulation of soybean seed protein mRNA levels. Proc. Natl. Acad. Sc.i USA. 83: 2123-2127.

Wang, F. and S.E. Perry. 2013. Identification of direct targets of FUSCA3, a key regulator of Arabidopsis seed development. Plant Physiol. 161: 1251-1264.

Wang, J., Y.C. Tse, G. Hinz, D.G. Robinson and L. Jiang. 2012a. Storage globulins pass through the Golgi apparatus and multivesicular bodies in the absence of dense vesicle formation during early stages of cotyledon development in mung bean. J. Exp. Bot. 63: 1367-1380.

Wang, X.D., Y. Song, M.B. Sheahan, M.L. Garg and R.J. Rose. 2012b. From embryo sac to oil and protein bodies: embryo development in the model legume *Medicago truncatula*. New Phytol. 193: 327-38.

Weber, H., L. Borisjuk and U. Wobus. 2005. Molecular physiology of legume seed development. Annu. Rev. Plant Biol. 56: 253-279.

White, D.A., I.D. Fisk and D.A. Gray. 2006. Characterisation of oat (*Avena sativa* L.) oil bodies and intrinsically associated E-vitamers. J. Cereal Sci. 43: 244-249.

Winichayakul, S., R.W. Scott, M. Roldan, J.H. Hatier, S. Livingston, R. Cookson, A.C. Curran and N.J. Roberts. 2013. *In vivo* packaging of triacylglycerols enhances Arabidopsis leaf biomass and energy density. Plant Physiol. 162: 626-639.

Yatsu, L.Y. and T.J. Jacks. 1972. Spherosome membranes: half unit-membranes. Plant Physiol. 49 : 937-943.

Zhou, X.R., P. Shrestha, F. Yin, J.R. Petrie and S.P. Singh. 2013. AtDGAT2 is a functional acyl-CoA:diacylglycerol acyltransferase and displays different acyl-CoA substrate preferences than AtDGAT1. FEBS Lett. 587: 2371-2376.

Zhu, Z., Y. Ding, Z. Gong, L. Yang, S. Zhang, C. Zhang, X. Lin, H. Shen, H. Zou, Z. Xie, F. Yang, X. Zhao, P. Liu and Z.K. Zhao. 2015. Dynamics of the lipid droplet proteome of the oleaginous yeast *Rhodosporidium toruloides*. Eukaryote Cell 14: 252-264.

Index

T - #0352 - 071024 - C396 - 234/156/18 - PB - 9780367782917 - Gloss Lamination